Welding

David Hoffman

Member, American Welding Society

Certified Welding Inspector

Certified CRAW Technician

Fox Valley Technical College

Kevin Dahle

Member, American Welding Society

Member, Artist Blacksmith Association of North America

Certified Welding Inspector

Fox Valley Technical College

David Fisher

Member, American Welding Society

Certified Welding Inspector

Fox Valley Technical College

Prentice Hall

Boston   Columbus   Indianapolis   New York   San Francisco   Upper Saddle River
Amsterdam   Cape Town   Dubai   London   Madrid   Milan   Munich   Paris   Montreal   Toronto
Delhi   Mexico City   São Paulo   Sydney   Hong Kong   Seoul   Singapore   Taipei   Tokyo

Editorial Director: Vernon R. Anthony
Acquisitions Editor: David Ploskonka
Development Editor: Dan Trudden
Editorial Assistant: Nancy Kesterson
Director of Marketing: David Gesell
Executive Marketing Manager: Derril Trakalo
Senior Marketing Coordinator: Alicia Wozniak
Marketing Assistant: Les Roberts
Project Manager: Maren L. Miller
Senior Managing Editor: JoEllen Gohr
Associate Managing Editor: Alexandrina Benedicto Wolf
Senior Operations Supervisor: Pat Tonneman

Operations Specialist: Laura Weaver
Senior Art Director: Diane Y. Ernsberger
Art Studio: Terrel Broiles
Cover and Interior Designer: Michael Fruhbeis
Cover Image: iStock
AV Project Manager: Janet Portisch
Full-Service Project Management: Kelly Keeler, PreMediaGlobal
Composition: PreMediaGlobal
Printer/Binder: R.R. Donnelley/Willard
Cover Printer: Lehigh-Phoenix Color/Hagerstown
Text Font: HelveticaNeue

Credits and acknowledgments borrowed from other sources and reproduced, with permission, in this textbook appear on the appropriate page within text.

Library of Congress Cataloging-in-Publication Data

Hoffman, David
    Welding / David Hoffman, Kevin Dahle, David Fisher.
        p. cm.
    Includes index.
    ISBN 978-0-13-234977-2
    1. Welding.   I. Dahle, Kevin, 1957-   II. Fisher, David, 1967-   III. Title.
TS227.H597 2012
671.5'2—dc22

                                2010013781

10 9 8 7 6 5 4 3 2

Prentice Hall
is an imprint of

www.pearsonhighered.com

ISBN 10:   0-13-234977-9
ISBN 13: 978-0-13234977-2

Contents

Today's students and instructors need to do more in less time than ever before. *Welding* was written from the ground up to be both comprehensive and concise. Each chapter is cleanly designed and covers exactly what students need to know to enter the workforce. The material is presented with striking images, easy-to-read reference charts and tables, and detailed step-by-step procedures. Because today's students and instructors do more with video and media than ever before, we developed *MyWeldingLab*, which features an extensive collection of instructor supplements, media, video presentations, and welding simulations to save instructors time and engage students in the classroom and anywhere else that they have Internet access.

STORY OF THE BOOK

The authors have experience both in industry and in the classroom. Their industry experience brings together knowledge of welding and manufacturing, welding inspection and quality control, power source design, troubleshooting, and customer service. Those collective industry experiences and more than 50 years of combined instructional expertise have inspired the writing of this textbook. As instructors, the authors wanted to produce an accessible book that covers welding in a clear and concise manner. They gave each major welding process its own chapter so they could easily be discussed in the classroom. Following the weld processes, their chapters focus on critical topics, such as safety, codes, destructive and nondestructive weld testing, welding symbols, welding metallurgy, and welding power sources. Most textbooks do not cover smaller diameter electrodes well. *Welding* does. The authors' goal was to create a text suitable for high schools and postsecondary schools, for industrial use, or by the home hobbyist, and they have done just that with *Welding*.

ORGANIZATION

First, the book orients students with basic information on welding careers and welding safety. Then, welding chapters follow, with uniquely detailed step-by-step weld setups and techniques. The authors devote an entire chapter to each of the common weld processes. Within those chapters they take the weld setup and break it down into its most important steps, providing students with instructions and photos. Students are visual learners, and this step-by-step format makes the material easier to learn and easier to reference later in the semester and even later in students' careers. The chapters cover important information with easy-to-read technical tables.

Next, the chapters turn to metallurgy and alloys, codes and costs, and other critical welding topics. Finally, the appendices cover additional topics, such as troubleshooting, conversion, and pipes and beams.

A SOLID FOUNDATION FOR LEARNING

CHAPTER LEARNING OBJECTIVES Each chapter begins with clearly stated objectives that enable students to focus on what they should achieve and know by the end of the chapter.

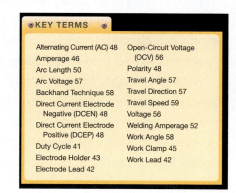

LEARNING OBJECTIVES

In this chapter, the reader will learn how to:

1. Determine how long the Shielded Metal Arc Welding process has been used in industry.
2. Operate the Shielded Metal Arc Welding process.
3. Select the type of power source and peripheral components that would work best at home, school, or in industry.
4. Identify and correctly use the various controls on power sources and their characteristics.
5. Set up for Shielded Metal Arc Welding by controlling amps, arc length (volts), travel speed, and work angles.
6. Adjust the power source to produce an arc that is best for the weld position, base metal, and desired outcome.
7. Select electrodes for the base metal used and the desired outcome.
8. Integrate process knowledge and technique, resulting in acceptable welds.

CHAPTER KEY TERMS Each chapter opens with a list of key terms with page references for the chapter. This provides for quick access and easy review by students.

KEY TERMS

Alternating Current (AC) 48
Amperage 46
Arc Length 50
Arc Voltage 57
Backhand Technique 58
Direct Current Electrode Negative (DCEN) 48
Direct Current Electrode Positive (DCEP) 48
Duty Cycle 41
Electrode Holder 43
Electrode Lead 42

Open-Circuit Voltage (OCV) 56
Polarity 48
Travel Angle 57
Travel Direction 57
Travel Speed 59
Voltage 56
Welding Amperage 52
Work Angle 58
Work Clamp 45
Work Lead 42

CHAPTER INTRODUCTION
These introductions provide a general overview of the chapter content. They are a preview of the essential content, which often includes historical information, that will be featured in the coming pages.

INTRODUCTION

The majority of the welding done in today's welding and metal fabrication shops is done with the four main arc welding processes: Gas Metal Arc Welding, Flux Cored Arc Welding, Shielded Metal Arc Welding, and Gas Tungsten Arc Welding. These processes and the equipment that is used to perform them are continuously being improved, pushing the limits of quality, craftsmanship, and production. Even with all their improvements, these processes are not always the best ones for every application. Other welding processes exist that have unique benefits that allow them to stand out and to fill niches that the four main welding processes cannot fill.

CHAPTER QUESTIONS/ASSIGNMENTS Great for study review and comprehension checks, these end-of-chapter questions help students assess their understanding of the material.

CHAPTER QUESTIONS/ASSIGNMENTS

MULTIPLE CHOICE

1. Which of the following is not an advantage of an Oxy-Fuel Welding system?
   a. Portability
   b. Does not require electricity
   c. Requires an inert shielding gas
   d. Can be used for welding, brazing, cutting, and

5. What is used to melt the base metal that is located between the tips when Resistance Welding?
   a. Vibrating energy
   b. High-temperature flame
   c. Electrical current flow
   d. DCEN welding arc

FILL IN THE BLANK

9. The four main arc welding processes are Gas Metal Arc Welding, _____, Shielded Metal Arc Welding, and Gas Tungsten Arc Welding.

10. Oxygen and _____ are the most popular combination for Oxy-Fuel Welding.

plasma welding power sources only provide a DC welding output.

15. A benefit of Plasma Arc Welding is that the electrode is _____ inside the torch.

16. _____ is the electrode of choice for Plasma

SAFETY Critical for every welding program, safety considerations/tables are clearly highlighted at the start of each of the welding process chapters.

SAFETY

TABLE 6-1 and **TABLE 6-2** show some basic safety considerations related to workplace environments. Table 6-1 shows the basic safety considerations for GTAW and Table 6-2 shows recommended lens shades for welding at varying amperage ranges.

Head	Body	Hands	Feet
Safety glasses Welding helmet	Flame-resistant wool or cotton pants and shirt	Leather welding gloves	High-top leather safety work boots

TABLE 6-1

Read "Chapter 2: Safety in Welding" for detailed discussion of safety related to workplace environment, workplace hazards, personal protective equipment, electrical consideration, gases and fumes, ventilation, fire prevention, explosion, and compressed cylinders and gases.

Arc Current	Suggested Shade
Less than 50	8
50–150	10
150–250	12
250–400	14

TABLE 6-2

STEP-BY-STEP PROCEDURES

Procedures including instruction, safety, and photos make material easy to understand.

GMAW Setup

1 Align and measure the tip gap (make sure the power is off).
CAUTION: Make sure power source is off whenever touching the tongs or tips.

2 Clean the tips and/or recondition the tips (make sure the power is off).

RICH ART THROUGHOUT

Striking art and photos are large and detailed to reach visual learners.

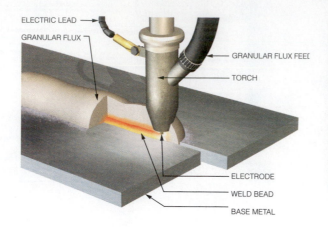

SUPPLEMENTS

Supplements are a crucial part of any book these days, and *Welding* includes a unique package for both students and instructors. To access supplementary materials online, instructors need to request an instructor access code. Go to **www.pearsonhighered.com/irc** to register for an instructor access code. Within 48 hours of registering, you will receive a confirming e-mail including an instructor access code. Once you have received your code, locate your text in the online catalog and click on the Instructor Resources button on the left side of the catalog product page. Select a supplement, and a login page will appear. Once you have logged in, you can access instructor material for all Prentice Hall textbooks. If you have any difficulties accessing the site or downloading a supplement, please contact Customer Service at **http://247.prenhall.com**.

PEARSON myweldinglab

MyWeldingLab (Text and *MyWeldingLab* Package ISBN-10: 0-13-248818-3) is an easy-to-use online study tool that provides robust assessment and personalizes course content to address your students' learning needs. Students can review their welding skills and knowledge online on their own time and at their own pace. *MyWeldingLab* tests students on chapter objectives and creates personalized study plans based on their results. Customized study plan tools include an e-book, interactive media exercises, and video tutorials.

The key component of *MyWeldingLab* is the Welding Lab Simulations. In these 3-D simulations, students are asked to set up various types of welds. They must choose and connect the components in the proper order. These fun, interactive simulations act as end-of-chapter capstones for each of the major welding processes.

MyWeldingLab saves instructors time by automatically grading homework and quizzes. It saves time in the lab because students can practice setups online. *MyWeldingLab* helps students by presenting the material in fun, visually striking interactive ways.

VIDEOS There are over 60 welding videos directly tied to the content in this book. These video presentations bring welding techniques to life and can be referenced by students at any time in *MyWeldingLab*. For easy access, they are referenced in the textbook with a video icon in the margin of the corresponding content.

VIDEO

INSTRUCTOR'S RESOURCE MANUAL The *Instructor's Resource Manual* (ISBN-10: 0-13-512231-6) includes chapter-by-chapter assistance through modifiable lesson plans, a sample syllabus, objectives, and answers to the end-of-chapter review questions.

POWERPOINT PRESENTATION The PowerPoint Presentation (ISBN-10: 0-13-614125-0) features a lecture outline format with key talking points. It includes videos and art from the textbook to enhance the classroom experience and to help instructors prepare for class.

MYTEST   This electronic question bank (ISBN-10: 0-13-234976-0) can be used to develop customized quizzes, tests, and/or exams.

INSTRUCTOR'S CUSTOMIZABLE WELDING LAB MANUAL (ONLINE)   The *Instructor's Customizable Welding Lab Manual* (ISBN-10: 0-13-159782-5) is an electronic resource that allows instructors to easily develop and customize labs for their students based on the equipment they have and the parameters they want to assign.

STUDENT'S WELDING LAB MANUAL (PRINT)   A printed *Student Welding Lab Manual* (ISBN-10: 0-13-159776-0) is also available. With more than 170 labs in all, this outstanding resource includes challenging lab exercises for the more common welds that students are likely to encounter.

ACKNOWLEDGEMENTS

We owe a special thank you to all the people who reviewed the manuscript and everyone who worked to make this book the best it can be.

- Ted Alberts, New River Community College
- Lawrence Bower, Blackhawk Technical College
- Sue Caldera, Clackamas Community College
- David Dillon, North Carolina Agricultural and Technical State University
- Jason Hill, Metropolitan Community College of Omaha, Nebraska
- Ken Paulus, Bismarck State College
- Nicholas Pinckney, Grand Rapids Community College
- Carlisle "Carl" Smith, West Virginia University, Parkersburg
- Christopher Staba, Alfred State College
- Mark Steinle, Casper College
- Alex Taddei, San Joaquin Delta College

We would also like to thank **Stephen Collins**, Maine Maritime Academy, and **Richard Rowe**, Johnson County Community College, for lending their considerable skill to authoring parts of *MyWeldingLab*.

David Hoffman
Kevin Dahle
David Fisher

part 1
INTRODUCTORY MATERIALS

Chapter 1

WELDING JOBS AND EMPLOYMENT SKILLS

myweldinglab

INTRODUCTION

Since the Bronze Age, welding has evolved from a simple science—where soft metals were bonded together by use of pressure—to the Iron Age—where heat and pressure were used to join iron. This method of joining metals is called *Forge Welding*. Blacksmithing evolved in the Middle Ages, and until the nineteenth century, metals were joined with these somewhat primitive forging methods. However, since the advent of oxyacetylene welding in 1836, welding has advanced into a field that includes more than 100 welding processes.

WHY WELDING?   **Welding** is the joining of materials produced by heating them to the proper temperature, with or without the application of pressure, and with or without filler metal or joining them together, with the application of pressure alone, without the use of filler metal. These methods create a part that is one continuous piece and can be used for joining virtually all commercially available metals. The welded part is in many ways superior to those made with other processes, such as bolting and riveting, for permanently joining materials. Welding is widely accepted as a more economically viable joining method than others. The bottom line is that it is cheaper and often easier to join structures by welding. For these reasons welding is the most common joining method.

WHAT ARE THE DEMANDS FOR WELDING EMPLOYMENT?   Welding jobs are plentiful. The U.S. Department of Labor's Bureau of Labor and Statistics estimates that nearly 450,000 welding jobs will be available in 2014. These statistics do not reflect the actual total of all welding-related jobs. The industry demands for welding-related jobs also include self-employed welders and consultants or other welding-related support careers, such as equipment manufacture and sales, research, and education. While the U.S. Department of Labor indicated a decline in the overall number of welders between 2000 and 2005, this decline has been offset by a high demand for new welders.

The current industry demand for welders appears to be growing with the number of welders reaching retirement age. This increase in demand is likely to continue as a growing number of the baby boomer generation—those born between 1946 and 1964—continue leaving the workforce. These workers have formed the backbone of the manufacturing industry, and as they leave the workforce, they leave behind increasing opportunities for new workers. The welding employment outlook has never been better.

WHAT TYPES OF JOBS ARE AVAILABLE FOR WELDING EMPLOYMENT?
The Bureau of Labor and Statistics surveyed the occupational employment and wages for the job category 51-4121, Welders, Cutters, Solderers, and Brazers, in May 2007. These jobs are for those who use manual welding, flame cutting, manual soldering, or brazing equipment to weld or join metal components or to fill holes, indentations, or seams of fabricated metal products.

Two interesting sets of data help to understand the surveyed welding jobs. **TABLE 1-1** brings together the industries with the highest levels of employment.

Industry	Employment	Hourly Mean Wage	Annual Mean Wage
Architectural and Structural Metals Manufacturing	47,260	$15.16	$31,530
Agriculture, Construction, and Mining Machinery Manufacturing	26,240	$15.59	$32,420
Motor Vehicle Body and Trailer Manufacturing	20,750	$14.48	$30,120
Commercial and Industrial Machinery and Equipment (except Automotive and Electronic) Repair and Maintenance	17,990	$16.11	$33,500
Other General Purpose Machinery Manufacturing	16,340	$16.01	$33,300

TABLE 1-1

Source: Bureau of Labor Statistics, U.S. Department of Labor, *Occupational Outlook Handbook, 2008–09 Edition*, Welding, Soldering, and Brazing Workers, on the Internet at http://www.bls.gov/oco/ocos226.htm

These broad groupings offer a diverse set of opportunities for welders. **TABLE 1-2** lists the top-paying industries for these welding occupations. As shown, a wide range of job fields exists for those who want an interesting career. Welders need to not only have welding skills, but they must also have knowledge about welding terms and weld **joint nomenclature** in order to communicate clearly in the workplace. Joint nomenclatures are technical terms related to welds and weld joints (**FIGURES 1-1** through **1-4**). For individuals who acquire adequate education and technical training, not only are there interesting jobs available, but there are also jobs that pay well and provide benefits worthy of a lifetime career as a welder.

Another occupational grouping is 51-4122, Welding, Soldering, and Brazing Machine Setters, Operators, and Tenders. This group of workers sets up, operates, or tends welding, soldering, or brazing machines or robots that weld, braze, or solder or heat treat metal products, components, or assemblies.[3] These jobs differ from those of category 51-4121 in that the workers set up and operate

Industry	Employment	Hourly Mean Wage	Annual Mean Wage
Motor Vehicle Manufacturing	360	$27.34	$56,870
Building Equipment Contractors	(8)	$23.34	$48,540
Local Government (OES designation)	80	$20.27	$42,150
Aerospace Product and Parts Manufacturing	700	$20.01	$41,620
Metal Ore Mining	50	$19.50	$40,570

TABLE 1-2

Source: Bureau of Labor Statistics, U.S. Department of Labor, *Occupational Outlook Handbook, 2008–09 Edition*, Welding, Soldering, and Brazing Workers, on the Internet at http://www.bls.gov/oes/current/oes514121.htm

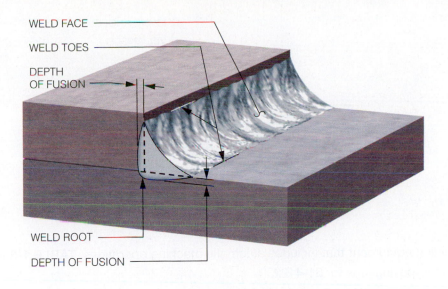

WELD FACE
WELD TOES
DEPTH OF FUSION
WELD ROOT
DEPTH OF FUSION

FIGURE 1-1 Fillet welds

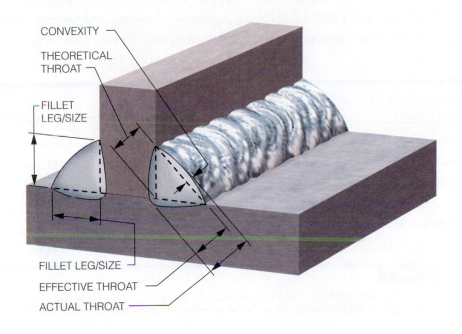

CONVEXITY
THEORETICAL THROAT
FILLET LEG/SIZE
FILLET LEG/SIZE
EFFECTIVE THROAT
ACTUAL THROAT

FIGURE 1-2 Convex fillet welds

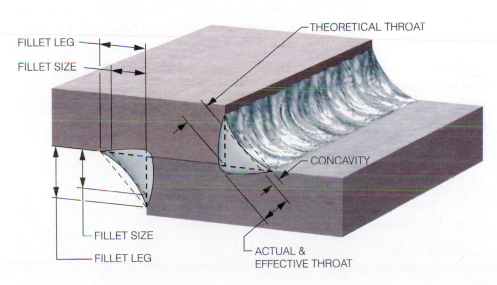

FILLET LEG
FILLET SIZE
THEORETICAL THROAT
CONCAVITY
FILLET SIZE
FILLET LEG
ACTUAL & EFFECTIVE THROAT

FIGURE 1-3 Concave fillet welds

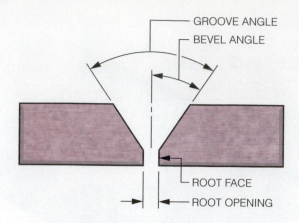

FIGURE 1-4 Groove welds

welding equipment that includes automatic machine operations. **TABLE 1-3** lists the top-paying jobs for 51-4122.

Keep in mind that the tables included here are examples and do not compile all the jobs related to the welding industry. Sales of welding equipment, welding alloys, and filler metals and welding inspection, welding engineering, and welding education offer other potential opportunities for those who seek careers in welding.

Industry	Employment	Hourly Mean Wage	Annual Mean Wage
Motor Vehicle Manufacturing	360	$27.34	$56,870
Building Equipment Contractors	(8)	$23.34	$48,540
Local Government (OES designation)	80	$20.27	$42,150
Aerospace Product and Parts Manufacturing	700	$20.01	$41,620
Metal Ore Mining	50	$19.50	$40,570

TABLE 1-3

Source: Bureau of Labor Statistics, U.S. Department of Labor, *Occupational Outlook Handbook, 2008–09 Edition*, Welding, Soldering, and Brazing Workers, on the Internet at http://www.bls.gov/oes/2007/may/oes514122.htm

Employability skills needed by and the benefits paid out to an individual welder are based on a broad range of acquired skills and knowledge. For example, an employer may require a welder to be qualified to weld in a specific **welding position**. Welding position relates to the physical orientation of the workpiece, weld joint, and weld being completed. The American Welding Society has standardized welding positions for welder qualification (**FIGURES 1-5 through 1-23**). A combination number/letter designation is assigned to welding positions:

- 1 = Flat
- 2 = Horizontal
- 3 = Vertical
- 4 = Overhead
- F = Fillet
- G = Groove

Welding Positions Pipe (Groove):

FIGURE 1-5 Flat groove position 1G (rotated)

FIGURE 1-6 Horizontal groove position 2G

FIGURE 1-7 Vertical groove position 5G

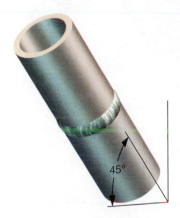

FIGURE 1-8 All position groove 6G

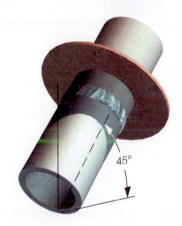

FIGURE 1-9 All position groove 6GR (T, K, & Y Joints)

Welding Positions Pipe (Fillet):

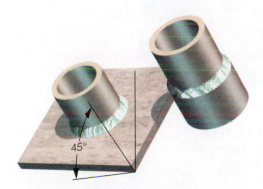

FIGURE 1-10 Flat fillet position 1F (rotated)

FIGURE 1-11 Horizontal fillet position 2F

FIGURE 1-12 Horizontal fillet position 2FR (rotated)

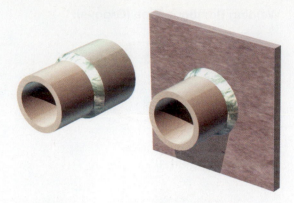

FIGURE 1-13 Vertical fillet position 5F (fixed position)

FIGURE 1-14 Overhead fillet position 4F

45°

FIGURE 1-15 Fillet multiple position 6F

Welding Positions Plate (Groove):

FIGURE 1-16 Flat groove position 1G

FIGURE 1-17 Horizontal groove position 2G

FIGURE 1-18 Vertical groove position 3G

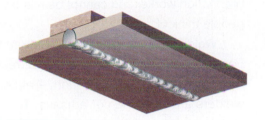

FIGURE 1-19 Overhead groove position 4G

Welding Positions Plate (Fillet):

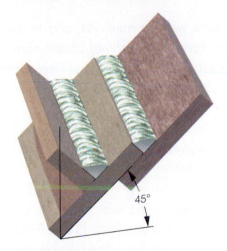

45°

FIGURE 1-20 Flat fillet position 1F

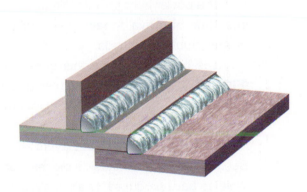

FIGURE 1-21 Horizontal fillet position 2F

FIGURE 1-22 Vertical fillet position 3F

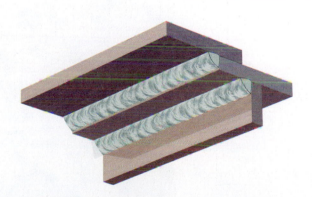

FIGURE 1-23 Overhead fillet position 4F

WELDERS, CUTTERS, SOLDERERS AND BRAZERS. THAT'S WELDING?

The jobs related to **TABLE 1-1**, **TABLE 1-2**, and **TABLE 1-3** are more than just means to acquiring income and security. Welding as an occupation provides challenging experiences for those who prepare adequately, and the ability to pull together a broad skill set can propel a welder into a wide range of careers. Career opportunities are available worldwide for those who choose welding (**FIGURE 1-24**). The field of welding encompasses more territory than that required to be a production worker and cannot be relegated to such brief titles as those that many people think of.

Welding has come a long way from those somewhat primitive processes of Forge Welding. Let's put aside the Bureau of Labor and Statistics job categories to make a broader observation. Welding may be considered a skill, but many welders are craftsmen or artisans. Their high degree of skill can be difficult to match. In many cases, these specialists will advance in their careers and their wares will be in demand not only at home but also abroad. Sometimes, our lives depend on the craft for safety or security. In other cases, welding applications may be a hobby or welding may be used as an art form, simply there for enjoyment or pleasure.

Some of the more primitive art forms used welding methods, and those same processes are used today. For example, Forge Welding with a hand hammer or a power hammer can join metals. The art of blacksmithing and forging is in a state of revival today by enthusiasts who appreciate working with once-nearly-forgotten methods.

It is hard to imagine this unrefined chunk of alternating layers of steel (**FIGURE 1-25**) cutting vegetables or being displayed as a piece or art. Three pieces of wrought iron are welded together with alternating layers of spring steel from an automotive leaf spring.

FIGURE 1-26 shows the layers after Forge Welding and folding three times. Because the original outside pieces were both wrought iron, folding and Forge Welding have produced 33 alternating layers of laminate steel. The piece is then countersunk with a drill to create a pattern and hammered flat to about 3/32 of an inch thick (4.8 mm).

FIGURE 1-24 Pipeline welding (Egypt). Photo courtesy of the ESAB Group, Inc.

FIGURE 1-25 Wrought iron/spring steel 5 layers

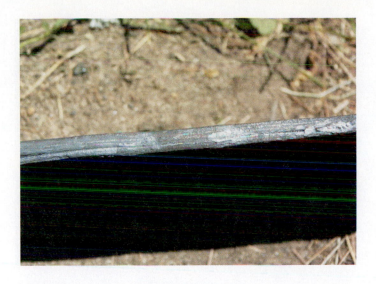

FIGURE 1-26 Forge-welded 33 layers

Finally, the forge-welded piece is cut to the desired shape, and finish-sanded with a very fine grit, and then sharpened, and an ebony handle is added to finish the piece (**FIGURE 1-27**).

Cloisonné, the art of combining enamel and copper, originated in China during the Ming Dynasty (1368–1644). Cloisonné, in part a marriage of welding methods and ancient art, is still in use today, although many people do not understand the connection to welding.

Copper alloys are formed into a desired shape: a vase, plate, bowl, or jewelry. Next, a process called filigree welding is used. Copper or brass alloys in the form of wire or flattened strips called *filigree* are soldered or glued to the copper form (**FIGURES 1-28** and **1-29**)—in this case, a vase.

The copper vessel and filigree are heated to 1652°F (900°C), and the materials are joined. Subsequent steps include filling the pattern with colored enamel, firing to set the enamel, polishing the surface on grinding wheels, cleaning with acid, and then dipping the object into liquid electroplating to add a layer of gold to the metal surfaces. The finished cloisonné is a fine piece of art (**FIGURE 1-30**).

FIGURE 1-27 Forge-welded mezaluna cutter

FIGURE 1-28 Cloisonné
filigree setting

FIGURE 1-29 Cloisonné filigree
complete

FIGURE 1-30 Cloisonné finished

Of course, these are not the common applications of welding, but they show that welding is a complex group of applications. Welding is so much a part of our everyday lives that most people are simply unaware of just how it impacts everything we do. If welding is not art, it certainly adds to the ambiance of our daily experience. How often do we use a map (**FIGURE 1-31**) or pass by an appealing display (**FIGURE 1-32**) and overlook the work of the welder.

Take a close look at **FIGURE 1-33**. What do you see? Do you see one weld? You see trucks and cars, a harbor, hundreds of shipping containers, cranes for loading containers on ships, and cranes for construction of the towers of the Stonecutters Bridge, which will soon become the longest cable-stayed bridge in the world. The steel bridge will stretch 1,018 meters (1,181 yards) and is supported by two towers constructed of concrete and stainless steel. How many pounds of weld metal do you suppose are used to create this picture?

FIGURE 1-31 City map
Photo courtesy of Hieptas Welding

FIGURE 1-32 Street scape
Photo courtesy of Hieptas welding

WHAT ARE THE SKILL SETS NECESSARY FOR WELDING EMPLOYMENT?

Obviously, a welder needs a broad knowledge of and skills in welding. Beyond good eyesight and depending on their areas of expertise, welders need to be physically capable of lifting significant loads, bending and flexing their bodies in confined work spaces, and enduring work environments that include heights (**FIGURE 1-34**)—all the while maintaining good dexterity and hand-eye coordination.

The qualifications welders need for employment are not limited to technical knowledge or manual skills and abilities related to welding. As welding technology advances and the labor force is exposed to computer-integrated applications and with expanding competition in the global marketplace, welders now more than ever must have advanced knowledge and training, such as in robotic welding (**FIGURE 1-35**). Comprehension of mathematic principles is critical to any manufacturing application. Welders must have a clear understanding of welding symbols and blueprint reading (**FIGURE 1-36**) in order to meet production demands that ensure quality and efficiency at work. Training and experience in sciences related to metals and electrical principles are also beneficial.

FIGURE 1-33 Container port, Stonecutters Bridge, Hong Kong

FIGURE 1-34 Welding heights
Photo courtesy of the ESAB Group, Inc.

FIGURE 1-35 Robotic welding
Photo courtesy of the ESAB Group, Inc.

DO WELDERS NEED SOFT SKILLS? The answer is yes! Welding employment encompasses a wide range of employment opportunities. The first image one might conjure when the term "welder" is mentioned may be an individual wearing a leather apron, a welding hood, and sparks flying in a small dark shop. That is the narrowest interpretation possible. The range of skill sets required stretches far beyond that image. In many cases, welders have to work in teams. A welding salesperson communicates with scores of people in a given day. A welding inspector, lead person, or foreman must communicate clearly with the welder on one hand and welding engineers and contractors or subcontractors on the other. All personnel at every level of welding expertise need to be competent in many ways.

For the welder, soft skills are those traits and behaviors an individual demonstrates in the workplace that influence work outcomes and work relationships with peers, customers, and employers. To gain an understanding of areas of competence beyond the manual skills that welders must have, data from a Wisconsin Technical College System (WTCS) survey can provide a true-life perspective of just what is expected by Wisconsin manufacturers. It is important to note that soft skills are not vital to Wisconsin manufacturers only.

As a result of a statewide conference, "Peak Performance for Wisconsin Manufacture," held in 2003, WTCS in conjunction with state manufacturers developed an Advanced Manufacturing Solutions (AMS) initiative. A purpose of this initiative is to prepare the workforce of the present and the future. Part of this initiative is a survey that identified 12 core skills that manufacturers deem critical to being a good employee (**FIGURE 1-37**).

Beyond technical skills such as welding, employers need workers with these 12 soft skills that can be broken down into 4 skill sets. For one, employees must have **productivity skills**. A good employee will work productively, must be able to follow directions, and functions in a manner that maintains a safe work environment. Another set of necessary proficiencies relate to **problem solving**. These include the ability to think critically, apply problem-solving strategies, and apply mathematical reasoning. **Team skills** are another necessary attribute employers seek when hiring. Employees must communicate clearly, listen effectively, and be

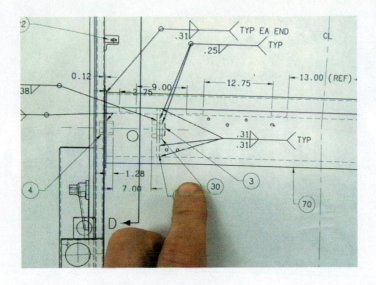

FIGURE 1-36 Weld symbols and print reading

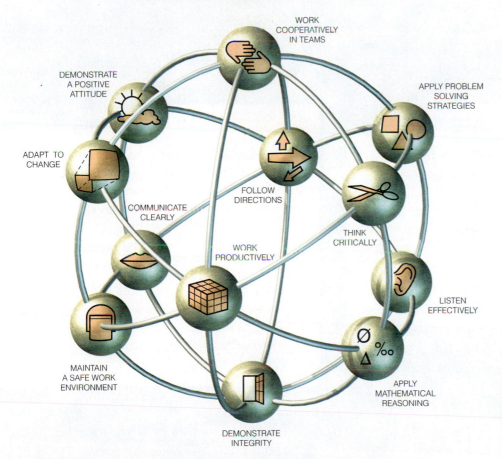

FIGURE 1-37 Critical core skills cluster
Source: Critical Core Skills Learning System, Wisconline.org, Fox Valley Technical College, Appleton, Wisconsin

capable of working cooperatively in teams. Finally, **adaptability skills** are necessary. Employers want workers who demonstrate integrity, can adapt to change, and demonstrate a positive attitude in the workplace.

A welder who masters these critical core skills in addition to technical training—both in theory and with hands-on methods—is a valuable link in the growing world of welding and manufacturing. Technical knowledge and productivity skills, problem-solving and team-building skills, and an ability to adapt to a constantly changing work environment will serve a welder just as much as he or she will serve the manufacturing community.

Chapter 2

SAFETY IN WELDING

PEARSON
myweldinglab

WHO IS RESPONSIBLE FOR A SAFE WORKPLACE? It is important for the welder to become familiar with safety concerns and hazard prevention. Several organizations have established guidelines for manufacturers and laborers to ensure a safe work environment. The American Welding Society has developed *Safety in Welding, Cutting, and Allied Processes*, an American National Standards Institute (ANSI)–approved publication, Z49.1. This standard covers all aspects of safety and health in a welding environment.

Another advancement of safety for workers was the establishment of the **Occupational Safety and Health Administration (OSHA)**. OSHA was created by Congress under the Occupational Safety and Health Act, signed by President Richard M. Nixon on December 29, 1970. Its mission is to prevent work-related injuries, illnesses, and deaths by issuing and enforcing rules (called standards) for workplace safety and health. Another outcome was the creation of the **National Institute for Occupational Safety and Health (NIOSH)** as a research agency whose purpose is to determine the major types of hazards in the workplace and ways of controlling them. *Occupational Safety and Health Standards*, Subpart Q, Welding, Cutting, and Brazing, standard 1910.252, General Requirements, lays out guidelines for safe welding practices.

OSHA has determined that employers must protect employees from potential workplace hazards through appropriate hazard control methods. OSHA's **Hazard Communication Standard (HCS)** requires development of a hazard communication program whereby all employers with hazardous chemicals in their workplaces must have labels and **Material Safety Data Sheets (MSDS)** for their exposed workers and train them to appropriately handle the materials.

An MSDS is a form containing data regarding the properties of a particular substance. An important component of workplace safety, it is intended to provide workers and emergency personnel with procedures for handling or working with that substance in a safe manner and includes information such as physical data (melting point, boiling point, flash point, etc.), toxicity, health effects, first aid, reactivity, storage, disposal, protective equipment, and spill-handling procedures. The exact format of an MSDS can vary from source to source.

A clear understanding of workplace hazards is a starting point for determining adequate safety protection. This must be followed by wearing the appropriate protective equipment.

THE HAZARDOUS MATERIALS IDENTIFICATION SYSTEM

HOW CAN AN EMPLOYER MEET HAZARD REQUIREMENTS? Another recent development, the **Hazardous Materials Identification System (HMIS)**, was developed by the National Paint & Coatings Association (NPCA) to help employers comply with OSHA's Hazard Communication requirements. The system is composed of a color-coding and numbering system, including labels or tags placed appropriately in the employee workspace to identify potential hazards and to inform workers of appropriate concerns and **personal protective equipment (PPE)** required. Personal protective equipment are safety protections required under specific workplace conditions.

HMIS provides clear, recognizable information to employees by standardizing the presentation of chemical information. With HMIS, color codes correspond to specific hazards, while numbers rate the degree of the hazard, and a letter designation indicates PPE required for material handling. Personal protective equipment may include any of the following examples:

- Safety glasses and other eye protection
- Ear protection (earplugs)
- Work gloves
- Safety shoes (steel-toed)
- Dust respirator
- Hard hat

Four color codes used by HMIS (**TABLE 2-1**) simplify identification of the types of hazards and what the colors represent.

In addition to the three color codes (**TABLE 2-2**) Red, Yellow, and Blue, a number designation signifies the degree of hazard level from minimal to extreme.

The white label designating PPE uses a letter designating the appropriate PPE employees should wear while handling the material (**TABLE 2-3**).

FIGURE 2-1 shows how the HMIS system can be used to create a labeling system for identifying hazards and personal protection. The example represents moderate health, minimal flammability, and slight reactivity risks. Recommended PPE are safety glasses, gloves, and an apron.

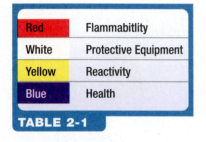

Red	Flammabitlity
White	Protective Equipment
Yellow	Reactivity
Blue	Health

TABLE 2-1

4	Extreme
3	Serious
2	Moderate
1	Slight
0	Minimal

TABLE 2-2

A	safety glasses
B	safety glasses and gloves
C	safety glasses, gloves, and an apron
D	face shield, gloves, and an apron
E	safety glasses, gloves, and a dust respirator
F	safety glasses, gloves, on apron, and a dust respirator
G	safety glasses and a vapor respirator
H	splash goggles, gloves, on apron, and a vapor respirator
I	safety glasses, gloves, and a dust/vapor respirator
J	splash goggles, gloves, on apron, and a dust/vapor respirator
K	airline hood or mask, gloves, full suit, and boots
L–Z	custom PPE specified by employer

TABLE 2-3

PERSONAL PROTECTIVE EQUIPMENT (PPE)

WHAT TYPES OF CLOTHING ARE RECOMMENDED FOR WELDING?

Ultraviolet (UV) rays from welding will degrade protective materials over time, as will molten weld spatter. Protective clothing recommended is flame-resistant wool or cotton material. While many inherently flame-resistant materials are available, they may be costly items. Many employers provide suitable clothing through uniform rental services. These clothes are washed and treated as necessary to render them flame-resistant.

Where those options are not practical, durable denim or canvas style (cotton duck) pants hold up fairly well. It is important to keep welding attire free of grease and oil. A quick change to coveralls may be helpful for workers who are exposed to grease and oil.

Because welding arc rays—visible, UV, and infrared—can cause damage to skin in the form of a burn, they must be avoided. Exposure of skin to UV rays results in **arc burn**. Arc burn results as the intensive UV and infrared rays from the welding arc burn the skin, typically a first-degree burn like that from exposure to sunlight, except it is a deeper burn. Much of the sun's rays are filtered by miles of Earth's atmosphere. The welding arc, being only inches away, is much more intense. To ensure proper protection the welder must always wear clothing that leaves no skin exposed to arc rays. The welder must button collars and sleeves.

Dark clothing is preferred over light colors that are more likely to reflect back arc rays. This is especially true for welding on highly polished materials, such as stainless steels and aluminum. It is also important to wear clothing with a tight-knit pattern that prevents passage of arc rays through the fabric.

In addition, the welder should wear leathers or a flame-resistant shop coat or apron for out-of-position welding and for high-amperage applications. Welding in vertical and overhead positions may result in slag, spatter, or molten metal dripping onto the welder. Leathers are preferred for durability and protection; however, flame-resistant jackets provide a reasonable amount of protection. **FIGURE 2-2** shows a welder with a flame-resistant jacket, wool welding hat, high-top leather safety shoes, and a passive-style welding helmet. **FIGURE 2-3** shows the more protective leather jacket, gauntlet-style welding gloves, insulating heat shield, and an auto-darkening welding helmet.

FIGURE 2-2 Proper personal protection prevents burns.

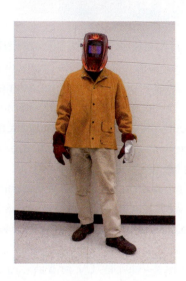

FIGURE 2-3 A leather jacket provides more protection from burns than a flame-resistant jacket.

ARE THERE ANY TYPES OF CLOTHING THAT SHOULD NOT BE WORN?

Synthetic materials such as nylon, polyester, and rayon should be avoided, as they tend to melt and can burn the skin. Care must also be taken to avoid wearing new clothes without washing them before use. Dyes and chemical treatments on new clothing may cause it to ignite. One case example is flannel shirts. A new flannel shirt may start on fire from a single spark, burning off the fuzzy, unwashed outer layer, leading to injury. Finally, do not wear frayed clothing, as it easily starts on fire.

ARE POCKETS SUITABLE FOR USE ON WELDERS' CLOTHING?

While some texts warn against wearing clothing with pockets, the greatest concern is what is actually in the pocket. For example, a butane lighter is an extreme hazard. Avoid having any flammable materials, such as flame lighters, paper, matches, or shop rags, in pockets because they are fire hazards.

WHAT TYPES OF PROTECTION ARE REQUIRED FOR HANDS AND FEET?

A wide range in styles of welding gloves is available for hand protection. For most welding applications, a gauntlet-style welding glove is preferred. Lighter-weight gloves may be suitable for low-amperage Gas Tungsten Arc Welding (GTAW) applications. An aluminized heat shield may be slipped over the welding glove for added protection from a more intense arc, such as with Flux Cored Arc Welding (FCAW) applications.

Footwear used must provide protection from burns and heavy metals and must also be durable enough to withstand the harsh welding environment. High-top boots offer the best protection because they reduce the chance of molten metal or slag from entering and burning the feet. Boots with rounded toes are best because molten metal rolls off them. Finally, steel-toed boots are usually required in welding environments where heavy materials and equipment are common.

WHAT TYPES OF HEAD PROTECTION ARE NEEDED? In addition to a welding helmet to protect the face, several other safety concerns exist. A welding hat made from flame-resistant material can prevent sparks from burning the head or singeing the hair. Any welding supplier will have a few choices available, ranging from the traditional welding hat to tie-style bandannas.

Furthermore, earplugs provide two kinds of protection. They protect the welder from possible noise damage, and they are also useful in preventing sparks from falling into the ear canal. Any time a welder has to work with the head positioned at an angle, sparks may enter the ear. This might be a concern when doing maintenance work—for example, under a vehicle.

WHAT EYE PROTECTION IS REQUIRED FOR WELDING? The first phase of eye protection in any welding environment is the use of approved safety glasses. Eyes must be protected from weld spatter and slag that are expelled during welding and cleaning operations, and safety glasses are an absolute necessity and should be worn under the welding helmet at all times. Safety glasses must also be worn underneath full-face shields for grinding and cutting shields during cutting operations.

The next phase of protection, and equally important, is the welding helmet. UV light rays produced by the welding arc can damage the welder's eyes. Exposure of the eyes to arc rays can result in a burning and blistering of the sensitive membrane of the eye, a condition called **arc flash**. The intensive UV and infrared rays from the welding arc burn the sensitive skin on your eyes, creating little blisters. Severe cases of arc flash may cause an eye infection and require medical attention. Arc flash is often not noticed immediately, but its symptoms are delayed, with the burning sensations occurring at a later time—often in the middle of the night. To eliminate this hazard, the welder wears one of a variety of welding helmets.

While many styles of helmets are available, the welder will need to choose either the old-style passive filter lens or a more modern auto-darkening lens or helmet. Lens filter shades protect the eyes from the welding arc in the same manner as sunglasses protect eyes from the sun, except welding filters are American National Standards Institute (ANSI) approved for welding. Never substitute sunglasses for safety glasses or use them for any welding operations. The passive filter lens is labeled with a number corresponding to the darkness of the shade, where a higher number equates to a darker shade. Auto-darkening lenses provide one shade or a range of shades.

FIGURE 2-4 Various protection devices for use during welding. On the left is a passive-style welding helmet, in the middle is a full-face shield for oxy-fuel torch work, and on the right is an auto-darkening welding helmet.

The advantages of the passive-style helmets are low cost and high durability compared to the more expensive auto-darkening models. There are several disadvantages to using the passive filter. The major drawback is that the welder cannot see through the lens before starting the arc. Another significant downside is the need to change the lens to a different shade, with considerable changes to lower or higher welding amperage ranges.

The auto-darkening filters offer several advantages. First, the welder can see the workpiece and work area for setup prior to starting the arc. This reduces the tendency of getting arc flash and increases productivity. Another major benefit of using the auto-darkening filter is the ability to change lens shades by simply selecting a shade at the touch of a button. **FIGURE 2-4** shows the passive style helmet (left), the full-face shield for light cutting (middle), and the auto-darkening helmet (right).

Drawbacks to auto-darkening helmets are that some styles are very lightweight and spot sensors may be blocked out, causing the lens to flash on and off. If you read the warning labels inside many auto-darkening helmets, you will find that they caution against using them for overhead welding. This is because heavy spatter or molten metal can land on the cover plate and may damage the filter. These styles have a limited durability standard. If a heavy and durable style is required, an auto-darkening module can be inserted into some standard welding helmets in place of a passive lens filter. Finally, the cost of auto-darkening helmets may not seem practical. However, in a production environment, the cost is easily recovered through increased productivity. For the student learner, the comfort and ease of use can speed learning and offset the cost factor.

Welding helmets with filter lenses protect welders from welding arc rays and from UV and infrared rays given off by Oxy-Fuel Welding, heating, and cutting operations. ANSI Z49.1:1999 recommends lens shades, as does OSHA. The lens shade required may be dependent on several factors, such as welding amperage, flame size and intensity, welding smoke, and the reflectivity of the metal being welded. While charts and tables (**TABLES 2-4, 2-5,** and **2-6**) are useful, the shade needed may vary from one person to another. Use the darkest shade possible that allows good visibility of the weld pool and arc.

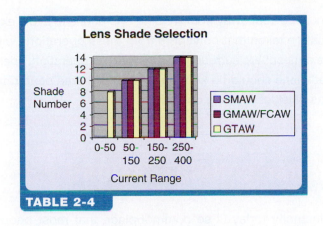

TABLE 2-4

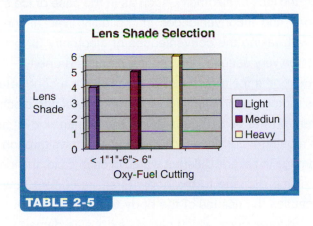

TABLE 2-5

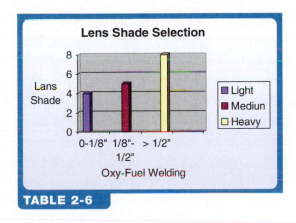

TABLE 2-6

WHAT OTHER UV CONSIDERATIONS MUST BE MADE? OSHA suggests that where the work permits, the welder should be enclosed in an individual booth painted with a finish of low reflectivity, such as zinc oxide (an important factor for absorbing UV radiations), and lamp black or shall be enclosed with noncombustible screens similarly painted. Booths and screens shall permit circulation of air at floor level. Workers or other persons adjacent to the welding areas shall be protected from the rays by noncombustible or flameproof screens or shields or shall be required to wear appropriate goggles.

Vapors from chlorinated hydrocarbons used for degreasing and other cleaning operations can turn to dangerous phosgene gas, a toxic gas used as a chemical weapon during World War I, when exposed to welding arc rays. These compounds must be kept away from welding operations to avoid exposure to arc rays. Care must be taken that no vapors from these operations will reach or be drawn into the atmosphere surrounding any welding operation. Trichloroethylene and perchlorethylene should be kept out of atmospheres penetrated by the UV radiation of gas-shielded welding operations.

ELECTRICAL CONSIDERATIONS

The use of electricity today is so commonplace that most people take it for granted. In fact, the loss of electricity, such as in the case of ice storms, causes major disruptions and inconveniences in the daily lives of the individuals it affects. Along with taking the conveniences of electricity for granted, we also forget that it can be very dangerous when not handled properly.

Electrical current is the movement of electrons in a conductor. The flow of current enables motors to turn, lights to light, and metal and electrodes to melt. The same current flow is very detrimental to the human body if any part of the body becomes a path for current flow. When current flows through body tissues, it can cause damage in several different ways. In each case, the longer the exposure, the greater the damage to the body. First, as current flows through the body, its flow causes the tissues of the body to heat up and burn. The burn is generally a third-degree burn, which causes extensive damage and prolonged healing time.

The second effect is to the body's nervous system. The heart, lungs, and muscles are controlled by the brain with electrical signals. These signals are conducted to the organs by the nervous system. When current is conducted through the body, the nervous system can be damaged and/or the organs can react incorrectly to the current flow. An example is when a person grabs an electrified conductor; the individual may not be able to let go of the conductor. In this case, the current flow has caused the muscles of the hand and arm to contract.

Fortunately, human skin when dry is not a very good conductor and has a high amount of electrical resistance. The top layer of our skin is called the *epidermis* and consists of a layer of skin cells with intermittent hair and sweat glands running through it to the surface. The skin cell layer has a very high resistance to current flow, especially in the harder, more callused skin of an adult. Underneath the epidermis lies the *dermis*, which consists of blood vessels, muscle, and nerve endings, all of which are very good conductors of electrical current. The epidermis acts as an insulating layer over the very conductive body underneath it.

WHAT ARE THE ELECTRICAL HAZARDS FOR THE WELDER?
Even though a welder is aware of the necessity of working in a dry environment, considerations must be made regarding ambient moisture and sweat. A dry epidermis has a high resistance value, but when the epidermis is wet, this resistance value begins to drop. Just being wet from water has a negative effect, but when the epidermis is covered with sweat, the resistance drops even more. The lower resistance value

of sweat is caused by the conductive amount of salt in the sweat and the flow of conductive sweat through the sweat glands, which runs through the epidermis and dermis layers. This is of special concern for a welder in that it is very common for a welder to become hot and sweaty on the job. Another concern for welders is in the course of doing their job, they can get cuts and abrasions on their skin. This type of damage also reduces the insulating level of the epidermis.

Because the skin is not always a great insulator, it is important to protect the body by adding additional layers of insulation. The same leathers, work clothes, boots, and gloves that protect the welder from heat, sparks, and infrared rays act as this layer.

Electrical safety when arc welding is even more important in that the welder is exposed to several different sources of electrical hazard. Sources of possible electrical hazard are the primary input power and the secondary welding output. Each one of these sources can cause injury or death. But when the proper safety precautions are followed, the chances of anything bad happening is greatly reduced—if not eliminated.

WHAT IS THE SIGNIFICANCE OF PRIMARY INPUT POWER?

Primary input power is used to operate the welding power source and electrical tools (grinders, drills, radios, etc). It is always in the form of alternating current (AC), and its voltage level can vary from 115 vac to 575 vac. People are exposed to primary power every day at home, but what makes it more dangerous when welding is the welding environment itself. At home, equipment is generally not wired to 460 vac three phase, tow motors are not running over cords, and sparks are not damaging insulation, as might be the case in any welding environment. Therefore, extra precaution must be taken in the welding environment.

The welding environment can be hard on equipment. It is important to avoid the use of damaged equipment and power cords. The insulation found on power cords is there to provide a protective layer between the welder and the electrical conductors inside the cable. If the insulation has become damaged, the cord must be repaired or replaced. Failure to do so can cause a shock when the welder touches the exposed conductor and any grounded surface.

WHAT IS THE SIGNIFICANCE OF PROPERLY GROUNDED CONNECTIONS?

Current flow of a primary electrical system is designed to flow back and forth through the electrical conductors or through a grounded conductor to earth ground. **FIGURE 2-5** shows a simple single-phase primary electrical circuit. The two conductors are used to power the welding power source. The grounded conductor is a safety conductor that is designed to only conduct current in case of emergency. Without ground, a damaged electrical cable on a power source or a tool such as a metal-encased grinder could allow electricity to pass through the worker. The ground is designed with this in mind. It provides a path for electrical current to follow rather than passing through the user.

FIGURE 2-6 shows how the ground conductor is used to safely conduct current to earth ground when one of the primary conductors touches the power source case. For this reason, all noninsulted cases of electrical equipment need to be grounded.

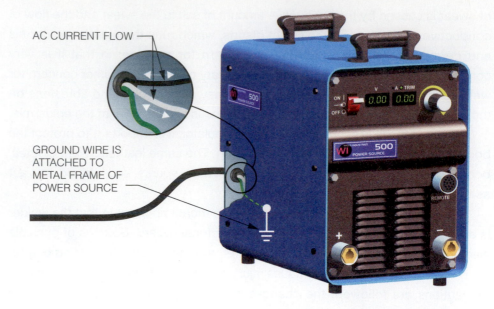

AC CURRENT FLOW

GROUND WIRE IS
ATTACHED TO
METAL FRAME OF
POWER SOURCE

FIGURE 2-5 A simple single-phase primary electrical circuit

AC CURRENT FLOW

WHITE WIRE IS
DAMAGED AND
SHORTED OUT
TO THE METAL
FRAME OF
POWER SOURCE

FIGURE 2-6 For safety, in case of electrical wire damage, all electrical equipment must be grounded.

SECONDARY WELDING OUTPUT

WHAT IS THE SIGNIFICANCE OF SECONDARY WELDING OUTPUT?

Secondary welding output is the current flow used to arc weld. The amount of current flow is controlled by the welding power source. This current flow is designed to flow from one welding output connection through the welding circuit and back to the other output connection. **FIGURE 2-7** shows current flow though a Gas Metal Arc Welding (GMAW) circuit. What causes a secondary welding circuit to become unsafe is when connections are not tight or conductors are in poor shape.

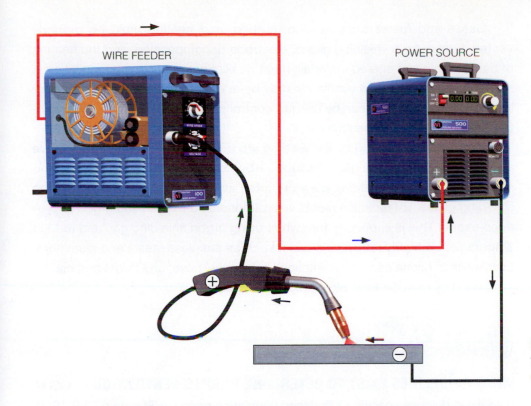

WIRE FEEDER

POWER SOURCE

FIGURE 2-7 Electrical current flow of a Gas Metal Arc Welding circuit

WHAT CAUTIONS ARE TAKEN REGARDING THE WELDING ENVIRONMENT?

All welding equipment should be manufactured according to appropriate standards. The National Electrical Manufacturers Association (NEMA) has developed standards for electric welding equipment. Consult the owner's manual or other manufacturer specification sheets to verify compliance. Follow manufacturer recommendations for installation and safe use of equipment.

Work and electrode leads and connections must be set up correctly and maintained to ensure that no electrical shock can occur. Electrical equipment must also be installed by a licensed electrical contractor by following guidelines of electrical codes. Electrical connections must be properly grounded.

Do not use electrode holders with damaged insulators or damaged cables. Dry clothing, leather work boots, and gloves must be worn at all times. Never operate equipment when damp conditions exist that can lead to electrical shock. When welding operations are required in damp areas, the welder must be protected from that damp environment by means such as an insulating platform or the use of insulated rubber boots.

GASES AND FUMES

WELDING SEEMS SMOKY. DOES SMOKE PRESENT A PROBLEM? The first defense against fumes and gases is for the welder to keep away from the **plume**. The plume is a column of heat that rises from the welding arc, and it contains particulate matter and vapors caused by the breakdown of filler metal, welding gas, and base metal constituents. The majority of fumes disintegrate as they are carried into the atmosphere by rising heat of the arc. While proper ventilation is important, keeping the welder's breathing zone out of the plume is critical.

Gases and fumes from welding, cutting, and heating must be properly vented. Fumes from shielding gases, electrode decomposition, and the heating of base metals and coated materials must be eliminated. If natural ventilation is not practical, mechanical ventilation may be acceptable to reduce exposure or local exhaust ventilation can be used to control shielding gas and fume concentrations in the welder's breathing zone.

Toxic gases produced by the welding arc tend to break down as they move away from the arc. These gases include carbon monoxide, nitrogen dioxide, and ozone. Ozone accumulations are most often associated with the use of argon shielding gases. UV radiation reacts with atmospheric oxygen and argon to produce ozone. This is especially true when using argon shielding gas and at high amperages on highly reflective metals, such as stainless steels and aluminum. Low levels of ozone can cause irritation to the eyes, nose, and throat and can be eliminated by use of a fresh air supply to the welder.

VENTILATION

WHAT GUIDELINES EXIST TO DETERMINE PROPER VENTILATION? OSHA
guidelines discuss specific air flow and ventilation needs in Standard CFR 1910. Fumes generated in the welding arc must be removed from the breathing zone by a suitable means of mechanical ventilation if welding is done:

- In a space of less than 10,000 cubic feet (284 m³) per welder
- In a room having a ceiling height of less than 16 feet (5 m)
- In confined spaces or where the welding space contains partitions, balconies, or other structural barriers to the extent that they significantly obstruct cross ventilation

Ventilation may be fixed-position enclosures as in a welding booth designed specifically for welding (**FIGURE 2-8**) or freely moveable hoods placed as nearly as possible to the welder.

FIGURE 2-8 This booth, specifically designed for arc welding, has a moveable exhaust ventilation at the table and overhead exhaust ventilation.

FIGURE 2-9 Moveable exhaust ventilation at the table

WHEN IS VENTILATION NOT ENOUGH PROTECTION?

When welding applications include fluorine compounds, zinc, lead, beryllium, cadmium, mercury, cleaning compounds, and cutting stainless steels, special precautions apply. In many cases, work on these special materials must be done by using local exhaust ventilation (**FIGURE 2-9**) or airline respirators. When in doubt as to what constitutes proper ventilation, see OSHA standards CFR 1910.252(c)(5) through (c)(12).

ANY CONCERNS FOR WELDING IN CONFINED SPACES?

Welding in confined spaces must be done in a manner that prevents accumulation of toxic materials or possible oxygen deficiency. Adequate ventilation must be provided. Combustion of electrodes, decomposition of fluxes, and the use of shielding gases produce hazardous fumes that will accumulate and are likely to displace air in the welder's breathing zone. Some form of air circulation and replacement is necessary.

In situations where adequate ventilation cannot be provided, airline respirators or hose masks must be used. These must be approved by the NIOSH under 42 CFR Part 84. NIOSH 42 CFR Part 84 also specifies special equipment for work in self-contained units. A full-facepiece, pressure-demand, self-contained breathing apparatus or combination full-facepiece, pressure-demand supplied-air respirator with auxiliary, self-contained air supply is used in areas immediately hazardous to life. This equipment must be fitted and cleaned regularly to prevent leaks and illness.

FIRE PREVENTION

WHAT CONSIDERATIONS ARE MADE FOR POSSIBLE FIRE HAZARDS?

Welding should be done in areas with established potential hazards related to welding in mind, and, because sparks from welding arcs can travel as far as

35 feet, fire prevention is a serious consideration. Fire hazards include sparks, molten metal and slag from welding, and cutting and grinding operations. Properly maintained fire extinguishers must be available. When work cannot be done in a shop designed specifically for welding, other steps must be taken. These steps include the assignment of a **fire watcher** to the work area. A fire watcher is a person who observes the work and work area during and for a half hour after welding and cutting operations are completed to prevent occurrence of a fire. Fire watchers have fire extinguishers available and are trained in their use. They are trained to be familiar with facilities for sounding an alarm if a fire should occur.

Sparks from welding can travel significant distances, and care should be taken to ensure that flammable or combustible materials are not within reach. OSHA 1910.252 (a) (2) (iii) (A) requires that a fire watcher is used under circumstances where a fire other than a minor one might occur and under the following circumstances:

- Appreciable combustible material, in building construction or contents, are closer than 35 feet (10.7 m) to the point of operation.
- Appreciable combustibles are more than 35 feet (10.7 m) away but are easily ignited by sparks.
- Wall or floor openings within a 35-foot (10.7 m) radius expose combustible materials in adjacent areas, including concealed spaces in walls or floors.
- Combustible materials are adjacent to the opposite side of metal partitions, walls, ceilings, or roofs and are likely to be ignited by conduction or radiation.

Use of written procedures, including a **hot work permit**, can be a great help in fire prevention where welding and cutting operations cannot be done safely. A hot work permit is a document that provides step-by-step procedures for workers to follow in areas where welding and cutting are not normally performed and is issued by the fire safety supervisor or other responsible party. A checklist of safety considerations is followed prior to approval of welding and cutting operations and also after welding and cutting operations are completed. Examples for a hot work safety checklist follow:
General precautions:

- Sprinklers are in service
- Welding and cutting equipment is in good repair.

Precautions within 35 feet of work:

- Floors are swept clean of combustibles.
- Combustible floors are wet down and covered with damp sand or fire-resistive sheets
- Flammable liquids are removed; other combustibles, if not removed, covered with fire-resistive sheets or materials.
- Explosive atmosphere is determined.
- Wall and floor openings are covered.

Work on walls and ceilings:

- Construction is noncombustible and without combustible insulation or covering.
- Combustibles are moved away from other side of wall.

Work on enclosed equipment:

- Enclosed equipment is cleaned of all combustibles.
- Containers are purged of flammable liquids and gases.

Fire watch:

- Fire watch will be provided during work, for at least 30 minutes after work, and during breaks in work.
- Fire watch is supplied with suitable fire extinguisher and charged small hose.
- Fire watch is trained in use of equipment and in alarm-sounding procedures.

EXPLOSION

WHAT SPECIAL PRECAUTIONS MUST BE TAKEN WHEN WELDING ON CONTAINERS OR ENCLOSURES?

Special care must be taken when welding on containers, whether they hold or have held combustibles or not. Tanks, barrels, or other containers that have held fuels, oils, tars, or other flammable materials might become flammable or explosive, or toxic vapors may form that can be harmful to the welder/cutter. These containers must be systematically cleaned to remove harmful elements completely before welding or cutting operations may begin. When hoses or pipes are connected to containers such as a drum or tank, they must be disconnected or closed off from the work completely.

Any hollow container when heated can burst or explode because of the expansion of air or gases it contains. For this reason, it is important that all hollow spaces and cavities on containers be vented to permit the escape of air or gases before preheating, cutting, or welding. **Purging** with inert gas is recommended in all cases to eliminate the possibility of combustion and therefore an explosion taking place. Purging refers to removing explosive atmosphere from a closed container by pushing it out with low-pressure concentrations of a gas, such as nitrogen, argon, or helium. Pumping in an inert gas such as argon forces out any air, oxygen, or other gases that may be present. Never attempt to cut or weld on containers without special training and following the recommendations of *Safety in Welding, Cutting, and Allied Processes*, ANSI Z49.1.

COMPRESSED CYLINDERS

HOW SHOULD COMPRESSED GAS CYLINDERS BE STORED?

Inside of buildings, compressed cylinders shall be stored in a well-protected, well-ventilated, dry location at least 20 feet (6.1 m) from highly combustible materials to prevent exposure of those materials to gases in case a leak occurs. Acetylene, propane, and others gases should be stored in a well-ventilated area away from oxygen and in the vertical position at a minimum distance of 20 feet (6.1 m) or by a noncombustible barrier at least 5 feet (1.5 m) high having a fire-resistance rating of at least one-half hour.

A suitable cylinder truck, chain, or other steadying device shall be used to keep cylinders from being knocked over while in use. Compressed gas cylinders shall be secured in an upright position at all times except, if necessary, for short periods of time while cylinders are actually being hoisted or carried.

HOW MIGHT A CYLINDER BE DAMAGED?
Care must be taken to prevent contact between compressed gas cylinders and flames, sparks, and electrical cables, which could damage them and lead to a cylinder failure. Inspect gas cylinders frequently. Any cylinder that leaks, has a bad valve, has damaged threads, has arc strikes on the cylinder must be identified, removed from service, isolated, and reported to the supplier in accordance with ANSI Z49.1.

HOW ARE CYLINDERS MOVED?
Cylinders should be transported with valve caps securely in place. Unless cylinders are firmly secured on a special carrier intended for this purpose, regulators shall be removed and valve protection caps put in place before cylinders are moved.

ARE THERE SPECIAL CONCERNS REGARDING FUEL GASES?
Acetylene cylinders are equipped with fusible plugs that melt below the boiling point of water (212°F). At times a cylinder may be exposed to weather conditions, such as sleet or snow. In order to use the equipment safely, it may be necessary to move the frozen, iced-up cylinder into a heated building to thaw and remove ice and snow to allow safe use of the equipment. Never use flame heat to thaw cylinders or use tools to remove ice or snow.

Acetylene tanks that have been lying on their side must be put upright for at least 15 minutes prior to use. Acetylene cylinders are not just a hollow cylinder with gas inside. Unlike other gas cylinders, they are filled with a porous material that stabilizes the gas at pressures up to 250 psi (1724 kPa). This porous material is saturated with acetone. If a cylinder has been lying on its side, the acetone will leak into the shoulder of the cylinder and must be put upright to allow the acetone to filter back into the cylinder—or else the acetone is likely to enter and damage the gas regulator.

Liquid gases such as liquid propane cylinders must not be transported or stored lying on their sides. They are supplied with a burst disc, which traps the excess gas pressure within the tank in case it is overfilled or subjected to high temperature gas expansion within the cylinder. If the cylinder is lying on its side and pressures increase too high, the burst disc might release liquid propane and cause the storage area to become saturated with explosive gas vapors.

HOW IS A COMPRESSED GAS CYLINDER SET UP?
High-pressure compressed gas cylinders must be handled with care. Cylinder caps must be secured firmly when in transport. Should a cylinder fall without the cap in place, the valve stem might break, and the high-pressure gas will propel the tank like a missile. Transport cylinders by chaining them to a cylinder cart. Furthermore, while in use,

FIGURE 2-10 Always stand on the opposite side of the valve when opening it.

cylinders should always be secured in the vertical position to the workstation or the welding machine cart.

The next step is to briefly crack open the cylinder valve, making sure that the valve opening is not pointing toward anyone. The operator stands opposite the valve opening (**FIGURE 2-10**). Do not stand in front of or behind the regulator. Cracking the valve prevents the possibility of dust or debris from the valve opening entering the regulator, which could lead to regulator failure and injury.

Always match the correct regulator to the cylinder and gas for which it was designed. Before attaching the regulator, check the cylinder valve stem opening and the regulator connections for damage. Attach the regulator. Back off the pressure-adjusting screw (counter-clockwise) before pressurizing the regulator. High-pressure cylinder valves, except fuel gas, should be opened all the way when in use and closed completely when not in use, with the pressure adjusting screw turned out. The fuel gas cylinder is opened from one-fourth to one-half turn. After pressurizing the regulator, turning the adjusting screw clockwise will increase the working pressure and the flow to the workpiece (**FIGURE 2-11**).

Before using any gas, check the system for leaks. With shielding gases like argon, carbon dioxide, or other mixed gases, a simple test involves only a couple

FIGURE 2-11 Working pressure on a shielding gas regulator.

of steps. First, pressurize the system, and set the appropriate gas flow. Then, shut off the cylinder valve completely, and watch the working pressure and the cylinder pressure needles for any movement. If a needle drops, there is a leak that must be corrected. Inspect hose and valve stem connections for damage. With fuel gases and oxygen, it is necessary to check for leaks with an approved leak-detecting solution. These can be acquired from a welding supplier.

CHAPTER QUESTIONS/ASSIGNMENTS

MULTIPLE CHOICE

1. What types of clothing materials are preferred for welding?
   a. Heavy dark polyester ✗
   b. Flame-resistant wool or cotton
   c. Synthetic composite fiber ✗

2. What lens shade is suggested for GTAW at current of 90 amps?
   a. #8
   b. #10
   c. #12
   d. #14

3. What lens shade is suggested for light Oxy-Fuel Cutting?
   a. #2
   b. #4
   c. #5
   d. #6

4. How far can sparks travel?
   a. 5 feet (1.5 meters)
   b. 20 feet (6.1 meters)
   c. 35 feet (10.7 meters)

5. How far apart should compressed oxygen and fuel gases be stored?
   a. 5 feet (1.5 meters)
   b. 20 feet (6.1 meters)
   c. 35 feet (10.7 meters)

6. To prevent formation of phosgene gas, what materials should be kept out of atmospheres penetrated by the UV radiation of gas-shielded welding operations?
   a. Argon and carbon dioxide
   b. Polyester and nylon
   c. Trichloroethylene and perchlorethylene
   d. Fluorine compounds, zinc, lead, and beryllium

7. Inside of buildings, compressed cylinders should be stored in a well-protected, well-ventilated, dry location at least how far from highly combustible materials?
   a. 5 feet (1.5 meters)
   b. 10 feet (3 meters)
   c. 20 feet (6.1 meters)
   d. 35 feet (10.7 meters)

8. Acetylene cylinders are equipped with what safety mechanism that melts below the boiling point of water (212°F/100°C)?
   a. Burst disc
   b. Fusible plugs
   c. Spark testers

FILL IN THE CHART

9. For the example shown, fill in the blanks for levels of risk for health, flammability, reactivity risks, and recommended PPE.
   a. Health ___2___
   b. Flammability ___0___
   c. Reactivity ___1___
   d. Recommended PPE ___M___

HEALTH	3
FLAMMABILITY	2
REACTIVITY	1
PERSONAL PROTECTION: M	

10. Workers or other persons adjacent to the welding areas shall be protected from the rays by noncombustible or flameproof _____ or _____ or shall be required to wear appropriate _____.

11. The first defense against fumes and gases is for the welder to keep away from the _____, a column of heat that rises from the welding arc.

12. Contained in the plume are _____ matter and _____ caused by the breakdown of filler metal, welding gas, and base metal constituents.

13. UV radiation reacts with atmospheric oxygen and _____ to produce ozone.

14. When welding applications include fluorine compounds, _____, lead, beryllium, cadmium, mercury, _____ compounds, and cutting stainless steels, special _____ may apply.

15. A danger in confined spaces is that the combustion of _____, the decomposition of fluxes, and the use of shielding _____ produces hazardous fumes that will accumulate and are likely to _____ air in the welder's breathing zone.

16. Welding should be done in areas with established potential hazards related to welding in mind, and because sparks from welding arcs can travel as far as _____, fire prevention is a serious consideration.

17. Tanks, barrels, or other containers that have held fuels, oils, tars, or other flammable materials might become _____ or explosive, or _____ vapors may form that can be harmful to the welder/cutter.

18. Never attempt to _____ or weld on _____ without special training and following the recommendations of *Safety in Welding, Cutting, and Allied Processes*, _____ Z49.1.

19. Unless cylinders are firmly secured on a special carrier intended for this purpose, _____ shall be removed and valve protection caps put in place before cylinders are moved.

20. After pressurizing a gas regulator, turning the adjusting screw _____ will increase the working pressure and the flow to the workpiece

21. What is an MSDS?

22. What is the purpose of HMIS?

23. List 6 PPE.

24. What protective apparel is worn over clothing?

25. What eye and head protection is required?

26. In many cases, work on special materials must be done by using local exhaust ventilation or airline respirators. When is this true?

27. What are the hazards from welding and cutting operations that could result in a fire?

Chapter 3

SHIELDED METAL ARC WELDING

LEARNING OBJECTIVES

In this chapter, the reader will learn how to:

1. Determine how long the Shielded Metal Arc Welding process has been used in industry.

2. Operate the Shielded Metal Arc Welding process.

3. Select the type of power source and peripheral components that would work best at home, school, or in industry.

4. Identify and correctly use the various controls on power sources and their characteristics.

5. Set up for Shielded Metal Arc Welding by controlling amps, arc length (volts), travel speed, and work angles.

6. Adjust the power source to produce an arc that is best for the weld position, base metal, and desired outcome.

7. Select electrodes for the base metal used and the desired outcome.

8. Integrate process knowledge and technique, resulting in acceptable welds.

KEY TERMS

Alternating Current (AC) 48

Amperage 46

Arc Length 50

Arc Voltage 57

Backhand Technique 58

Direct Current Electrode Negative (DCEN) 48

Direct Current Electrode Positive (DCEP) 48

Duty Cycle 41

Electrode Holder 43

Electrode Lead 42

Open-Circuit Voltage (OCV) 56

Polarity 48

Travel Angle 57

Travel Direction 57

Travel Speed 59

Voltage 56

Welding Amperage 52

Work Angle 58

Work Clamp 45

Work Lead 42

myweldinglab

Shielded Metal Arc Welding (SMAW) has been the stalwart of the structural welding industry for decades. While the first coated electrode rods were developed in the early 1900s, their use did not become popular until after World War I. Advances in coated electrodes were explored, and in 1929 they were made available to the public by the Lincoln Arc Welding company.

While many structural parts are fabricated in shops with semiautomatic welding equipment, SMAW is primarily used onsite for erection of structural members and piping systems. Portability and the ability to weld without the use of external gas supplies that are not suitable for welding in open environments make SMAW the logical choice for many applications.

Shielded Metal Arc Welding equipment is simpler to use and set up than equipment for gas-shielded welding processes, such as Gas Metal Arc Welding, Flux Cored Arc Welding, and Gas Tungsten Arc Welding. Electrical circuitry and components are generally less complex and cheaper. No external gas supplies, gauges, regulators, gas cylinders, or peripheral equipment are needed. Motor generator power sources can be transported to work sites on truck beds, and some small power supplies that operate on 110 volt circuits can even be carried to the work site.

Shielded Metal Arc Welding is a joining process where heat for welding is generated between a consumable electrode and the workpiece (**FIGURE 3-1**). The electrode is coated with a flux that, as it disintegrates, provides a gas shield that protects the molten weld pool and, upon cooling, produces a slag that blankets the weld.

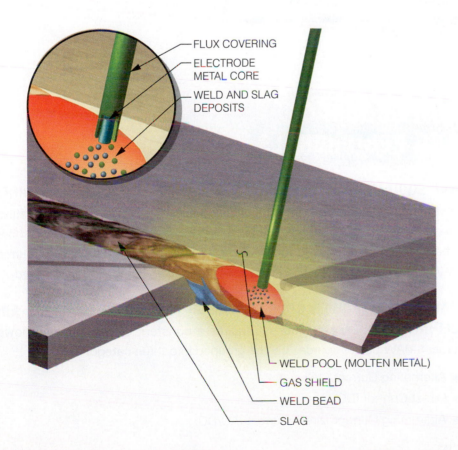

FLUX COVERING
ELECTRODE METAL CORE
WELD AND SLAG DEPOSITS

WELD POOL (MOLTEN METAL)
GAS SHIELD
WELD BEAD
SLAG

FIGURE 3-1 A breakdown of the Shielded Metal Arc Welding process

TABLE 3-1 shows the basic safety considerations for SMAW. **TABLE 3-2** shows recommended lens shades for welding at varying amperage ranges.

Head	Body	Hands	Feet
Safety Glasses	Flame-Resistant Wool or Cotton Pants and Shirt	Gauntlet-Style Welding Gloves	High-Top Leather Safety Work Boots
Welding Helmet	Flame-Resistant Jacket		
Welding Hat	Leathers		

TABLE 3-1

Read "Chapter 2: Safety in Welding" for a detailed discussion of safety related to workplace environment, workplace hazards, personal protective equipment, electrical consideration, gases and fumes, ventilation, fire prevention, explosion, compressed cylinders, and gases.

FIGURE 3-2 Basic equipment required for Shielded Metal Arc Welding

Arc Current	Suggested Shade
Less than 50	—
50–150	10
150–250	12
250–400	14

TABLE 3-2

POWER SOURCE AND PERIPHERALS

WHAT EQUIPMENT IS NEEDED FOR SHIELDED METAL ARC WELDING?

Shielded Metal Arc Welding uses one of the most simple equipment configurations available when compared to most other arc welding processes. **FIGURE 3-2** shows the basic equipment required for Shielded Metal Arc Welding: a constant current power source, electrode lead and holder, work lead and clamp, and power cord.

WHAT TYPE OF POWER SOURCE IS USED FOR SHIELDED METAL ARC WELDING?
The SMAW process requires a constant current welding power source. SMAW power sources can be grouped into three categories:

- Alternating Current (AC)
- Direct Current (DC)
- Alternating Current/Direct Current (AC/DC)

Within these categories are four different technology levels:

- Magnetic Amplifier (Mag Amp)
- Moveable Shunt
- Solid State
- Inverter

More in-depth information on the power sources for Shielded Metal Arc Welding, including engine-driven generators and the peripheral equipment, can be found in Chapter 15.

HOW DO I KNOW WHAT POWER SOURCE TO SELECT?

In order to select the right power source for the job, several factors must be considered. These factors include the required welding current and amperage range, arc on time or **duty cycle**, and incoming power (single-phase or 3-phase). Duty cycle is the amount of time the power source is designed to operate in a 10-minute time frame. A 20% duty cycle means the power source is designed to operate without damage from overheating 2 minutes in a 10-minute time frame. Not all duty cycle ratings are the same. Some manufacturers rate their duty cycle based on 32°F (0°C) at 0% relative humidity. Others rate their welding power sources at 85°F (29°C) at 50% relative humidity. Still others rate their welders at 104°F (40°C) at 95° relative humidity. This helps explain why welders at discount stores are so inexpensive compared to name-brand welders.

A welding power source must be selected that has sufficient amperage output and duty cycle to perform to the work requirements. The next consideration is for the type of output current to the work that is desired. While simple alternating current (AC) output welding power sources are less expensive than those that produce direct current (DC) output, operating characteristics are generally less favorable. A choice can be made among welding power sources that can supply AC, DC, or AC/DC outputs. It is important to note here that a welder may be restricted to the use of a specific output characteristic based on restrictions imposed by written welding procedure specification.

Finally, a consideration is made based on incoming power availability of single-phase or 3-phase current. Single-phase current is standard for residential construction and is the logical choice for home hobbyist and small shops. The choice of 3-phase is based on the efficiency of 3-phase current and the ability to recover the additional costs involved with its installation.

ALONG WITH THE POWER SOURCE, WHAT ELSE IS NEEDED TO START WELDING?

Whether using the magnetic amplifier, moveable shunt, solid state, or inverter-type power source, the selection of other equipment needed is fairly simple. The various parts include:

- Power Source
- Work Clamp
- Power Cord
- Welding Electrode Holder
- Welding Leads (cable)

FIGURE 3-3 Single-phase 230V plug

FIGURE 3-4 3-phase 460V plug

POWER CORD   Power cords are selected based on input power supplied to the welder. Incoming power may be single-phase or 3-phase, and this determines the receptacle configuration. **FIGURE 3-3** shows a typical single-phase, and **FIGURE 3-4** shows a 3-phase plug for Shielded Metal Arc Welding power input. All electrical connections should be installed by a qualified electrician in accordance with an established electrical code.

WELDING LEADS   Two electrical cables are required to complete the circuit from the power source: one to the electrode and another to the workpiece. These are called the **electrode lead** and the **work lead**. The electrode lead routes current between the power source and the electrode holder, while the work lead carries current between the power source and the workpiece. Lead cables have a core composed of a number of copper wire strands bound together and encased in reinforced insulating material and then coated with rubber or plastic.

The size of the welding leads required is dependent on two factors: the maximum amperage for welding and the distance from the power source to the work. Higher amperages require larger diameter leads to allow adequate current flow. There is also a voltage drop over greater distances to the work. In order to compensate for these losses, it is critical to select welding leads large enough to accommodate electrical needs. Work and electrode leads should both be the same size. They are sized according to their diameters. The standard for wire sizes is American Wire Gauge.

Weld Type	Amps	50 Feet	100 Feet	150 Feet	200 Feet	300 Feet	400 Feet
Manual or semiauto-matic weld (up to 60% duty cycle)	75	6	6	4	3	2	1
	100	4	4	3	2	1	1/0
	150	3	3	2	1	2/0	3/0
	200	2	2	1	1/0	3/0	4/0
	250	2	2	1/0	2/0	4/0	
	300	1	1	2/0	2/0		
	350	1/0	1/0	3/0	3/0		
	400	1/0	2/0	3/0	4/0		
	450	2/0	3/0	4/0			
	500	3/0	3/0	4/0			

TABLE 3-3

TABLE 3-3 shows recommended lead sizes for varying distances to the work at required maximum welding currents. The larger numbers correspond to smaller diameter cable. In Table 3-3, the smallest cable size, 6, can be used at low settings of 75 amps at distances of 50–100 feet. Cable sizes increase in diameter in the following order: 6, 5, 4, 3, 2, 1, 1/0, 2/0, 3/0, 4/0, 5/0, and 6/0. Sizes 6–1 are identified as gauge, while the larger sizes, 1/0–6/0, are identified as aught. A 0 size is 1 aught (1/0), a 00 is 2 aught (2/0), 000 is 3 aught (3/0), etc. **FIGURE 3-5** shows 2 lead cable sizes: The smaller on the left is 4 gauge, and the larger one on the right is 1 gauge.

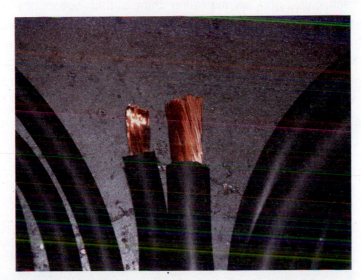

FIGURE 3-5 A 4-gauge wire (left) compared to a 1-gauge wire (right)

TABLE 3-4 shows voltage drops, per 100 feet, for different lead cable sizes. Larger-sized welding leads can be used to minimize voltage drop. When selecting leads, both voltage drop and amperage loss should be taken into consideration to ensure required operating parameters can be maintained.

WELDING ELECTRODE HOLDER The **electrode holder** is a clamping device attached to the electrode lead used to hold the electrode in position for welding. The electrode holder is composed of several parts: copper alloy contacts or jaws,

Amps	#2	#1	#1/0	#2/0	#3/0	#4/0
50	1.0	0.7	0.5	0.4	0.3	0.3
75	1.3	1.0	0.8	0.7	0.5	0.4
100	1.8	1.4	1.2	0.9	0.7	0.6
125	2.3	1.7	1.4	1.1	1.0	0.7
150	2.8	2.1	1.7	1.4	1.1	0.9
175	3.3	2.6	2.0	1.7	1.3	1.0
200	3.7	3.0	2.4	2.0	1.5	1.2
250	4.7	3.6	3.0	2.4	1.8	1.5
300		4.4	3.4	2.8	2.2	1.7
350			4.0	3.2	2.5	2.0
400			4.6	3.7	2.9	2.3
450				4.2	3.2	2.6
500				4.7	3.6	2.8
550					3.9	3.1
600					4.3	3.4
650						3.7
700						4.0

TABLE 3-4

Voltage drop/100 feet of lead vs. cable size (American Wire Gauge)

insulating covers to prevent contact with the user or work surfaces, a spring-loaded clamping mechanism, and an insulating handle that covers the work lead connection to the electrode holder. **FIGURE 3-6** shows a partially disassembled electrode holder (top) that has had its insulating parts removed and a fully assembled one (bottom).

The electrode holder, attached to the electrode lead, brings electrical contact to the electrode from the power source. The electrode holder jaws have grooves machined in them to allow for holding the electrode at various different

FIGURE 3-6 A partially disassembled electrode holder (top) and a fully assembled one (bottom)

angles. This allows the welder to choose an angle that provides comfortable access to the weld based on weld joint design and the welding position of the work (**FIGURE 3-7** and **FIGURE 3-8**).

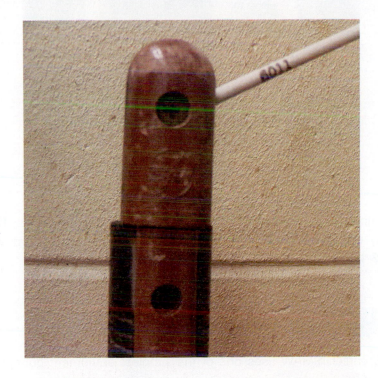

FIGURE 3-7 Grooves cut in the electrode holder allow the electrode to be securely positioned at an angle.

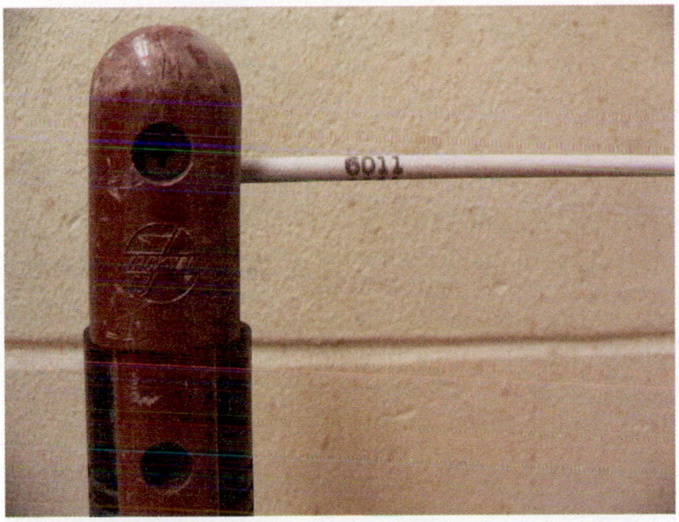

FIGURE 3-8 Grooves cut in the electrode holder allow the electrode to be securely positioned at 90°.

The end of the lead, opposite the electrode holder, is attached to the power supply by bolting a copper lug onto the power source terminal or by the use of a DINSE adapter inserted into the terminal (**FIGURE 3-9**). The DINSE adapter is the most convenient method to use. This is because the ease of connecting and disconnecting the leads with the DINSE—in order to change current characteristics— is favorable over having to use a wrench as required to change leads from DCEP to DCEN with the lug and nut type connection.

WORK CLAMP The **work clamp** is a device attached to the work lead that makes the physical connection to the workpiece from the power source. Like the electrode lead, it is connected to the power source with a copper lug or a DINSE adapter.

FIGURE 3-9 The end of
the lead may be bolted (left)
or attached by the use of a
DINSE adapter.

Work clamps can be spring-loaded, screw-down, or magnetic types. The most
commonly used work clamp is a spring-loaded type similar to that shown in
FIGURE 3-10.

CONTROLS AND CHARACTERISTICS

As stated earlier, SMAW uses one of the simplest equipment configurations.
Because of this simple fact, after polarity is selected, only two possible control
choices remain: amperage and arc control. **Amperage** is set on the power
source. In some cases, power sources also have an arc control function, which is
used to provide a more forceful arc for open-root welding on pipe and plate.
Other names for arc control are "dig" and "arc force."

HOW IS AMPERAGE CONTROLLED?   With SMAW, amperage is controlled by
adjusting the amperage dial (**FIGURE 3-11**) on the power source. Amperage
controls are typically dialed in increments or shown as a digital readout.
Increasing amperage increases the melting rate of the electrode and the wetting
out of a weld pool, making a wider weld.

FIGURE 3-10 The most
commonly used work clamp is
a spring-loaded type.

FIGURE 3-11 Amperage is
adjusted by a dial on the power
source.

FIGURE 3-12 If equipped on a power source, arc control is dialed in increments or given as a digital readout.

HOW IS ARC CONTROL SET?   Setting the arc control is accomplished by adjusting a dial on the front of the power source (**FIGURE 3-12**). Arc control is set to a percentage setting from 0%–100%. The set percentage corresponds to a percent of the power source amperage setting.

SMAW SETUP

There are a number of variables the welder has to select or set for proper welding control. Prior to welding, the welder selects the:

- Electrode
- Polarity
- Amperage
- Arc Control

HOW ARE ELECTRODES SELECTED FOR WELDING?   Electrode selection is very critical. Unless a written welding procedure specification exists that specifies the electrode to be used, several considerations must be made for selecting an electrode. The first criterion for selecting an electrode is to ensure that the filler metal is compatible with the base metal alloy type. Electrode composition, diameter, coating type, and thickness determine the electrode operating parameters. Other factors considered when choosing an electrode for welding include base metal thickness, joint design, and the welding work position. **FIGURE 3-13** displays the 5 common weld joints.

   When welding, it is critical that an electrode be selected that, when the weld is completed, will match the base metal properties and strength. For each base metal type—ferrous alloys, such as carbon steels, stainless steels, and cast iron, or nonferrous metals, such as aluminum or nickel—it is necessary to match appropriately the electrode's chemical composition compatibility with the base metal. This will ensure that the resulting weld properties, such as strength, ductility, and corrosion resistance, are sufficient.

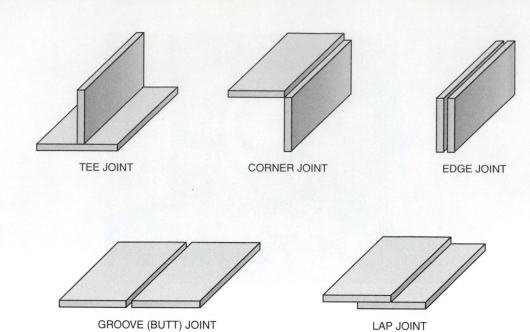

FIGURE 3-13 The 5 common weld joints

TEE JOINT

CORNER JOINT

EDGE JOINT

GROOVE (BUTT) JOINT

LAP JOINT

Where base metal thickness is concerned, larger diameter electrodes can help provide the higher amperage capacity and heat input necessary for welding thicker base metals. Select an electrode diameter that provides adequate amperage for fusion within its operating range. Smaller diameter electrodes work better on thin metals, and larger diameter electrodes work better on thicker metals.

The final determining factors in electrode selection are joint design and welding position. On the initial pass on open-root groove joints, an electrode (i.e., EXX10 or EXX11) with a cellulose-based coating is usually preferred. These electrodes produce deep-penetrating characteristics and a weld pool that freezes rapidly, suitable for root passes on pressure pipe and boiler repair. These welds are required to have root reinforcement, which is a desirable amount of weld buildup on the backside of the weld. Where welding position is concerned, small diameter electrodes work better for vertical and overhead welding, in part because the weld volume of the heated weld pool is easier to control. Electrode diameters greater than 5/32 inches (4 mm) are not practical for vertical and overhead welding. Not all electrode types can be used for welding in the vertical and overhead positions. When welding in these positions, an electrode that can be used in all positions is required.

WHAT ARE THE EFFECTS OF CHANGING POLARITIES? **Polarity** is the
selection of electrical lead settings for welding current. Polarity selection is critical with SMAW and determines the direction of electron or current flow. The influence of electron flow has significant impact on the heating of base metals and the electrode. Electrons flow from negative to positive, as shown in **FIGURE 3-14**.

The three polarity choices for welding are **direct current electrode positive (DCEP)**, **direct current electrode negative (DCEN)**, and **alternating current (AC)**. DCEP has, in the past, been referred to as reverse polarity, and DCEN has been referred to as straight polarity. With direct current settings, current flows in one direction. With DCEP, the electrode lead is connected to the positive terminal

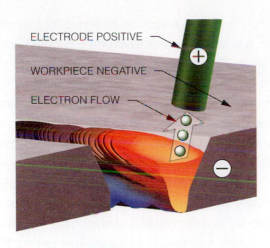

FIGURE 3-14 When arc welding, electrons flow from negative to positive.

on the power source. With DCEN, the electrode lead is connected to the negative terminal on the power source. With alternate current settings, the direction of current alternates from DCEP to DCEN in cycles called hertz.

Factors affected by the selection of polarity are varied. They include ease of arc starting, arc stability, metal transfer to the base metal, and the amount of weld penetration. DC is usually preferred over AC. The primary reason for this is that with DC, the electron flow moves continuously in one direction, resulting in a more stable arc than can be achieved with AC. Starting and maintaining the arc is also easier when welding with DC. While many electrodes can be operated by using either AC or DC, DC is almost always preferred because of its smooth operating characteristics and the ease of arc starting, even at lower amperages.

With DCEP settings, approximately two-thirds of the welding heat is present at the positive side of the circuit (**FIGURE 3-15**). This higher heat on the electrode provides sufficient arc force and localized heating of the base metals, resulting in deeper weld penetration, which is usually a desirable outcome on thicker base metals. Filler metal transfer to the base metal is smoother with DCEP than can be achieved with DCEN or AC.

With many electrodes, DCEN produces a more globular transfer than does DCEP. With DCEN, because only one-third of the welding heat is on the electrode, there are poorer penetration and fusion characteristics, making its use undesirable in most circumstances. DCEN is often recommended for welding on thin sheet metals. However, using electrode negative to reduce penetration should be considered as a

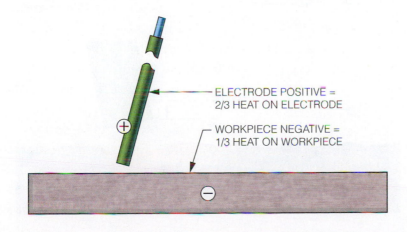

FIGURE 3-15 With DCEP, approximately two-thirds of the welding heat is present at the positive side.

last resort and only used with an electrode designed for use with DCEN. A better approach is to select a smaller diameter electrode and classification that are recommended for sheet metals. Some electrodes can be run on any polarity, so the choice of DCEN on thin sheet with the right electrode may be suitable. An E6012 electrode, which is designed for use with DCEN, is a good choice for welding thin metals.

AC, as its name implies, has electron flow that changes direction at regularly occurring intervals. While DC welding has many advantages, there are a couple of situations that warrant the use of AC. The first is low-cost AC-output only power supplies. Low-cost power supplies can be purchased that simply transform incoming high voltages to safer working voltages and transform incoming low amperage service to amperages suitable for welding operations. This eliminates the costs of circuitry that would otherwise increase power source prices. The second advantage of using AC for welding is that it reduces effects of arc blow on magnetic materials during welding. Arc blow is the deflection of the welding arc from its intended path due to magnetic forces imposed.

WHAT ARE THE CAUSES OF ARC BLOW, AND HOW IS IT CONTROLLED?

Arc blow is sometimes a problem encountered when using DC. When using DC, a magnetic effect of current flowing in one direction can become strong and cause the arc to wander. Using AC is often recommended for reducing the effects of arc blow. Because the current flow reverses continuously in AC, magnetic forces cannot build up in one direction and deflect the arc. However, in many situations, changing the current type or polarity is not an option due to existing welding procedure specification limitations.

There are other reasonable means for reducing arc blow. Use a sequential approach, starting with the easiest possible remedy and working through all possibilities until arc blow is resolved. The welder can reduce arc blow by decreasing the arc length and moving the electrode travel angle from 10°–15° to a 0° angle. **Arc length (FIGURES 3-16** and **3-17)** is the distance from the end of the electrode to the weld pool. Once arc blow disappears, the correct arc length and lead angle can be re-established. Other methods used to reduce effects of arc blow include:

- Have the work lead as close as possible to the weld
- Relocating the work lead clamp
- Welding away from the work lead
- Welding toward large tack welds
- Using the back-step sequence

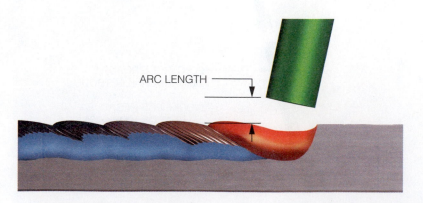

ARC LENGTH

FIGURE 3-16 Arc length is the distance from the end of the electrode to the weld pool.

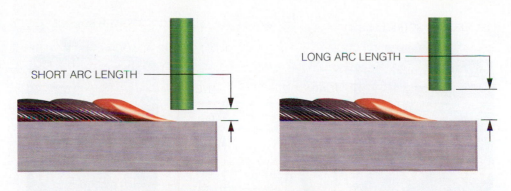

SHORT ARC LENGTH

LONG ARC LENGTH

FIGURE 3-17 Arc length

HOW IS POLARITY SET ON THE POWER SOURCE?

How to change or set polarity is determined by the type of SMAW power source available. Some DC inverter-type power sources first require that you select the process (**FIGURE 3-18**). If it is a DC-only power source, there is no polarity control on the power source. The welder simply attaches the electrode and work leads for the desired polarity. An example of the terminal setup for a DC power source is shown in **FIGURE 3-19**. A change between DCEP and DCEN is made by reversing the electrode and work leads at the power source terminal connections (**FIGURE 3-20**).

On solid state, AC/DC power sources, polarity is typically controlled with a switch (**FIGURE 3-21**) rather than physically changing the electrode leads at the

FIGURE 3-18 Some power sources require the selection of the welding process.

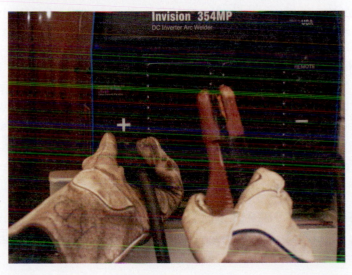

FIGURE 3-19 A terminal setup for a DC.

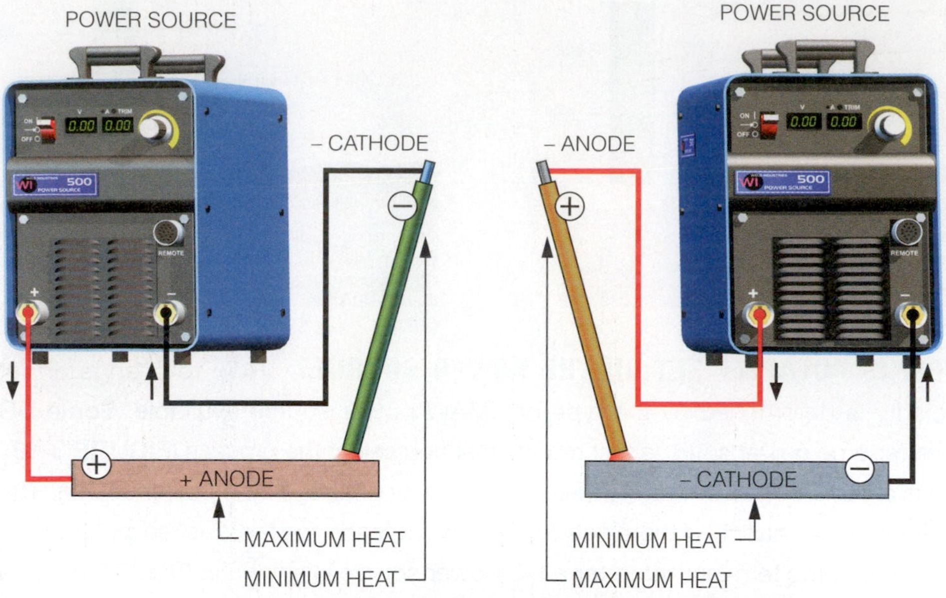

POWER SOURCE

POWER SOURCE

− CATHODE

− ANODE

+ ANODE

− CATHODE

MAXIMUM HEAT
MINIMUM HEAT

MINIMUM HEAT
MAXIMUM HEAT

FIGURE 3-20 Change between DCEP and DCEN may be made by reversing the electrode and work leads.

FIGURE 3-21 On solid state AC/DC power sources, polarity is typically controlled with a switch.

power source. The polarity control often allows the welder to choose polarities between DCEN and DCEP or to select AC. Many inverter and solid state power sources that are used for both Shielded Metal Arc Welding and Gas Tungsten Arc Welding (GTAW) set the polarity based on the welding process selection. When using SMAW, the electrode is positive for DC, and when using GTAW, the setting is electrode negative.

HOW IS AMPERAGE SET?　**Welding amperage** is the amount of the current output to the electrode and therefore the welding arc. It is increased or decreased by adjusting the amperage control on the front of the power source. The required welding amperage setting depends on several factors, including base metal thickness, electrode diameter and type, and the welding position. Increasing amps melts the electrode faster and produces more heat input into the base metal, which can improve fusion and increase penetration.

WHAT IS THE ARC CONTROL SETTING?　Not all power sources have an arc control. Arc control is used to prevent electrodes from shorting and sticking to the

base metal. It is commonly used for open-root welding of plate and pipe, with electrodes covered by cellulose-based fluxes, such as the E6010 and E6011 electrodes. Setting the arc control is accomplished by adjusting a control on the front of the power source. Typically, the control is set in increments from 0–100. The settings relate to a percent increase in the actual power source amperage settings when the electrode arc length is decreased. If the power source is set to operate at 100 amps and the arc control setting is at 30, then 30 additional amps is supplied at the instance where shorting would otherwise occur with a very short arc length. The additional amps produce a digging effect into the base metal. The amount of arc control needed varies with welder experience or preference.

The arc control percentage setting controls the amount the welding current increases when the arc voltage decreases with shorter arc lengths. With no arc control, the electrode may stick to the base metal, causing the arc to extinguish. Using arc control, the power source senses the low arc voltage and increases the welding current by the arc control setting percentage. In turn, the current increase causes the electrode to melt back, increasing the arc voltage. Once the arc voltage and arc length increase, the arc control turns off, and current returns to the amount set on the amperage control.

It is important to use arc control sparingly. Too high a setting makes it difficult to maintain a short arc length, as is required in most SMAW applications. For this reason, it is not used with other electrodes, such as the E7012, E7013, E7014, and E7018 electrodes. The use of arc control for all practical purposes is limited to electrodes in the F3 filler metal group, such as the E6010 and E6011. These electrodes and their uses are discussed in detail in the electrodes section.

SMAW Setup

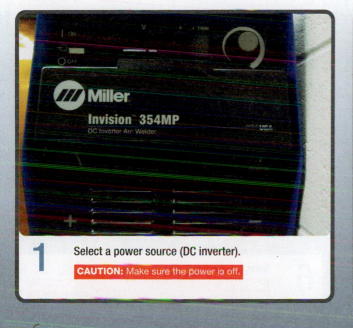

1 Select a power source (DC inverter).
CAUTION: Make sure the power is off.

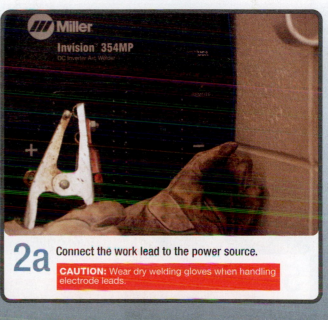

2a Connect the work lead to the power source.
CAUTION: Wear dry welding gloves when handling electrode leads.

2b **CAUTION:** Inspect electrical cables (leads). Never use damaged electrical cables.

3 Clamp the work lead to the work or table.

4a Connect the electrode lead to the power source.

4b **CAUTION:** Never use a damaged electrode holder.

5 Hang the electrode holder safely.

CAUTION: Always remove the electrode before hanging it up. Never lay the electrode holder on a metal surface, as it may become electrically alive.

6 Select an electrode for welding.

CAUTION: Never remove electrodes from a rod-holding oven without wearing gloves. Low-hydrogen electrodes are stored at over 250°F (121°C)

7 Turn on the power source.

8 Set the amps.

9 Set the arc control (dig).

10 Put on your personal protective equipment before welding.

11 Insert an electrode into the electrode holder. Squeeze the handle, insert the metal end of the electrode, and then release the grip.

OTHER THAN ADJUSTING THE POWER SOURCE AND SELECTING AN ELECTRODE, WHAT ELSE CAN THE WELDER DO TO AFFECT THE WELD?

There are a number of variables the welder has control over after the arc is initiated:

- Arc Length (voltage)
- Electrode Work Angle
- Electrode Travel Angle
- Direction of Travel
- Travel Speed
- Electrode Manipulation

HOW IS CORRECT ARC LENGTH DETERMINED? Arc length is not difficult to maintain if the electrode used is running at the proper amperage settings. The electrode diameter has a direct correlation to arc length. Larger diameter electrodes run at higher current ranges; consequently, a longer arc length is maintained. Using the electrode diameter as a reference, maintain an arc length equal to or slightly less than the electrode diameter. Keep in mind that electrode coatings affect the appearance of the arc and therefore the appearance of the arc length. Electrodes

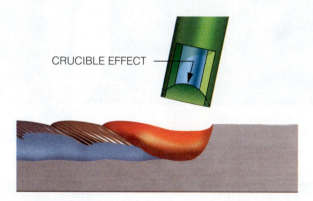

CRUCIBLE EFFECT

FIGURE 3-22 Electrodes with iron powder produce a crucible effect.

with iron powder coatings have a less visible arc because the electrode melts up slightly inside the coating; this is referred to as the crucible effect (**FIGURE 3-22**).

Too short of an arc length causes the electrode to short out and stick to the metal being welded. If too long an arc length is maintained, the arc will wander erratically, causing undesirable spatter and decreased penetration. The weld bead will be wide and irregular in shape.

VIDEO

WHAT IS VOLTAGE, AND HOW IS VOLTAGE ADJUSTED WITH SMAW?

Voltage is the electrical force that causes or pushes the flow of electrons through the welding leads. For this reason, it is also known as electrical potential or pressure. There are two types of voltage to be concerned about. The first is referred to as **open-circuit voltage (OCV)**. OCV is the voltage measured across the welding terminals before the arc is started. **FIGURE 3-23** shows an OCV of 83.1 volts for an inverter power source. When first turned on, a SMAW power source provides zero amperage and a typical OCV of around 60V–90V. The

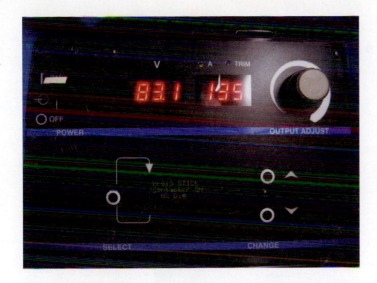

FIGURE 3-23 Power source shows OCV.

amperage remains zero until an arc is initiated for welding. After welding is started, amperage increases to the amperage set on the power source. At the same time, voltage drops from the OCV to about 18V upon electrode contact to the workpiece and then changes to a normal operating voltage for common small-diameter electrodes, between 20V–24V, depending on the arc length. Increasing the arc length greater than the electrode diameter results in excessively high arc voltage, and the arc becomes erratic. An even longer arc length extinguishes the arc.

Power sources used in shipyards operate at an OCV between 7 and 10 volts. The use of this low OCV is to compensate for the possible hazard encountered in the event the power source is submerged in water. The operating voltage is the typical 20V–24V. Arc starting is more difficult with the lower open-circuit voltages.

Arc voltage, the voltage in the welding circuit as measured at the welding arc, is a function of arc length. There is no voltage setting on the welding power source. Voltage has a proportional relationship with arc length. Welding arc voltage is the primary variable controlled by arc length, which influences the fluidity or the wetting out of the weld onto the base metal workpiece. Increasing the arc length increases the arc voltage and fluidity of the weld pool, while decreasing the arc length decreases the arc voltage and fluidity of the weld pool. Another way to look at effects of voltage is that increasing the arc length (voltage) makes a flatter, wider weld bead.

V I D E O

HOW SHOULD THE ELECTRODE BE POSITIONED IN RELATION TO THE BASE METAL?

The electrode position is described in several ways. The common electrode angles are often referred to as the travel angle and the work angle. Using the correct electrode angles is critical to making good quality welds. Together, these angles affect the bead shape, the position on the base metals, and the depth of penetration or fusion onto the base metals.

The **travel angle** is the electrode angle in relation to the direction of travel determined by the angle between a line perpendicular to the weld axis and the electrode axis (**FIGURE 3-24**). **Travel direction** refers to the path the electrode and weld pool progresses along the weld joint. The travel angle for most applications ranges approximately 5°–15° from perpendicular to the base metal axis. This range of angle gives both optimum penetration to the base metal and

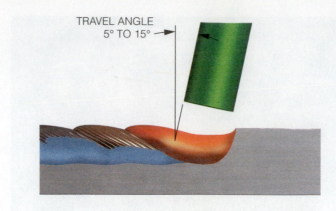

TRAVEL ANGLE
5° TO 15°

FIGURE 3-24 Travel angle is measured from the vertical.

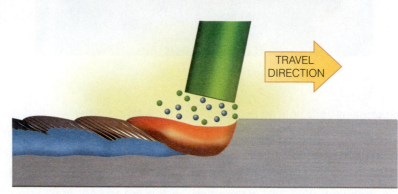

TRAVEL DIRECTION

FIGURE 3-25 Travel direction is from left to right for a right-handed person and from right to left for a left-handed person.

optimum weld flow, resulting in a uniform bead shape. Too steep of an angle reduces penetration and causes the weld bead to pile high onto the base metal, resulting in too convex a bead and a slightly V-shaped ripple pattern.

Travel direction also has significant effects on weld bead shape and penetration. A drag or **backhand technique**, as shown in **FIGURE 3-25**, with the electrode pointing opposite the direction of travel, provides for a narrow, convex weld bead with deep penetration. Using the backhand technique with correct travel angle is important. This method is always employed with Shielded Metal Arc Welding positions other than for vertical up-welding. Vertical up-welding is sometimes referred to as uphill welding progression, while vertical down-welding is sometimes called downhill welding.

The **work angle** is a variable angle of the electrode in an adjacent position to the workpiece surfaces. The work angle is the angle between the electrode axis and a line perpendicular to the workpiece surfaces. The work angle controls the flow of the weld edges onto the workpiece. For groove welds in flat and overhead positions, the work angle is perpendicular to the work. Single-pass fillet welds on T-joints and lap joints (**FIGURE 3-26**) have 45° angles for welds made in the flat and vertical positions. This results in a fillet weld that flows equal distances onto each base metal of the weld joint. If the electrode angle is not 45°, the fillet weld may be uneven and undercut at the weld toes may result (**FIGURE 3-27**).

To ensure even flow of the weld bead onto both metals being welded, it is important to watch the weld pool to see that it flows evenly onto both base metals. Gravity, magnetic forces, and the welding position impact considerably the flow of weld metal. Therefore, the electrode angle must be adjusted continuously to compensate for these external forces, taking care to control the flow of filler metal to ensure proper weld bead profiles.

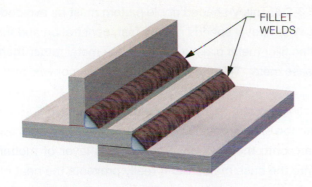

FILLET WELDS

FIGURE 3-26 Fillet welds

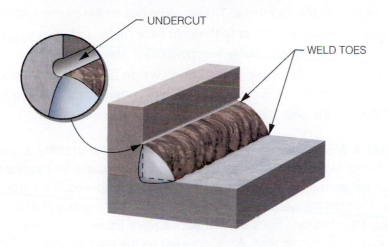

UNDERCUT

WELD TOES

FIGURE 3-27 Incorrect work angle could result in discontinuities

Travel speed is a welder-controlled variable that has a high degree of influence on weld bead shape, penetration, and fusion. Travel speed is a measure of the distance the weld bead progresses along the weld joint in inches per minute. The welder controls the travel speed, taking care to ensure that the electrode stays at the leading edge of the weld pool. If this step is followed, the welding arc will sufficiently heat the base metal, and the arc force mixes the filler material with the melted base metal. A correct travel speed produces a uniformly-shaped oval weld pool and a weld bead with a slightly oval ripple pattern (**FIGURE 3-28**). The weld profile will be slightly convex.

Too fast a travel speed occurs when the electrode leads the weld pool slightly. In this case, the electrode travels too fast to sufficiently heat and melt the base metal. The weld will be too convex, with an elongated V-shaped ripple

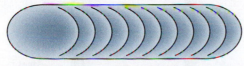

DESIRABLE WELD POOL AND BEAD SHAPE

UNDESIRABLE WELD POOL AND BEAD SHAPE

FIGURE 3-28 Correct travel speeds are essential to proper bead shape.

pattern (**FIGURE 3-28**). This V-shaped ripple pattern must be avoided on steel and aluminum alloys. When travel speeds are too fast, penetration and fusion are poor at best. The molten filler metal piles up on the base metal rather than melting and mixing with the base metal. This creates a high, narrow, convex, and ropy-looking weld bead.

Just as traveling too fast reduces penetration and fusion, so too can traveling too slow. With too slow a travel speed, the molten weld pool rolls out ahead of the welding arc onto the cold base metal. This layer of molten metal may appear to melt into the base metal, but it really prevents the heat of the welding arc from melting the base metal. As a result, there is no mixing and melting of the base metals with the filler metal. The arc impinges on the molten weld pool as the molten filler metal flows out onto the cold base metal, reducing penetration and resulting in poor fusion. Many inexperienced welders think that slow travel speeds and big welds equate with strong welds. This is a poor assumption to make, and the opposite is true. A bigger weld often leads to inadequate fusion and defects in the weld that can result in a weld failure.

VIDEO

STARTING A WELD   Two common techniques used for starting a weld are the scratch and tap methods. Either technique is acceptable as long as arc strikes are not left on the base metal after the weld is completed. As a rule, it is good practice to start the welding arc by tapping or scratching ahead of the weld start, bringing the electrode back so that subsequent travel results in the weld bead burning through the arc start. Arc strikes left on the base metal cause stress risers that may lead to cracking. Before attempting to start the arc, it is imperative that the amperage setting is correct for the electrode being used and the position the weld is to be made in.

VIDEO

The tapping method is accomplished by holding the electrode perpendicular to the base metal, tapping the end onto the base metal and lifting the electrode to the correct arc length (**FIGURE 3-29**). Begin at start position point A by holding the electrode perpendicular to the base metal. Bring the electrode

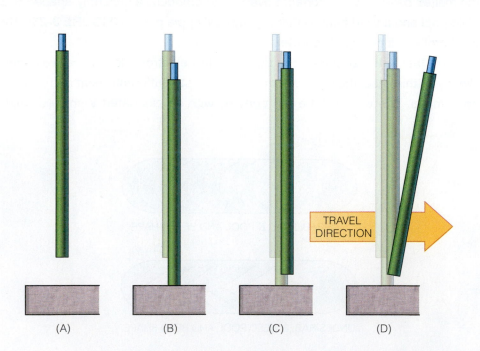

FIGURE 3-29 Tapping method for striking an arc

(A)   (B)   (C)   (D)

TRAVEL DIRECTION

straight down onto the base metal (position B), and bounce it off the base metal to the correct arc length (position C). Then, adjust to the correct travel angle, position D and begin travel down the base metal surface. Bouncing the electrode implies that the electrode rebounds off the plate in a sharp, quick manner. If the amperage is set correctly, it is easy to maintain a steady arc length. The arc length should be no greater than the diameter of the electrode core wire. Lifting the electrode too high extinguishes the arc, while too slow of a tap causes the electrode to stick to the base metal. The electrode sticking to the base metal must be avoided, since this may result in arc strikes on the base metal that can cause localized hardening and must be ground off the surface for code welding.

Using the scratch technique to start the arc involves striking the electrode against the base metal in the same way a dry wooden match is struck on a matchbox. An important factor with the scratch method is to avoid dragging the electrode across the base metal, which is a common problem for the beginning welder and leaves undesirable arc scratches on the base metal. To start the arc, the end of the electrode must be glanced off the surface of the base metal in a shortly swung arc (**FIGURE 3-30**). Start with a proper travel angle at position A, swing the electrode scratching the base metal at position B, and lift to position C, initiating the arc. Then, bring the arc back to the start position D, and begin travel down the base metal surface.

V I D E O

Starting an arc with a partially used electrode can be difficult. With some electrodes, such as the E7018, slag from the flux coating solidifies on the end of the electrode. Before restarting the electrode, this slag can be broken off by removing the electrode from the holder and bluntly tapping it directly on the base metal, after which it can then be used for welding, or by scratching or tapping the electrode on a nonconducting surface, such as a cement floor. With some other electrodes, such as the E7024, the core wire melts back conically inside the flux. These electrodes should restart easily.

V I D E O

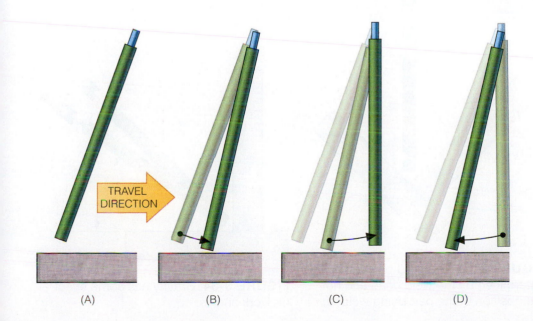

(A) (B) (C) (D)

TRAVEL DIRECTION

FIGURE 3-30 Scratch technique for starting an arc

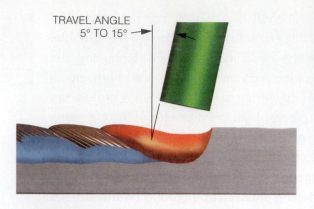

TRAVEL ANGLE
5° TO 15°

FIGURE 3-31 Most welding positions use a 5°–15° angle in the direction of travel.

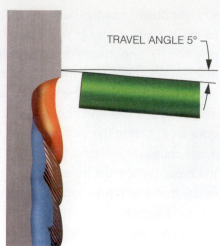

TRAVEL ANGLE 5°

FIGURE 3-32 Vertical welding positions use a 0°–5° angle in the direction of travel.

ELECTRODE MANIPULATION Once the arc is initiated, the electrode is held in position briefly, allowing the welding arc to melt the base metal and to flow to an appropriate width before travel begins. At this point, a backhand travel angle between 5°–15° (**FIGURE 3-31**) is used for all welding positions except vertical up-welding, where a slight forehand angle of 5°–10° (**FIGURE 3-32**) is usually used.

In addition to travel angle, the work angle must be adjusted according to the joint type and welding position. For welding in the flat and vertical positions, the electrode is positioned perpendicular to the workpiece, as shown in **FIGURE 3-33**. The effects of gravity help make the weld bead flow evenly at both the weld toes. Fillet welds are best deposited by using a stringer bead technique with the electrode centered at the joint interface.

Horizontal position welds are affected by gravitational forces and often require adjusting the work angle to compensate for these effects. **FIGURE 3-34**

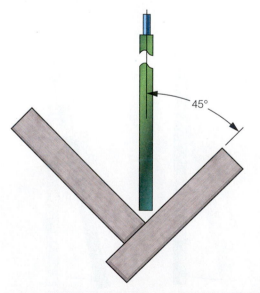

45°

FIGURE 3-33 The weld position is determined by the weld bead and not the position of the part being welded.

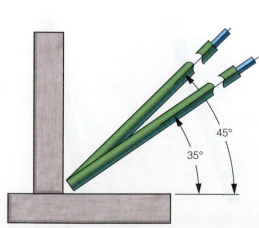

45°

35°

FIGURE 3-34 Gravity has some effect on work angles.

shows the possible range of work angle variation that may be used to offset gravitational forces on a horizontal fillet weld. The welder must observe the flow of the weld pool on the weld joint and adjust the electrode work angle as needed to ensure even flow onto both base metal surfaces.

FIGURE 3-35 shows the approximate work angle for overhead position fillet welds. Overhead position fillet welds also require compensating for gravitational effects. In this case, the electrode must be angled more toward the top base metal, forcing the weld pool to flow evenly onto both base metal surfaces. Care must also be taken to maintain a short arc length, as the filler metal will be attracted to the heated base metal.

Finally, a choice between two basic electrode manipulation concepts exists. They are the stringer bead technique and the weave bead technique. In either case, the electrode position must be maintained at the leading edge of the weld pool. When using any weave technique, the welder must ensure that the weld pool is not allowed to roll ahead of and lead the arc. The choice between the stringer or weave techniques is often imposed on the welder as defined in a welding procedure specification. Some welding codes, such as AWS D1.1/D1.1M Structural Welding Code—Steel, set limits on maximum weld size, layer thickness, and weld layer width based on electrode size and welding positions, and these limits must be observed. In all cases, groove welds must be filled completely to be flush to the base metal surface, with up to 1/8" face reinforcement.

A stringer bead is deposited with little or no side-to-side motion. This technique is favorable for use on fillet welds in the horizontal and overhead welding positions. Weave beads in these welding positions are susceptible to undercut on the weld toe. When larger welds are required, a multiple pass fillet may be employed. **FIGURE 3-36** displays both a single pass and a multiple pass stringer bead profile. A weld pass is a single layer of weld deposited longitudinally within the weld joint.

Vertical and flat position fillet welds can use either stringer or weave techniques. In these welding positions, gravitational forces aid in minimizing undercutting at the weld toes if the electrode is made to pause or dwell briefly at the outer limit of both sides of the weave pattern. It is imperative that the electrode dwells at the weld toes in vertical up-welding to ensure that undercutting does not result.

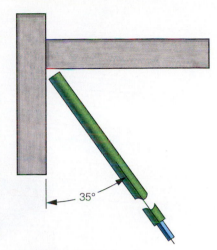

FIGURE 3-35 Angle more toward the top plate for overhead welding.

V I D E O

V I D E O

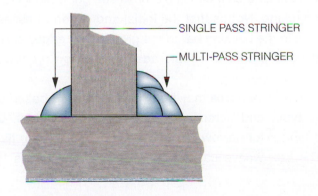

SINGLE PASS STRINGER

MULTI-PASS STRINGER

FIGURE 3-36 A multipass weld may be a better choice than a single stringer in some cases.

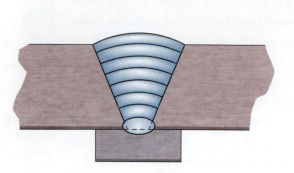

FIGURE 3-37 Care must be taken when welding weave beads to obtain proper fusion.

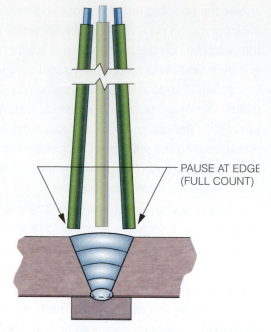

PAUSE AT EDGE (FULL COUNT)

FIGURE 3-38 Technique for weaving

V I D E O

Weave beads can be used on root passes in groove welds and for filling groove welds (**FIGURE 3-37**) in the flat and vertical positions. When using weave techniques, absolute care must be taken to control the weld pool as it flows out onto the base metal or previous weld deposit. At the outer edge of the weave pattern, the welder must angle the electrode slightly outward and dwell (pause), causing the weld pool to flow out and eliminate possible undercutting at the weld toe (**FIGURE 3-38**). It can be helpful when learning weave techniques for the welder to count and to develop a rhythm, keeping control of travel speed and dwell time. Using a count such as 1 & 2 & 1 & . . . where the numbers 1 and 2 carry a full count at the weld edges, while the & carries half a count as the electrode moves across the weld center.

Stringer beads are preferred for groove welds in the horizontal and overhead positions. **FIGURE 3-39** displays a multiple-pass stringer sequence used to fill an open-root groove weld. For groove welds used to fill thick weld joints, strongbacks may be used to minimize distortion or pulling the plates out of alignment. A block sequence or cascade sequence may also be used for filling heavy grooves as a method of controlling distortion.

FIGURE 3-40 illustrates the multiple-pass stringer bead weld sequence for a groove with backing. Notice that the initial and second passes use enough side motion to tie in the backing bar to the groove root edges and to produce a somewhat flat bead profile to enable adequate fusion on subsequent weld bead passes.

Numerous weave patterns may be used for welding based on weld position, electrode type, and personal preference of the welder. Crescent and Z patterns are typical for narrow fill deposits in grooves or for flat and vertical fillet welds. The box weave can be used on the cover pass to finish off a

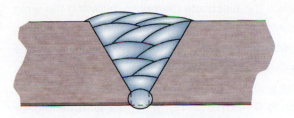

FIGURE 3-39 Stringer beads are preferred for groove welds in the horizontal and overhead positions.

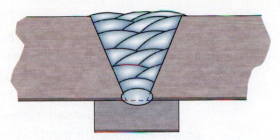

FIGURE 3-40 Weld beads must not only be tied to each other but also to the base plate.

multiple-pass weld. These patterns (**FIGURE 3-41**) are most suitable for use with electrodes that have titania (rutile) and iron powder coatings, including the low-hydrogen types.

V I D E O

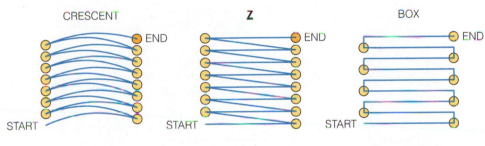

○ INDICATES PAUSE AT EDGE OF WEAVE PATTERN

FIGURE 3-41 Various weave welding techniques

Electrodes with cellulose coatings, such as E6010 and E6011 classifications (F3), are often used with one of several distinct weave motions. A straight step whipping technique, also called stitching, is often used on root passes of open-root groove welds used for pipe and plate groove designs and on fillet welds. **FIGURE 3-42** illustrates the straight step whipping technique. The arc is initiated, and the electrode is held in position (travel angle), allowing the weld pool to flow out, forming an oval shape. The electrode is then lifted upward and

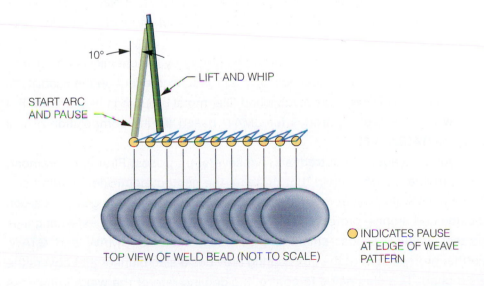

10° LIFT AND WHIP

START ARC AND PAUSE

TOP VIEW OF WELD BEAD (NOT TO SCALE)

○ INDICATES PAUSE AT EDGE OF WEAVE PATTERN

FIGURE 3-42 The whipping technique

VIDEO

slightly forward (approximately 2 times the electrode diameter), allowing the weld pool to cool slightly. Then, the electrode is swung back to the travel angle, advanced about one-half of the electrode diameter, and then paused to allow the weld pool to flow out evenly to an oval shape. This step-and-pause sequence is carried out through the length of the weld.

VIDEO

Two other significant techniques (**FIGURE 3-43**) are used with cellulose-coated electrodes. A tightly weaved circular manipulation is used at times. This technique works well on groove welds, including square grooves, as well as for rounding corners on outside corner joints. The other technique, the triangle technique, can be used for making fillet welds in the vertical position with an upward weld progression, particularly when large welds are required.

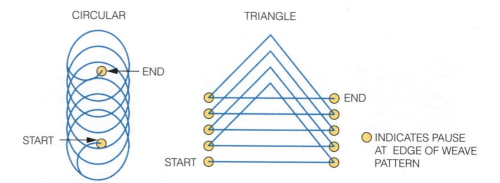

FIGURE 3-43 The circular and triangle welding techniques

ELECTRODES

Filler metals for SMAW are composed of a metal alloy core covered by a flux coating. The core wire carries the current for welding sufficiently enough to melt the base metal and mixes with melted base metal to form the weld. These SMAW electrodes are manufactured in accordance with AWS filler metal specifications that establish procedures and criteria for the manufacture of electrodes. Several specification standards (**TABLE 3-5**) exist for the manufacture of metal alloy electrodes for SMAW.

Electrode classifications are influenced greatly by the chemical composition of the electrode coatings. Coatings have a pronounced effect on the usability of the electrode. The AWS has established filler metal groupings F-1 through F-4 for low-carbon steel electrodes for SMAW based on the composition of flux coatings (**TABLE 3-6**).

Flux coatings on electrodes serve several purposes. First and foremost, they provide the gas shield that protects the heated electrode, welding arc, and molten weld pool from atmospheric contamination. These gases—carbon dioxide and argon—protect the weld arc in the same way that external gases protect welds with gas-shielded processes, such as GMAW and GTAW. Further protection from the flux coating provides a slag blanket that covers the weld bead. This slag helps to control the cooling rate of the weld, influences

AWS Filler Metal Specification	Electrode Alloy Type
AWS A5.1	Carbon steels
AWS A5.3	Aluminum and aluminum alloys
AWS A5.4	Corrosion-resistant steel (stainless steels)
AWS A5.5	Low alloy steels
AWS A5.6	Copper and copper alloys
AWS A5.11	Nickel and nickel alloys
AWS A5.15	Cast irons

TABLE 3-5

Source: **American Welding Society (AWS) Welding Handbook Committee,** *2004 Welding Handbook*, 9th ed., Vol. 2, Welding Processes. **Miami: American Welding Society**

AWS A5.1 Filler Metal Groupings for Shielded Metal Arc Welding

Grouping	Characteristic	Coating Type	Electrodes
F-1	High deposition, shallow to medium penetration	High iron oxide iron powder	E7024, E7027, E7028, E6020, E6022, E6027
F-2	Shallow penetration	High titania	E6012, E6013, E6019, E7014
F-3	Deep penetration	Cellulose	E6010, E6011
F-4	Low hydrogen, medium penetration	Low hydrogen	E7015, E7016, E7018, E7048

TABLE 3-6

the weld bead shape, and holds the weld metal in place for vertical and over-head procedures.

Electrode coatings also contain ionizing elements that help to initiate and stabilize the arc. While all mild or low-carbon steel core wires are composed of the same composition, the flux coatings are sometimes used to add alloying elements, such as chromium, nickel, and manganese, to influence the mechanical properties of the weld deposit. Finally, iron powder is added to some electrode coatings to improve deposition rates.

In summary, the electrode coatings have these purposes:

- Provide shielding gas
- Help to control the cooling rate of the weld
- Influence weld bead shape
- Help to hold molten metal in place for vertical and overhead welding
- Help to initiate and stabilize the arc
- Add alloying elements to influence weld mechanical properties

WHAT DO THE NUMBERS FOR EACH ELECTRODE MEAN?

Electrodes are classified and identified by a filler metal classification number painted onto the electrode (**FIGURE 3-44**). Carbon steel electrodes manufactured under AWS A5.1/A5.1M—Specification for Carbon Steel Electrodes for Shielded Metal

FIGURE 3-44 AWS has devised a numbering system for electrode identification.

Arc Welding have 4 digits, while low-alloy steel electrodes manufactured under AWS A5.5/A5.5M—Specification for Low-Alloy Steel Electrodes for Shielded Metal Arc Welding include 4 or 5 digits and a suffix indicating alloy composition.

For carbon steels, the electrode classification starts with the letter E, which indicates electrode, followed by 4 digits. After the letter E, the first 2 digits indicate the approximate minimum tensile strength of the filler metal ($\times$ 1000 psi) in the as-welded condition. Carbon steel electrodes have 62,000 pounds or 72,000 pounds per square inch (psi) minimum tensile strengths. The next-to-last digit indicates the welding position capabilities for electrode use (**TABLE 3-7**). The final digit represents the chemical composition of the electrode coating, which greatly influences electrode characteristics and usability.

Characteristics and usability of the electrode refers to factors such as recommended electrical polarity settings, welding position capabilities, and weld penetration into the base metals. Data from Table 3-6 indicates that electrodes with high iron oxide or iron powder coatings (F1 group) provide high deposition rates for high production work and have limited position capabilities as indicated by the next-to-last number (2) in their classification. Electrodes with titania coatings (F2 group) produce shallow penetration more suited to thin materials and sheet metal welding applications. The cellulose based electrodes (F3 group) provide deep penetration characteristics suited to open-root complete joint penetration (**FIGURE 3-45**) applications, such as root welds on pipe systems.

Next-to-Last Digit Indicates Welding Position				
AWS Classification	**Flat**	**Horizontal**	**Vertical**	**Overhead**
EXX1X	Yes	Yes	Yes	Yes
EXX2X	Yes	Fillet	No	No
EXX3X	Yes	No	No	No
EXX4X	Yes	Yes	Down	Yes

TABLE 3-7

COMPLETE JOINT PENETRATION

FIGURE 3-45 Complete joint penetration

Finally, the low-hydrogen group of electrodes (F4 group) used in structural welding applications are kept in rod-holding furnaces to keep moisture from the coatings. It is the coating that determines the welding positions used (Table 3-7). Usability is also somewhat defined by polarity selection or current type, as indicated in Tables 3-9 and 3-10.

AWS D1.1/D1.1M Structural Welding Code—Steel sets limitations on low-hydrogen electrodes (F4 group) atmospheric exposure and requires that once a box of electrodes is opened, the electrodes must be stored in a rod-holding oven at a minimum temperature of 250°F (121°C). The E7018 classification can have an allowable atmospheric exposure time of 4 hours. Filler metals with added suffixes may vary in their allowable exposure times. Some changes have taken place with the AWS classification numbering system. To avoid the confusion of developing an entirely new classification system, low-hydrogen electrodes may include suffixes to indicate special hydrogen and moisture content qualifications. The standard E7018, for example, may appear, with additions, as one of the following:

E7018R	E7018-1	E7018-1R	E7018H4
E7018-1H4	E7018H4R	E7018-1H4R	E7018H8
E7018H8R	E7018-1H8	E7018-1H8R	E7018H16
E7018H16R	E7018-1H16	E7018-1H16R	E7018HZR

R: The addition of the letter R indicates that the electrode may tolerate a humid environment of 80°F for up to 9 hours without the coating absorbing more than 0.4% moisture. This is twice the length of time a typical E7018 can tolerate moisture without having to put it back in an electrode furnace for reconditioning.

1: The standard A5.1 electrode has a satisfactory notch toughness value down to at least 20 foot-pounds at −20°F (27 J at −29°C). The addition of the −1 indicates that the electrode has a satisfactory average minimum notch toughness value down to 20 foot-pounds at −50°F (27 J at −46°C).

H or **HZ**: This is used to designate an electrode that will meet an average diffusible hydrogen level Z. The Z designator is usually 4, 8, or 16. The numbers are the maximum average diffusible hydrogen in mL(H_2) per 100 g of weld metal deposited.

Using a common low-hydrogen electrode, the E70184HR classification can be broken down as follows (**FIGURE 3-46**):

E	=	Electrode
70	=	Minimum tensile strength × 10,000 (psi)
1	=	Weld position—all
8	=	Characteristics—low hydrogen, iron powder, current type
H4	=	Hydrogen less than 4 ml/100 g of deposited weld metal
R	=	Meets requirements of absorbed moisture tests

FIGURE 3-46

The position number for the most commonly-used electrodes is 1 (all-position welding), followed by the less frequent use of 2 (flat positions and horizontal fillet). However, there are other numbers, 3 and 4, that may be on the electrode, but these are rarely used. Table 3-7 shows the position number and the welding positions electrodes can be welded in.

For low-alloy electrodes, the classification starts with the letter E, which indicates electrode, followed by 4 or 5 digits. Following the letter E, the first 2 or 3 (low-alloy types) digits indicate the approximate minimum tensile strength of the filler metal ($\times$ 1000 psi) in the as-welded condition. The next digit indicates the welding position capabilities for electrode use. The final digit represents the chemical composition of the electrode coating, which greatly influences electrode characteristics and usability. A suffix is then added to signify the chemical composition of the weld metal in the as-welded condition.

A suffix is necessary information for matching the correct low-alloy filler metal with a compatible base metal alloy. **TABLE 3-8** lists the low-alloy suffixes and quantifies the chemical composition.

Suffix	C	Mn	Si	Ni	Cr	Mo	V
A1	0.12	0.40–0.65	0.40–0.80			0.40–0.65	
B1	0.12	0.90	0.60–0.80		0.40–0.65	0.40–0.65	
B2L	0.05	0.90	0.80–1.00		1.00–1.50	0.40–0.65	
B2	0.12	0.90	0.60–0.90		1.00–1.50	0.40–0.65	
B3L	0.05	0.90	0.80–1.00		2.00–2.25	0.90–1.20	
B3	0.12	0.90	0.60–0.80		2.00–2.25	0.90–1.20	
B4L	0.05	0.90	1.00		1.75–2.25	0.40–0.65	
B5	0.07–0.15	0.40–0.70	0.30–0.60		0.50–0.60	1.00–1.25	0.05
B6	0.05–0.10	1.00	0.90	0.40	4.00–6.00	0.45–0.65	
B8	0.05–0.10	1.00	0.90	0.40	8.00–10.50	0.85–1.20	
B9	0.08–0.13	1.25	0.30	1.00	8.00–10.50	0.85–1.20	0.15–0.30
C1	0.12	1.20	0.60–0.80	1.00–2.75			
C2	0.12	1.20	0.60–0.80	3.00–3.75			
C3	0.12	0.40–1.25	0.80	0.80–1.10	0.15	0.35	0.05
D1	0.12	1.25–1.75	0.60–0.80			0.25–0.45	
D2	0.15	1.65–2.00	0.60–0.80			0.25–0.45	
G		1.00 Min	0.80 Min	0.50 Min	0.30 Min	0.20 Min	0.10 Min
			Specifies to military specifications				
M	0.10	0.60–2.25	0.60–0.80	1.40–2.50	0.25–1.50	0.25–5.50	0.05
		Explanations Single values indicate maximum amounts of elements. Percentages vary with electrode classifications (see AWS A5.5).					

TABLE 3-8

WHAT ARE THE CORRECT CURRENT SETTINGS FOR SMAW ELECTRODES?

Current settings vary with electrode diameters and with changes in flux-coating composition. **TABLES 3-9** and **3-10** give probable usage ranges for common carbon steel electrodes. Where the welder is set within these ranges depends on base metal thickness and the welding position. Thinner base metals are welded

Electrode Diameters and Current Ranges

	Current Type	3/32"	1/8"	5/32"
EXX10	DCEP	40–80	65–130	90–165
EXX11	AC/DCEP	50–90	75–125	110–160
EXX12	AC/DCEN	50–95	70–130	125–200
EXX13	AC/DC	50–95	75–125	120–165
EXX14	AC/DC	65–95	110–140	130–200
EXX15	DCEP	75–110	90–145	120–210
EXX16	AC/DCEP	70–100	80–130	120–170
EXX18	AC/DCEP	75–110	90–145	120–210
EXX18M	DCEP	75–110	90–145	120–210

TABLE 3-9

Carbon steel all welding-position-capable electrodes

Carbon Steel High-Deposition Electrodes

	Current Type	3/32"	1/8"	5/32"	3/16"	1/4"
EXX22	AC/DC		110–150	150–170		
EXX24	AC/DC	60–120	110–150	170–220	220–300	320–420
EXX27	AC/DCEN			170–230	230–300	310–440
EXX28	AC/DCEP			170–250	220–310	270–400

TABLE 3-10

Carbon steel limited position capability electrodes

with lower amperage settings, as are welds made in the vertical up-welding progression. The welder must select a current that provides an adequate melt-off rate and penetration while maintaining an ability to control the weld pool. Each welding electrode manufacturer recommends operating parameters for each electrode type. It is critical to limit welding to parameters within that range.

Stainless steel electrodes are classified by AWS, and their classification numbers are the same as the base metal classifications for their identification by AISI, SAE, and ASTM standards. Current type depends on the electrode coating composition. Electrode coatings are composed of either lime or titania. Electrodes with the lime coating use the 15 suffix and run on DCEP, while the electrodes with titania coatings use the 16 suffix and are used with either DCEP or AC. Due to superior arc starting and arc stability, DCEP is preferred. **TABLE 3-11** gives recommended current ranges for stainless steel electrodes.

The welding parameters given in this chapter should suffice for welding steels and stainless steels. On thicker materials, for the electrode diameters listed, only a slight increase in amperage may be needed as well as additional passes to complete the desired weld size. Those wishing to use larger diameter electrodes should consult electrode manufacturers to produce optimum results. For welding other base metals and for using hard surfacing alloys, it is best to follow specific welding amperage parameter tables for each electrode manufacturer.

Stainless Steel Electrodes					
	Current Type	3/32"	1/8"	5/32"	3/16"
E308L	Lime (15 suffix)	45–85	55–120	65–170	160–205
E309	coatings use	45–85	55–120	65–170	160–205
E310	DCEP and	50–75	70–100	100–145	
E312	titania	50–75	70–125	65–170	
E316	(15 suffix)	45–85	55–120	65–170	160–205
E347	coatings use DCEP or AC	60–90	80–130	140–170	

TABLE 3-11

CHAPTER QUESTIONS/ASSIGNMENTS

MULTIPLE CHOICE

1. What lens shade is recommended for SMAW at 120 amps?
   a. #9
   b. #10
   c. #11
   d. #12

2. What welding lead size is recommended for welding at 150 amps at 300 feet?
   a. 1
   b. 1/0
   c. 2
   d. 2/0

3. How much is the voltage drop for the welding example given in question #2?
   a. 1.7 volts
   b. 2.8 volts
   c. 4.2 volts
   d. 4.4 volts

4. What welding control has the greatest effect on penetration?
   a. Amperage
   b. Arc control
   c. Duty cycle
   d. Voltage

5. What welding control reduces the chance of an electrode sticking to the base metal with open-root joint designs?
   a. Amperage
   b. Arc control
   c. Duty cycle
   d. Voltage

6. What is the correct arc length for SMAW?
   a. Between 5–15 degrees
   b. Slightly less than the electrode diameter
   c. The base metal thickness

7. What are the indications that the arc length is too long?
   a. A narrow and high weld bead
   b. The arc will wander, causing too much spatter.
   c. The electrode turns red, ruining the flux.

8. What is the travel angle for most applications?
   a. approximately 0°–5°
   b. approximately 5°–15°
   c. approximately 30°–45°
   d. perpendicular to the weld joint

9. In what situation is a forehand travel angle used?
   a. In the flat position
   b. In the horizontal position
   c. In the vertical-up position
   d. In the overhead position

10. When welding, where should the electrode position be in relation to the weld pool?
    a. The electrode position must be kept ahead of the weld pool.
    b. The electrode position must be maintained at the leading edge of the weld pool.
    c. The electrode position must trail behind the weld pool.

11. What is the AWS filler metal specification for carbon steel?
    a. AWS A5.1
    b. AWS A5.3
    c. AWS A5.4
    d. AWS A5.5

12. The size of the welding leads required depends on two factors: the _____ amperage for welding and the _____ from the power source to the work.

13. The electrode holder jaws have _____ machined into them to allow for holding the electrode at various different angles. This allows the welder to choose (an) _____ that provides comfortable access to the weld joint based on welding joint design and the welding position of the work.

14. Amperage is the measure of _____ flow or movement of _____ from the power source through the welding cable and electrode holder to the electrode.

15. Increasing the arc length _____ the arc voltage and _____ of the weld pool, while decreasing the arc length _____ the arc voltage and fluidity of the weld _____.

16. A _____ travel speed produces a uniformly-shaped _____ weld pool and a weld bead with a slightly oval ripple pattern.

17. With too fast a travel speed, the weld will be too _____, with an elongated _____ ripple pattern.

18. What three factors are considered when selecting a welding power source?

19. What basic equipment is required for SMAW?

20. How is a work clamp attached to the work?

21. What four variables can the welder select before welding begins?

22. What factors determine the operating parameters of the electrode?

23. What is the significance of a SMAW filler metal electrodes chemical composition?

24. What factors determine the selection of SMAW filler metal electrodes?

25. What are the advantages of using DCEP?

26. When might DCEN be useful?

27. What two reasons may justify the use of AC?

28. Other than using AC, how else can arc blow be reduced?

29. What are the effects of an improper work angle on a fillet weld?

30. In addition to the work angle, what other factors affect weld metal flow?

31. What are the results of too fast a travel speed?

32. What are the results of too slow a travel speed?

33. What are the two methods used for starting the welding arc?

34. Where should the weld arc be initiated in relation to the weld start?

35. What problem may be encountered by using poor techniques when using weaving techniques?

36. What are three common weave techniques used with cellulose-coated electrodes?

37. Flux coatings on electrodes serve what purposes?

38. Describe what the numbers and letters represent for an E6011 electrode.

Chapter 4

GAS METAL ARC WELDING

LEARNING OBJECTIVES

In this chapter, the reader will learn how to:

1. Recount the history and development of Gas Metal Arc Welding (GMAW).

2. Follow safety protocols when welding with the Gas Metal Arc Welding process.

3. Select the type of power source and peripheral components that would work best at home, school, or in industry.

4. Identify and correctly use the various controls on power sources and their characteristics.

5. Properly set up for Gas Metal Arc Welding by adjusting volts, amps, travel speed, work angles, and electrode extension.

6. Adjust the power source to produce the mode of transfer that is best for the weld position, base metal, and desired outcome.

7. Select the correct electrode for the base metal used and the desired outcome.

8. Select the correct shielding gas depending on the metal and the desired mode of transfer.

KEY TERMS

Wire Feed Speed (WFS) 77

Contact Tip 77

Drive Rolls 83

Liner 85

Bird's Nest 88

Spool Drag 88

Electrode Extension (stick-out) 94

Forehand Technique 96

Short-Circuit Transfer 99

Axial-Spray Transfer 99

Slope 101

Inductance 103

Globular Transfer 104

Transition Current 105

Pulse-Spray Transfer 105

Current Density 108

Carbide Precipitation 117

Flow Rate 122

Surface Tension 123

Deoxidizers 123

Ionization Potential 129

INTRODUCTION

Gas Metal Arc Welding (GMAW) is a welding process where an electrode wire is continuously fed from an automatic wire feeder through a conduit and welding gun to the base metal, where a weld pool is created. If a welder is controlling the direction of travel and travel speed the process is considered semi-automatic. The process is fully automated when a machine controls direction of travel and travel speed, such is in the case of robotics.

Automatic welding was introduced in 1920 and utilized a bare electrode wire operated on direct current and arc voltage as the basis of regulating the feed rate. Automatic welding was invented by P. O. Nobel of the General Electric Company. The GMAW process we recognize today was successfully developed at Battelle Memorial Institute in 1948 under the sponsorship of the Air Reduction Company. This development used a gas shielded arc similar to the gas tungsten arc but replaced the tungsten electrode with a continuously fed electrode wire. One of the basic changes that made the process more usable was the small-diameter electrode wires and the constant-voltage power source. The high deposition rate led users to try the process on steel. In 1953, Lyubavskii and Novoshilov announced the use of welding with consumable electrodes in an atmosphere of carbon dioxide gas (CO_2). The CO_2 shielding gas immediately gained favor because it utilized equipment developed for inert GMAW but could now be used for economically welding steels. The development of the short-circuit arc variation, which was also known as Micro-wire, short-arc, and dip transfer welding appeared late in 1958 and early in 1959. Short-circuit transfer allowed all-position welding on thin materials and soon became the most popular of the GMAW process variations.

GMAW is a common welding process used in conditions where air current is not too severe to be a detriment to the external shielding gas used to protect the weld pool. External shielding gases vary in composition, allowing the electrode wire to change mode conversions—provided proper voltage and amperage parameters are used. Shielding gas composition also affects bead appearance, fusion, and operator appeal.

A variety of electrode wires are used, including steel, stainless steel, aluminum, composites, and silicon bronze. Electrode diameters range from very small—allowing fusion on thin sheet metal—to larger diameters, where increased deposition rates and depth of root penetration are beneficial.

Manufacturers of power sources fill user needs from weekend hobbies and crafts to sophisticated microprocessor applications, which automatically recognize and link to robots. The characteristics of GMAW allow the process to be, arguably, the easiest welding process to learn how to use. However, not following proper techniques and essential variables can lead to defective welds.

SAFETY

TABLE 4-1 shows the basic safety considerations for GMAW. **TABLE 4-2** shows recommended lens shades for welding at varying amperage ranges.

Head	Body	Hands	Feet
Safety glasses	Flame-resistant wool or cotton pants and shirt	Gauntlet-style welding gloves	High-top leather safety work boots
Welding helmet	Flame-resistant jacket		
Welding hat	Leathers		

TABLE 4-1

Read "Chapter 2: Safety in Welding" for a detailed discussion of safety related to workplace environment, workplace hazards, personal protective equipment, electrical consideration, gases and fumes, ventilation, fire prevention, explosion, compressed cylinders, and gases.

Arc Current	Suggested Shade
Less than 50	—
50–150	10
150–250	12
250–400	14

TABLE 4-2

POWER SOURCE AND PERIPHERALS

WHAT MAKES UP THE GAS METAL ARC WELDING SETUP? The basic equipment needed to start welding with GMAW is:

- Power Source
- Wire Feeder
- Welding Gun (or Torch)
- Regulator/Flow Meter

The GMAW process requires a constant voltage (also called a constant potential) welding power source. We can loosely group the GMAW power source into three categories:

- Heavy Industrial (460V)
- Light Industrial (230V)
- Home Hobbyist (115V)

The type or category of power source determines the need for other accessories. For heavy-industrial machines, everything must be specified and ordered (**FIGURE 4-1**). The various parts include:

- Power Source
- Wire Feeder
- Drive Rolls
- Shielding Gas Hose
- Regulator/Flow Meter
- Coolant Circulator (when using a water-cooled gun)
- Welding Gun (Torch)
- Contact Tips
- Nozzle
- Work Clamp and Cord
- Power Cord

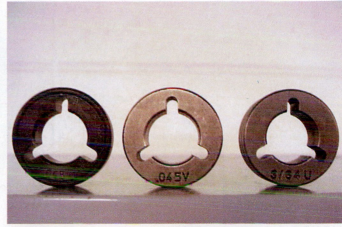

The power source, the main component, is a DCCV (direct current constant voltage) welder. The wire feeder is a device that feeds the electrode wire. While the power source controls the voltage output, increasing or decreasing the **wire feed speed (WFS)** on the wire feeder increases or decreases the welding amperage.

Manufacturers package small home hobby welders with everything needed to start welding. The wire feeder and power source are encased as a single unit. The welding gun has the gas diffuser and a contact tip attached. The **contact tip** is made of copper where the electrode wire travels through and where electricity energizes the electrode. A few extra contact tips are usually supplied. Drive rolls, power cord, and work clamp and cable are also supplied. Often, a small spool of self-shielded flux-cored electrode, welding helmet, gloves, and safety glasses are included. For solid wire, a separate shielding gas cylinder must be purchased separately.

The next size up from the 115V power supplies is a light industrial welder. These welders can be found in automotive repair shops, machine shops, farms, and in the home of a serious hobbyist. Generally, these welders are packaged similar to the smaller 115V units, operate on 230V, and have larger amperage output and duty cycle. Note the power cord plugs in **FIGURES 4-2** and **4-3**.

WHICH POWER SOURCE IS BEST FOR HOME OR BUSINESS?

A heavy-industrial power source is one that is going to be used for manufacturing on a daily basis. The heavy-industrial power source also operates on 3-phase power,

FIGURE 4-2 115-volt GMAW power source

POWER CORD

requiring special power line installation. For the manufacturer, this is not an issue. The manufacturer makes up the cost difference between how much more the welder costs and the savings in labor due to the higher duty cycle and increased savings in power usage with 3-phase power.

For machine shops, automotive repair, farms, and other similar businesses not using the welder on a continual basis, the choice is made on the benefits versus the purchase price. Schools are another establishment where a choice between a heavy-industrial and a light-industrial power source is made. Some schools run intense programs and may benefit from 3-phase heavy industrial power sources or may wish to have students train on equipment similar to that which they will experience in industry. The choice of power supplies is a decision between costs and needs.

For home use, the choice is between the 230V "light industrial" and the 115V "home hobby" size welder. A middle ground power source exits with the same physical size of the 120V power sources, but is available to operate on 230V (**FIGURE 4-4**). The difference between those and the light-industrial models is output capability and a lower duty cycle decreasing the purchase price.

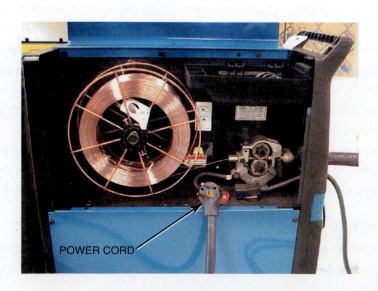

FIGURE 4-3 Large-size 230-volt GMAW power source

POWER CORD

POWER CORD

FIGURE 4-4 Small-size
230-volt GMAW power source

CONTROLS AND CHARACTERISTICS

WHAT TYPES OF CONTROLS DO POWER SOURCES HAVE? There are mainly two controls that all Gas Metal Arc Welders have: voltage and amperage. Voltage is set on the power source, and the wire feed speed controls the amperage. The larger heavy industrial power sources may have an inductance control, also known as arc control, pinch effect, or arc force. Some heavy industrial power sources may also have slope control.

WHAT TYPES OF VOLTAGE CONTROLS DOES A POWER SOURCE HAVE?

VOLTAGE CONTROL (POTENTIAL) Power sources used for GMAW, as mentioned earlier, are constant voltage. Increasing or decreasing the voltage control on the power source increases or decreases the arc voltage. There are two types of voltage controls on the power source: Range/Tap and Full Range Potentiometer.

1. *Range/Tap:* The Range/Tap control is commonly found on home hobbyist and light industrial power sources. This style of voltage control is less expensive and simpler to use. It is very straightforward and eliminates confusion when setting the welding voltage (**FIGURE 4-5**).

   The operator adjusts the welding voltage by selecting one of a number of preset voltage settings on the front of the power source. For example, Range/Tap 1 may equal 17 volts DC, Range/Tap 2 may equal 18 volts DC, and so on. Most manufactures provide a detailed parameters chart showing recommended voltage range/tap settings (**FIGURE 4-6**). The operator need only search in one column to match the type of welding they will be doing—that is, aluminum, flux core, or steel—along with the type of shielding gas used and then find the thickness of material they are going to weld in a row at the top of the chart. The intersection of this row and column will provide the operator with a tap setting and a wire feed speed.

   The disadvantage to the Range/Tap control is that there is no ability to fine-tune voltages between settings. Ranges/taps cannot be adjusted during welding. If the Range/Tap switch is adjusted when welding, damage

FIGURE 4-5 Adjustable wire feed speed with tapped voltage

can occur to the contact points of the Range/Tap switch. If the Range/Tap switch is accidentally placed in between a tap, no voltage is available, and the machine will not produce an arc.

2. **Full Range Potentiometer:** The Full Range Potentiometer allows the operator to set welding voltage from the lowest to highest value of the power source's voltage range (see **FIGURE 4-7**). This style of control enables the operator to fine-tune voltage. Adjustment of voltage can be done before or during welding. If the voltage is too high in relation to the wire feed speed, the puddle may appear to wet out and fuse to the base metal, but there will be insufficient penetration. If the voltage is too low in relation to the wire feed speed, the weld bead will appear ropy, with incomplete fusion at the weld toes.

FIGURE 4-6 Tapped voltage settings chart inside the power source

			Suggested Welding Settings (GMAW)			
	Wire Diameter (inch)	Operator Control Settings	Material Thickness			
Wire Type			1/4 in.	3/16 in.	1/8 in.
Mild Steel 75% Argon 25% CO$_2$	.023	Tap	—	—	3
		Feed Range	—	—	80
	.030	Tap	4	4	3
		Feed Range	75	70	60
	.035	Tap	4	3	3
		Feed Range	55	55	50
E71 T-GS Flux Core	.030	Tap	4	4	3
		Feed Range	70	60	50
	.035	Tap	4	4	3
		Feed Range	100	50	40

FIGURE 4-7 Adjustable voltage and amperage settings dial

The advantage to the Full Range is the ability to fine-tune the welding arc over a wider range of material thicknesses and joint designs. Adjustability is helpful if material fit-up is less than ideal.

HOW IS THE WIRE FEED SPEED SET?

WIRE FEED SPEED (WFS) The wire feed speed controls the amperage. An increase in wire feed speed increases the amperage while welding, and decreasing the wire feed speed decreases the amperage while welding. The wire feed speed control is found on power supplies that are equipped with a built-in wire feed system. On the large-industrial power sources, the wire feed system is separate from the power source; the wire feed control is found on the wire feeder.

Wire feeders provide a consistent supply of electrode wire to the welding arc. The function of the wire feed speed control is to allow the operator to adjust the wire feed speed setting from the lowest to the highest speeds of the wire feed range. All wire feeders have a listed wire feed range and a Full Range Potentiometer–style wire feed speed control. Turning the wire feed speed control knob clockwise increases the wire feed speed, and turning the wire feed speed control knob counterclockwise decreases the wire feed speed. The Full Range Potentiometer usually has a numbering system from 1 to 10 or 10 to 100. These numbers are just references and do not indicate the actual wire feed speed. Top wire feed speeds, when the control knob is turned to maximum, are usually 750 to 800 inches per minute (300 to 350 mm/second). In this scenario, each increment, 1, 2, 3, or 10, 20, 30, etc., will equal 75 or 80 inches per minute (30 to 35 mm/second). See **FIGURE 4-5**. Wire feed speed is also displayed digitally on feeders so equipped (see **FIGURE 4-8**).

WHERE IS THE SLOPE CONTROL SET ON A POWER SOURCE? Slope control is generally not found on modern power sources, but it can be found as a control or as a selectable tap on some large-industrial power sources. On these

FIGURE 4-8 Digital voltage and amperage settings dials

power sources, slope control is found on the front panel, along with the voltage control. They are adjusted in the same manner as voltage control. The selectable tap type is accessed by removing a side panel of the power source. A series of taps are available by loosening a set screw and moving a heavy wire from one tap block to another (**FIGURE 4-9**). All power sources have some slope. Most power sources have a fixed amount of slope and cannot be adjusted.

WHERE IS AN INDUCTANCE OR PINCH EFFECT CONTROL ON A POWER SOURCE?

Inductance is also called arc control, pinch effect, or arc force. There are several different versions of inductance control, ranging from a fixed value to a Full Range Potentiometer. Like slope, all welding power sources designed for GMAW short-circuit transfer welding have some inductance. The home hobbyist and light-industrial power sources have a fixed amount of inductance and cannot be adjusted. Many large-industrial welding power sources have a Full Range Potentiometer inductance control. Turning the control clockwise increases the inductance, while turning the control counterclockwise decreases the inductance. A Full Range Potentiometer inductance control is pictured in **FIGURE 4-10**.

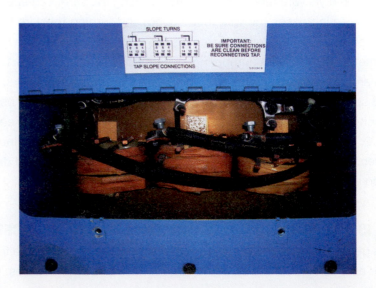

FIGURE 4-9 View of manual tapped slope inside of a power source

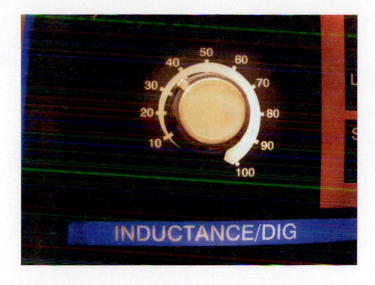

FIGURE 4-10 Adjustable inductance pinch effect dial

IS THERE A DIFFERENCE IN DRIVE ROLLS?
There are generally three types of **drive rolls** (see **FIGURE 4-11**):

- V-Groove Drive Rolls (**FIGURE 4-12**)
- U-Groove Drive Rolls (**FIGURE 4-13**)
- Knurled Drive Rolls

V-groove drive rolls are generally used for steel and stainless steel electrode wires. U-groove drive rolls are generally used for aluminum electrode wires. And knurled drive rolls are generally used for tubular composite electrode wires. The drive rolls pictured in **FIGURE 4-11** are installed on the heavy-industrial separate wire feeder and, in some cases, the light-industrial all-in-one power sources. In most home hobbyist power sources that also contain the wire feeder, the drive rolls are a combination. The combination configuration often has a knurled bottom drive roll and a smooth roller for the top. Drive rolls on these machines are intended to be used on a variety of electrodes.

WHAT OPTIONS ARE AVAILABLE FOR WELDING TORCHES (WELDING GUNS)?
There are many options for welding torches. Welding torches are included on most integrated power source/wire feeder systems, such as the

FIGURE 4-11 From left to right, the knurled drive roll, the V-groove drive roll, and the U-groove drive roll

FIGURE 4-12 The V-groove drive roll and a closeup view

light-industrial or home hobbyist power sources. Considerations for welding torches include:

- **Duty cycle and welding range.** If there is a great deal of arc-on time, duty cycle becomes very important to prevent the torch getting very hot in the welder's hand and burning up the welding torch. Torches designed for a maximum of 150 amps will not last long if the power source is set consistently in the 200–300 amperage range for spray transfer arc welding.

- **Replacement parts.** If a contact tip takes a long time to change, this time can add up as cost in a heavy-industrial setting. Some contact tips have fine thread, some have quick disconnect tips, and others have no threads at all and are slip fit or are similar to a collet-type fit. Nozzles also can be threaded or slip fit. Insulators are either a separate unit that the nozzle slips over or are integrated as part of the nozzle (see **FIGURE 4-14**).

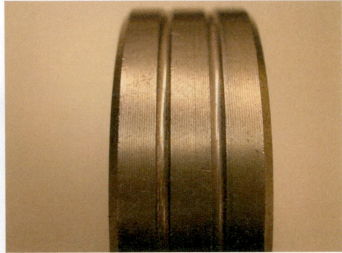

FIGURE 4-13 The U-groove drive roll and a closeup view

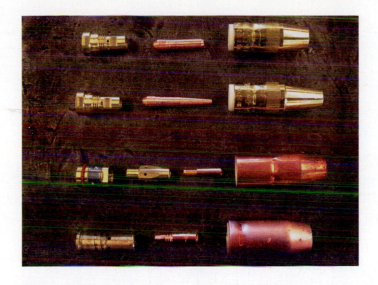

FIGURE 4-14 Various types of gas diffusers, contact tips, and nozzles

Torches can be air- or water-cooled. There are two advantages of water-cooled torches. They weigh less than the same welding range air-cooled torch, and they stay cooler during extended welding periods. The disadvantage is that a water-cooled torch is more expensive and requires a water circulator. In addition, operators of water-cooled torches must take care not to place the whip of the torch leading back to the water circulator on hot metal, which could melt a hole in the coolant hose.

After the electrode wire is fed past the drive rolls, the wire travels through the torch via a **liner**. This liner is composed of various materials, including steel, composite, and nylon, depending on the electrode wire. The liners made of steel and composite are also know as conduits. Aluminum electrodes, for example, feed more freely through a nylon liner but will be damaged with steel electrodes. Stainless steel electrode wires feed more smoothly through a composite conduit liner than the typical conduit used for steel electrode wires.

HOW IS A GAS REGULATOR OR FLOW METER SELECTED?

Regulator manufacturers have several different styles of regulators. Cost and features distinguish the styles. With most integrated power source/wire feeder systems, such as the light-industrial or home hobbyist power source, the regulator is included. Considerations for regulators include:

- Degree of precision control of gas flow
- Display of gas flow actual or preset
- Capability of being used with different gas mixes

SINGLE-STAGE FLOW GAUGE Gas flow is adjusted by turning a T-handle on the front of the regulator body. They are equipped with a flow gauge and outlet orifice rather than a flow meter to measure gas flow (see **FIGURE 4-15**).

SINGLE-STAGE FLOW METER Gas flow is adjusted by a thumb screw at the top of the float tube. Flow meter regulators provide efficient and accurate gas regulation (see **FIGURE 4-16**).

TABLE 4-3 indicates equipment and accessories that are often included with a welding power source purchase. It can be noted that the buyer of a heavy industrial power source is required to select and purchase are all peripherals separately.

FIGURE 4-15 Single-stage gas flow gauge

FIGURE 4-16 Flow meter regulators

Power Source Guide Chart (what is included in purchase)			
Power Source	**Industrial**	**Light Industrial**	**Home Hobbyist**
Wire Feeder	No	Yes	Yes
Drive Rolls	No	Yes	Yes
Shielding Gas Hose	No	Yes	Yes
Regulator/Flow Meter	No	Yes	Yes
Work Clamp and Cord	No	Yes	Yes
Power Cord	No	Yes	Yes
Welding Gun (torch)	No	Yes	Yes
Welding Helmet	No	No	Yes – 120V Unit
Spool of Wire	No	No	Yes – 120V Unit

TABLE 4-3

HOW IS THE WELDING POWER SOURCE SET UP FOR WELDING? There are a number of variables the welder has control over prior to welding. Prior to welding, the welder selects and adjusts:

- Type of Shielding Gas and Flow Rate
- Electrode
- Drive Rolls
- Contact Tips
- Nozzle
- Polarity
- Electrode and Spool Tension
- Amperage (Wire Feed Speed)
- Voltage

The type and amount of shielding gas is also critical to producing a sound weldment. The type of shielding gas is primarily determined by the electrode selected and the mode of transfer desired. The amount of shielding gas, or flow rate, is also determined by the electrode selected and the mode of transfer desired but is also determined by other factors, such as draft.

Electrode selection is critical. For the most part, the electrode must match the base metal properties being used. Depending on the type of electrode, the diameter, and the mode of transfer, the voltage and wire feed speed settings can be drastically different.

VIDEO

Drive roll selection depends on the electrode selected. V-groove drive rolls are used with steel and stainless steel electrodes; U-groove drive rolls are used with aluminum electrode wire; and knurled drive rolls are used with composite electrode wires.

Contact tips must match the electrode wire selected. A 0.035 inch (0.8 mm) contact tip is required for a 0.035 inch (0.8 mm) steel, stainless steel, or composite electrode. However, for aluminum, a 3/64 inch (1.2 mm) electrode requires a 0.052 inch (1.3 mm) contact tip. The larger contact tip for aluminum is required because the amount of expansion of the aluminum electrode wire is considerably more than steels. The use of a smaller size contact tip for aluminum may cause the electrode to seize during a start/stop weld sequence. Note that the contact tip manufactured for aluminum may have 3/64 AL stamped on the contact tip, along with the 0.052/1.2 mm size (see **FIGURE 4-17**).

Nozzles, as shown in **FIGURE 4-18**, come in various sizes and shapes. The size and shape depend on intended use. A needle-like taper nozzle works well on T-joints and hard-to-reach places. A larger, more robust nozzle lends itself to high voltage/high amperage welding.

Polarity selection is critical with GMAW and influences the direction of current flow. Direct current electrode positive (DCEP) is the preferred setup, yielding sufficient localized heating of both filler and base metals. This provides quality bead characteristics and superior weld penetration. Direct current electrode

VIDEO

FIGURE 4-17 Contact tip for aluminum electrode wire

VIDEO

VIDEO

negative (DCEN) produces a globular transfer and poor fusion properties that make its use undesirable. DCEN is not used with GMAW electrodes. Other welding processes may use DCEN and have been switched for their use. Always check to ensure proper polarity prior to welding.

Electrode tension is set by the drive rolls. If the tension is set too loose, the electrode wire feeds erratically and often sticks to the contact tip. Too much tension, and the electrode wire may form a **bird's nest** around the drive rolls if feeding is constricted at either the contact tip or the gun/whip are tangled (see **FIGURE 4-19**). The spool nut must have enough tension to stop the spool from spinning and unraveling the electrode wire after the trigger on the welding gun is disengaged but not too much tension as to cause excessive drag on the electrode wire being fed through the drive rolls. This is known as **spool drag**.

Welding amperage (current), as mentioned earlier, is increased or decreased by adjusting the wire feed speed (WFS). However, how much or little is determined by several factors, including base metal thickness, electrode diameter,

FIGURE 4-18 Nozzles of various sizes. The larger nozzles are designed to handle higher amounts of amperage

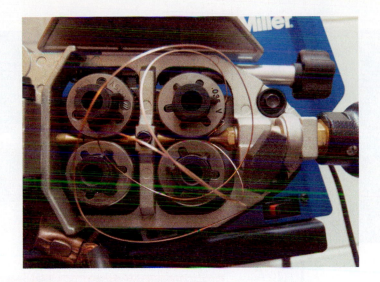

VIDEO

FIGURE 4-19 A bird's nest may result if an obstruction occurs at the contact tip and the drive roll tension is set too high

and the desired mode of metal transfer. Increasing WFS will increase both amps and penetration. Decreasing WFS will decrease both amps and penetration. Excessive WFS causes electrode stubbing off the plate.

Arc voltage is adjusted on the GMAW power supply. Welding arc voltage is the primary variable for controlling arc length (wire tip-to-work distance) and the fluidity or wetting out of the weld onto the base metal workpiece. Increasing voltage increases arc length and fluidity of the weld pool. Decreasing voltage decreases arc length and fluidity of the weld pool.

Adjusting the voltage to produce the proper arc length and bead characteristic for the desired mode of metal transfer is critical. Higher voltages produce a longer arc length and a flatter weld bead profile (see **FIGURE 4-20**). Too high a voltage for any WFS causes the electrode to burn back to the contact tip, while insufficient voltage causes difficult arc starting, spatter, or stubbing of the electrode on the base metal. The GMAW constant voltage power source automatically regulates arc length to compensate changes in WFS, electrode drag in the liner, or changes in arc length.

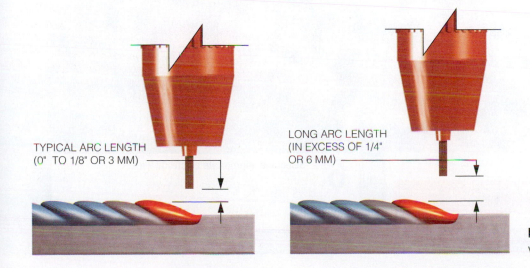

TYPICAL ARC LENGTH (0" TO 1/8" OR 3 MM)

LONG ARC LENGTH (IN EXCESS OF 1/4" OR 6 MM)

FIGURE 4-20 Short arc length versus long arc length

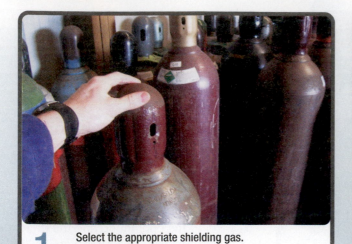

1 Select the appropriate shielding gas.

2 Secure the shielding gas bottle.

CAUTION: Failure to secure the shielding gas bottle could result in the bottle tipping over, breaking off the stem, and propelling it like a rocket.

3 Remove the cap.

4 Crack the cylinder.

CAUTION: Do not stand facing the valve opening.

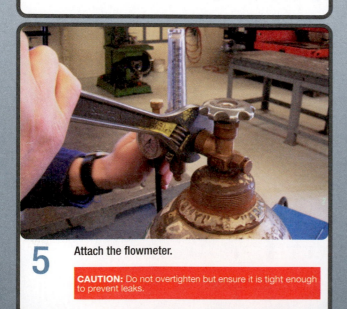

5 Attach the flowmeter.

CAUTION: Do not overtighten but ensure it is tight enough to prevent leaks.

6 Select the appropriate electrode wire.

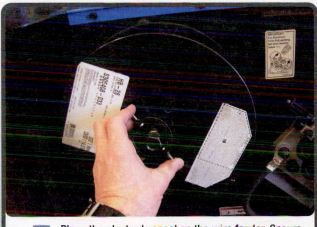

7 Place the electrode spool on the wire feeder. Secure the wire spool.

CAUTION: Failure to secure the wire spool could result in the spool falling off.

8 Select the appropriate drive rolls.

9 Obtain the contact tip to match the electrode filler metal selected.

10 Install the drive rolls.

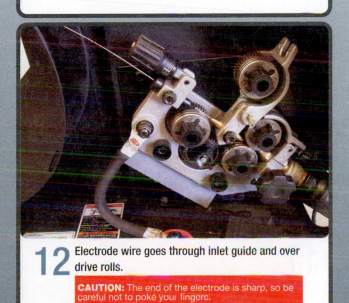

11 Check the polarity.

12 Electrode wire goes through inlet guide and over drive rolls.

CAUTION: The end of the electrode is sharp, so be careful not to poke your fingers.

13 Plug in the power source to the proper electrical receptacle.

CAUTION: Electricity can be dangerous. Always be careful when plugging and unplugging electrical receptacles. Check for frayed or damaged electrical cords.

14 Adjust the drive roll tension.

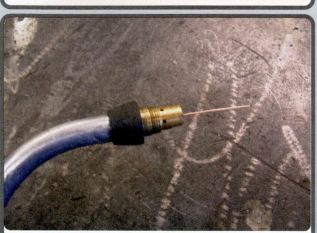

15 Feed the electrode through the welding whip and gun.

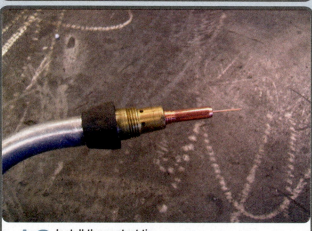

16 Install the contact tip.

17 Install the nozzle.

18 Test the drive roll tension.

CAUTION: Never test drive roll tension with your hands; sharp electrode wire could pierce your gloves.

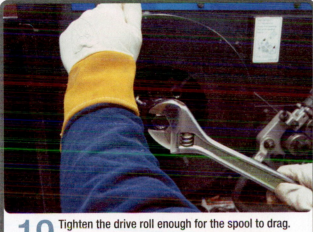

19 Tighten the drive roll enough for the spool to drag.

CAUTION: If the tension is too tight, the electrode wire may bind. If the tension is too loose, the wire may unravel.

20 Open the cylinder valve.

CAUTION: Never stand in front of the regulator when opening the cylinder.

21 Adjust the flowmeter.

22 Check for shielding gas leaks. O-rings should not be exposed but should be pressed tight against the drive roll housing. **CAUTION:** Never check for shielding gas leaks by holding the welding gun *toward* your ear/head.

23 Attach the work lead.

24 Adjust the voltage and wire feed speed (amperage).

OTHER THAN ADJUSTING THE POWER SOURCE, CAN THE WELDER DO SOMETHING TO AFFECT THE WELD? There are a number of variables the welder can control after the arc is initiated. After the arc is initiated, the welder has control over:

- Electrode Extension (stick-out)
- Electrode Work Angle
- Electrode Travel Angle
- Direction of Travel
- Travel Speed
- Electrode Oscillation

One way to affect changes in the arc and weld puddle is by changing the electrode extension. **Electrode extension** (stick-out) is the distance the electrode sticks out from the contact tip either prior to or while welding (see **FIGURE 4-21**). As the GMAW gun is moved away from the work, the result is a greater electrode extension and a slight reduction in the welding arc current. Required electrode extension varies with electrode diameters and metal transfer modes. There are definite effects of changing electrode extension. While voltage and WSF remain constant, there is a reduction of current with an increase in electrode extension. Typical electrode extension for short-circuit welding is approximately 3/8 inch (9.5 mm) and 5/8 inch (16 mm) for spray transfer welding.

Another way to affect the weld puddle is to change the electrode work angle and the direction of travel. Electrode work angle refers to the GMAW gun position in relation to the workpiece. Proper welding angles are critical for making satisfactory welds. Travel angle refers to the angle of the welding gun in the direction of travel. The electrode position in **Figure 4-22** is at a 45° angle between the top and bottom base metal and about 10°–15° in the direction of travel. The travel angle in **FIGURE 4-22** is for right-hand welders. A left-hand

V I D E O

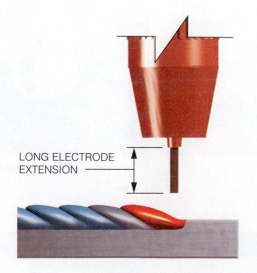

LONG ELECTRODE EXTENSION

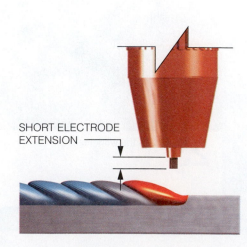

SHORT ELECTRODE EXTENSION

FIGURE 4-21 Short and long electrode extension

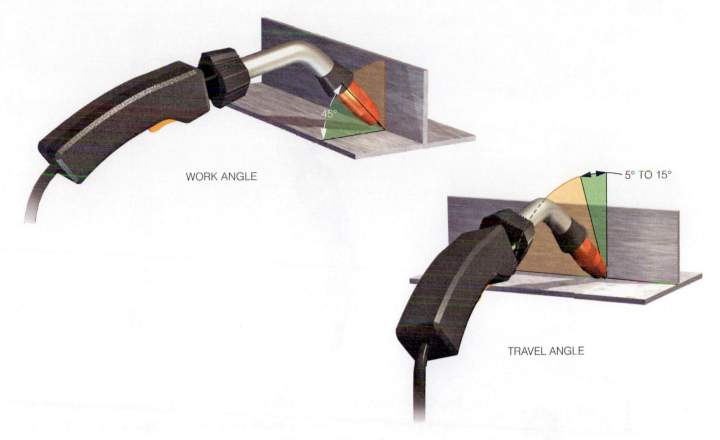

WORK ANGLE

5° TO 15°

TRAVEL ANGLE

FIGURE 4-22 Work and travel angles on a fillet weld (T-joint)

welder would tilt the welding gun 10°–15° to the left. If the welding gun is more vertical, for example, the weld penetrates more of the base member and has insufficient fusion to the upright member.

On groove joints the work angle is perpendicular to the base metal or at a 90° angle and about 10°–15° in the direction of travel (see **FIGURE 4-23**).

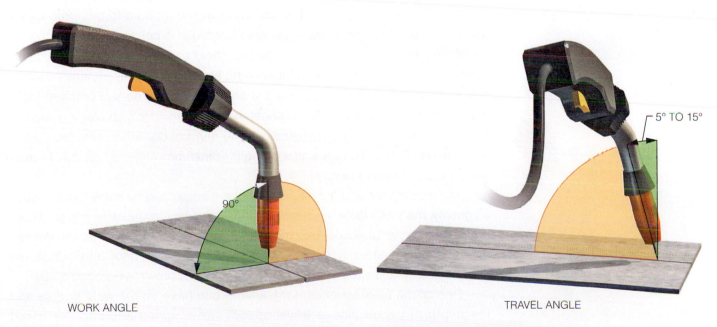

WORK ANGLE

5° TO 15°

TRAVEL ANGLE

FIGURE 4-23 Work and travel angles on a square groove weld (butt joint)

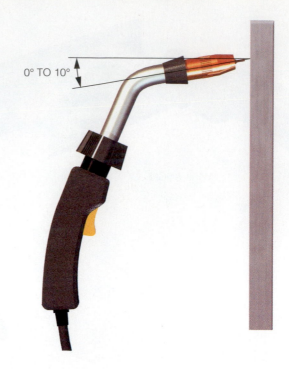

FIGURE 4-24 Travel angle for a vertical square groove weld (butt joint)

0° TO 10°

V I D E O

Direction of travel changes slightly when welding in the vertical position to 0°–10° (see **FIGURE 4-24**). As the angle increases, penetration decreases, and the weld profile flattens. The desired outcome and the welder's skill determine the best work angles.

Direction of travel has a significant affect on weld bead shape, penetration, and efficiency. A backhand technique, also referred to as a drag or pull technique as shown in **FIGURE 4-25**, with travel from left to right, results in the most penetration; the weld torch/gun is directed back into the weld opposite the direction of travel. The technique also produces a narrower, more convex weld bead. The backhand method provides deeper penetration on heavier sheet metal, such as 3/16 inch (4.8 mm) thick. The **Forehand Technique**, also referred to as a push technique as shown in **FIGURE 4-25**, with travel from right to left, produces a flatter weld and less penetration; the weld torch/gun is pointed in the direction of travel. This technique is suitable for thinner sheet metals. Forehand technique is preferred for spray transfer and for welding aluminum because it lends itself to better shielding gas coverage over the solidifying weld metal. Forehand technique with short-circuit transfer on sheet materials provides high welding speeds, flat bead profiles, adequate penetration, and minimal distortion. The decrease in penetration afforded by forehand welding is sometimes preferred on thinner metals to prevent burning through.

V I D E O

The welding arc and molten weld pool are influenced by many factors, such as gravity, magnetic force, and weld position. Work angles must be adjusted continuously in order to ensure that the weld pool flows out evenly into the desired direction. The welder must always observe the weld pool and adjust the gun/workpiece angle to position the weld pool for optimum weld bead quality.

One of the greatest effects the welder can have on the weld puddle and the depth of penetration is travel speed. Travel speed refers to the speed the weld bead travels down the workpiece in inches/minute (mm/second).

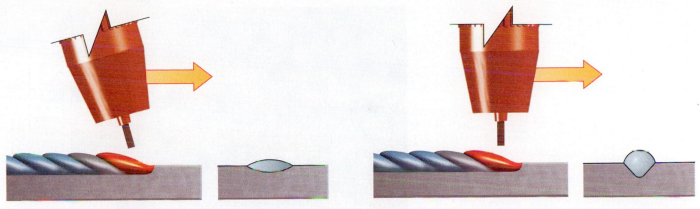

PUSH/FOREHAND TECHNIQUE

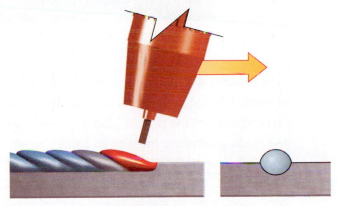

DRAG/BACKHAND TECHNIQUE

FIGURE 4-25 Weld bead shape, penetration, and efficiency can be affected by the direction of travel

The welder controls travel speed, taking care to ensure that the electrode stays in the leading edge of the weld pool. The welding arc will sufficiently heat the base metal, and the arc force mixes the filler material with the melted base metal.

Too fast a travel speed will not heat the base metal enough for adequate penetration and fusion to occur. The molten filler metal will pile up on the base metal rather than melt and mix with the base metal, eliminating the chance for fusion to occur. This creates a high, convex, and ropy appearing weld bead. The bead tends to have a ripple pattern that is long and v-shaped.

Too slow a travel speed also prevents the arc from heating the base metal. The weld pool actually forms an insulating layer between the arc and the base metal. The arc impinges on the molten weld pool as the molten filler metal flows out onto the cold base metal, resulting insufficient penetration, insufficient fusion, and cold lap (see **FIGURE 4-26**). Producing a large weld bead does not guarantee a stronger weldment. The opposite is often the case. The impinging on the weld pool is why most failures occur on welds, as the weld lifts away from the base metal. Also, it is a reason why welders fail to pass a qualification welding test. Too slow a travel speed results in incomplete penetration to the base metal.

Oscillation and travel speed often go hand in hand. While oscillation, or the movement of the electrode in a pattern that is side to side or circular in motion

V I D E O

FIGURE 4-26 Too slow a travel speed or incorrect weld parameters could cause cold lap

V I D E O

(see **FIGURE 4-27**), influences the shape of the puddle, two things must be kept in mind. The first is electrode location. It is imperative to keep the electrode at the leading edge of the puddle, and a circular motion does not promote staying on the leading edge. As mentioned earlier, an electrode that impinges on the weld pool will have a definite and negative effect on weld metal penetration into the base metal. The second consideration in oscillation is overwelding. Excessive oscillation, such as a side-to-side motion, leads to slower travels speeds, which not only affect weld penetration if not done properly by not staying at the leading edge of the puddle but also increase weld metal size, which leads to overwelding, distortion, increased thermal and residual stresses, and increased labor cost. For most GMAW applications, the welder will produce fewer discontinuities if welding is done in a straight line with the backhand or forehand technique (see **FIGURE 4-28**).

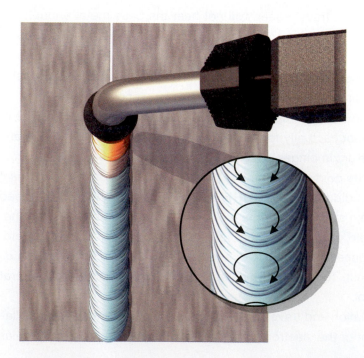

FIGURE 4-27 Making a weld bead by using oscillation will increase bead size but will decrease weld penetration

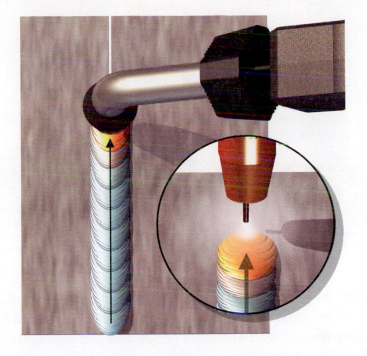

Starting an arc is not difficult. The trigger is pulled on the welding gun to initiate the arc. If the arc is not initiated, it generally means that the work lead connection is poor or is not connected at all. Oxidation present on the end of the electrode may cause the arc to stumble. Cutting the tip off the electrode prior to starting the arc ensures arc initiation. To aid in arc initiating further, cut the tip of the electrode at an angle, as opposed to perpendicular (see **FIGURE 4-29**).

FIGURE 4-29 Cutting the electrode wire at an angle will aid in initiating the arc

MODES OF METAL TRANSFER

WHAT ARE THE MODES OF METAL TRANSFER?   Metal transfer refers to how filler metal is deposited to the base metal to form the weld bead. The common modes of metal transfer are **short-circuit**, **globular**, **axial-spray**, and **pulsed-spray** transfers. The mode of metal transfer is determined by many mitigating factors:

- Base Metal Type
- Filler Metal Composition
- Electrode Diameter
- Polarity
- Arc Current
- Arc Voltage/Arc Length
- Shielding Gas Composition
- Welding Position

Short-circuit transfer occurs when filler metal is deposited from the electrode by short-circuiting to the workpiece surface. Amperage and voltage settings, along with proper shielding gas selection, determine if welds are produced with short-circuit transfer. Short circuit can be easily understood as it is related to what occurs in a circuit of an electrical fuse. If too much current (amperage) is

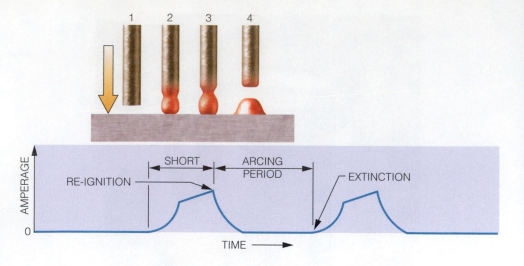

FIGURE 4-30 An electrical view of the short-circuit cycle

introduced to the circuit, the fuse overheats and blows. The same principles occur when short-circuit welding. The difference is that the electrode (fuse) is being continually fed, and the short circuit occurs over and over again, depositing the electrode on the base metal as weld.

On steels, typical voltages range from 16–20 volts. The filler metal is deposited when the electrode heats up and shorts to the workpiece, more than 100 times in a second, with typical short circuits in the 150–180 times per second range. There is no visible open arc length when using short-circuit current unless observed in slow-motion photography.

Short-circuit transfer operates at relatively low voltages and amperage ranges. **FIGURE 4-30** shows the rise in amperage that causes the short-circuit cycle.

1. At this step, the trigger is pressed on the welding gun, and the wire feeds.
2. The electrode contacts the work, voltage drops, and amperage increases.
3. Magnetic forces pinch the electrode while the amperage peaks and voltage increases.
4. The electrode melts off, and the arc opens. There is a separation of weld deposit from electrode, and the open arc period begins. The arc length increases, and the weld pool becomes more fluid and wets out to the base metal.

The low operating parameters of short-circuit transfer provide sufficient heat to weld sheet materials up to 3/16 inches (4.8 mm) thick. Welding in all positions is also easily accomplished. Small electrode diameters from 0.024 inches (0.6 mm) to 0.035 inches (0.9 mm) are easier to weld. Larger electrode diameters require higher electrical parameters more suitable for globular, spray transfer, or pulse-spray transfer methods. Welding with larger electrode diameters increases the difficulty of successfully welding the thinner gauge steel without burn-through and other discontinuities. It is easier to produce a weld free of defects when using smaller electrode diameters on thin metals. Large electrode diameters increase productivity.

Due to low heat input, short-circuit transfer is not recommended for plate materials, as penetration and fusion capabilities are limited. While it can be used

FIGURE 4-31 A small lapse in welder technique and concentration could lead to weld defects, such as lack of fusion

V I D E O

successfully on plate thickness, it must be pointed out that the skill level and concentration level of the welder must be significantly increased. A high degree of concentration and a higher degree of welder skill are necessary to produce sound welds. A small lapse in welder technique and concentration could lead to weld defects, such as incomplete fusion (see **FIGURE 4-31**) and poor penetration. Groove joint designs can help to overcome these issues for welding plate.

WHAT IS SLOPE, AND IS IT ONLY EFFECTIVE WITH SHORT-CIRCUIT WELDING?

Most power sources have a preset slope and cannot be adjusted by the welder. However, as stated earlier, some industrial power sources may be equipped with a variable slope control or tapped slope terminals. Slope is most effective when using the short-circuit transfer. As stated, typical short circuits occur about 150–180 times per second. The number of turns of slope determines the number of short circuits per second. **Slope** can be explained by examining the representative volt/amp curves, found in **FIGURE 4-32**.

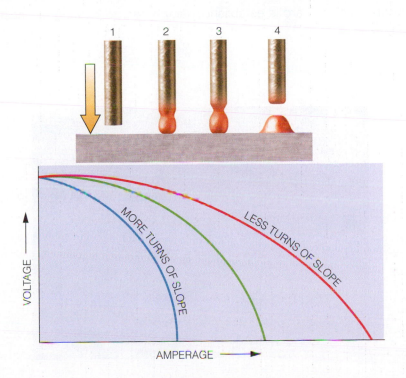

FIGURE 4-32 The amount of available amperage to pinch off the electrode wire (steps 2, 3, and 4) depends on the turns of slope

All three curves have the same maximum voltage and are smooth. The difference among these slopes can be observed as welding amperage is increased. Remember, the rise in amperage causes the short-circuit cycle.

The three given slopes provide different responses when the electrode wire short-circuits to the base metal. Because a constant voltage power source is designed to maintain a preset welding voltage, the short-circuiting of the electrode causes the power source to increase its welding amperage until the short-circuit deposit is removed and the welding voltage returns to its preset value.

With flat slope (less turns of slope), there is no limiting the short-circuit current. Thus, the high incoming amperage causes the wire to be pinched off rapidly and violently, resulting in poor arc starts and excessive spatter (**FIGURE 4-33**).

For steels, about 8 turns of slope limits short-circuit current. This limited short-circuit amperage allows the electrode wire to be pinched off at a slower rate, resulting in smoother starts and less spatter. This volt/amp curve is typical of a fixed slope found in most power sources manufactured today. It provides acceptable arc performance for the short-circuit and spray transfer arc welding of mild steel and aluminum. Because of the poor conductivity of stainless steel, the curve would still allow too much short-circuit amperage, resulting in difficult arc starts and spatter.

The steeper slope (more turns of slope) reduces the available short-circuit current compared to those curves of the flatter slopes (less turns of slope). A slope setting at 10 turns is beneficial when welding stainless steel, reducing both popping starts and spatter. However, it results in poor performance for welding aluminum and mild steel. Too steep a slope reduces short-circuit currents so low that the electrode may just pile up on the base metal.

If the slope is set too steep for the other modes of transfer—i.e., globular, spray, or pulse spray—the electrode often stumbles prior to establishing the arc. If the slope is set too flat in these other modes of transfer, arc initiation is difficult, and the electrode wire often snaps back to the contact tip and may fuse to the contact tip. Once the arc is established, slope has no bearing on these other modes of transfer.

V I D E O

FIGURE 4-33 A slope set too flat may cause excessive spatter, as in this case on stainless steel

WHAT ROLE DOES INDUCTANCE PLAY IN THE SHORT-CIRCUIT CYCLE?

While slope controls how many short circuits occur per second, **inductance** controls the rise in amperage (current) between the time the electrode contacts the base metal and then pinches off. The higher the inductance is set, the longer the arcing period. Both short-circuit cycles shown in **FIGURE 4-34** illustrate the same amount of time. As the inductance increases, the rise in amperage is impeded, thus increasing the amount of time the electrode wire is in contact with the base metal while decreasing the amount of time the open arc occurs, which increases the puddle fluidity.

Provided the voltage and amperage are set correctly and if the inductance is set correctly, the puddle will be fluid with little spatter. Voltage and amperage (wire feed speed) are the primary parameters to set. Inductance can be thought of as the final stage to fine-tune the arc. If the inductance is set too high, the rise of amperage is impeded, and the electrode will have poor arc starting stability. The electrode wire will stumble, and the operator will feel the welding gun push back.

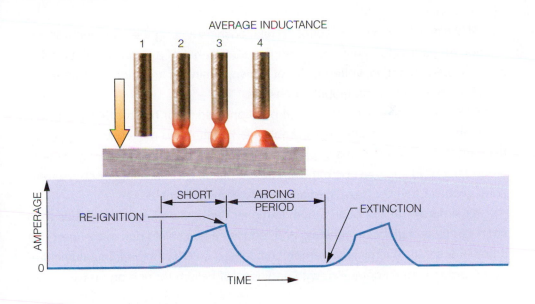

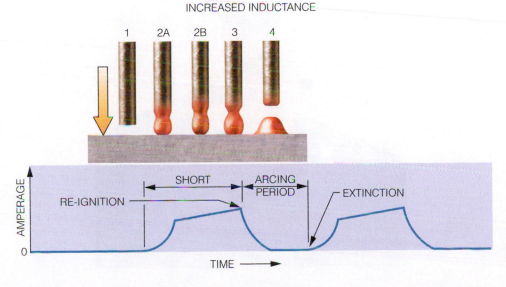

FIGURE 4-34 An electrical view of inductance of pinch effect. Steps 2, 2a, 2b, and 3 are the short, and step 4 is the arcing period

If the inductance is set too low, the rise of amperage is not impeded, and the short-circuit cycle is fast and violent, producing a great deal of spatter. In worse cases, the short circuit is so violent that the electrode wire snaps back and fuses to the contact tip. Today's modern power sources have a usable range of inductance and have eliminated the problem areas at the extreme high and low ends.

For carbon steels, 30% inductance is sufficient to reduce spatter and provide good wetting at the weld edges. Inductance settings for stainless steels are set significantly higher in order to reduce spatter, and a 50% setting is desired. The higher inductance tends to ball the end of the electrode, which must be cut before restarting the arc.

VIDEO

WHAT IS GLOBULAR TRANSFER?   Globular transfer (FIGURE 4-35) occurs with increased arc voltages in the same amperage range as the short-circuit transfer. The increased arc voltage melts the electrode back away from the weld pool. The electrode forms a balled end approximately 2 times the electrode diameter, and filler metal is deposited when gravity causes the molten droplet to detach from the electrode and fall across an open gap, causing an explosive deposit. The welder will observe an open arc or gap between the electrode and the weld pool.

FIGURE 4-36 shows the volt/amp relationships among the common transfer modes. Amperage for the globular range parallels that of a short-circuit transfer, and an adjustment in voltage, up or down, changes the arc characteristic between short-circuit and globular transfers.

Globular transfer is limited to flat position welding or horizontal fillet welds. The highly erratic arc creates undesirable weld spatter and makes welding in vertical and overhead positions impractical. Spatter often sticks to the base metal around the weld and must be removed before painting or finishing. For these reasons, globular transfer is avoided as a mode of metal transfer.

Axial-spray transfer occurs both at higher currents and higher voltages than short-circuit and globular transfers and requires a gas shield rich in the inert gas argon. For steels, the minimum argon percentage is 80%. For welding aluminum, 100% argon is most common, with additions of helium for use on heavier plate

FIGURE 4-35 Globular transfer

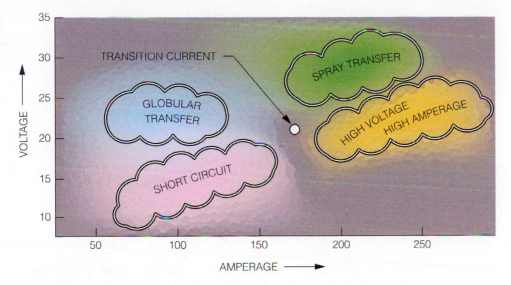

FIGURE 4-36 Area of modes of transfer

materials. Helium should not exceed 50% for spray transfer to occur when aluminum welding.

The point where one mode of transfer converts to another mode is called the **transition current**. Note the transition current in **FIGURE 4-36**. Increasing current beyond the transition current and increasing voltage allows for spray transfer to occur. Higher voltages cause an increase in arc length greater than that of globular or short-circuit transfers. Small individual droplets are transferred across the open arc to the workpiece at rates from 150–200 drops/second. Electromagnetic forces carry the droplets across the arc gap. The open arc is visible to the welder and looks like the lower half of an hourglass (**FIGURE 4-37**).

V I D E O

High voltage and amperage settings of the spray transfer arc and the fluidity of the weld pool provide for deep penetration, flat wide bead contours, and high welding speeds. These characteristics are suitable to welding on thicker plate materials. Excellent bead profiles and elimination of spatter are factors that make use of spray transfer desirable. Due to weld puddle fluidity, spray transfer welding is limited to use in the flat position for grooves and flat and horizontal positions for fillet welds.

Pulse-Spray Transfer combines high heat inputs of the spray transfer arc with slightly lower currents near the globular transfer range to provide a balanced average current low enough to allow for metal transfer in all positions. This is accomplished by pulsing the arc current between a high peak current in the spray transfer range and a lower background current (see **FIGURE 4-38**.)

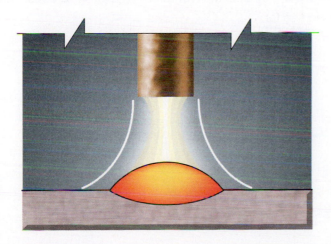

FIGURE 4-37 Spray transfer

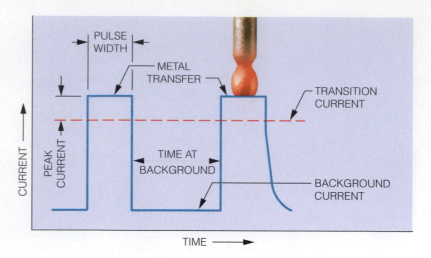

FIGURE 4-38 An electrical schematic of pulse-spray transfer

Increasing or decreasing times at each current level can change the waveform pulses between peak and background currents. For ease of use, many newer power sources have preprogrammed output waveforms for filler metal electrode types and shielding gas combinations to utilize this mode of metal transfer. This simplifies electrical setup for the welder to adjustments of wire feed speed and arc length. In some instances, welding engineers within a business make customized programs specific to company needs.

The main advantages of pulsed-spray transfer in steels and stainless steels over conventional spray transfer that are it can be used for welding in all positions and the low fume output. Pulsing can also be used on thin materials with large diameter filler metals at high welding speeds. The lower overall heat input reduces distortion and size of the heat affected zone. Stainless steels benefit from lower heat input that reduces the chances of inter-granular corrosion occurring and considerably reduces distortion.

VIDEO

Aluminum electrodes should never be short-circuit-welded and are almost always used in the spray transfer mode. However, the advantage of pulse-spray welding with aluminum electrodes is the ability to use larger diameters. Oxides are a bane of aluminum electrodes. Larger electrode diameters have a greater cross-sectional area of clean (non-oxide) electrode compared to smaller diameters (**FIGURE 4-39**). It may not be possible to weld on 1/8 inch (3 mm) aluminum sheet metal using a 1/16 inches (1.6 mm) diameter electrode in the spray transfer mode. However, using pulse spray, the task is easily accomplished.

TABLE 4-4 summarizes the variation in GMAW process metal transfer modes discussed so far.

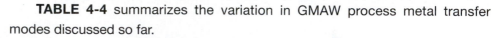

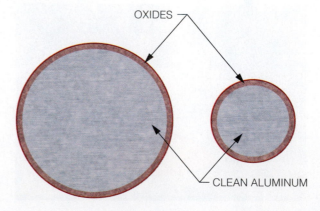

FIGURE 4-39 A cross-section of an aluminum electrode wire. Oxide contamination forms a ring around the electrode wire

Variations in the GMAW Process							
	Short Circuit	**Spray Transfer**	**Pulse Spray**	**Voltage**	**Amperage**	**Inductance**	**Slope**
Plate Steel	Not recommended	Flat position and horizontal on fillets	All thicknesses; all positions	Above 22 volts	Set to match volts	Low	≈ 8 turns
Sheet Metal Steel	Performs well in all positions	Caution— burn-through	All thicknesses; all positions	Below 22 volts	Set to match volts	≈ 30% SC	≈ 8 turns
Stainless	Performs well in all positions	On plate in flat position and horizontal on fillets	All thicknesses; all positions	Lower for short circuit; higher for spray	Set to match volts	≈ 50% SC	Steep
Aluminum	Not recommended	All thicknesses; all positions	All thicknesses; all positions	Typically above 20 volts	Set to match volts	Low	Flat
Electrode Extension	1/4–1/2 inches	1/2–3/4 inches	3/8–5/8 inches	—	—	—	—

TABLE 4-4

ELECTRODES

WHAT MUST BE CONSIDERED WHEN CHOOSING AN ELECTRODE FOR GMAW? When choosing an electrode, one must consider arc stability, solidification rate, mechanical properties, deposition rate, base metal compatibility, and parameter settings. Parameter settings for GMAW electrodes depend on:

- Diameter of the Electrode
- Mode of Transfer
- Shielding Gas Composition
- Welding Position
- Manual or Automated

Other considerations may also include corrosion resistance, wear resistance, and if the metal is to be preheated, postheated, or have no heat treatment. Color match is also a consideration, especially when welding aluminum.

WHAT SIZES OF STEEL ELECTRODES ARE USED FOR GMAW? The GMAW process uses a consumable, automatically fed wire electrode and is considered a semiautomatic welding process when welded by hand. Electrode diameters used in GMAW typically range from 0.024 inches (0.6 mm) to 0.062 inches (1.6 mm). However, electrode diameters are manufactured from as small as 0.020 inches (0.5 mm) to as large as 1/8 inch (3 mm).

As a rule, electrode size selection is based on base metal thickness, welding position, and mode of metal transfer. Larger electrodes can obviously provide higher welding speeds and the higher amperages desired for spray transfer welding on heavier base metals. Beyond that, a few considerations should be made.

For one, there has been a shift recently toward the smaller electrode diameters when semiautomatic welding. The smaller the electrode diameter, the easier it is to operate by hand. For the home hobbyist, it may make sense to run a small diameter electrode for two reasons: It is easier, and it consumes more of the electrode per inch of weld produced. This is a good thing because many home hobbyists do not weld enough to use up an entire roll of electrode wire before it oxidizes. Rusty steel, oxidized aluminum, and pitted electrodes cause many difficulties. Oxidized electrodes create weld discontinuities and wire-feeding problems.

However, when automating a process, it is often advantageous to run as large an electrode diameter as possible to increase productivity. Automation circumvents welder discomfort due to larger electrode diameters requiring more amperage and producing more heat.

Electrodes must be used within a reasonable range of voltage and wire feed speed setting. Each electrode diameter is capable of carrying a maximum amount of current that may actually be less than the power source output can provide. While a welder/operator might want to turn up the power source and feeder to provide maximum welding speeds and penetration, it is often problematic to run at excessively high welding wire feed speeds. This is in part because the electrode reaches a saturation point where **current density** is exceeded and can carry no more increase in current.

For example, when using 0.045 inches (1.1 mm) diameter carbon steel, GMAW electrode saturation may occur around 450–475 amps, or approximately 680 IPM WFS (290 mm/sec), depending on the power supply. An increase in WFS carries no more current, but more filler metal is deposited on the base metal. The travel speed must be increased to maintain the same size weld, and incomplete fusion occurs on thicker materials. With excessive travel speed, the arc will outpace the shielding gas, resulting in porosity. Finally, at the saturation current with elevated WFS, oxidation of the electrode is limited, and deoxidizer chemistry and mechanical properties of the weld metal are altered.

Many industrial wire feeders are limited to 750 IPM (320 mm/sec). To reach the current density of 0.045 inches (1.1 mm), the feeder would have to be maxed out. With proper electrode extension, it is difficult to produce a proper single pass weld bead size at those high wire feed speeds.

HOW ARE ELECTRODES FOR GMAW CATEGORIZED?
Electrodes are categorized in a number of different ways. They include:

- Filler Numbers or F-Numbers
- Chemical Analysis
- AWS Specification
- AWS Classification

Categorization divides and separates electrodes. All steel and stainless electrodes fall into the F-6 grouping. Aluminum electrode wires fall into the F-21, F-22, or F-23 grouping. If an electrode is proprietary and has not been categorized, it can be specified by chemical analysis.

Specifications for GMAW electrodes falls into different categories based on electrode compositions, such as steel, aluminum, stainless steel, etc. The AWS specifications for GMAW wires are as follows:

Steel	=	A5.18
Stainless Steel	=	A5.9
Aluminum	=	A5.10
Copper Alloys	=	A5.7
Magnesium	=	A5.19

Classifications of electrodes differ according to their composition and are generally similar to that of the base metal. Alloying and deoxidizing elements are added to the electrode to minimize porosity in the weld and to assure satisfactory weld metal mechanical properties. The most common deoxidizing elements used in steel electrodes are manganese, silicon, and aluminum.

Solid electrodes for steel GMAW are classified under the AWS A5.18 specification (**FIGURE 4-40**). The AWS classification is ERXXS-X. A typical example would be ER70S-3 or ER70S-6.
Classifications for steel electrode wires are:

- ER70S-2
- ER70S-3
- ER70S-4
- ER70S-5
- ER70S-6
- ER70S-7
- ER70S-GS

Each letter and number of a specific electrode classification has significant meaning. **FIGURE 4-41** conveys this meaning.

Some modifications have been made to the AWS classification numbering system. In order to avoid confusion, AWS added suffixes to its original classification system. An example is ER70S-3H4. The addition of H, or HZ, is used to designate

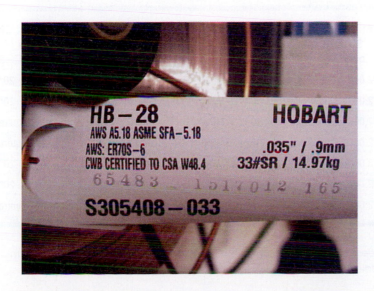

FIGURE 4-40 The AWS and ASME classification and specifications are often found on a tag attached to the spool of electrode wire

FIGURE 4-41

AWS Classification of Solid Steel Electrodes		
ER70S-X		
ER	=	Electrode rod
70	=	Minimum tensile strength × 1000
S	=	Solid electrode wire
X	=	Chemical composition and shielding gas

that an electrode meets an average diffusible hydrogen level Z. The Z designator is usually 4. The number is the maximum average diffusible hydrogen in $mL(H_2)$ per 100 g of weld metal deposited.

Another common electrode classification may have an additional suffix containing a letter/number that specifies the type of alloy. The electrode ER80S-D2, for example, contains the additional alloying elements nickel and molybdenum. **TABLES 4-5** and **4-6** show the major alloying elements by % weight and mechanical properties of the common filler metal classifications respectively.

Major Alloying Elements—% By Weight (AWS Specifications)						
AWS Classification	**Carbon**	**Manganese**	**Silicon**	**Titanium**	**Zirconium**	**Aluminum**
E70S-2	0.07	0.90–1.40	0.40–0.70	0.50–0.15	0.02–0.12	0.05–0.15
E70S-3	0.06–0.15	0.90–1.40	0.45–0.70	—	—	—
E70S-4	0.07–0.15	1.00–1.50	0.65–0.85	—	—	—
E70S-5	0.07–0.19	0.90–1.40	0.30–0.60	—	—	0.50–0.90
E70S-6	0.07–0.15	1.40–1.85	0.80–1.15	—	—	—
E70S-7	0.07–0.15	1.50–2.00	0.50–0.80	—	—	—
E70S-G	No Chemical Requirements					

TABLE 4-5

Data compiled from AWS specifications

AWS Classification	Current Electrode Polarity	Min. Tensile Strength (psi / MPa)	Min. Yield Strength (psi / MPa)	% of Elongation (2 in/50 mm)	Impact Test Charpy V (20 ft. lbs = 27 J) (−20°F = −29°C)
ER70S-2	DCEP	72,000 (500 MPa)	60,000 (415 MPa)	22%	20 @ −20°F
ER70S-3	DCEP	72,000 (500 MPa)	60,000 (415 MPa)	22%	20 @ −20°F
ER70S-4	DCEP	72,000 (500 MPa)	60,000 (415 MPa)	22%	Not Required
ER70S-5	DCEP	72,000 (500 MPa)	60,000 (415 MPa)	22%	Not Required
ER70S-6	DCEP	72,000 (500 MPa)	60,000 (415 MPa)	22%	20 @ −20°F
ER70S-7	DCEP	72,000 (500 MPa)	60,000 (415 MPa)	22%	20 @ −20°F
ER70S-G	DCEP	72,000 (500 MPa)	60,000 (415 MPa)	22%	Not Required

TABLE 4-6

Adapted from Cary, Howard B., *Modern Welding Technology*, 4th edition, © 1998. Reproduced by permission of Pearson Education, Inc., Upper Saddle River, New Jersey.

WHAT ARE THE INDIVIDUAL FILLER METAL CHARACTERISTICS OF STEEL ELECTRODES FOR GAS METAL ARC WELDING?

ER70S-2   This classification covers filler metal electrodes containing small amounts of titanium, zirconium, and aluminum in addition to the normal deoxidizing elements manganese and silicon. These wires are commonly referred to as triple deoxidized wires. They produce sound welds in all types of carbon or mild steels and are especially suited for welding carbon steels that are rusty or have mill scale on the surface. The ERXXS-2 electrodes work well when encountering porosity and high sulfur contents in base metals. Weld integrity varies with the amount of oxides on the surface of the steel. They work well in the short-circuiting mode for out-of-position welding. An application for the ER80S-D2 is on chrome-moly steels, such as 4130, when preheat and postheat treatment is not feasible. Although most application of the ER80S-D2 wire is with GTAW, it is also applicable for the GMAW process.

ER70S-3   Electrodes of this classification contain a relatively low percentage of deoxidizing elements (silicon and manganese); however, they are one of the most widely used GMAW wires. They produce welds of fair quality when used on all types of steels, especially if using Argon-O_2 or Argon-CO_2 as a shielding gas. Straight CO_2 can be used, but parameters should remain in short-circuit mode. The use of straight CO_2 is not recommended when welding with high voltage and high amperage. When CO_2 shielding gas is used, in the high welding currents (high heat input), welds produced may not meet the minimum tensile and yield strengths of this specification. High heat input using straight CO_2 will burn out the relatively low percentage of deoxidizing elements in this electrode wire. A better choice when using straight CO_2 and high welding parameters is the ER70S-7 or the more popular ER70S-6 with its higher percentage of deoxidizers.

ER70S-4   Containing slightly higher silicon and manganese contents than the E70S-3, these filler metals produce weld metal of higher tensile strength. Primarily used for CO_2 shielding gas applications where a higher degree of deoxidization is necessary.

ER70S-5   The electrodes in this classification contain aluminum as well as silicon and manganese as deoxidizers. The addition of aluminum allows these wires to be used at higher welding currents with straight CO_2 shielding gas. However, it is not recommended for out-of-position, short-circuit transfer because of high puddle fluidity. This electrode can be used for welding rusty or dirty steels with only a slight loss of weld quality.

ER70S-6   Electrodes in this classification contain the highest combination deoxidizers in the form of silicon and manganese. This allows them to be used for welding all types of carbon steel, even rimmed steels, by using CO_2 shielding gas. They produce smooth, well-shaped beads, and are particularly well-suited for welding sheet metal. This filler metal is also usable for out-of-position welding with short-circuit transfer. It is the best choice for when welding on rusted, scaled steel, or even oily surfaces. The weld quality depends on the degree of surface impurities. This electrode wire also displays excellent weldability for argon/carbon dioxide mixtures in the short-circuit, axial-spray, and pulse-spray modes of transfer. This wire may be used for high current, high deposition welding by also using argon mixed with 5–10% oxygen. When compared to the ER70S-3 series, there is a noticeable difference in how smooth the welding arc performs.

ER70S-7   This electrode is similar to the ER70S-3 classification, but it has a higher manganese content, which provides better wetting action and bead appearance. The tensile and yield strengths are slightly higher, and welding speed may be increased compared to ER70S-3. This filler metal is usually recommended for use with Argon-O_2 shielding gas mixtures, although Argon-CO_2 and straight CO_2 may be used. The weld metal will be harder than that of the ER70S-3 types but not as hard as an ER70S-6 deposit. Like the ER70S-6 electrode wire, there is a noticeable difference of arc smoothness during welding.

ER70S-G   This classification may be applied to solid electrodes that do not fall into any of the preceding classes. It has no specific chemical composition or shielding gas requirements but must meet all other requirements of the AWS A5.18 specification.

The welding parameters given in this chapter for steel, stainless steel, and aluminum should suffice for the home hobbyist, high school, trade school, technical college, or many industrial settings. On thicker materials, for the electrode diameters listed, only a slight increase in voltage, amperage, and wire feed speed may be needed as well as additional passes to complete the desired weld size. Those wishing to use larger electrode diameters should consult electrode manufacturers to produce optimum results. **TABLES 4-7** and **4-8** display welding parameters for short circuit transfer and for spray transfer welding applications.

GMAW-S (Short Circuit) Parameters – STEEL Shielding Gas – 85% Argon/15% CO_2 Gas Flow – 25–35 CFH (12–17 LPM) CTWD – ≈ 3/8 inch (≈ 9.5 mm) Position – All									
Thickness	**26–22 Gage (0.6–0.8 mm)**		**20–18 Gage (1.0–1.3 mm)**		**16–14 Gage (1.6–2.0 mm)**		**12–10 Gage (2.4–3.4 mm)**		**7 Gage (5 mm)**	
Diameter, inch (mm)	.024 (0.6)	.030 (0.8)	.024 (0.6)	.030 (0.8)	.030 (0.8)	.035 (0.9)	.030 (0.8)	.035 (0.9)	.035 (0.9)	.045 (1.1)
Voltage	12–14	14–16	14–16	14–16	16–18	16–18	17–19	17–19	18–20	18–20
Amperage	25–35	35–45	45–55	45–55	70–90	90–110	115–145	125–155	145–175	155–185
WFS IPM (mm/sec)	120–140 (50–60)	80–100 (35–45)	145–175 (60–75)	100–120 (45–50)	155–195 (65–85)	145–175 (60–75)	270–330 (115–140)	200–250 (85–105)	260–320 (110–135)	170–210 (70–90)

TABLE 4-7

Note: CFH = cubic feet per hour (LPM = liters per minute)   CTWD = contact to work distance
     IPM = inches per minute (mm = millimeters)   WFS = wire feed speed

	GMAW (Axial Spray) Parameters – STEEL Shielding Gas – 90% Argon/10% CO_2 Gas Flow – 30–40 CFH (14–19 LPM) CTWD – ≈ 3/4 inch (≈ 19 mm) Position – Flat Butt or Flat/Horizontal Fillets									
Thickness	3/16 inch (5 mm)		1/4 inch (6.4 mm)		5/16 inch (8 mm)		3/8 inch (9.5 mm)		1/2 inch (1.3 mm)	
Diameter, inch (mm)	.035 (0.9)	.045 (1.1)	.035 (0.9)	.045 (1.1)	.035 (0.9)	.045 (1.1)	.045 (1.1)	.052 (1.3)	.045 (1.1)	.052 (1.3)
Voltage	22–25	22–26	22–26	23–29	23–27	25–31	27–33	27–33	29–35	29–35
Amperage	175–215	220–270	180–220	235–285	185–225	270–330	300–370	325–395	330–410	390–470
WFS IPM (mm/sec)	335–415 (145–175)	245–305 (105–130)	360–440 (150–185)	295–355 (125–150)	385–465 (165–195)	335–415 (145–175)	425–525 (180–220)	335–415 (145–175)	515–635 (215–270)	435–535 (185–225)

TABLE 4-8

Note: CFH = cubic feet per hour (LPM = liters per minute)   CTWD = contact to work distance
IPM = inches per minute (mm = millimeters)   WFS = wire feed speed

WHAT ARE COMPOSITE ELECTRODES?

Composite electrodes, also known as metal-cored electrodes, are tubular electrodes with an inner core filled with metal particles. They deliver excellent welder appeal with good arc stability to produce slag free and spatter-free welds. The weld bead has a general appearance of a weld made with solid wire in the spray transfer mode but with the added benefit of being able to weld in the overhead and vertical down positions. Because metal-cored electrode wires produce higher current density, along with increased deposition and a less turbulent weld pool, the penetration pattern is more uniform and reduces the chance of cold lap as compared to solid steel electrodes. The penetration benefit is slightly superior to that of a solid wire in the spray transfer mode.

Composite steel electrodes have the same AWS specification as solid steel electrodes: A5.18. Composite electrodes require an external shielding gas. Unlike solid steel electrodes, composite electrodes shielding gas requirements are designated in the AWS classification with a C or M. **FIGURE 4-42** illustrates the use of the classification for composite filler metal electrodes. **Classifications** for steel electrode wires are:

- E70C-3C
- E70C-3M
- E70C-6C
- E70C-6M
- E70C-G

The 70 may be replaced with other tensile strengths designations. Common tensile strengths designations besides 70 are 80, 90, 100, and 110.

The electrode classification may also have one or more additions, such as E70C-6M H4. The addition of H, or HZ, is used to designate an electrode that will meet an average diffusible hydrogen level Z. The Z designator is usually 4 or 8. The number is the maximum average diffusible hydrogen in $mL(H_2)$ per 100 g of weld metal deposited. **TABLE 4-9** displaces the mechanical properties of composite filler metal electrodes.

FIGURE 4-42

AWS Classification of Composite Steel Electrodes		
E70C-3M		
E	=	electrode
70	=	minimum tensile strength × 1000
C	=	composite electrode wire
3 or 6	=	chemical composition
M	=	mixed gas of argon/carbon dioxide
C	=	carbon dioxide

AWS Classification	Current Electrode Polarity	Minimum Tensile Strength	Minimum Yield Strength (psi)	% of Elongation (2 inches/50 mm)	Impact Test Charpy V (ft. lbs/Joule)
E70C-3M	DCEP	80,000 psi (550 MPa)	65,000 psi (450 MPa)	31	47 @ −20°F (64 @ −29°C)
E70C-3C	DCEP	75,000 psi (520 MPa)	62,000 psi (430 MPa)	29	25 @ −20°F (40 @ −29°C)
E70S-6M	DCEP	80,000 psi (550 MPa)	70,000 psi (485 MPa)	30	50 @ −20°F (68 @ −29°C)
E70C-6C	DCEP	80,000 psi (550 MPa)	65,000 psi (450 MPa)	29	45 @ −20°F (61 @ −29°C)
E70C-G	DCEP	72,000 psi (500 MPa)	60,000 psi (415 MPa)	22	20 @ −20°F (27 @ −29°C)

TABLE 4-9

Note: Strengths, elongation, etc., are approximate depending on the percentage of shielding gas blend used.

Classification for composite electrodes vary with changes to their chemical composition. **TABLE 4-10** shows their categorization based on percent of major alloying elements.

For the E70C-3M and E70C-6M electrodes, a minimum shielding gas requirement is 75% argon, with a maximum of 95% argon. The balance mixture is CO_2. Check with electrode manufacturers for best blends to use with a particular composite electrode.

Standard electrode diameters range from 0.035 inches (0.9 mm) to 3/32 inches (2.4 mm). This range makes them useful from thin gauge metal to heavy plate. They are applicable for manual (semiautomatic) welding to fully automatic or robotic applications. Welding parameters for common carbon steel composite electrode diameters are given in **TABLE 4-11**.

Major Alloying Elements—% by Weight (Approximate Weld Metal Chemistry)					
AWS Classification	Carbon	Manganese	Silicon	Phosphorous	Sulfur
E70C-3X	0.02	1.56	0.90 Max.	—	—
E70C-6X	0.07	1.00–1.50	0.90 Max.	0.008–0.015	0.011
E70C-G	No Chemical Requirements				

TABLE 4-10

GMAW-C (Composite) Parameters – STEEL Shielding Gas – 90% Argon/10% CO_2 Gas Flow – 30–40 CFH (14–19 LPM) CTWD – < 3/4 inch ($\approx$ 19 mm) Position – Flat/Horizontal and Vertical Down				
Thickness	**All Thicknesses**			
Diameter, Inch (mm)	.035 in (0.9 mm)	.045 in (1.1 mm)	.052 in (1.3 mm)	1/16 in (1.6 mm)
Voltage	26–32	26–32	26–32	26–32
Amperage	180–220	225–275	270–330	315–385
WFS IPM (mm/sec)	490–600 (210–255)	360–440 (150–155)	315–385 (135–165)	270–330 (115–140)

TABLE 4-11

Note: Travel speeds and deposition rates are twice that of short-circuiting.

CFH = cubic feet per hour (LPM = liters per minute) CTWD = contact to work distance

IPM = inches per minute (mm = millimeters) WFS = wire feed speed

WHAT ARE THE ADVANTAGES AND DISADVANTAGES OF COMPOSITE ELECTRODES?
There are a number of advantages of metal-cored electrodes. Advantages include:

- Higher in deoxidizers for improved performance on mill-scaled plate
- Better wetting action than solid wire to minimize cold lap
- Good bead appearance
- Low fume levels that are comparable to solid wires
- Slag-free welds
- Low to non-existent spatter
- Higher deposition rates for faster travel speeds and increased productivity
- Improved fusion at the toes for better weld quality
- Deeper root penetration at the root of the weld
- Good weld efficiencies up to 96%

The greatest disadvantage associated with metal-cored wires is the higher initial cost for the plain carbon steel varieties. Because of wire manufacturing methods—i.e., rolled with metal powder in the center of a tube—they are inherently more expensive than solid wires. In addition, the new generation of metal-cored wires requires special production technology to assure low fume generation rates while maintaining high levels of weld quality. These generally result in higher manufacturing and selling costs for basic wires of this type.

WHAT ELECTRODES ARE USED FOR STAINLESS STEEL?
The AWS classification for stainless steel is different than for steel wires. Stainless steel filler wires are designated like the ANSI or ASTM base metal designations. For example, typical ANSI/ASTM base metal designations for stainless steel are 308, 316, 316L, and 320. The AWS electrode classification would be ER308, ER308L, ER316, ER316L, etc.

WHAT IS THE SPECIFICATION/CLASSIFICATION FOR STAINLESS STEEL?
The AWS specification for solid stainless steel electrode wire is AWS A5.9. An example of the AWS specification is ER308L (**FIGURE 4-43**).

AWS Classification of Solid Stainless Steel Electrodes		
ER308-L		
ER	=	electrode rod
308	=	electrode composition
L	=	additional requirements change in original alloy

FIGURE 4-43

Additional requirements that may be added to the classification are:

- L = Lower carbon content; generally no more than 0.03% carbon
- H = Limited to the upper range on the carbon content
- Mo = Molybdenum added—has pitting resistance and increased strength

Filler metal electrodes are classified based on variations in their chemical compositions (**TABLE 4-12**). These changes in composition alter their usability characteristics. In addition to impacting physical properties of the weld admixture, altering the filler metal electrode chemical composition impacts mechanical properties as indicated in **TABLE 4-13**. Filler metal specification and classification are provided by electrode manufacturers on product labeling (**FIGURE 4-44**).

WHAT ARE THE INDIVIDUAL FILLER METAL CHARACTERISTICS FOR STAINLESS STEEL?

ER308 This electrode is a general welding electrode and used for welding stainless steels of similar composition. It is one of the most common stainless electrodes used for a variety of applications.

ER308L This electrode is preferred for use on the same materials as the ER308. The chromium and nickel content is the same as the ER308 but

AWS Chemical Composition Requirements of Solid Stainless Steel Electrodes (%)								
AWS Classification	**Carbon**	**Chromium**	**Nickel**	**Molybdenum**	**Manganese**	**Silicon**	**Phosphorous**	**Copper**
ER308	0.08 Max	19.5–22.0	9.0–11.0	0.75 Max	1.0–2.5	0.30–0.65	0.03 Max	0.75 Max
ER308H	0.04–0.08	19.5–22.0	9.0–11.0	0.50 Max	1.0–2.5	0.20–0.65	0.03 Max	0.75 Max
ER308L	0.03 Max	19.5–22.0	9.0–11.0	0.75 Max	1.0–2.5	0.30–0.65	0.03 Max	0.75 Max
ER308LSi	0.03 Max	19.5–22.0	9.0–11.0	0.75 Max	1.0–2.5	0.65–1.00	0.03 Max	0.75 Max
ER309	0.12 Max	23.0–25.0	12.0–14.0	0.75 Max	1.0–2.5	0.30–0.65	0.03 Max	0.75 Max
ER309L	0.03 Max	23.0–25.0	12.0–14.0	0.75 Max	1.0–2.5	0.20–0.65	0.03 Max	0.75 Max
ER309LSi	0.03 Max	23.0–25.0	12.0–14.0	0.75 Max	1.0–2.5	0.65–1.00	0.03 Max	0.75 Max
ER316	0.08 Max	18.0–20.0	11.0–14.0	2.00–3.00	1.0–2.5	0.30–0.65	0.03 Max	0.75 Max
ER316H	0.04–0.08	18.0–20.0	11.0–14.0	2.00–3.00	1.0–2.5	0.30–0.65	0.03 Max	0.75 Max
ER316L	0.03 Max	18.0–20.0	11.0–14.0	2.00–3.00	1.0–2.5	0.30–0.60	0.03 Max	0.75 Max
ER316LSi	0.08 Max	18.0–20.0	11.0–14.0	2.00–3.00	1.0–2.5	0.65–1.00	0.03 Max	0.75 Max
ER347	0.08 Max	19.0–21.5	9.0–11.0	0.75 Max	1.0–2.5	0.30–0.65	0.03 Max	0.75 Max
ER410	0.12 Max	11.5–13.5	0.60 Max	0.75 Max	0.60 Max	0.50 Max	0.03 Max	0.75 Max

TABLE 4-12

Note: Ni = Nickel added—high temperature strength, corrosion resistance, and added ductility

AWS Classification	Current Electrode Polarity	Minimum Tensile Strength	Minimum Yield Strength	% of Elongation (2 inches/50 mm)	Impact Test Charpy V (ft. lbs / Joule)
ER308	DCEP	85,000 psi (585 MPa)	58,000 psi (380 MPa)	36	96 @ RT (130 @ RT)
ER309	DCEP	87,000 psi (600 MPa)	59,000 psi (410 MPa)	40	100 @ RT (135 @ RT)
ER316	DCEP	86,000 psi (595 MPa)	57,000 psi (395 MPa)	36	82 @ RT (110 @ RT)
ER347	DCEP	90,000 psi (620 MPa)	59,000 psi (410 MPa)	42	112 @ RT (150 @ RT)
ER410	DCEP	79,000 psi (545 MPa)	44,000 psi (305 MPa)	25	—

TABLE 4-13

Typical Mechanical Properties

Note: Strengths, elongation, etc., are approximate depending on percentage of shielding gas blend used. RT = room temperature

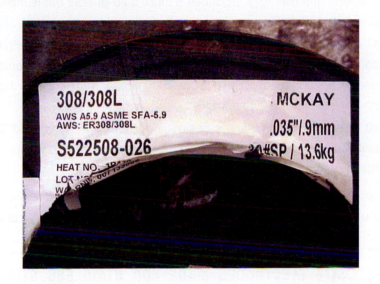

FIGURE 4-44 The AWS and ASME classification and specifications of this stainless steel electrode wire are attached to the spool

with a lower carbon content, as designated by the L. Lower carbon content reduces any possibility of **carbide precipitation** and the inter-granular corrosion that can occur. Carbide precipitation is the chemical reaction of carbon with chromium to form chromium carbides, which do not have resistance to corrosion. They are formed during welding with slow cooling if the base metal is held in the temperature range of 800°F (425°C) to 1600°F (870°C) for prolonged periods.

ER308LS This electrode not only has low carbon content but also has higher silicon content than the previous two designated by the letter S. Higher silicon levels improve the wetting characteristics of the weld metal. However, if the dilution of the base metal is extensive, the higher silicon can cause greater crack sensitivity.

ER309 This electrode is used for welding comparable base metal (309). This electrode is one of the most versatile stainless electrodes and is recommended for welding stainless steel to mild steel, and stainless steel to low alloy steels, and for joining heat-treatable stainless steels when heat treatment is not possible.

ER309L The lower carbon content increases resistance to inter-granular corrosion.

ER310 Most frequently used to weld base metals of similar composition. It is also recommended in many applications that ER309 is also recommended.

ER316 This electrode is used for welding 316 and 319 stainless steels. It contains additional amounts of molybdenum for high temperature service and for increased pitting corrosion resistance. It is a general welding electrode used similar to ER308 but is more costly.

ER316L This electrode filler wire has lower carbon content than the previous example, as designated by the letter L. It is less susceptible to inter-granular corrosion caused by carbide precipitation.

ER316LHS This electrode filler wire not only has lower carbon content than the ER316 filler wire but also has high silicon content, as designated by the HS. This filler wire produces a more fluid puddle but is more susceptible to cracking.

ER347 This electrode wire has fewer problems with carbide precipitation because niobium (colombium) or niobium along with traces of tantalum is added as a stabilizer. It is used for welding similar metals where high temperature strength is required.

ER410 Used for welding alloys of similar composition. Also used for overlays on carbon steels to resist corrosion, erosion, or abrasion. Usually requires preheat and postheat treatments.

Stainless steel electrodes can be used for both short circuit transfer and for spray transfer applications. Welding parameters for short circuit and spray transfers are provided in **TABLES 4-14** and **4-15**.

ARE COMPOSITE ELECTRODES MADE FOR STAINLESS STEEL TOO?

Composite electrodes are also manufactured in stainless steel and deliver excellent welder appeal with good arc stability to produce slag-free and spatter-free welds. Electrodes are generally formulated to run in the spray transfer mode with a

GMAW-S (Short Circuit) Parameters – STAINLESS STEEL							
Shielding Gas: Argon/Nitrogen/CO_2 Blend or Argon/Helium/CO_2 Blend							
Gas Flow – 30–40 CFH (14–19 LPM)							
CTWD – ≈ 3/8 inch (≈ 9.5 mm)							
Position – All							
Thickness	**20–18 Gage** **(1.0–1.3 mm)**		**16–14 Gage** **(1.6–2.0 mm)**		**12–10 Gage** **(2.4–3.4 mm)**		**7 Gage** **(5 mm)**
Electrode Diameter, Inch (mm)	.030 (0.8)	.035 (0.9)	.030 (0.8)	.035 (0.9)	.030 (0.8)	.035 (0.9)	.035 (0.9)	.045 (1.1)
Voltage	14–18	14–18	14–18	15–18	16–20	16–20	17–21	17–21
Amperage	80–100	120–150	90–110	125–155	115–145	145–175	160–200	160–200
WFS IPM (mm/sec)	245–305 (105–130)	270–330 (115–140)	270–330 (115–140)	295–355 (125–150)	450–550 (190–235)	360–440 (155–185)	425–525 (180–225)	295–355 (125–150)

TABLE 4-14

Note: CFH = cubic feet per hour (LPM = liters per minute)   CTWD = contact to work distance
IPM = inches per minute (mm = millimeters)   WFS = wire feed speed

GMAW (Axial Spray) Parameters – STAINLESS STEEL
For 1/4 inch (6.4 mm) and under – Shielding Gas: Argon/Nitrogen/CO_2 Blend or Argon/Helium/CO_2 Blend
For 5/16 inch (84 mm) and over – Shielding Gas: 98% Argon/2% O_2
Gas Flow – 30–40 CFH (14–19 LPM)
CTWD – ≈ 3/4 inch (≈ 19 mm)
Position – Flat Butt and Flat/Horizontal Fillets

Thickness	3/16 inch (5 mm)		1/4 inch (6.4 mm)		5/16 inch (8 mm)		3/8 inch (9.5 mm)		1/2 inch (1.3 mm)	
Electrode Diameter, Inch (mm)	.030 (0.8)	.035 (0.9)	.030 (0.8)	.035 (0.9)	.035 (0.9)	.045 (1.1)	.035 (0.9)	.045 (1.1)	.035 (0.9)	.045 (1.1)
Voltage	23–27	23–27	23–27	23–27	23–27	23–29	23–27	23–29	23–27	23–29
Amperage	150–190	215–265	150–190	215–265	190–230	225–270	195–235	230–280	200–240	235–285
WFS IPM (mm/sec)	540–660 (230–280)	495–605 (210–255)	540–660 (230–280)	495–605 (210–255)	425–525 (180–220)	290–360 (120–155)	450–550 (190–235)	305–375 (130–160)	495–605 (210–255)	325–395 (140–170)

TABLE 4-15

Note: CFH = cubic feet per hour (LPM = liters per minute)   CTWD = contact to work distance
IPM = inches per minute (mm = millimeters)   WFS = wire feed speed

98% argon and 2% oxygen mixed shielding gas. These electrodes exhibit very high deposition rates and faster travel speeds than solid stainless steel electrodes.

The composite electrodes have the same AWS specification as solid stainless steel electrode wires: A5.9. The AWS classification also mirrors the stainless steel solid electrodes and the previously discussed steel composite electrodes (**FIGURE 4-45**). Some examples are:

- EC308L Si
- EC309L Si
- EC316L Si

Filler metal electrodes are classified based on variations in their chemical compositions (**TABLE 4-16**). **TABLE 4-17** lists welding parameters for some common diameter filler metal electrodes.

AWS Classific ation of Composite Stainless Steel Electrodes

EC308-L Si		
EC	=	electrode composite
308	=	electrode composition
L	=	lower carbon content
Si	=	higher silicon content

FIGURE 4-45

AWS Classification	Carbon Max	Manganese	Silicon	Nickel	Chromium	Molybdenum	Iron
EC308L Si	0.03	1.50–1.75	0.75–0.85	10.00	20.00	0.25	Balance
EC309L Si	0.03	1.50– 1.75	0.75–0.85	13.00	24.00	0.25	Balance
EC316L Si	0.03	1.50– 1.75	0.75–0.85	12.00	18.00	2.50	Balance

TABLE 4-16

Major Alloying Elements—% by Weight (Approximate Weld Metal Chemistry)

GMAW-C (Composite) Parameters – STAINLESS STEEL Shielding Gas – 98% Argon/2% O_2 Gas Flow – 30–40 CFH (14–19 LPM) CTWD – ≈ 3/4 inch (≈ 19 mm) Position – Flat/Horizontal and Vertical Down				
Thickness	**All Thicknesses**			
Electrode Diameter, Inch (mm)	.035 in (0.9 mm)	.045 in (1.1 mm)	.052 in (1.3 mm)	1/16 in (1.6 mm)
Voltage	23–27	23–27	23–27	23–29
Amperage	180–220	225–275	270–330	315–385
WFS IPM (mm/sec)	495–605 (210–255)	360–440 (150–155)	315–385 (135–165)	270–330 (115–140)

TABLE 4-17

Note: Travel speeds and deposition rates are twice that of short-circuiting.

CFH = cubic feet per hour (LPM = liters per minute)   CTWD = contact to work distance
IPM = inches per minute (mm = millimeters)   WFS = wire feed speed

As industrial needs have changed and new alloys are introduced, the development of composite wires has also increased. It can be expected that this area will expand and grow in the years to come.

WHAT IS SPECIAL ABOUT ALUMINUM ELECTRODE WIRE?

Unlike steel and stainless steel electrodes, composition of the electrode for aluminum is generally not similar to that of the base metal. Aluminum welding is basically an alloying welding process. Using the same base material as the filler metal leads to cracks on some base metals. Using a filler metal specifically formulated to match the base metal is essential to quality welds. Solid electrodes for aluminum GMAW fall into the AWS A5.10 specification. The AWS classification is ERXXXX. An example of an AWS aluminum filler metal electrode classification is an ER4043 (**FIGURE 4-46**). As with other filler metal electrodes, product labeling identifies the AWS and ASME specification and classification (**FIGURE 4-47**).

WHAT ARE THE MORE COMMON FILLER METALS USED TO WELD ALUMINUM, AND WHAT ARE THEIR CHARACTERISTICS?

The two most common aluminum filler metals are ER4043 and ER5356. Some companies prefer the ER4047, which is very similar to the ER4043 except with higher silicon content that improves puddle fluidity. A complete base metal/filler metal comparison chart is found in Chapter 10. **TABLE 4-18** gives welding parameters for spray transfer welding of various metal thicknesses with common diameter electrodes.

ER4043 This electrode is used for general-purpose welding. Its main alloying element is silicon, ranging from 4.5–6%. The addition of silicon

AWS Classification of Solid Aluminum Electrodes		
ER4043		
ER	=	electrode rod
4043	=	composition

FIGURE 4-46

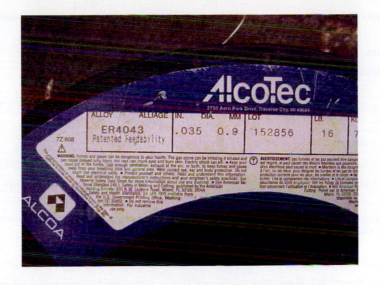

FIGURE 4-47 The AWS and ASME classification and specifications of this aluminum electrode wire is attached to the spool

GMAW (Axial Spray) Parameters – 4043/5356 ALUMINUM Shielding Gas – 100% Argon Gas Flow – 35–45 CFH (17–21 LPM) CTWD – ≈ 3/4 inch (≈19 mm) Position – All								
Thickness	1/8 in – 3/16 in (3 mm – 5 mm)		1/4 in – 5/16 in (6.5 mm – 8 mm)		3/8 in – 7/16 in (9.5 mm – 11 mm)		1/2 in and up (13 mm and up)	
Electrode Diameter, Inch (mm)	.035 (0.9)	3/64 (1.2)	3/64 (1.2)	1/16 (1.6)	3/64 (1.2)	1/16 (1.6)	3/64 (1.2)	1/16 (1.6)
Voltage	19–23	20–24	21–25	22–26	23–27	23–29	24–30	25–31
Amperage	110–120	115–145	125–155	160–200	190–230	200–240	205–255	250–310
WFS IPM (mm/sec)	315–385 (135–165)	225–275 (95–115)	270–330 (115–140)	150–190 (65–80)	335–415 (145–175)	200–240 (85–100)	425–525 (180–220)	225–275 (95–115)

TABLE 4-18

Note: Travel speeds and deposition rates are twice that of short-circuiting.

CFH = cubic feet per hour (LPM = liters per minute) CTWD = contact to work distance
IPM = inches per minute (mm = millimeters) WFS = wire feed speed

lowers the melting temperature of the filler and promotes a free-flowing puddle. ER4043 wires have good ductility and high resistance to cracking during welding.

ER5356 This electrode has improved tensile strength due to the major alloying element of magnesium, ranging from 4.5–5.5%. The ER5356 has good ductility but only average resistance to cracking during welding. This filler metal is often used as a substitute for ER4043 on 6061T-X.

SHIELDING GASES

WHAT SHIELDING GASES OR COMBINATIONS OF SHIELDING GASES ARE USED IN GMAW? The shielding gases used in GMAW are presented in **TABLE 4-19**.

Gas	Molecular Weight	Remarks
Carbon Dioxide	44.010	Reactive shielding gas for ferrous metals, providing deep weld penetration.
Argon	39.940	As the amount increases transition current decreases.
Oxygen	32.000	Increases puddle fluidity.
Nitrogen	28.016	Increases corrosion resistance in stainless steels.
Helium	4.003	High ionization potential increases penetration when blended with argon in nonferrous metals.

TABLE 4-19

Oxygen, argon, and nitrogen gases are obtained from the atmosphere and cooled and liquefied. Because each gas has a different boiling point, they can be separated and purified. Carbon dioxide is manufactured by various methods of burning, production of chemicals, or fermentation. Helium is found and extracted from natural gas. Purification of all shielding gases is essential and must be at least 99% pure.

WHAT KINDS OF PROPERTIES AND CHARACTERISTICS DO SHIELDING GASES EXHIBIT?

Shielding gases are necessary for GMAW to protect the weld pool from atmospheric contamination. Without this protective shield, the weld metal will exhibit porosity and embrittlement. Depending on the choice of shielding and the **flow rate** of the shielding gas, there is a pronounced effect on:

- The way the weld puddle wets out
- The depth of penetration and bead profile (see **FIGURE 4-48**)
- The width of the weld puddle
- The characteristics of the arc
- The mode of metal transfer that can be achieved
- The end mechanical properties of the weld metal
- The weld metal efficiency
- The cleaning action on the base metal

It is important to set the proper flow rate, depending on the shielding gas density (proportional to its molecular weight), to obtain an adequate shielding without wasting gas, creating discontinuities, or changing the mechanical properties of the weld. Insufficient flow rate will cause porosity and a brittle weld metal (see **FIGURE 4-49**). Excessive flow rate will also cause porosity by forcing the shielding gas into the molten weld pool. Depending on the shielding gas density, the flow rate will vary. A flow rate of 35 CFH (17 LPM) for argon may compare to 100 CFH (47 LPM) of a gas that is primarily helium because of the density difference.

FIGURE 4-48 Effects on weld profile with various shielding gases

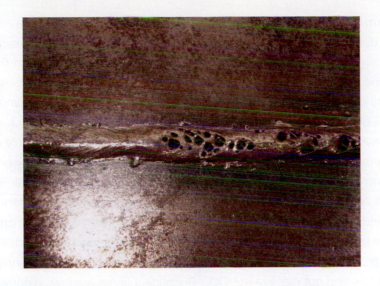

FIGURE 4-49 Air current, wind, and improper shielding gas flow settings can cause porosity

Selection of the type of shielding gas can reduce **surface tension**. Surface tension can be reduced by the use of argon gas because it is a cleaning gas. However, surface tension can be best eliminated by cleaning the mill scale off steel surfaces. Surface tension can be recognized when the weld puddle does not fuse to the base metal and appears to separate (cold lap).

WHAT SHIELDING GASES ARE USED TO WELD CARBON STEEL?

Shielding gas composition and flow rates are critical components of a welding procedure. While some gases and gas combinations are more common than others, it is important to select a gas based on need rather than on what is popular. Consider choices for welding carbon steels:

Carbon Dioxide (CO_2): This is a common shielding gas for any of the ERXXS-X electrodes. Carbon dioxide is composed of 72% oxygen and 29% carbon. It is the least-expensive shielding gas to purchase for welding plain carbon steel with the GMAW process. It is a reactive gas that produces deep penetration, a rough bead profile, and more spatter than the blended shielding gases used for GMAW. It is also the least efficient shielding gas to use for steel electrodes. Depending on the quantity of welds produced, a blended shielding gas, with its greater efficiency, compensates for the initial purchase price and is less expensive to use in the long run.

Short-circuit transfer is the only suitable mode of transfer that is achieved when using carbon dioxide. Globular transfer may also be achieved. However, with the decreased efficiency of globular transfer coupled with being able to only weld in the flat butt and flat/horizontal fillet weld positions, globular transfer is impractical.

At elevated arc welding parameters (higher heat input), the reactive carbon dioxide gas burns out **deoxidizers** (alloying elements) that are added to steel electrodes. This can cause the weld metal to be more brittle, leading to service failure. If high voltage/high amperage settings are used with CO_2, an electrode with higher amounts of deoxidizers—i.e., use ER70S-6 as opposed to ER70S-3—is a better choice. Also, at these elevated-welding parameters, an increased diffusion of carbon into the weld metal results. An increase in carbon into the weld metal may result in excessive weld metal hardness and brittleness.

Argon/Carbon Dioxide: When two shielding gases are mixed, they may also be called dual blends or binary blends. There are many advantages for using a mixture of argon and carbon dioxide. By optimizing the amount of CO_2 in the argon mixture, the fluidity of the weld puddle can be controlled to give good bead shape in a variety of welding positions. It allows for good control and speed when in flat, horizontal, or vertical welding positions (up or down). Because Argon/CO_2 mixtures provide an arc that remains more stable when welding over light mill scale or residual oil, there is a significant reduction of the possibility of weld porosity occurring. Also, by increasing the percentage of CO_2 in a mixture, there is a greater tendency to remove some material contamination in advance of the arc, which can improve overall weld quality, particularly when coated steels are used. As the amount of CO_2 in the mixture is increased, penetration will broaden and become less finger-like. This provides for greater tolerance to poor fit-up and mismatch. Finally, argon is an inert gas and not reactive like CO_2. Because it is not reactive, oxidizers in the filler metal—i.e., electrodes—are not burned out at elevated operating parameters.

Argon/CO_2 mixtures offer versatility. With the correct blend, solid wire can be used in the short arc, spray, or pulsed-spray transfer modes. Most sources agree that the correct blend must exceed 80% argon in order to achieve axial-spray or pulse-spray transfer. Below 80% argon, only short-circuit or globular transfer is achieved.

In comparison, argon/oxygen mixtures are restricted to spray transfer. All argon mixtures significantly reduce fumes. There may be a reduction in fume levels by as much as 25%–50% when using an argon/CO_2 mixture as opposed to using 100% CO_2.

Any mixture of the two shielding gases can be created. For example, 93/7, 60/40, and 50/50 mixtures can be purchased if desired. The suppliers of shielding gas will mix any percentage the customer requires. However, when mixing special orders, the price of the special mixture will be reflected.

The more common mixtures are:

- 75% argon/25% CO_2
- 80% argon/20% CO_2
- 85%argon/15% CO_2
- 90% argon/10% CO_2
- 95% argon/5% CO_2

The weld in **FIGURE 4-50** was short-circuited by using 85% argon and 15% CO_2 shielding gas.

The weld in **FIGURE 4-51** was made with the spray transfer mode by using argon/CO_2 (85/15).

Argon/Oxygen: There are advantages and disadvantages to using an argon/oxygen mixture. The advantages include:

- The addition of oxygen to argon lowers the spray transition current and thus allows spray arc transfer at lower average currents than argon/CO_2 mixtures.
- Oxygen reduces the surface tension of the weld pool and enhances its flow characteristics. On thin material, travel speeds can be increased when using a spray transfer mode due to the lower voltage required.

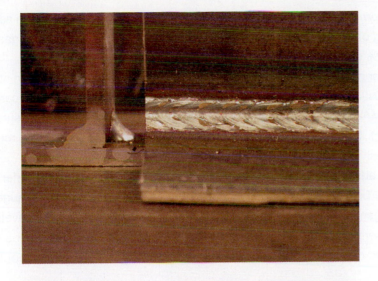

FIGURE 4-50 An etched cross-section of GMAW steel with the short-circuiting transfer

FIGURE 4-51 An etched cross-section of GMAW steel with the spray transfer

- Because spray transfer can be achieved at lower voltage levels, it is possible to use a larger wire size with argon/O_2 mixtures than is possible with argon/CO_2 in a similar application and still maintain fine droplet transfer.
- If argon/O_2 is compared with argon/CO_2 using solid wire, fumes are reduced by 15%.

However, there are some disadvantages to the use of argon/O_2:

- There can be problems with undercutting and incomplete fusion due to bead rolling at the bottom edge of fillet welds. The very fluid weld puddle makes bridging gaps more difficult. In addition, there is a greater tendency for crater cracking due to bead shape and puddle fluidity.
- To achieve stable spray transfer and minimum weld porosity, the material surfaces must be clean with no scale or residual oil film. Arc instability is a common cause of weld defects and irregular weld bead shape. The power supply slope must be steep for acceptable starting. Arc voltage and tip-to-work distance must be carefully controlled to minimize spatter, manage bead shape, and attain good arc stability. This requires a higher level of operator skill and attention. In addition, there must be little or no variation in travel speed as burn-through may result on thinner material.

Argon/Carbon Dioxide/Oxygen: Mixing argon/CO_2/O_2 is known as a tri-mix shielding gas or a ternary blend. A common mixture of this tri-mix gas is: 90% argon, 8% carbon dioxide, and 2% oxygen.

Other mixtures of tri-mix gases exist for welding on plain carbon steels, but there is usually only a slight deviation in the percentages. This mixture can be used in the GMAW process with all modes of metal transfer (spray, pulsed-spray, short-circuiting). It produces good arc characteristics and excellent weld mechanical properties.

The advantage of this mixture is its ability to use any metal transfer mode to shield carbon steel and low-alloy steel of all thicknesses. On thin gauge metals, the oxygen constituent assists arc stability at very low current levels (30–60 amps), permitting the arc to be kept short and controllable. This helps minimize excessive melt-through and distortion by lowering the total heat input into the weld zone.

The disadvantage of this mixture is expense. It also offers no real advantage over dual blend shielding gases of argon/CO_2 capable of all modes of metal transfer already discussed.

Straight, or 100%, argon is never used for shielding when welding on plain carbon steels. At least 1% of either carbon dioxide or oxygen is needed to stabilize the arc. With 100% argon, the transfer mode jumps around from short-circuit transfer, spray transfer, and globular transfer.

TABLE 4-20 shows a Sheilding Gas Comparison Chart for Carbon Steels.

WHAT SHIELDING GASES ARE USED TO WELD STAINLESS STEEL?

Traditionally, there have been two types of shielding gases for stainless steel: one tri-mix shielding gas for a short-circuit transfer and one dual-mixed shielding gas

Shielding Gas Comparison Chart—Carbon Steels				
Shielding Gas	**Short Circuit**	**Spray/Pulse**	**Flow—CFH**	**Characteristics**
100% CO_2	Yes	No	15–25	Deep penetrating
75 Ar/25 CO_2	Yes	No	25–35	Operates well at higher parameters
80 Ar/20 CO_2	Yes	Lower Limit	25–35	Marginal at spray transfer parameters
85 Ar/15 CO_2	Yes	Yes	25–35	A good choice for those wanting all transfer modes with the higher energy of CO_2 Upper limit for FCAW if choosing one shielding gas for everything
90 Ar/10 CO_2	Yes	Yes	25–40	A good choice for those wanting all transfer modes Works great with metal-cored composite electrodes
95 Ar/5 CO_2	Upper Limit	Yes	30–40	Mixture more tolerant to mill scale and a more controllable puddle than an argon-oxygen mixture
95 Ar/5 O_2	No	Yes	30–40	Addition of oxygen to argon lowers the transition current
98 Ar/2 CO_2	No	Yes	30–40	Good penetration, high efficiency, pleasing appearance
90 Ar/8 CO_2/2 O_2	Yes	Yes	25–40	Ability to use any metal transfer mode to shield carbon steel and low-alloy steel of all thicknesses

TABLE 4-20

for a spray or pulse-spray transfer. However, there are several shielding gas combinations that allow all three modes of metal transfer.

Helium/Argon/Carbon Dioxide: The common mixtures for short-circuit welding are:

- 90% helium, 7.5% argon, and 2.5% CO_2
- 90% helium, 8% argon, and 2% CO_2

There is little difference between these two mixtures. Both exhibit excellent characteristics. When appropriate voltage, amperage, and travel speeds are employed, the weld bead exhibits a rainbow of blue and gold colors. When excessive parameters are travel speeds are too slow, the bead will appear much darker, taking on a blackish color. One drawback to these two mixtures is that they are limited to the short-circuit transfer mode only. Another drawback is the world's limited supply of helium. It has been predicted that the world's supply of helium will be exhausted before the year 2020.

Carbon dioxide content is kept low to minimize carbon absorption and assure good corrosion resistance, especially in multipass welds. For every 5% CO_2 added to argon, there is an increase of 2 points (0.02%) of carbon added to the weld metal. Any increase in carbon to stainless steel is significant, and additions of more than 5% CO_2 should be avoided whenever possible. The argon and carbon dioxide additions provide good arc stability and depth of fusion. The high helium content provides significant heat input to overcome the sluggish nature of the stainless steel weld pool. It also provides increased puddle fluidity for good bead shape and faster travel speeds.

Argon/Oxygen: Two shielding gases that may be used for spray and pulse-spray transfers in stainless steels are:

- 98% argon/2% O_2
- 99% argon/1% O_2

These mixtures are used for spray transfers on stainless steels. One percent oxygen is usually sufficient to stabilize the arc and improve the droplet rate and bead appearance. As the oxygen increases, the weld puddle exhibits a better wetting action. Weld mechanical properties and corrosion resistance of welds made with either oxygen additions are similar. However, bead appearance will be darker and more oxidized as the percentage of oxygen increases.

Tri-Mixed Shielding Gases: There are many different tri-mixed shielding gases for stainless steel. The combinations vary in percentages; however, the major gas is argon. This combination works well in the short-circuit, spray transfer, and pulse-spray transfer modes. Some mixtures are:

- Argon/helium/CO_2
- Argon/nitrogen/CO_2
- Argon/CO_2/hydrogen

Argon/Helium/Carbon Dioxide: Note that the shielding gas listed first is the main constituent. The argon mixture for this shielding gas may range from

60–80%; helium from 15% to near 40%; and the balance of CO_2 from 1%–5%. This tri-mix provides a high welding speed, a broad weld with a flat crown and good color match, reduced porosity, and excellent alloy retention with good corrosion resistance. This mixture can also be used for the robotic welding of stainless steel. The argon and helium with controlled additions of CO_2 will produce an excellent weld bead appearance with minimum spatter. Higher amounts of argon, near the 80% range, allow better arc starting characteristics. Also, the higher percentage of argon reduces the transition current, allowing spray and pulse-spray transfers to occur at lower parameters. As mentioned previously, the disadvantage of this mixture is the world's limited supply of helium.

Argon/Nitrogen/Carbon Dioxide: This mixture exhibits very good arc stability, low levels of welding fume, improved color match, very good short-circuiting performance with minimal spatter, and very good performance in a pulsed-spray transfer, with good bead shape and optimized travel speed. It produces excellent quality welds when joining light gauge material with a short-circuiting transfer.

With its controlled CO_2 content, this mixture can be utilized in most austenitic stainless steel applications, particularly where weld metal carbon control is required. The addition of nitrogen enhances arc performance by increasing its stability, improving weld penetration, and reducing distortion in the welded part. It also assists in maintaining weld metal nitrogen levels for materials such as duplex stainless steels where such chemistry control is critical to maintaining micro-structural integrity and increased corrosion resistance. The weld in **FIGURE 4-52** was made with an argon/nigtrogen/CO_2 shielding gas. Voltages are generally 2 to 3 volts less when welded on the same thickness compared to helium/argon/CO_2 blends. The axial-spray transfer weld in **FIGURE 4-53** was made with argon/helium/CO_2 shielding and appropriate welding parameters.

Argon/CO_2/Hydrogen: This mixture is designed for all-position welding of austenitic stainless steels by using axial-spray and pulsed-spray transfers.

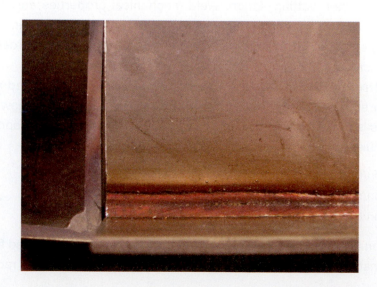

FIGURE 4-52 An etched cross-section of GMAW stainless steel with the short-circuiting transfer

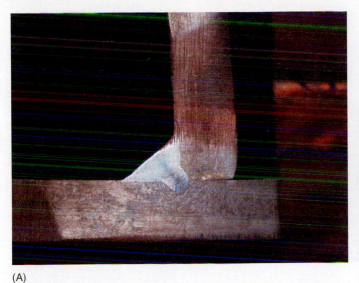
(A)

(B)

FIGURE 4-53 (A) An etched cross section of a GMAW stainless steel weld made with the spray mode of transfer. Note the deep root penetration in the etched cross section. (B) A GMAW stainless steel weld made with the spray mode of transfer

WHAT SHIELDING GASES ARE USED TO WELD ALUMINUM?
Basically, there are only two shielding gases used for welding with aluminum electrode wire with the GMAW process: argon and helium. Where argon works well on thin gauge metal, the addition of helium requires higher voltage parameters, making it more suitable for welding heavier plate. A blended shielding gas of argon/helium requires at least 50% argon in order to achieve spray transfer. Common shielding gases are:

- 100% argon
- 75% argon/25% helium
- 50% argon/50% helium

The only suitable modes of transfer with GMAW with aluminum electrode wires are the axial-spray and pulse-spray transfers. Short circuit should only be used on the thinnest of metals where, even with the smallest electrode diameter, burn-through occurs when using spray transfer and only if pulse spray is not an option. Because of aluminum's ability to dissipate heat readily, short circuit does not allow proper penetration and fusion. Cold lap and insufficient penetration are two defects that often occur when short-circuiting on aluminum (see **FIGURE 4-54**) compared to the spray transfer weld in **FIGURE 4-55**.

100% Argon: Argon provides the best arc starting characteristic compared to argon/helium mixtures because of its lower **ionization potential**. The ionization potential, or ionization energy, of a gas atom is the energy required to strip it of an electron. That is why a shielding gas such as helium, with only 2 electrons in its outer shell, requires more energy (higher voltage parameters) for welding. The ionization potential of a shielding gas also establishes how easily an arc will initiate and stabilize. A low ionization potential means the arc will start relatively easy and stabilize quite well. A high ionization potential has difficulty initiating and may have difficulty keeping the arc stable. Argon is also a cleaning gas that

FIGURE 4-54 The short-circuit transfer on aluminum does not allow adequate penetration due to aluminum's ability to dissipate heat. The lack of root penetration can be seen in the etch cross section in this weld

FIGURE 4-55 When the spray transfer mode is used on aluminum, root penetration is greatly increased. Spray transfer on aluminum can be performed in all welding positions

helps cut through some of the oxides that form on aluminum. Oxides on aluminum form immediately, even after cleaning, and continue to build up as time goes on. These oxides melt at approximately 3600°F (1980°C) as opposed to the aluminum base metal, which melts around 1250°F (675°C). For maximum penetration and fusion, oxides should be removed prior to welding. There are a variety of methods for cleaning the base metal, ranging from chemicals, such as acetone, to mechanical grinding. The axial-spray transfer aluminum weld in **FIGURE 4-55** was made with 100% argon.

Argon/Helium: As the helium content increases, voltage parameters and travel speeds are increased. The increase in helium makes it easier to achieve greater penetration on heavier material, such as 1/2 inches (13 mm) plate to 1 inch (25 mm) and thicker. Although an increase in helium does increase voltage parameters compared to argon, 50% helium is the maximum amount before spray transfer is no longer achievable. A mixture of half argon and half helium is suitable for high-speed welding on material under 3/4 inches (19 mm) thick.

Other Mixtures: As with other metals, shielding gases for aluminum are not restricted to the three listed. Any percentage of argon/helium can be mixed to satisfy the requirements of the customer.

MULTIPLE CHOICE

1. When was the GMAW process we recognize today developed?
   a. 1920
   b. 1948
   c. 1953
   d. 1958

2. What is the suggested lens shade for FCAW?
   a. 5–8
   b. 8–10
   c. 10–12
   d. 12–15

3. What are the three common incoming voltages for GMAW power sources?
   a. 24V, 50V, and 110V
   b. 110V, 200V, and 300V
   c. 120V, 240V, and 560V
   d. 115V, 230V, and 460V

4. In order to weld with GMAW, which of the following components is *not* needed?
   a. Portable cart
   b. Power source
   c. Wire feeder and welding torch
   d. Shielding gas

5. What two controls are commonly found on GMAW power sources?
   a. Voltage and wire feed speed
   b. Voltage and inductance
   c. Wire feed speed and inductance
   d. Inductance and slope

6. How is voltage usually adjusted on GMAW power sources?
   a. By adjusting the control on the wire feeder
   b. By adjusting the control on the power source
   c. By adjusting the arc length
   d. By adjusting the amount of stick-out

7. How is amperage adjusted on GMAW power sources?
   a. By adjusting the control on the wire feeder
   b. By adjusting the control on the power source
   c. By adjusting the arc length
   d. By adjusting the amount of stick-out

8. What does slope control?
   a. Slope impedes the rise in voltage when the electrode wire makes contact with the base metal.
   b. Slope controls the amount of available voltage when the electrode wire makes contact with the base metal.
   c. Slope controls the amount of available resistance when the electrode wire makes contact with the base metal.
   d. Slope controls the amount of available amperage when the electrode wire makes contact with the base metal.

9. What does inductance control?
   a. Inductance impedes the rise in voltage when the electrode wire makes contact with the base metal.
   b. Inductance controls the amount of available voltage when the electrode wire makes contact with the base metal.
   c. Inductance impedes the rise in current when the electrode wire makes contact with the base metal.
   d. Inductance controls the amount of available resistance when the electrode wire makes contact with the base metal.

10. What type of drive roll is used when welding steel?
    a. V-groove drive roll
    b. U-groove drive roll
    c. Knurled drive roll

11. How is duty cycle measured?
    a. By the amount of time the welding power source can operate without overheating in a 1-hour period
    b. By the amount of time the welding power source can operate without overheating in a 30-minute period
    c. By the amount of time the welding power source can operate without overheating in a 10-minute period

12. What does a shielding gas flow meter measure?
    a. Cylinder pressure in psi (kPa)
    b. Cylinder pressure in CFH (LPM)
    c. Flow rate to the welding nozzle in psi (kPa)
    d. Flow rate to the welding nozzle in CFH (LPM)

13. What is the type of current and polarity used in GMAW?
    a. DCEN
    b. DCEP
    c. ACEN
    d. ACEP

FILL IN THE BLANK

14. _____ is the distance from the base metal to the end of the electrode wire, whereas _____ is the distance from the contact tip to the end of the electrode wire.

15. The typical electrode extension for GMAW is _____ for short-circuit transfer and _____ for spray transfer welding.

16. _____ refers to the gun position in relation to the workpiece, whereas _____ refers to the angle of the welding gun.

17. Work angles are _____ on fillet welds and _____ on groove welds.

18. Travel angle, regardless of the direction of travel, is _____ to _____.

19. Backhand welding is also referred to as the drag or _____ technique; it provides _____ weld penetration than the other technique.

20. Forehand welding is also referred to as the _____ technique; it provides better gas coverage for _____ transfer and pulse-spray welding.

21. Other than the forehand/backhand welding techniques _____ has the most influence on the weld puddle and depth of penetration.

22. The electrode should be _____ in relation to the weld puddle while welding.

23. The modes of metal transfer for GMAW are _____, spray transfer, pulse-spray, and globular transfer.

24. List an advantage and a disadvantage for each mode of metal transfer.

25. List an advantage and a disadvantage for using small electrode diameters.

26. Match the AWS specification with the GMAW electrode.
    a. Mild steel   __ A5.19
    b. Stainless steel   __ A5.7
    c. Aluminum   __ A5.18
    d. Copper alloys   __ A5.9
    e. Magnesium   __ A5.10

27. Match the AWS classification with the steel GMAW electrode.
    a. ER   __ Tensile strength × 10,000
    b. 70   __ Alloy changes
    c. S   __ Chemical composition
    d. X (2,3,4,5,6, or 7)   __ Electrode rod
    e. H, HZ, etc.   __ Solid

28. Match the AWS classification with the steel composite GMAW electrode.
    a. EC   __ Electrode composite
    b. 70   __ Chemical composition
    c. X (3 or 6)   __ Tensile strength × 10,000
    d. M, C, etc.   __ Shielding gas requirements

29. Match the AWS classification with the stainless steel GMAW electrode.
    a. ER   __ Electrode composition
    b. XXX (308, 316, etc.) __ Alloy changes
    c. L, LS, etc.   __ Electrode rod

30. Match the AWS classification with the stainless steel composite GMAW electrode.
    a. EC   __ Alloy changes
    b. XXX (308, 316, etc.) __ Electrode composition
    c. L, LS, etc.   __ Electrode composite

31. Match the AWS classification with the aluminum GMAW electrode.
    a. ER   __ Electrode composition
    b. 4043, 5356, etc.   __ Electrode rod

SHORT ANSWER

32. Which are the most common mild steel GMAW electrodes, and what deoxidizing elements do they have in common?

33. Explain an advantage and a disadvantage of using carbon dioxide as a reactive gas.

34. At what percentage can spray transfer be obtained with argon on steel and aluminum, and what effect does argon have on the transition current?

35. What effect does the addition of oxygen have in a blended shielding gas?

36. List the blends of shielding gas used on stainless steel electrodes.

37. Helium is a popular addition to shielding gases used for stainless steel. What problem exists for helium?

38. What advantage does the addition of nitrogen have to a blended shielding gas for stainless steels?

39. What are the two most common shielding gases, either straight or blended, for aluminum? Write an advantage of each.

Chapter 5
FLUX CORED ARC WELDING

PEARSON

myweldinglab™

INTRODUCTION

Flux Cored Arc Welding (FCAW) is more or less a cross between Gas Metal Arc Welding (GMAW) and Shielded Metal Arc Welding (SMAW). FCAW is similar to GMAW because an electrode is automatically fed from a spool of wire through a welding gun when activated. FCAW is similar to SMAW because a slag is formed on the surface of the weld bead. FCAW has operator appeal because it is easier to achieve complete fusion to adjoining base metal members than many other welding processes. FCAW performs well on gauge thickness with smaller diameter steel electrodes and well on plate thickness with medium to large diameter electrodes. When an external shielding gas is required, the Flux Cored Arc Welding process shares the same issues with air current or drafts as does GMAW. However, some electrode classifications of FCAW are **self-shielded**, meaning they use no external shielding gas because the inner core fluxes provide all the necessary shielding and can be used outdoors. The process generates a considerable amount of smoke in comparison with some other welding processes, and the comfort level for the operator is lower due to the typical increased voltage and amperage requirements for this process.

FCAW was developed and made its industrial debut in the 1950s. The electrode is tubular, with a metal sheath encapsulating a core of flux, alloying elements, and various ingredients that when exposed to the intense heat of an electrical arc generate a shielding gas for protecting the weld pool. Initially developed as an alternative to SMAW, only the self-shielded FCAW process was produced. The FCAW process increases welding speeds and productivity. Increased productivity was a key reason for its use in the construction of the Sears Tower in Chicago. The building was welded using Lincoln Electric's self-shielded Innershield flux-cored electrode.

Later, the composition of the internal flux was altered, and an external shielding gas was added. The combination of using the internal cored fluxes and an external shielding gas is called **gas-shielded**, sometimes also referred to as dual-shielded. The gas-shielded FCAW process generates less fumes and spatter than the self-shielded electrode, increasing efficiency and productivity but limiting the use to indoor areas where air movement is not a concern. The construction of the John Hancock Center in Chicago is an example where gas-shielded E70T-1 flux-cored electrode was used to weld the box sections in the building's frame.

Today, sales of gas-shielded FCAW far outnumber the sales of the self-shielded variety. They produce low-hydrogen quality welds on plate thicknesses in structural applications and increase productivity over manual welding methods.

SAFETY

TABLE 5-1 shows the basic safety considerations for FCAW. **TABLE 5-2** shows recommended lens shades for welding at varying amperage ranges.

Head	Body	Hands	Feet
Safety glasses	Flame-resistant wool or cotton pants and shirt	Gauntlet-style welding gloves	High-top leather safety work boots
Welding helmet	flame-resistant jacket		
Welding hat	Leathers		

TABLE 5-1

Read "Chapter 2: Safety in Welding" for detailed discussion of safety related to workplace environment, workplace hazards, personal protective equipment, electrical consideration, gases and fumes, ventilation, fire prevention, explosion, and compressed cylinders and gases.

Arc Current	Suggested Shade
Less than 50	—
50–150	10
150–250	12
250–400	14

TABLE 5-2

POWER SOURCE AND PERIPHERALS

WHAT EQUIPMENT IS NEEDED FOR FCAW? The basic equipment needed to start welding with FCAW is:

- Power Source
- Wire Feeder
- Welding Gun (or Torch)
- Regulator/Flow Meter (when gas-shielded flux core is used)

WHAT TYPE OF POWER SOURCE IS USED FOR FCAW? The FCAW process requires a constant voltage (constant potential) welding power source. The FCAW power sources can be grouped into three categories:

- Heavy Industrial (460V)
- Light Industrial (230V)
- Home Hobbyist (115V)

Flux-cored electrode diameters typically range from 0.030 inch (0.76 mm) to 1/8 inch (3.2 mm). The size of the power source limits the size of the electrode used. The smaller (115V) power sources can only handle the smaller electrode diameters, and the larger (460V) power sources can handle the larger electrode diameters.

WHEN SELECTING A POWER SOURCE, IT LOOKS LIKE EVERYTHING LISTED FOR GMAW IS APPLICABLE FOR FCAW. IS THERE ANY DIFFERENCE? The same constant voltage power source and peripheral equipment used for the GMAW process is also used for the FCAW process. The smaller home hobbyist

FIGURE 5-1 Knurled drive rolls and an enlarged view of the knurled V-groove

welder utilizes the smaller diameter flux-cored electrodes due to the limitations of available voltage, while the heavy-industrial machines can handle both smaller and larger electrode diameters. The part of the equipment requiring the most attention is the drive rolls. Flux-cored electrodes are stiffer and are more difficult to feed to the contact tip. When knurled drive rolls are used, the drive roll tension can be reduced, eliminating the possibility of flattening the electrode, which presents feeding problems. **Knurled drive rolls** have serrations in the "v" portion, gripping the electrode for a more consistent feed. Because flux-cored electrodes may slip when V-groove drives rolls are installed, welders usually overtighten drive roll pressure and crush the electrode wire.

Knurled drive rolls on heavy-industrial and light-industrial power sources are usually easily changed to accommodate various electrode diameters used in Flux Cored Arc Welding (**FIGURE 5-1**). Many home hobbyist power sources have drive rolls not quickly or easily changed. Also, many home hobbyist power sources are sold with a roll of self-shielded flux-cored electrode. Because drive rolls are not easily changed and self-shielded flux-cored electrodes are included in the home hobbyist package, one drive roll is often knurled (**FIGURE 5-2**).

FIGURE 5-2 Home hobbyist–type power sources often come with one drive roll knurled for use with solid electrode wire and flux-cored electrode wire.

WHAT TYPE OF POWER SOURCE IS USED FOR HOME OR BUSINESS? The considerations for purchasing a heavy-industrial machine for FCAW are the same as those for GMAW. For a machine shop, automotive repair, farms, and other similar businesses not using the power source on a continual basis, a better choice is the light-industrial power source. Trade schools and technical colleges, depending on their needs, often choose between heavy-industrial or light-industrial power sources.

Duty cycle is more of a concern when choosing a power source for Flux Cored Arc Welding than for GMAW. Duty cycle is rated on how many minutes the power source can operate continuously in a 10-minute period. Think of this as a window of time the welder can operate without overheating. DC constant voltage heavy-industrial welding power sources are usually rated at or near 100% duty cycle. Light-industrial power sources' duty cycles vary but are less than 100%. Flux-cored electrodes operate at higher welding parameters than when welding with GMAW-S because the demands on the power source are greater.

While the definition of duty cycle is concise, the way in which duty cycle is rated from one manufacturer to another varies. For example, manufacturer A rates duty cycle at 104°F (40°C) at 98% relative humidity, while manufacture B rates duty cycle at 32°F (0°C) at 0% relative humidity. On a typical day, power source B does not hold up as well as power source A. The difference in rated duty cycle is a primary reason why some power sources cost more than others. The difference in duty cycle is also one criterion distinguishing a heavy-industrial from a light-industrial or a home hobbyist power source.

CONTROLS AND CHARACTERISTICS

DO CONTROLS FOR FCAW DIFFER FROM GMAW?
Setup for GMAW includes voltage and wire feed speed (WFS) and may include inductance and slope. FCAW utilizes only the voltage and wire feed speed (WFS) controls. Voltage is set on the power source and the wire feed speed controls the amperage. While inductance controls the rise in amperage as an electrode necks down on a GMAW short-circuit cycle, FCAW operates with an open arc. Inductance has no influence on amperage when an open arc is used. Slope also has no influence on the electrode when an open arc is used. The only effect inductance or slope has when Flux Cored Arc Welding occurs if either is set at one extreme or the other. An excessive inductance setting or steep slope causes the electrode to stumble when initiating the arc. An extremely low inductance setting or flat slope causes the electrode to snap back and fuse to the contact tip.

WHAT EFFECT DOES INCREASING OR DECREASING THE VOLTAGE HAVE?
VOLTAGE CONTROL: POTENTIAL Power sources used for FCAW are constant voltage. Increasing or decreasing the voltage control on the power source increases or decreases the arc voltage. Welding arc voltage is the primary variable for controlling the fluidity or wetting out of the weld onto the base metal workpiece. Increasing the voltage increases the fluidity of the weld pool. Decreasing the voltage decreases the fluidity of the weld pool. A decrease in both amperage and voltage decreases penetration and fusion to the base metal

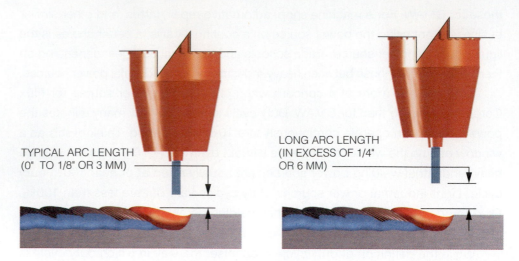

TYPICAL ARC LENGTH
(0" TO 1/8" OR 3 MM)

LONG ARC LENGTH
(IN EXCESS OF 1/4"
OR 6 MM)

FIGURE 5-3 Maintain a shorter arc length, as shown on the left.

and subsequent weld layers while increasing the possibilities of slag inclusions and internal porosity.

Proper arc length (wire tip-to-work distance) is associated with adjusting the voltage. However, voltage for FCAW is relatively consistent regardless of the electrode diameter. If the voltage is set correctly and the arc length is excessive, increase the wire feed speed to remedy the situation. The longer the arc, the more difficult it is to control the weld pool. While an open arc is desirable, a shorter arc length provides excellent penetration with good wetting characteristics and minimal spatter (**FIGURE 5-3**).

WHAT EFFECT DOES INCREASING OR DECREASING THE WIRE FEED SPEED HAVE?

WIRE FEED SPEED (WFS) The same power source and wire feed speed (WFS) control is used for FCAW and GMAW. An increase in WFS increases the amperage while welding, and decreasing the WFS decreases the amperage while welding. Increasing WFS also increases weld penetration, while decreasing WFS decreases penetration. Decreasing WFS also increases spatter due to poor electrode separation to the weld pool.

The WFS control is found on power supplies equipped with a built-in wire feed system. On the large industrial power sources, the WFS is separate from the power source; the wire feed control is found on the wire feeder. The Full Range Potentiometer usually has a numbering system from 1 to 10 or 10 to 100 (see **FIGURE 5-4**). These numbers are references and do not indicate the actual WFS. When the control knob is turned to maximum, WFS is usually 750 to 800 IPM (inches per minute) or 19 to 20.3 MPM (meters per minute) and can exceed 800 IPM. In this scenario, each increment—1, 2, 3, or 10, 20, 30, etc.,—equals 75 or 80 IPM (1.9 or 2 MPM).

Modern wire feeders feature a digital readout for WFS. This feature is convenient when written welding parameters and welding procedures provide a WFS range as opposed to or in conjunction with an amperage range (see **FIGURE 5-5**).

While WFS is a variable set by the welder, the power source is an electrical machine with the two major variables: amperage and voltage. Whether monitoring

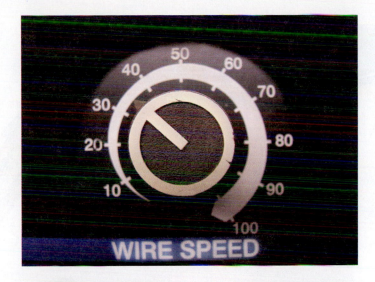

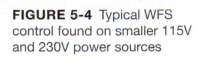

FIGURE 5-4 Typical WFS control found on smaller 115V and 230V power sources

FIGURE 5-5 Digital readouts found on some 230V and many 460V machines

parameters for production purposes with an analog or digital meter displayed on the power source or by a certified multimeter for welder qualification purposes, it is amperage and voltage being verified to ensure quality. This is an important distinction to keep in mind, as the welder is not just setting WFS.

DO WELDING TORCHES FOR FCAW DIFFER FROM GMAW (WELDING GUNS)?

Gas-shielded Flux Cored Arc welding parameters, regardless of electrode diameter, are typically higher than those for gas metal arc short-circuit welding parameters, and a larger welding gun is required. Welding guns are rated on operating amperage. An example of a very light welding gun is a 100 amperage rating, whereas a light to medium welding gun has a 200 amperage rating. A heavy-duty welding gun is likely be rated in the 300 to 400 amperage range. Handheld welding guns can be purchased with a 600 amperage rating. However, welding amperages in the 600 range, even if water-cooled, can be quite uncomfortable for the operator, and automating the process is a desirable option.

Weld torch components increase in size for higher welding parameters. Using components designed for light-gauge GMAW short-circuiting welding at flux-cored parameters can overheat and destroy them.

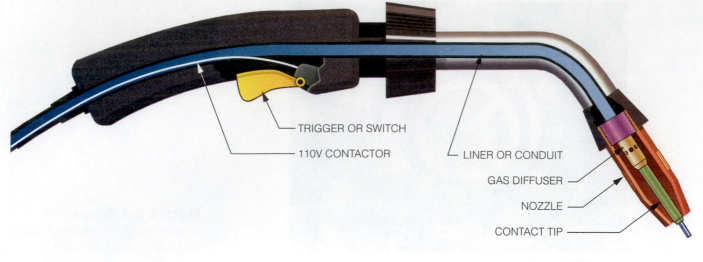

TRIGGER OR SWITCH

110V CONTACTOR

LINER OR CONDUIT

GAS DIFFUSER

NOZZLE

CONTACT TIP

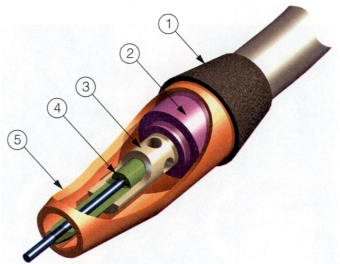

FIGURE 5-6 The top illustration shows a cutaway of a typical wire feed torch. The bottom illustration shows the (1) neck, (2) insulator, (3) gas diffuser (4) contact tip, and (5) nozzle.

In addition to welding guns, work lead cabling and work lead clamps should be heavier than if the power source duties were regulated to lower voltage GMAW operations. How much heavier is determined by the amount of amperage produced and how much the welding power source is in operation (duty cycle). **FIGURE 5-6** shows a cutaway of a typical torch and nozzle end.

IS THERE ANY DIFFERENCE IN GAS REGULATORS OR FLOW METERS IN FCAW OR GMAW? Manufactures of regulators sell several different styles. Their cost and features distinguish the styles. Flow meters and gas regulators used for GMAW can also be used for FCAW.

FCAW SETUP

HOW IS THE WELDING POWER SOURCE SET UP FOR WELDING? There are a number of variables the welder has control over prior to welding. Prior to welding, the operator selects the:

- Type of Shielding Gas and Flow Rate
- Electrode

- Drive Rolls
- Contact Tips
- Nozzle
- Polarity
- Electrode and Spool Tension
- Amperage (wire feed speed)
- Voltage

Shielding gas for carbon steels is typically either 100% CO_2 or an argon/CO_2 blend. Typical flow rates for gas-shielded Flux Cored Arc Welding are from 30 to 45 CFH (14 to 21 LPM). The flow meter can be read from the top, middle, or bottom of the ball in the float tube, since it is only an approximation and will vary due to length of the hose and restrictions in the gun. As the diameter of the nozzle at the end of the welding gun increases or decreases, the force of shielding gas increases or decreases to the weld pool. It is only essential that the weld pool is adequately blanketed by the protective shielding gas (**FIGURE 5-7**).

FIGURE 5-7 Shielding gas flow need only be enough to adequately blanket and protect the weld pool.

Electrodes for FCAW are a hollow tube filled with fluxes and alloying elements. When carefully examined, a line can be seen where the electrode was rolled; it can be split to see the flux inside (**FIGURE 5-8**). Electrode selection is very critical. Some flux-cored electrodes operate on electrode positive, while others operate on electrode negative. Some flux-cored electrodes are designed for single-pass welding only, while others are designed for both single and multiple-pass welding. Some flux-cored electrodes are manufactured to weld with no shielding gas, while others use 100% CO_2 only, an argon/CO_2 shielding gas only, or the electrode manufacturer has designated either 100% CO_2 or an argon/CO_2 shielding gas.

Drive rolls, as previously stated, are usually knurled. The harder flux cored electrode wires are not like solid steel electrodes, which have a copper cladding that will flake off while feeding. If a V-groove drive roll is used on flux-cored electrode wires, they may have to be tightened to the point of crushing the hollow wire in order to produce reliable feeding. The longer the whip and conduit liner and the smaller the electrode diameter, the more problematic feeding becomes using V-groove drive rolls with flux-cored electrode wires.

FIGURE 5-8 A flux-cored wire split to show the electrode is hollow

FIGURE 5-9 Contact tips for FCAW, such as the top two, are generally more heavy duty than contact tips (bottom) used for GMAW.

Contact tips (**FIGURE 5-9**) must match the electrode wire selected. A 0.045 inch (1.1 mm) contact tip is required for a 0.045 inch (1.1 mm) flux cored electrode wire diameter. Although the same contact tip used for GMAW-S may be used, it is advisable to use a more robust contact tip for Flux-Cored Arc Welding. The higher welding parameters of Flux Cored Arc Welding place a higher demand on components used. **FIGURES 5-9** and **5-10** illustrate the differences in the heavy-duty components of FCAW compared to those required for the lower output parameters of short-circuit transfer GMAW.

Nozzles, as with contact tips, should also be more robust than nozzles used in GMAW-S. A light-duty nozzle can quickly deteriorate due to the higher welding parameters used with FCAW.

Polarity selection is critical with FCAW. The FCAW electrode selected determines if the terminals on the power source must be electrode positive or electrode negative. A typical heavy-industrial power source has a positive (+) and a negative (–) terminal where the electrode and work cables are connected. Home hobbyist power sources and some light-industrial power sources have taps. They

FIGURE 5-10 Nozzles for welding guns, such as the top two, are generally more heavy duty than nozzles used for GMAW, such as the bottom one.

must be switched for electrode positive (reverse polarity) or electrode negative (straight polarity). See **FIGURE 5-11**.

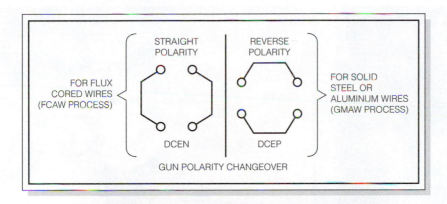

FIGURE 5-11 A home hobbyist–type power source shows a schematic view of how the polarity needs to be connected on the power source.

V I D E O

   Electrode tension is set by the drive rolls. If the tension is set too loose, the electrode wire feeds erratically and often sticks to the contact tip. Too much tension, even with knurled drive rolls, may crush the hollow flux-cored electrode wires. Although generally not an issue with large diameter electrodes wires, small diameter electrode wires with excessive tension form a bird's nest around the drive rolls when feeding. Also, the spool nut must have enough tension to stop the spool from spinning and unraveling the electrode wire after the trigger on the welding gun is disengaged but not too much tension as to cause excessive drag on the electrode wire being fed through the drive rolls.

Welding amperage (current) is increased or decreased by adjusting the wire feed speed (WFS). The range of the WFS depends on the flux-cored electrode classification and diameter. For the same amperage range, smaller electrode diameters have higher wire feed speeds and larger electrode diameters have lower wire feed speeds.

Arc voltage is adjusted on the constant potential power supply. Typical voltage settings for semiautomatic operations are 24 to 25 volts with (DCEP) electrode diameters from 0.035 inches (0.9 mm) to 1/16 inch (1.6 mm). Voltages can be higher or lower, but 24 to 25 are typical for the diameters listed. When the voltage and WFS are set correctly, most flux-cored electrodes produce a spray-like metal transfer in all welding positions.

FCAW Setup

1 Select the appropriate shielding gas for gas-shielded electrode wires or no shielding gas for self-shielded electrode wires.

2 Secure the shielding gas bottle for gas-shielded electrode wires ONLY!

CAUTION: Failure to secure the shielding gas bottle could result in the bottle tipping over, breaking off the stem, and propelling like a rocket.

3 Remove the cap for gas-shielded electrode wires ONLY!

4 Crack the cylinder for gas-shielded electrode wires ONLY!

CAUTION: *Do not* stand facing the valve opening.

5 Attach the flowmeter for gas-shielded electrode wires ONLY!

CAUTION: Do not over-tighten, but secure tight enough to prevent leaks.

6 Select the appropriate electrode wire.

7 Place the electrode spool on the wire feeder and secure the wire spool.

CAUTION: Failure to secure the wire spool could result in the spool coming off the wire feeder.

8 Select the appropriate drive rolls.

9 Obtain the contact tip to match the electrode filler metal selected.

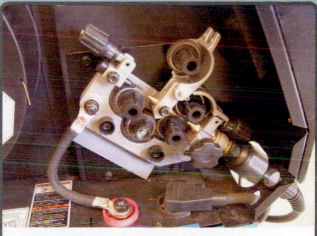

10 Check the polarity: Some FCAW electrodes require DCEP; others require DCEN.

11 Install the drive rolls.

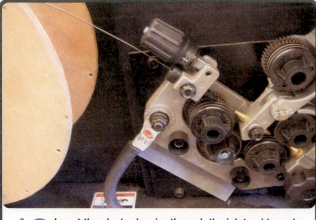

12 Insert the electrode wire through the inlet guide and over the drive rolls.

CAUTION: The end of the electrode is sharp. Be careful not to poke your fingers.

13 Plug in the power source to the proper electrical receptacle.

CAUTION: Electricity can be dangerous. Always be careful when plugging and unplugging electrical receptacles. Check for frayed or damaged electrical cords.

14 Adjust the drive roll tension.

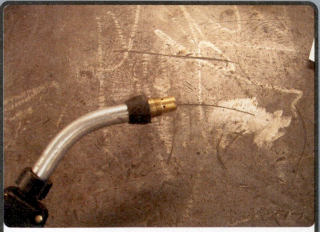

15 Feed the electrode through the welding lead and gun.

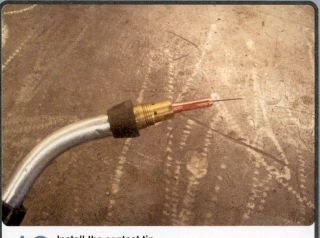

16 Install the contact tip.

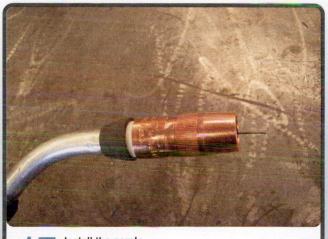

17 Install the nozzle.

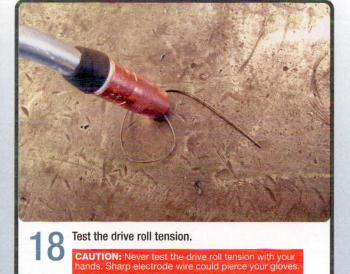

18 Test the drive roll tension.

CAUTION: Never test the drive roll tension with your hands. Sharp electrode wire could pierce your gloves.

19 Tighten enough to produce drag.

CAUTION: Undertightening can cause the wire to unravel.

20 Open the cylinder valve for gas-shielded electrode wires ONLY!

CAUTION: Never stand in front of the regulator when opening the cylinder.

21 Adjust the flowmeter for gas-shielded electrode wires ONLY!

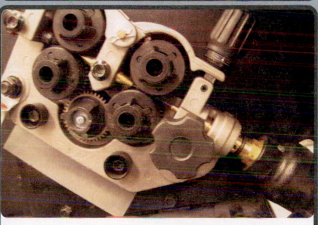

22 Check for shielding gas leaks. O-rings should not be exposed but pressed tight against the drive roll housing.

CAUTION: Never check for shielding gas leaks by holding the welding gun *toward* your ear/head.

23 Attach the work lead.

24 Adjust the voltage and amperage (wire feed speed).

TECHNIQUE

OTHER THAN ADJUSTING THE POWER SOURCE, CAN THE WELDER DO SOMETHING TO AFFECT THE WELD? There are a number of variables the welder has control of after the arc is initiated. After the arc is initiated, the welder has control over:

- Electrode Extension (stick-out)
- Electrode Work Angle
- Electrode Travel Angle
- Direction of Travel
- Travel Speed
- Electrode Manipulation

Electrode extension (stick-out) is the distance the electrode extends from the contact tip either prior to or while welding. (**FIGURE 5-12**). Typical electrode extension for flux-cored welding is 1/2 inch to 1 inch (13 mm to 25 mm).

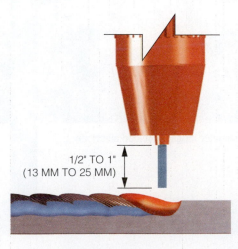

1/2" TO 1"
(13 MM TO 25 MM)

FIGURE 5-12 Electrode extension, or stick-out, for FCAW is generally 1/2 inch to 1 inch

Electrode work angle refers to the FCAW gun position in relation to the workpiece. Proper welding angles are critical for making satisfactory welds. Travel angle refers to the angle of the welding gun in the direction of travel. The electrode position in **FIGURE 5-13** is at a 45° angle between the top and bottom base metal and about 10° to 15° in the direction of travel. The travel angle in **FIGURE 5-14** is for right-handed welders. A left-handed welder would tilt the welding gun 10° to 15° to the left. If the welding gun is more vertical, for example, the weld penetrates more of the base member and has insufficient fusion to the upright member.

FIGURE 5-13 Typical work and travel angles on a fillet weld

FIGURE 5-14 Typical work and travel angles on a butt joint

On groove joints, the work angle is perpendicular to the base metal or at a 90° angle and about 10° to 15° in the direction of travel (**FIGURE 5-14**).

Direction of travel has a significant effect on weld bead shape, penetration, and efficiency. The backhand technique, also referred to as the drag or pull technique, positions the weld torch/gun angle back into the weld pool opposite the direction of travel. A backhand technique produces a more convex weld bead

FIGURE 5-15 The photo on the left illustrates the backhand, or drag, technique, while the photo on the right illustrates the forehand, or push, technique.

VIDEO

with deeper penetration than the forehand technique. The forehand technique, sometimes referred to as a push technique, provides a flatter weld profile, less penetration, and higher welding speeds. The weld torch/gun for the forehand technique is pointed in the direction of travel. The optimum angle for both backhand and forehand ranges from approximately 10°–15° (**FIGURE 5-15**).

Direction of travel changes slightly when welding in the vertical position to 0°–10° (**FIGURE 5-16**). As the angle increases, penetration decreases, and the weld profile flattens. The desired outcome and the welder's skill determine the best work angles.

Travel speeds can greatly affect the weld puddle and the depth of penetration. The welder controls travel speed, taking care to ensure the electrode stays at the leading edge of the weld pool. Travel speed, in any arc welding operation, affects penetration more than any other welding variable (within reason).

When travel speeds are too slow, the electrode impinges on the molten weld pool as the molten filler metal flows out onto the cold base metal, resulting in

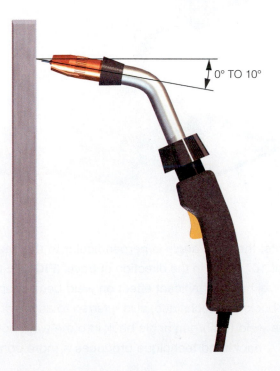

0° TO 10°

FIGURE 5-16 Typical work and travel angle for vertical welding

insufficient penetration and an incomplete of fusion. Excessive travel speed results in skips in the weld. Proper travel speeds sufficiently heat the base metal, and the arc force mixes the filler material with the melted base metal, producing the best results of fusion, penetration and productivity.

Travel manipulation—and weaving in particular—is usually unnecessary and often undesirable with Flux Cored Arc Welding. Oscillating makes it more difficult to maintain the electrode at the leading edge of the weld pool.

Starting an arc is not difficult. The trigger is pulled on the welding gun to initiate the arc. If the arc is not initiated, it generally means the work lead connection is poor or is not connected at all. Slag present on the end of the electrode causes the arc to stumble. Cutting the tip off the electrode prior to starting the arc ensures arc initiation. To aid in the arc initiating further, cut the tip of the electrode at an angle as opposed to perpendicular.

WHAT TYPE OF METAL TRANSFER DOES FCAW PRODUCE?
Metal transfer refers to how the molten filler metal leaves the electrode and transfers across the arc to the base metal, creating the weld pool (**FIGURE 5-17**). Unlike GMAW—which necks down, balls up, or pinches from a solid electrode—the tubular flux-cored electrode's outer metal sheath melts in a less than uniform fashion and is propelled across an open arc to the base metal. From the welders perspective, the metal transfer appears spray-like when the weld parameters are set correctly for some of the flux-cored electrode classifications and appears globular-like with other flux-cored electrode classifications. Both self-shielded and gas-shielded electrodes exhibit similar characteristics. The self-shielded electrode, however, produces more fumes and weld metal spatter than the gas-shielded electrode. Manufacturers endeavor to reduce the amount of fumes produced, as this is a concern when determining a welding process.

V I D E O

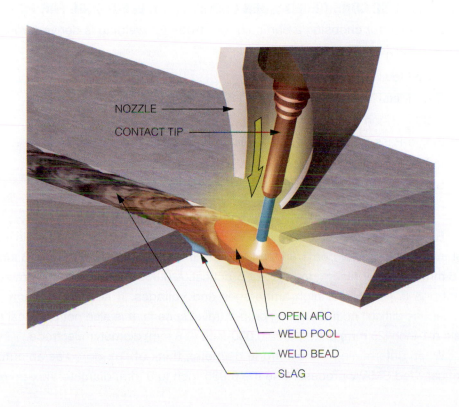

NOZZLE

CONTACT TIP

OPEN ARC

WELD POOL

WELD BEAD

SLAG

FIGURE 5-17 Cross-sectional view of the Flux Cored Arc Welding process

ELECTRODES

WHAT SIZES OF STEEL ELECTRODES ARE USED FOR FCAW?
The FCAW process uses a consumable, automatically fed tubular electrode and is considered a semiautomatic welding process when welded by hand. Electrode diameters used in FCAW range from 0.030 inches (0.8 mm) to 1/8 inch (3.2 mm). Welding parameters in this text do not exceed 1/16 inch (1.6 mm). Welder discomfort due to the heat produced by the electric arc escalates as the electrode diameter increases. Electrode diameters up to and including 1/16 inch (1.6 mm) are easily welded in all positions. When automating, it is advantageous to run larger electrode diameters to increase productivity.

There are several points of interests about flux-cored electrodes worth noting. For example, most classifications of flux-cored electrodes do not easily allow open-root welding. A backing weld from another welding process, a backing plate, or back-gouging and back-welding are used to create an acceptable root pass. Another issue is the shelf life for most flux-cored electrodes, especially the low-hydrogen type. Flux-cored electrode wires exceeding their shelf life pick up moisture, begin to rust, and develop an erratic arc when placed back in service. No adjustment in weld parameters corrects the problem. With high humidity and when electrode wires sit on the shelf for at least 6 months, the erratic arc starts to become evident. Schools training with flux core often experience this when the electrode wire sits on a shelf over the summer when courses do not run. Sealing the spool with an industrial food sealer, for example, prevents the electrode wire from absorbing moisture when not in use for a prolonged period of time.

WHAT MUST BE CONSIDERED WHEN CHOOSING AN ELECTRODE FOR FCAW?
Consideration for choosing a particular electrode for welding is derived from a number of factors. They include:

- Base Metal Composition
- Base Metal Thickness
- Weld Position
- Single-Pass or Multiple-Pass Welding
- Indoor or Outdoor Operation
- Shielding Gas Composition for Gas-Shielded Flux-Cored Electrodes
- Finished Weldment Structural Requirements

Base metal thickness determines the electrode diameter selected. While it is not impossible to weld on 1/8 inch (3.2 mm) thick base metal with a 1/16 inch (1.6 mm) diameter electrode, it is not practical. A 0.035-inch (0.9 mm) diameter electrode is capable of high amperages and voltages. If set to maximum, a welder has difficulties maintaining proper travel speeds. It is also not practical to weld a 1-inch (25 mm) plate with a 0.030-inch (0.8 mm) diameter electrode.

When determining the electrode diameter, think of flux-cored as an automatically fed SMAW process. Use the 0.035-inch (0.9 mm) diameter flux-cored electrode where the 3/32-inch (2.4 mm) shielded metal arc electrode is used; the 0.045-inch (1.1 mm) and 0.052-inch (1.3 mm) diameter flux-cored electrode where the 1/8-inch (3.2 mm) shielded metal arc electrode is used; and the 1/16-inch (1.6 mm) diameter flux-cored electrode where the 5/32 inch (4 mm) shielded metal arc electrode is used.

Weld position determines if a particular flux-cored electrode can be used. Some flux-cored electrodes are designed to weld in all positions, while others are designed to weld in flat-groove and flat and horizontal fillets only. It is easier to produce a sound weldment with flux-cored than most other welding processes. A 0.035-inch (0.9 mm) and 1/16-inch (1.6 mm) diameter electrode both produce a satisfactory weld on a 1/4-inch (6.5 mm) plate in the flat and horizontal position. However, even for the electrodes designed to weld in all positions, the 1/16-inch (1.6 mm) diameter is a bit excessive in the vertical-up and overhead welding positions on the 1/4-inch (6.5 mm) plate.

For weld position, use the same analogy as previously stated with the SMAW electrodes. Both the 1/16-inch (1.6 mm) diameter flux-cored electrode and the 5/32-inch (4 mm) diameter SMAW electrode are the practical limit for vertical-up and overhead welding.

Single-pass or multiple-pass welding needs to be established prior to selecting a flux-cored electrode. Some flux-cored electrodes are designed for multiple-pass welding, while others are designed for single-pass welding only. The difference between the two is the amount of alloying elements within the tubular electrode. **Over-alloying** occurs if multipass welding occurs with an electrode designed for single-pass welding only. Over-alloying increases tensile strength but decreases ductility and could result in premature in-service weld failure. A thorough understanding of the AWS electrode classification is essential to avoid problems.

Indoor or outdoor operations are accomplished with flux-cored electrodes. Self-shielded flux-cored electrodes are welded indoors or outdoors. Proper ventilation is more of a concern with self-shielded flux-cored electrodes producing more fumes than gas-shielded flux-cored electrodes. Gas-shielded flux-cored electrodes must avoid air movement or drafts, which disrupts proper atmospheric protection created by the external shielding gas.

Many self-shielded flux-cored electrodes operate on direct current electrode negative (DCEN). All gas-shielded flux-cored electrodes operate on direct current electrode positive (DCEP). It is essential to check the manufacturer's specifications or cross-reference the AWS classification in a chart to ensure proper polarity is connected.

Shielding gas composition for gas-shielded flux-cored electrodes alters the metallurgical structure of the weld metal depending on which shielding gas is selected. Carbon dioxide (CO_2) is a reactive gas that burns deoxidizers from the weld pool. To ensure the proper amount of deoxidizers are maintained, an additional amount of deoxidizers are added in the flux when designating a flux-cored electrode to use 100% CO_2 only. Using a blended argon/CO_2 shielding increases tensile strength, decreases ductility, and affects impact properties. Conversely, when an electrode is manufactured to use only a blend of argon/CO_2, the amount of deoxidizers is kept to a minimum. Using a 100% CO_2 shielding gas with an electrode designed to run with blended shielding gas causes arc deterioration, decreases tensile strength, and affects impact properties. Additional letters in the AWS classification numbering system for flux-cored electrodes designate shielding gas requirements.

WHAT CATEGORIES DO FLUX-CORED ELECTRODES FALL INTO?

Flux-cored electrodes are identified by Filler Number Group, Specification, and Classification. The Filler Number Group for all flux-cored electrodes is F6. The Specification for FCAW electrodes are:

	ASME	AWS
■ Mild Steel FCAW Electrodes	SFA 5.20	A5.20
■ Low-Alloy FCAW Electrodes	SFA 5.29	A5.29
■ Stainless Steel FCAW Electrodes	SFA 5.22	A5.22

WHAT IS THE AWS CLASSIFICATION FOR FCAW?

MILD STEEL FLUX-CORED ELECTRODES   Gas-shielded and self-shielded electrodes are classified similar to other welding processes except for the second numerical digit. Rather than designating tensile strength as 72,000, 82,000, etc., only the first numerical digit is used—i.e., 7, 8, 9, etc. All flux-cored electrodes with an E70T-X or E71T-X classification have a minimum tensile strength of 72,000 psi (500 MPa). The second numerical digit is either a 0 (zero) or 1 (one) to indicate the position the electrode can be welded (**FIGURE 5-18**).

Some changes have taken place with the AWS classification numbering system for flux-cored electrodes. To prevent confusion by developing a whole new numbering system, AWS added to the original classification system. An extended numbering classification for E71T-1, as an example, is E71T-1MJH8.

M: Shielding gas mixture of 75% to 80% argon, with the remainder of CO_2

J: Included if the impact properties are tested at −40°C

H: Used to designate an electrode meeting an average diffusible hydrogen level. The designator is usually 4, 8, or 16. The numbers are the maximum average diffusible hydrogen in $mL(H_2)$ per 100 g of weld metal deposited.

On the base classification E71T-1, the last numerical digit indicates characteristics, such as shielding gas requirements and the electrode polarity, and indicates if the electrode is designed to run as a single-pass or multiple-pass weld. The chemical composition is not as evident as the other characteristics. The E71T-1 and E70T-5, for example, consist of the same chemical composition as an E7018 electrode; both are low-hydrogen electrodes and are used in similar situations.

AWS Classification of Steel Flux-Cored Electrodes E70T-X or E71T-X		
E	=	electrode
7	=	minimum tensile strength × 10,000 (approximate)
1	=	weld position[1] – 0 or 1
T	=	tubular electrode wire
X	=	gas type/usability/performance capabilities

[1]0–Indicates flat-groove and flat/horizontal fillets only
1–Indicates all positions

FIGURE 5-18

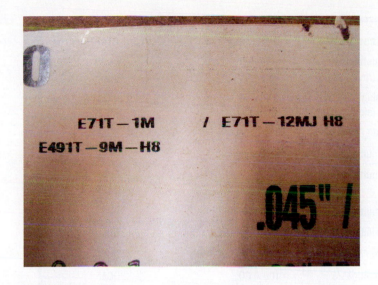

FIGURE 5-19 Product label indicating E71T-1M electrode classification

For flux-cored welding electrodes there is a letter designation indicating the use of a specific type of gas. An example is the difference between the E71T-1M and the E71T-1C. If the electrode requires a gas mixture such as 75/25, the M is added as with the E71T-1M. If the electrode requires a 100% carbon dioxide shielding, the C is added as with the E71T-1C. Finally, if no letter is added as with E71T-11, then no shielding gas is required. The electrode wire in **FIGURE 5-19** is designated for use with mixed gas argon/CO_2 only. **TABLE 5-3** outlines requirements for mild steel flux-cored electrodes.

EXXT-1: This electrode is characterized by a spray transfer, low spatter, high deposition rate, and flat bead with minimum slag covering. The excellent penetration this electrode produces requires a back plate, back weld, or backing weld on open-root joints. The E71T-1 electrode produces excellent out-of-position welds. Electrodes in this classification are low hydrogen and can be substituted where EXX18 shielded metal arc electrodes are used.

EXXT-2: This electrode has the same characteristics as the EXXT-1 but has higher percentages of manganese, silicon, or both. It is intended for single-pass welding because multipass welding leads to over-alloying. The electrode in this classification tolerates rust, mill scale, or other particulates much better than the EXXT-1 electrode.

EXXT-3: This electrode is designed to weld single pass on sheet metal and 1/4-inch plate only. They are primarily designed for flat, horizontal, and slight vertical-down positions. The electrode in this classification produces a spray transfer.

EXXT-4: This electrode is designed for low penetration, making it excellent for poor joint fit-up. The electrode is designed for very high travel speeds and has very low sulfur content that resists hot-cracking. The electrode in this classification produces a globular transfer not lending itself to out-of-position welding.

EXXT-5: This electrode has superior properties and resists hot-cracking compared to the EXXT-1 electrode but has less operator appeal. It is designed to operate in flat and horizontal positions only. The electrode in this classification produces a globular transfer.

EXXT-6: This electrode produces deep penetration with easy slag removal but is designed to operate in flat and horizontal positions only. The electrode in this classification produces a spray-like transfer.

AWS Classification	Shielding Type	Current and Polarity
EXXT-1 (multipass)	100% CO_2	DCEP
EXXT-1M (multipass)	Argon/CO_2	DCEP
EXXT-2 (single pass)	100% CO_2	DCEP
EXXT-2M (single pass)	Argon/CO_2	DCEP
EXXT-3 (single pass)	Self-Shielded	DCEP
EXXT-4 (multipass)	Self-Shielded	DCEP
EXXT-5 (multipass)	100% CO_2	DCEP
EXXT-5M (multipass)	Argon/CO_2	DCEP
EXXT-6 (multipass)	Self-Shielded	DCEP
EXXT-7 (multipass)	Self-Shielded	DCEN
EXXT-8 (multipass)	Self-Shielded	DCEN
EXXT-9 (multipass)	100% CO_2	DCEP
EXXT-9M (multipass)	Argon/CO_2	DCEP
EXXT-10 (single pass)	Self-Shielded	DCEN
EXXT-11 (multipass)	Self-Shielded	DCEN
EXXT-12 (multipass)	100% CO_2	DCEP
EXXT-12M (multipass)	Argon/CO_2	DCEP
EXXT-13 (multipass)	Self-Shielded	DCEN
EXXT-14 (multipass)	Self-Shielded	DCEN
EXXT-G[1] (multipass)	(Check with supplier)	
EXXT-GS[2] (single pass)	(Check with supplier)	

TABLE 5-3

Gas-shielding and polarity requirements
Mild steel flux-cored electrodes: specification A5.20

[1, 2]The G and GS are proprietary compositions.

Source: American Welding Society (AWS) Welding Handbook Committee, 2004, **Welding Handbook**, 9th ed., Vol. 2, Welding Processes, Miami: American Welding Society

EXXT-7: This electrode resists cracking by producing very low sulfur in the weld metal. Depending on the arc transfer droplet, this electrode can be welded in various positions. When the droplet is large, it is welded in flat and horizontal positions only. When the droplet is decreased, the electrode can be used in welding in all positions. The electrode in this classification produces a spray-like transfer.

EXXT-8: This electrode produces welds having good notch toughness at low temperature and high resistance to cracking. It is designed to operate in all positions. The electrode in this classification produces a spray-like transfer.

EXXT-9: This electrode has essentially the same characteristics as the EXXT-1 electrode but with improved impact properties.

EXXT-10: This electrode is designed to operate in flat, horizontal, and slight vertical-up positions. The electrode in this classification produces a spray-like transfer.

EXXT-11: This electrode is not designed to weld on thickness greater than 3/4 inches. The electrode in this classification produces a smooth spray transfer.

EXXT-12: This electrode has essentially the same characteristics as the EXXT-1 electrode but with better impact properties. The electrode in this classification has decreased tensile strength to increase toughness.

EXXT-13: This electrode works well on open-root joints in all weld positions. The electrode in this classification produces a short-circuit-like transfer.

EXXT-14: This electrode is designed to weld on sheet metal in all welding positions. It welds especially well on galvanized, aluminized, or other coated steels. The electrode in this classification produces a smooth spray transfer.

AWS Classification of Low-Alloy Flux-Cored Electrodes		
E90T1-D2		
E	=	electrode
9	=	minimum tensile strength $\times$ 1000
0	=	weld position[1] — 0 or 1
T	=	tubular electrode wire
1	=	gas type/usability/performance capabilities
-D2	=	changes to original alloy[2]

[1]0– Indicates flat-groove and flat/horizontal fillets only
 1 – Indicates all positions

[2]A – carbon/molybdenum   K – Various alloys, K1 through K9
 B – chromium/molybdenum   Ni – nickel
 D – manganese/molybdenum   W – weathering
 G – general

FIGURE 5-20

LOW-ALLOY FLUX-CORED ELECTRODES Flux-cored electrodes are almost as versatile as the SMAW electrodes. There are a number of low-alloy tubular electrodes for a variety of situations. The AWS classification is similar to the mild steel flux-cored electrode classification but has an addition of a suffix, indicating a change to the original alloy (**FIGURE 5-20**). Low-alloy flux-cored electrodes have tensile strength classifications from E80TX-X to E120TX-X.

The characteristics of the low-alloy flux-cored electrodes are the same as the mild steel flux-cored electrodes; the EXXT-1 mild steel flux-cored electrodes have the same characteristics as the EXXT1-X flux-cored electrodes. The exception is the EXXT5-X flux-cored electrode, which operates on DCEP or DCEN depending on manufacturer's recommendations (**TABLE 5-4**). The difference in the low-alloy

AWS Classification	Gas Shielding	Current and Polarity
EXXT1-X (multipass)	100% CO_2 argon/CO_2	DCEP DCEP
EXXT4-X (multipass)	Self-Shielded	DCEP
EXXT5-X (multipass)	100% CO_2 argon/CO_2	DCEP or DCEN DCEP or DCEN
EXXT6-X (multipass)	Self-Shielded	DCEP
EXXT7-X (multipass)	Self-Shielded	DCEN
EXXT8-X (multipass)	Self-Shielded	DCEN
EXXT11-X (multipass)	Self-Shielded	DCEN
EXXTX-G (multipass)[1]	(Check with supplier)	

TABLE 5-4

Gas shielding and polarity requirements
Low-alloy steel flux-cored electrodes: specification A5.29

[1]The G is a proprietary composition.

Source: American Welding Society (AWS) Welding Handbook Committee, 2004, **Welding Handbook**, 9th ed., Vol. 2, Welding Processes, Miami: American Welding Society

AWS Classification of Stainless Steel Flux-Cored Electrodes		
E308LT1-1		
E	=	electrode
308	=	type or grade of stainless steel
L	=	changes to original alloy[1]
T	=	tubular electrode wire
1	=	weld position[2]
–1	=	indicates shielding gas, if any[3]

[1]L – has lower carbon content
H – limited to upper range of carbon content
Mo – molybdenum added for resistance to pitting, increasing creep strength and ferrite
Ni – nickel added to increase corrosion resistance, temperature strength, and ad toughness
Cb – columbium added to prevent corrosion
Ti – titanium added to prevent corrosion

[2]0 – flat-groove and flat/horizontal fillets only
1 – all positions
3 – self-shielded

[3]1 – CO_2
4 – argon/CO_2

FIGURE 5-21

flux-cored electrodes is the suffix of A, B, D, K, Ni, and W (**FIGURE 5-21**). The alloying elements are added to the flux-cored electrodes to weld various low-alloy base metals.

STAINLESS STEEL FLUX-CORED ELECTRODES The classification of stainless flux-cored electrodes is similar to other welding processes for classifying stainless steel electrodes; therefore, the ANSI system is employed (i.e., 308, 316, 410, etc.). Alloying elements are noted following the ANSI designation; see **FIGURE 5-21** and **TABLE 5-5**.

Examples of self-shielded stainless steel flux-cored electrodes include:

- E308LT0-3
- E309LT0-3
- E316LT0-3
- E410NiMoT0-3

AWS Classification	Gas Shielding	Current and Polarity
EXXXTX-1 (multipass)	100% CO_2 Argon/CO_2	DCEP
EXXXTX-3 (multipass)	None	DCEP
EXXXTX-4 (multipass)	Argon/CO_2	DCEP
EXXXTX-5 (multipass)	100% Argon	DCEP
EXXXTX-G[1] (multipass)	(Check with supplier)	

TABLE 5-5

Gas shielding and polarity requirements

Stainless steel flux-cored electrodes: specification A5.22

[1]The G is a proprietary composition.

Source: American Welding Society (AWS) Welding Handbook Committee, 2004, **Welding Handbook**, 9th ed., Vol. 2, Welding Processes, Miami: American Welding Society.

Examples of gas-shielded stainless steel flux-cored electrodes include:

- E308T1-1
- E309LMoT1-4
- E316LT1-4
- E410NiMoT1-1

Most of the stainless steel flux-cored electrodes are manufactured to operate in the flat/horizontal position, T0, or all positions, T1. Additives such as columbium and titanium are listed as corrosion preventatives.

There are hard-surfacing flux-cored electrodes available from various manufactures. If a hard-surfacing flux-cored electrode is desired, check with the manufacturer for specific applications.

WHAT ARE THE GENERAL WELD PARAMETERS OF FLUX-CORED ELECTRODES?

The welding parameters given in this chapter for flux-cored electrodes should suffice for the home hobbyist, high school, trade school, technical college, or many industrial settings. The parameters listed in **TABLE 5-6** through **TABLE 5-10** were established using the more common flux-cored electrodes. However, the parameters given should be adequate regardless of the electrode classification; any mild steel flux-cored electrode—regardless of the classification—should operate correctly for the parameters listed in **TABLE 5-6**, and any stainless steel flux-cored electrode—regardless of the classification—should operate correctly for the parameters listed in **TABLE 5-9**, etc. When welding in the flat/horizontal position only, an increase in parameters increases productivity. Those wishing to use larger electrode diameters should consult electrode manufacturers to produce optimum results.

Mild Steel and Low-Alloy FCAW Parameters Shielding Gas – 75% Argon/25% CO_2 Gas Flow – 30–40 CFH (14–19 LPM) CTWD – ≈3/8"–3/4" (≈9.5 mm–19 mm) Position – All			
Thickness	All Thicknesses		
Diameter	.035" (0.9 mm)	.045" (1.1 mm)	.052" (1.3 mm)	1/16" (1.6 mm)
Voltage	24–26	24–27	24–27	24–28
Amperage	170–200	190–250	200–275	225–330
WFS	450–600 IPM (190–255 mm/sec)	305–450 IPM (130–190 mm/sec)	245–400 IPM (105–170 mm/sec)	180–330 IPM (75–140 mm/sec)

TABLE 5-6

Note: CFH = cubic feet per hour (LPM = liters per minute)   CTWD = contact to work distance
      IPM = inches per minute (mm = millimeters)   WFS = wire feed speed

Mild Steel and Low-Alloy FCAW Parameters
Shielding Gas – 100% CO_2
Gas Flow – 25–35 CFH (12–17 LPM)
CTWD – ≈3/8''–3/4'' (≈9.5 mm–19 mm)
Position – All

Thickness	All Thicknesses			
Diameter	.035'' (0.9 mm)	.045'' (1.1 mm)	.052'' (1.3 mm)	1/16'' (1.6 mm)
Voltage	25–27	25–28	25–28	25–29
Amperage	170–200	190–250	200–275	225–330
WFS	450–600 IPM (190–255 mm/sec)	305–450 IPM (130–190 mm/sec)	245–400 IPM (105–170 mm/sec)	180–330 IPM (75–140 mm/sec)

TABLE 5-7

Note: CFH = cubic feet per hour (LPM = liters per minute) CTWD = contact to work distance
IPM = inches per minute (mm = millimeters) WFS = wire feed speed

Mild Steel FCAW Parameters
Shielding Gas – None
CTWD – ≈1/4''–1/2'' (≈9.5 mm–13 mm)
Position – All

Thickness	All Thicknesses			
Diameter	.030'' (0.8 mm)	.035'' (0.9 mm)	.045'' (1.1 mm)	1/16'' (1.6 mm)
Voltage	14–18	14–19	15–20	15–20
Amperage	35–140	75–150	80–220	110–270
WFS	100–225 IPM (45–95 mm/sec)	100–270 IPM (45–115 mm/sec)	75–200 IPM (35–85 mm/sec)	50–125 IPM (25–50 mm/sec)

TABLE 5-8

Note: CFH = cubic feet per hour (LPM = liters per minute) CTWD = contact to work distance
IPM = inches per minute (mm = millimeters) WFS = wire feed speed

Stainless Steel FCAW Parameters
Shielding Gas – 75% Argon/25% CO_2
Gas Flow – 35–45 CFH (17–21 LPM)
CTWD – ≈3/8''–3/4'' (≈9.5 mm–19 mm)
Position – All

Thickness	All Thicknesses		
Diameter	.035'' (0.9 mm)	.045'' (1.1 mm)	1/16'' (1.6 mm)
Voltage	24–28	24–28	28–32
Amperage	110–130	160–200	240–280
WFS	280–375 IPM (120–160 mm/sec)	275–380 IPM (115–160 mm/sec)	230–290 IPM (100–125 mm/sec)

TABLE 5-9

Note: CFH = cubic feet per hour (LPM = liters per minute) CTWD = contact to work distance
IPM = inches per minute (mm = millimeters) WFS = wire feed speed

Stainless Steel FCAW Parameters Shielding Gas – 100% CO_2 Gas Flow – 30–40 CFH (14–19 LPM) CTWD – ≈3/8"–3/4" (≈9.5 mm–19 mm) Position – All			
Thickness	All Thicknesses		
Diameter	.035" (0.9 mm)	.045" (1.1 mm)	1/16" (1.6 mm)
Voltage	25–29	25–29	29–33
Amperage	110–130	160–200	240–280
WFS	280–375 IPM (120–160 mm/sec)	275–380 IPM (115–160 mm/sec)	230–290 IPM (100–125 mm/sec)

TABLE 5-10

Note: CFH = cubic feet per hour (LPM = liters per minute) CTWD = contact to work distance
IPM = inches per minute (mm = millimeters) WFS = wire feed speed

SHIELDING GASES

WHAT SHIELDING GASES OR COMBINATIONS OF SHIELDING GASES ARE USED IN FCAW?

The shielding gases used in FCAW with mild steel, stainless steel, or low-alloy electrodes are:

- Carbon Dioxide
- Argon/Carbon Dioxide

The two most common shielding gases for all flux-cored electrodes are 100% CO_2 or a blend of 75% argon/25% CO_2. Some flux-cored electrodes are manufactured to use only carbon dioxide; other electrodes are manufactured to use only the blend of argon/CO_2. The maximum amount of argon in a blend is usually 80%. Some variations within an electrode classification allow as much as 85% argon. Fabricators have successfully passed procedure qualification tests, with mixtures of 90% argon/10% CO_2. The increase in the percentage of argon allows for more of the deoxidizers to remain in the weld metal, increasing weld metal tensile strength. The increase in tensile strength results in a decrease in ductility.

Stainless steel flux-cored electrodes use a mixture of argon/CO_2, 100% CO_2, or no shielding gas. Using 100% CO_2 burns out some of the deoxidizers and adds carbon to the weld metal—but not enough to prove detrimental to the weld metal chemistry. Using no shielding gas on stainless steel flux-cored electrodes designed to do so loses some deoxidizers and picks up a significant amount of nitrogen. The addition of nitrogen stabilizes austenite and reduces ferrite content in the weld metal. In the past, a blend of argon/O_2 was used with certain classifications of stainless steel electrodes; however, the use of this blend did not prove useful, and the idea was abandoned.

An understanding of the electrode classification is essential in matching the correct shielding gas with the electrode to avoid under- or over-alloying. The initial decision whether to choose straight carbon dioxide or a blended shielding

FIGURE 5-22 The use of carbon dioxide (left photo) provides deeper weld metal penetration but produces more spatter than when using a mixed shielding gas of argon and carbon dioxide (right photo).

gas is based on cost and operator appeal. The cost of a cylinder of carbon dioxide is not as expensive as an equal amount of an argon/carbon dioxide blend. However, argon/carbon dioxide promotes greater weld metal efficiency; less of the electrode is lost due to spatter than straight carbon dioxide. In a situation where a good deal of weld metal is produced in a given day, the argon/carbon dioxide mixture is less expensive in the long run due to its greater efficiency. The weld on the left in **FIGURE 5-22** was made with carbon dioxide shielding gas, and the weld on the right was made with an argon/carbon dioxide shielding gas.

The second reason for choosing which shielding gas to use is operator appeal. The argon/carbon dioxide shielding gas generally has greater operator appeal because of a smoother arc, and it generates less spatter. In the vertical-up progression, more welding technique is needed when using straight carbon dioxide shielding gas to prevent undercutting at the toes of the weld. The argon/carbon dioxide blend is more fluid and wets out better, making it easier to prevent undercutting (**FIGURE 5-23**). The argon in a blended shielding gas is also known as a cleaning gas, creating acceptable welds when the base metal preparation, mill scale, or rust is not optimum. Never use 100% CO_2 with an electrode classification designed for use with mixed gas.

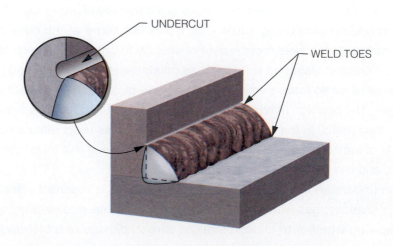

FIGURE 5-23 Undesirable undercut at the weld toes

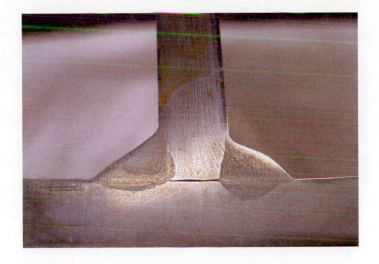

FIGURE 5-24 Note the deeper penetration of the carbon dioxide gas–shielded weld on the right compared to the weld on the left made using a mixed gas of argon and carbon dioxide.

A positive effect of carbon dioxide is it generates more energy, increasing penetration into the base metal more than the argon/carbon dioxide blend does. However, the Flux Cored Arc Welding process, regardless of the shielding gas choice, produces excellent base metal penetration. The weld on the left in **FIGURE 5-24** was made with argon/carbon dioxide, and the weld on the right was made with 100% carbon dioxide.

CHAPTER QUESTIONS/ASSIGNMENTS

MULTIPLE CHOICE

1. The _____ and the John Hancock Center are two out-standing buildings constructed using the FCAW process.
   a. Empire State Building   b. Sears Tower
   c. Eiffel Tower   d. Epoch Center

2. The suggested lens shade for FCAW is
   a. 3 to 5.   b. 8 to 10.
   c. 10 to 14.   d. 12 to 16.

3. The type of drive rolls installed when using flux-cored electrodes is
   a. V-groove.   b. U-groove.
   c. knurled.   d. smooth.

4. How is duty cycle measured?
   a. By the amount of time the welding power source can operate without overheating in a 10-minute period
   b. By the amount of time the welding power source can operate without overheating in a 1-hour period
   c. By the amount of time the welding power source can operate without overheating in a 30-minute period

5. Are duty cycle ratings uniform from various manufacturers?
   a. Yes   b. No

6. Does inductance have the same effect on the arc in FCAW as it does in GMAW?
   a. Yes   b. No

7. If the arc length becomes excessive
   a. there is no noticeable effect.
   b. the weld depth penetration is increased.
   c. the arc length is increased.
   d. the arc become unstable, making it more difficult to produce a consistent weld with decreased penetration and more spatter.

8. How is amperage adjusted on FCAW power sources?
   a. By adjusting the control on the wire feeder
   b. By adjusting the control on the power source
   c. By adjusting the arc length
   d. By adjusting the amount of stick-out

9. Why is it recommended that the contact tips be larger when welding with FCAW electrodes?
   a. Because the electrode diameters are typically larger
   b. Because typical voltage and amperage parameters are higher and produce more heat
   c. Because the nozzle diameter is typically larger

10. Which variable is *not* controlled by the welder when *setting up* for FCAW?
    a. Voltage
    b. Amperage (wire feed speed)
    c. Electrode selection
    d. Polarity
    e. Shielding gas selection
    f. Electrode extension

11. Polarity for FCAW, depending on the electrode, may be DCEP or _____.
    a. DCEN
    b. ACHF
    c. DCRP

12. Inside the hollow electrode tube is varying amount of fluxes and _____.

13. The two most common shielding gases used for FCAW are a blend of argon and carbon dioxide or _____.

14. The typical amount of shielding gas for FCAW is between _____ and _____ CFH.

15. The variables the welder controls *while welding* are electrode extension, work angle, _____, directions of travel, and electrode manipulation.

16. The typical electrode extension for FCAW is _____ to _____.

17. Work angles are _____ on fillet welds and _____ on groove welds.

18. Another name for backhand welding is _____ or drag, and for forehand, it is _____.

19. Travel angle, regardless of the direction of travel, is _____ to _____.

MATCHING

20. Match the AWS specification with the FCAW electrodes.
    a. Mild Steel _____ A5.22
    b. Low-Alloy Steel _____ A5.29
    c. Stainless Steel _____ A5.20

21. Match the AWS classification of steel FCAW electrodes.
    a. E _____ Tensile Strength $\times$ 10,000
    b. 7 _____ Tubular
    c. 0 or 1 _____ Usability and Performance
    d. T _____ Electrode
    e. X (1, 2, 3, 4, etc.) _____ Useable Electrode Position
    f. M, J, or H _____ Shielding Gas, Impact Properties, Diffusible H_2

SHORT ANSWER

22. Where should the electrode be in relationship to the weld puddle while welding?

23. What happens if an electrode manufactured to use 100% CO_2 is welded with an argon/CO_2 mixture?

24. Explain what happens to the electrode wire if it is exposed while in storage for an extended period of time.

25. List the considerations when selecting a FCAW electrodes.

26. Why is it important to consider the factors listed in the previous question?

27. Which mild steel electrode is for single-pass welding only?

28. What is the correct polarity for an E71T-1 electrode.

29. What changes in the electrode classification for low-alloy electrodes?

30. How is stainless steel electrode classification different from mild steel and low alloy flux-cored electrodes?

31. Which two types of shielding gases are typically used with flux-cored electrodes?

32. What is the maximum allowable amount of argon in a blended shielding gas?

33. Which shielding gas or mixture promotes greater efficiency and produces a more-fluid weld puddle?

Chapter 6

GAS TUNGSTEN ARC WELDING

LEARNING OBJECTIVES

In this chapter, the reader will learn how to:

1. Use the Gas Tungsten Arc Welding process safely.
2. List what types of power sources and other peripheral components make up the Gas Tungsten Arc Welding process.
3. Identify the characteristics and correctly use the various controls on power sources.
4. Determine which polarity is best for the base metal used and desired outcome.
5. Select the correct electrode for the base metal used and the desired outcome.
6. Select the correct shielding gas.
7. Properly set up for Gas Tungsten Arc Welding by adjusting the filler addition amount, travel speed, work angles, and electrode-work distance.

KEY TERMS

Air-Cooled Torch 169	Inert Gas 181
Water-Cooled Torch 169	Asphyxiant 182
Torch Body 172	Venire Control 183
Cup 172	Postflow 184
Collet 172	Preflow 184
Collet Body 172	Cathotic Etching 186
Back Cap 173	High-Frequency 188
Tungsten Electrode 176	Autogenous Weld 203

PEARSON
myweldinglab

Gas Tungsten Arc Welding (GTAW) is an extremely useful welding process that uses a nonconsumable tungsten electrode to conduct welding current. Base metal and arc gap resistance to current flow creates heat, causing the base metal to melt, forming a molten weld pool. The weld can then be finished by blending base metal pools together or by manually adding filler metal. **FIGURE 6-1** shows a typical GTAW torch and welding arc. In the past, this welding process was called TIG welding. TIG stands for Tungsten Inert Gas. While the acronym TIG is not entirely inaccurate, GTAW is the universally acceptable acroynym for this welding process.

The GTAW process is like other common welding processes in that work angle, lead angle, and other such variables are similar. However, it is quite different in many respects from other common welding processes. For example, GTAW usually establishes an arc via a high-frequency electrical circuit that jumps from the tungsten electrode to the grounded base metal. Generally, once an individual acquires proficiency in one welding process, there is a carryover of skills to other welding processes. However, the GTAW process differs enough that an entirely new skill set must be attained in order to master the process.

Once competence is achieved, it is easy to see a whole array of potential for use of this process. GTAW can be used to melt any metal and it can make welds in most. Very thin metals not practical with other welding processes are commonly welded with the GTAW process. The more competent a person becomes with the GTAW process, the more useful the process becomes—with seemingly endless possibilities.

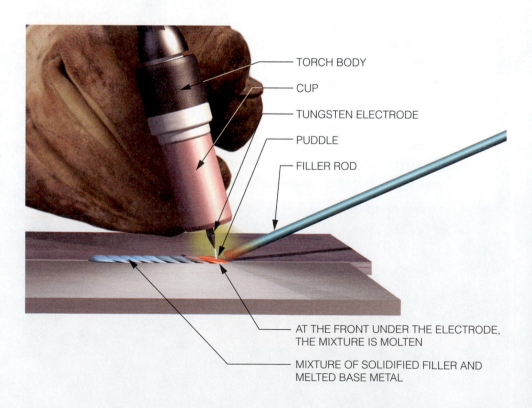

TORCH BODY

CUP

TUNGSTEN ELECTRODE

PUDDLE

FILLER ROD

AT THE FRONT UNDER THE ELECTRODE, THE MIXTURE IS MOLTEN

MIXTURE OF SOLIDIFIED FILLER AND MELTED BASE METAL

FIGURE 6-1 The GTAW process

SAFETY

TABLE 6-1 and **TABLE 6-2** show some basic safety considerations related to workplace environments. Table 6-1 shows the basic safety considerations for GTAW and Table 6-2 shows recommended lens shades for welding at varying amperage ranges.

Head	Body	Hands	Feet
Safety glasses Welding helmet	Flame-resistant wool or cotton pants and shirt	Leather welding gloves	High-top leather safety work boots

TABLE 6-1

Read "Chapter 2: Safety in Welding" for detailed discussion of safety related to workplace environment, workplace hazards, personal protective equipment, electrical consideration, gases and fumes, ventilation, fire prevention, explosion, and compressed cylinders and gases.

Arc Current	Suggested Shade
Less than 50	8
50–150	10
150–250	12
250–400	14

TABLE 6-2

POWER SOURCE AND PERIPHERALS

Equipment for GTAW is similar to that used for Shielded Metal Arc Welding (SMAW). Most GTAW power sources can also be used for SMAW. However, GTAW uses more periphery equipment, including gas lines for the externally supplied shielding gas required. **FIGURE 6-2** shows a typical GTAW setup.

The basic equipment needed to start welding with GTAW is:

- Power Source
- Welding Torch
- Work Lead
- Regulator/Flow Meter
- Remote Current Control
- Tungsten Electrode
- Water Circulator
- Shielding Gas
- Filler Wire

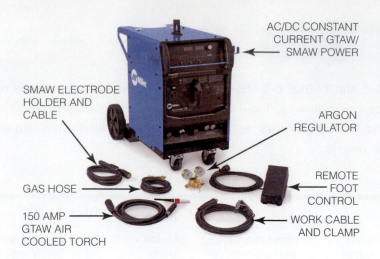

AC/DC CONSTANT CURRENT GTAW/ SMAW POWER

SMAW ELECTRODE HOLDER AND CABLE

ARGON REGULATOR

GAS HOSE

REMOTE FOOT CONTROL

150 AMP GTAW AIR COOLED TORCH

WORK CABLE AND CLAMP

FIGURE 6-2 The GTAW power source and accessories
Source: Photo courtesy of Miller Electric Manufacturing Co.

In its most basic form, all that is required for GTAW is a power source, a welding torch, and a regulator/flow meter. The type and configuration of metal, amount of time welding, welding amperage level, and working environment often determine how a GTAW system will be set up.

For example, a piping contractor installing stainless steel piping in a dairy may choose to limit the system to just a power source, an air-cooled torch, and a regulator/flow meter. When installing stainless steel piping, the contractor must have the ability to easily move equipment around; thus, the system is kept light and simple. Because the piping is stainless steel and the "walking the cup" technique is used, the need for a remote amperage control is eliminated.

When GTAW welding is done in a production or fabrication environment, a welder may choose to add components to the system. For example, a fabrication shop that is welding aluminum radiator components would utilize a remote foot control in order to compensate for the aluminum's changing melting rate. A lighter and more flexible water-cooled torch and a water circulator are chosen in order to reduce the welder's wrist fatigue.

WHAT TYPE OF POWER SOURCE IS USED FOR GTAW?
The GTAW process requires a constant current welding power source. GTAW power sources can be grouped into two categories:

- DC: Direct Current
- AC/DC: Alternating Current/Direct Current

Within these categories, there are three different technology levels. Throughout this chapter, the power sources will be referred to as:

- Magnetic Amplifier (Mag Amp)
- Solid State
- Inverter

Much more in-depth information can be found on the power sources for GTAW and the peripheral equipment in Chapter 15.

WHAT OPTIONS ARE AVAILABLE FOR WELDING TORCHES?
A GTAW torch is the conductor of welding current and the transmission line for shielding gas. The conduction of welding current and the heat of the welding arc will cause

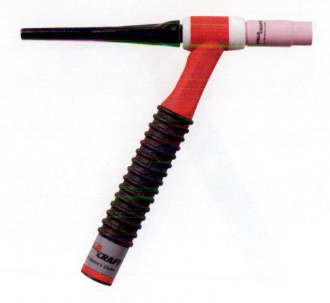

FIGURE 6-3 Air-cooled GTAW torch with standard back cap
Source: Photo courtesy of Weldcraft.

the torch to become hot. If the heat is not removed, the torch will be damaged; thus, torches must be cooled. There are two basic methods of cooling GTAW welding torches:

- Air cooled
- Water cooled

An **air-cooled torch** uses the flow of the argon shielding gas to help cool the torch. Three standard models of air-cooled torches are available: 9 (125 amp), 17 (150 amp), and 26 (200 amp). **FIGURE 6-3** shows a 150 amp air-cooled torch. In order to keep the torches from becoming overheated, the torch body and power cable are larger in size and diameter than water-cooled types.

The benefits of air-cooled torches are:

- Lower cost
- Simple setup (no need for coolant or water circulator)
- One cable
- Increased torch body size and cable diameter can withstand daily wear and tear of more abusive environments
- Can have built in manual gas valve. Used when the power source does not have a built in gas solenoid. **FIGURE 6-4** on the next page shows a 150 amp air-cooled torch with a built-in manual valve.

The drawbacks of air-cooled torches are:

- Increased weight
- Decreased flexibility of welding cable
- Will become hot to the touch when used at or near the maximum amperage range

A **water-cooled torch** uses the flow of shielding gas and a continuous flow of liquid coolant to cool the torch. For water-cooled torches, two standard models are available; the 20 (250 amp) and 18 (350 amp). Welding current range may be reduced when welding in AC; always consult the owner's manual for specifications.

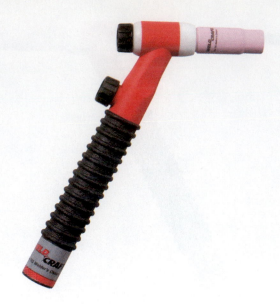

FIGURE 6-4 Air-cooled GTAW torch with short back cap and hand flow control valve
Source: Photo courtesy of Weldcraft.

FIGURE 6-5 shows a 250 amp water-cooled torch. Because of the continuous flow of coolant, the torch body and welding cable can be smaller in size and diameter. The benefits of water-cooled torches are:

- Decreased weight
- Increased flexibility of the power cable
- Smaller size can allow welding in smaller spaces
- Can help reduce strain and fatigue of the welder's wrist and arm

The drawbacks of water-cooled torches are:

- More expensive
- Requires coolant and a water circulator. Because of its smaller diameter welding cable and torch body, the torch will be damaged in a very short time if the coolant fails to circulate.

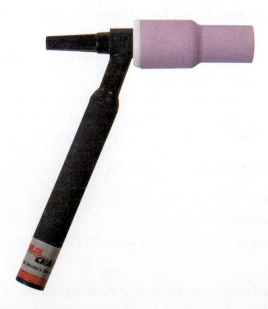

FIGURE 6-5 Water-cooled GTAW torch
Source: Photo courtesy of Weldcraft.

Three lines run between the power source and torch body: power cable, coolant return line, and gas hose. **FIGURE 6-6** shows the three lines of a

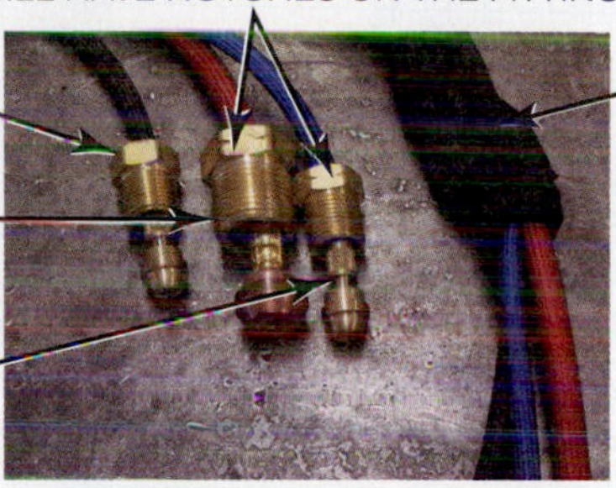

NOTE: LINES THAT CARRY WATER WILL BE LEFT-HANDED
THREADS AND WILL HAVE NOTCHES ON THE FITTINGS

ARGON LINE

POWER CABLE/
WATER-IN LINE

WATER
RETURN LINE

CABLE
COVER

FIGURE 6-6 Coolant and shielding gas lines for a water-cooled GTAW torch

water-cooled torch. It is recommended that the cables be contained inside a cable cover to prevent damage to the cables.

When determining which style of torch is best for an application, it is recommended that the following questions be answered:

- What will be the average welding amperage?
- How many hours of welding per day?
- What metals will be welded? AC used for aluminum and magnesium will produce more heat at the torch.
- Does the GTAW setup need to be portable?
- How much do you want to spend on the setup?

WHAT COMPONENTS MAKE UP A GTAW TORCH?

Air- and water-cooled torches are very similar in function and components. Both have two main functions: to conduct welding current and to carry shielding gas.

Notice that there are more similarities than there are differences. Both torches have a power cable, torch body, cup/nozzle, collet, collet body, and back cap (tail). **FIGURE 6-7** shows the components of a welding torch. The differences between the two are that the water-cooled components are smaller in size, and the water-cooled torch has three separate lines running between the power source and torch body.

POWER CABLE The power cable for the air-cooled torch has an electrical conductor contained in a rubber hose. The inside diameter of the hose is big enough for the conductor, with spare room for the shielding gas to flow. As the shielding gas flows past the conductor, it helps carry away heat produced when electrical current flows through the conductor.

BACK
CAP

TORCH
BODY

COLLET

COLLET
BODY

CUP
NOZZLE

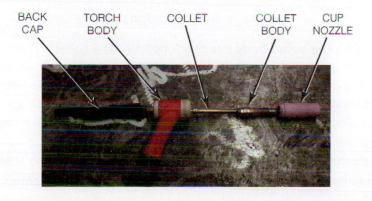

FIGURE 6-7 Parts of the GTAW torch

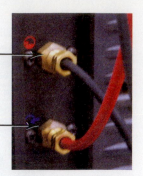

COOLANT IN: THIS HOSE IS THE COOLANT RETURN LINE (NOTE LEFT-HAND FITTINGS)

COOLANT OUT: HOSE RUNS TO POWER CABLE ADAPTER; THUS THE LOWEST TEMPERATURE COOLANT IS DELIVERED TO THE POWER CABLE (NOTE LEFT-HAND FITTINGS)

FIGURE 6-8 Recommended water flow for a water-cooled torch

The water-cooled torch consists of three different lines: the power cable, the coolant line, and the gas hose. The power cable has an electrical conductor contained in a rubber hose. The inside diameter of the hose is big enough for the conductor, with spare room for the liquid coolant to flow. As the coolant flows from the circulator, the coolant flows around the conductor and carries away heat produced from current flow. **FIGURE 6-8** shows the recommended water flow for a water-cooled torch. The coolant return line is the return path for coolant after it has flowed through the power cable and torch body. The gas hose provides the path for shielding gas to the torch body.

TORCH BODY The **torch body** consists of a brass body covered by a nonconductive coating (rubber or plastic). The design of the head provides the correct angle for the welder when welding. Models are available that allow the head angle to be flexible, thus allowing the welder to customize the head angle.

CUP The **cup** funnels shielding gas over the tungsten tip and welding puddle. It is made of ceramic and other heat-resistant materials that can withstand extreme temperatures of 11,000°F (6,100°C). Cups are available in different lengths and diameters.

COLLET Current needs to be conducted from the torch body to the tungsten electrode. Because the tungsten is designed to be adjusted and removed, there is no fixed connection in which current is conducted. In the GTAW torch, current is conducted through the **collet**. The collet is designed to push against the inside wall of the torch body and squeeze around the tungsten, holding it in place. It has a long fluted shape that is designed to provide as much surface contact area as possible in order to conduct current.

It is very important that the collet size matches the size of the tungsten being used. The tungsten will not fit in a collet that is too small, and the surface contact area of a collet that is too large will be reduced. Smaller tungsten may also fall out of the torch due to the lack of holding pressure.

COLLET BODY The **collet body** has two functions. The collet is pushed against the inside of the collet body, which forces the collet to tighten around the tungsten. It also diffuses the shielding gas through the cup to the weld. Like the collet, it is important that the collet body be the correct size. There are two types of collet bodies: standard and gas lens. **FIGURE 6-9** shows both types of collet bodies.

The standard collet body has 4 to 5 holes through which the shielding gas is diffused into the cup. As gas flows through the holes and enters the cup,

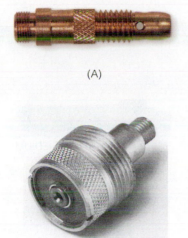

(A)

FIGURE 6-9A Standard collet body

Source: Photo courtesy of Weldcraft.

(B)

FIGURE 6-9B Gas lens collet body

Source: Photo courtesy of Weldcraft.

it begins to swirl around the inside of the cup. It continues to swirl as it leaves the cup and comes in contact with the weld pool and base metal. This swirling action causes turbulence that can pull atmosphere into the shielding gas if the gas pressure is too high causing contamination. This contamination leads to discoloration and porosity in the finished weld.

Shielding gas that flows through a gas lens does not swirl. Instead of flowing through 4 to 5 holes, shielding gas flows through layers of a mesh screen. The screen diffuses the gas into a smooth-flowing solid column. The gas column envelops the weld puddle and base metal, providing a pure environment of shielding gas. **FIGURE 6-10** shows the gas flow difference between a standard collet body and a gas lens.

GAS FLOW FROM
A GAS LENS

(A)

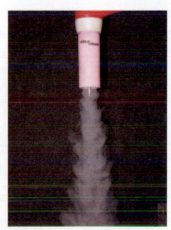

GAS FLOW FROM A
STANDARD COLLET BODY

(B)

FIGURE 6-10 Note how the gas lens (left) controls the flow of shielding gas compared to a standard collet body

Source: Photos courtesy of Weldcraft.

BACK CAP The **back cap** has two functions. First, it puts pressure against the back of the collet, pushing the collet against the collet body. The second function is to seal the back of the torch body, forcing the shielding gas through the collet body. The back cap is also called a tail.

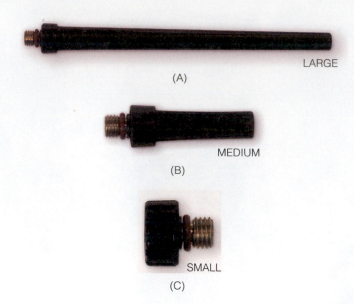

(A)   LARGE

(B)   MEDIUM

(C)   SMALL

FIGURE 6-11 Back caps come in a variety of sizes and styles for a given application.
Source: Photos courtesy of Weldcraft.

The back cap is available in three different lengths **(FIGURE 6-11)**. The long cap (4.5 inches/115 mm) is designed to allow a full-length piece of tungsten to be used in the torch. Unfortunately, the long length can get in the way when welding in confined spaces or included angles. The medium cap (2.75 inches/70 mm) and short cap (button cap) (1 inch/25 mm) allow the torch to be placed in confined spaces and included angles.

WHY ARE THERE TWO GAUGES ON WELDING REGULATORS?

The first gauge displays the internal pressure of the gas cylinder. This allows the welder to recognize how much gas is left in the cylinder. Pressure is displayed in kPa (kilopascals) and psi (pounds per square inch). **FIGURE 6-12** shows a welding regulator.

The second gauge displays the amount of gas flow used to shield the weld. It is displayed in CFH (cubic feet per hour) and L/min (liters per minute).

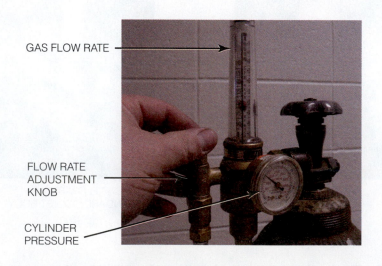

GAS FLOW RATE

FLOW RATE ADJUSTMENT KNOB

CYLINDER PRESSURE

FIGURE 6-12 A regulator displays cylinder pressure, and shielding gas flow to the GTAW torch is adjusted with the flow meter

HOW DOES A REMOTE CURRENT CONTROL OPERATE?

A remote current control has two functions. The first is to turn the welding arc on and off. Second, it is used to control the level of the welding current.

FIGURE 6-13 A foot control allows variation in the amount of available current to the welding arc.
Source: Photo courtesy of Miller Electric Manufacturing Co.

The most popular remote current control is a foot control **(FIGURE 6-13)**. The foot control is relatively easy to use and separate from the welding torch. It is well-suited for welding done on a bench where the welder can sit or stand. The ease of operation makes the foot control ideal for training new welders.

The foot pedal performs two tasks. It turns the welding output on and off. When the pedal is first depressed, a built-in switch closes. The closing of the switch signals the power source to turn on its welding output. When the pedal is returned to its top position, the switch is opened, and the welding output stops.

To control the welding current level, as the pedal is depressed, the resistance potentiometer changes. As the resistance changes, a voltage signal to the power source also changes. The power source reacts to this voltage change and in turn changes the welding output. The maximum range of the remote control is determined by the amperage setting on the front panel of the power source.

ARE THERE OTHER TYPES OF REMOTE CURRENT CONTROLS BESIDES THE FOOT CONTROL?

Because of its size and the fact that the foot control has to be kept on the floor, the control does not function well when the welder has to weld inside or on top of the weldment. An example of this is in the manufacturing of tank trailers. In this case, welders are very seldom welding at a bench. Most often, they are on a lift or ladder or inside, under, or on top of the trailer. The remote current control needs to be smaller and lighter. The control of choice is called a remote hand control **(FIGURE 6-14)**. These controls operate exactly the same as a foot control except they are in a smaller form, which allows them to be attached to the torch handle.

When choosing a hand control, make sure that the control does the following:

- Easily attaches to the torch handle
- Allows movement of the dial or slide control without moving the torch. This is very important because unwanted movement of the torch will show in the finished weld bead.
- Adjusts without causing excessive fatigue to the fingers or wrist.

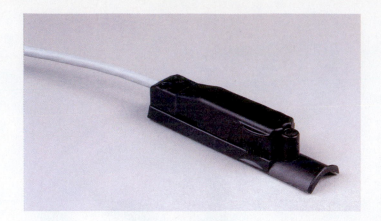

FIGURE 6-14 A hand-controlled remote control is preferred by some welders over a foot-controlled unit for varying amperage.

Source: Photo courtesy of Miller Electric Manufacturing Co.

WHAT DOES THE TUNGSTEN ELECTRODE DO? The GTAW process is different from the other common arc welding processes in that the electrode is not consumed. The temperature created around the welding arc can reach an average of 11,000°F (6,100°C); thus, any material that is used as a conductor must not only be conductive but also have a very high melting temperature. Tungsten is used as the conductor of welding current because it has the highest melting temperature of all pure metals at 6192°F (3422°C). With its ability to withstand high temperatures, the **tungsten electrode** is considered to be nonconsumable.

WHY ARE TUNGSTEN ELECTRODES SHARPENED TO A POINT? Control over the welding arc is one of the strengths and benefits of the GTAW process. When the tungsten electrode is sharpened to a taper, the arc is more controllable, resulting in narrower, deeper-penetrating welds. Tungsten is always sharpened when welding with the direct current electrode negative (DCEN) polarity. This applies to magnetic amplifier, solid-state, and inverter power sources. **FIGURE 6-15** shows sharpened tungsten.

The most popular way to produce the taper is by forming it with a grinding wheel. No matter what type of sharpener is used, it is important that the taper be produced in the correct way. When tungsten is ground, lines are left in the taper

FIGURE 6-15 An electrode with a taper creates an arc, which is more controllable, resulting in a narrower, deeper-penetrating weld.

of the tungsten. The welding current follows these lines toward the base metal. For this reason, the lines need to run toward the point of the taper. When the tungsten is sharpened incorrectly, the lines run around the diameter of the tungsten. The welding arc can be misdirected by the lines, resulting in a loss of control over the welding arc. **FIGURE 6-16A** shows a tungsten electrode being correctly sharpened using a belt grinder and **FIGURE 6-16B** shows the resulting grinding lines running toward the tip. **FIGURE 6-17A** shows a tungsten electrode being incorrectly sharpened using a belt grinder and **FIGURE 6-17B** shows the resulting grinding lines running around the tip.

Ceriated, thoriated, and lanthanated tungstens consist of pure tungsten alloyed with a percentage of cerium oxide, thorium oxide, or lanthanum oxide, respectively. The addition of the oxide allows the tungsten to withstand higher

VIDEO

CORRECT POSITION

(A)

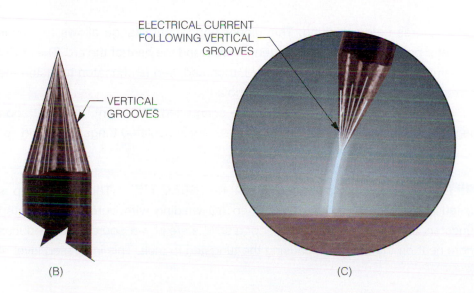

ELECTRICAL CURRENT FOLLOWING VERTICAL GROOVES

VERTICAL GROOVES

(B)

(C)

FIGURE 6-16 Grind marks from sharpening must run parallel with the electrode.

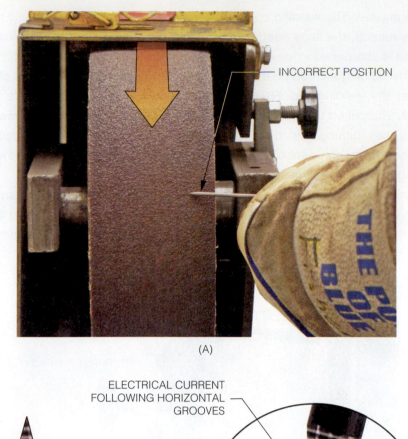

INCORRECT POSITION

(A)

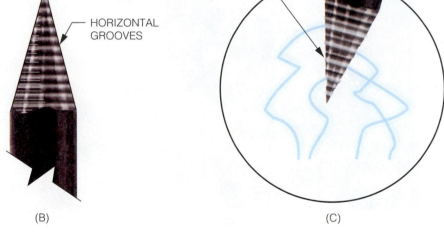

ELECTRICAL CURRENT
FOLLOWING HORIZONTAL
GROOVES

HORIZONTAL
GROOVES

(B)

(C)

FIGURE 6-17 An unstable arc may result if grind marks from sharpening run perpendicular to the electrode.

temperatures before melting. This higher temperature range allows the small diameter point of sharpened tungsten to withstand the heat of the arc. The ability to maintain their shape makes ceriated, thoriated, and lanthanated tungsten the correct choice for welding in the DCEN polarity.

If pure or zirconiated tungstens are sharpened to a point, they will soon begin to melt and form a ball. This is why pure or zirconiated tungsten should not be used to weld with DCEN polarity.

WHY DO SOME TUNGSTENS HAVE A ROUNDED TIP? The rounded tip is called a ball, and it is only used when AC welding with pure or zirconiated tungsten, using a magnetic amplifier or a solid-state power source. AC produces more heat on the tungsten, causing the tungsten to melt. This increased level of heat is caused by the direct current electrode positive (DCEP) portion of AC. Magnetic amplifier and solid-state power sources do not have the ability to reduce the heat enough to stop the melting. Because the tungsten is going to melt, welders have learned to weld with the ball formed on the tip of the tungsten **(FIGURE 6-18)**.

The drawback of a rounded shape of balled tungsten is that the welding arc wanders around the entire surface of the ball. The bigger the ball, the more the arc wanders. **FIGURE 6-19** shows an excessively balled tungsten electrode.

HOW IS A BALLED TIP FORMED?
The ball is created in one of two ways. With one method, the polarity switch is placed in the DCEP position, and the tungsten is blunted to a flat tip. An arc is started, and as the welding amperage is increased, a ball will form on the tip. After the ball is formed, the polarity switch is moved to AC. It is important that the ball size be limited to no more than 1.5 times the diameter of the tungsten. Excessive melting will negatively affect the control of the welding arc.

Another way to ball the tungsten is with the polarity switch in the AC position and the tungsten blunted flat. An arc is started on a piece of aluminum of the same thickness as the aluminum being welded. The benefit of this method is that the ball will only become as big as the welding amperage will produce. If the tungsten is the correct diameter, the ball will be smaller, and wandering will be reduced.

When welding AC with a magnetic amplifier or a solid-state power source, it is important not to use ceriated, thoriated, or lanthanated tungsten. These tungstens are made to handle higher temperatures, but when the temperature becomes too high, the tungsten will not melt into a ball. Instead, the tip will form into multiple small balls or crowns. **FIGURE 6-20** shows a crowned 2% ceriated tungsten electrode. The welding arc will wander more than if just one ball was formed. The individual balls can fall into the puddle, causing a tungsten inclusion, which is considered a defect in the weld and may need to be removed and rewelded. The pure and zirconiated tungstens are preferred for AC welding with magnetic amplifier and solid-state power sources.

WHY CAN SHARPENED TUNGSTEN BE USED WHEN WELDING ALUMINUM WITH AN INVERTER POWER SOURCE?
With the inverter power source, the tungsten is sharpened and then the point is blunted. Because of their extended balance control range, inverter power sources have the ability to reduce the amount of heat on the tungsten tip. The reduction of heat allows ceriated, thoriated, and lanthaniated tungstens to be used to weld with AC. There is more heat than when welding with the DCEN polarity, so the small diameter tip of a point will melt. This melting can cause the arc to wander. In order to eliminate the wandering, the tip is removed by blunting it flat. **FIGURE 6-21** shows a blunted 2% ceriated tungsten electrode.

Using sharpened and blunted tungsten gives the welder greater control over the welding arc. This greater control allows the finished weld to have a smaller size and deeper penetration than a weld made with balled tungsten.

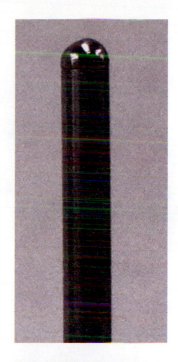

FIGURE 6-18 When welding on aluminum and using a pure tungsten electrode, a ball is preferred.

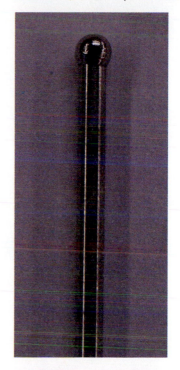

FIGURE 6-19 A large ball at the end of the electrode create a larger, wandering arc.

VIDEO

FIGURE 6-20 Tungsten electrodes other than pure or 1% zirconium often will not be able to form a ball.

FIGURE 6-21 A blunt sharpened tungsten electrode may aid in preventing the arc from wandering.

WHEN IS A WATER CIRCULATOR NEEDED? The only time a water circulator is needed is when using a water-cooled torch. The water circulator is a closed-loop system that circulates coolant through the torch to keep it from being damaged by the heat **(FIGURE 6-22).**

When water-cooled torches were first developed, welders often cooled the torches with water directly from the building's water supply. The water cooled the torches but caused several problems. Besides being a waste of water, minerals

FIGURE 6-22 A water circulator for a water-cooled GTAW torch.
Source: Photo courtesy of Miller Electric Manufacturing Co.

and other hard particles in the water can cause high-frequency current to be conducted to earth ground. High frequency current is used to start and stabilize the arc. Without it, the arc becomes hard to start or inconsistent. Minerals and particles can also build up in the torch and cause overheating or failure. Also, cold tap water running through the torch while the arc is off can cause condensation inside of the torch, which can cause porosity.

WHAT TYPE OF COOLANT SHOULD BE USED IN THE WATER CIRCULATOR?

There are three issues that restrict what type of coolant can be used in the water circulator. Because of minerals and particles in water, using the wrong type of coolant can cause the loss of high-frequency current and cause restrictions in the torch. If the coolant is not treated with an algaecide, algae can grow, which can cause restrictions in the torch.

Because of these issues, it is recommended that a water circulator be filled with the circulator manufacturer's recommended coolant or distilled water. If distilled water is used, the water must be changed often to stop the buildup of algae. Do not use any other antifreeze.

WHAT SHIELDING GASES ARE USED FOR GTAW?

The selection of the type of shielding gas is pretty straightforward for the GTAW process. In most situations, pure argon is used. The typical flow rate is from 15 to 35 cubic feet per hour (cfh) (7 to 16 liters per minute (L/min)). **FIGURE 6-23** shows a cylinder of argon.

The other choice for shielding gas is helium. Generally, helium is never used alone but is mixed with argon. The use of helium has the advantage of transferring more heat into the weld than argon. For this reason, helium is often used when welding thicker plates. Helium is similar to argon in that it is inert, but it differs in that it is a lighter than air, and it has a higher ionization level. What this means is that flow rates need to be higher for helium. The higher ionization level causes difficulty when starting the arc and instability when welding with AC.

WHY IS ARGON USED?

Argon is the perfect shielding gas for the GTAW process because it is inert and heavier than air. **Inert gas** is a gas that does not react with the welding arc, molten puddle, or finished weldment. The benefit of

FIGURE 6-23 100% argon is the most common shielding gas for GTAW.

the argon being heavier than air is that when the argon flows out of the torch, it will fall down over the weld and push out all the other atmospheric gases. Unlike the GMAW and FCAW processes, argon can be used as the shielding gas for all the metals the process is able to weld.

The drawback of a gas being heavier than air is that it is an **asphyxiant**. If inhaled asphyxiant gas will cause oxygen to be pushed out of the lungs, causing a person to suffocate, resulting in injury or death. Care must always be taken when using argon in a confined space in which there is a limited amount of fresh air.

HOW IS THE GTAW FILLER WIRE DIFFERENT FROM THE FILLER USED FOR THE OTHER COMMON ARC WELDING PROCESSES?

The filler rod used for GTAW is sold in the form of 3 feet (91.4 cm) lengths. The rods are generally called cut lengths or sticks. They are available in diameters from 0.10 inches (0.25 mm) to 1/4 inches (6.4 mm). **FIGURE 6-24** shows a box of 3/32-inch (2.38 mm) ER70S-2 mild steel filler rod.

Unlike GMAW and FCAW, the filler rod in the GTAW process is added manually to the weld puddle. Manually adding filler provides several advantages and disadvantages.

Advantages:

- Complete control over the amount of filler added to the weld puddle
- Greater control over penetration; the addition of filler can be delayed until the base metal melts
- Easier to control finished weld profile

Disadvantages:

- Deposition rates are slower than other arc welding processes
- Greater welder skill level is needed

In the GMAW and FCAW processes, the filler wire is also the electrode for the process. This means that the filler wire is conducting the welding current. Because the filler wire has a small diameter, it cannot conduct the welding current without melting. Thus, the filler wire melts and is added to the weld.

MOST FILLER RODS WILL HAVE THEIR CLASSIFICATION ON EACH INDIVIDUAL ROD

FIGURE 6-24 The filler rod classification is stamped on one end.

The GTAW process is different in that the filler rod is not the electrode. Because it is not the electrode, there is no welding current flowing through the wire. The filler wire for GTAW is melted by the heat of the molten weld puddle.

CONTROLS AND CHARACTERISTICS

There are many controls used on GTAW power supplies. Understanding these controls is essential to making good welds and getting the best performance from the power supply possible.

The standard controls found on most GTAW power sources are:

- Amperage
- Postflow/Preflow
- Polarity
- Balance
- High Frequency
- Lift Arc
- AC Frequency
- AC Wave Form
- Independent Amperage Control

Some controls are only found on one or two types of power sources; this will be noted in the body of the text.

AC frequency, AC wave form, and independent amperage control are only available with inverter power supplies.

WHAT EFFECT DOES INCREASING AND DECREASING THE WELDING AMPERAGE HAVE?

As welding amperage increases, the more the base metal melts. This increased melting directly affects the amount of penetration and the width of the puddle.

One unique feature of most GTAW power sources is the ability to control the welding amperage remotely. Unlike GMAW, FCAW, and SMAW, the welder has the ability to change welding amperage while welding. This gives the welder more adaptability than is found with the other arc welding processes. Control of the amperage is generally done by using a remote foot pedal or hand control. The name for this style of remote amperage control is called a **venire control**. A venire control's maximum amperage is controlled by the welding amperage setting on the power source.

How venire control works is that the welder sets the amperage control on the power source. This setting is the maximum amperage range of the remote control. For example, the welder sets the amperage at 146 amps. One hundred forty-six amps is now the maximum of the remote. The range of the remote is now from 1 to 146 amps. When the welder varies the remote, the change in welding amperage will be in much smaller increments. Smaller increments in change allows for finer control over the welding amperage. **FIGURE 6-25** shows how the amperage setting on the power source limits the range of the remote control.

FIGURE 6-25 While a remote control may vary the amount of the amperage to the tungsten electrode, the amperage cannot exceed the present amperage set on the power source.

Source: Photo courtesy of Miller Electric Manufacturing Co.

HOW IMPORTANT IS THE POSTFLOW?

Postflow is a timed gas flow at the end of a weld. This gas flow benefits the weld in two ways. One is to protect the weld puddle, filler rod, and tungsten electrode from the contaminating atmosphere. This protection is needed until the molten weld puddle and molten filler rod tip solidify. The second is to help cool the very hot tungsten electrode.

The rule of thumb for setting postflow is: for every 10 amps of welding current, there should be 1 second of gas flow. Thus, a weld of 130 amps should have a timed flow of around 13 seconds.

Remember that the flow of shielding gas is very important. It needs to protect the molten weld puddle, the molten tip of the filler rod, and the tungsten electrode from the contaminating effects of the atmosphere. After the welding has stopped, the puddle and filler rod tip generally need to be protected for 3 seconds for most metals. After 3 seconds, the remaining gas flow is needed to protect the tungsten.

In order to lower costs, some welders try to reduce the postflow time below the recommended 1 second/10 welding amps. If the postflow time is insufficient, the tungsten will turn black and can begin to degrade and lose shape. **FIGURE 6-26** shows tungsten contaminated by the atmosphere. This will cause defects in the finished weld.

Too much postflow time can also cause problems. First of all, it is a waste of money. Second and most important, argon can be dangerous when used in a confined space or any situation in which the amount of available fresh air is limited. Always remember to keep the postflow time at the lowest time needed to produce defect-free welds when welding in confined spaces.

WHEN DO YOU NEED A PREFLOW OF SHIELDING GAS?

Preflow is a timed amount of gas flow that happens before the start of welding. The benefit of this gas flow is that it will purge the welding torch and welding area of atmosphere, reducing the chance of contamination.

Generally, 0.2 seconds preflow time is adequate for most situations. However, there are situations in which it is beneficial to have more preflow time. Welding inside a recessed port requires more preflow in order to purge the port of trapped atmosphere.

FIGURE 6-26
Postflow shielding gas protects the electrode from contamination.

Polarity	Welding Process	% of Heat on Electrode	Metals Welded with GTAW
Straight Polarity	GTAW SMAW	30%	steel, stainless steel, titanium, copper, brass, bronze, nickel
Reverse Polarity	SMAW	70%	Only used to prepare (ball) the tip of pure or zirconiated tungsten
Alternating Current (AC)	GTAW SMAW	Depends on the balance control setting	aluminum, magnesium

TABLE 6-3

WHAT POLARITY SHOULD I USE TO WELD DIFFERENT METALS?

As mentioned in the beginning of the chapter, there are two different types of GTAW power sources: DC and AC/DC. If you have a DC-only power source, there is no polarity control on the power source. You can change between DCEN, also called straight polarity, and DCEP, also called reverse polarity, by reversing the torch and work leads.

An AC/DC power source has a polarity control in the form of a switch. This control allows the welder to choose from DCEN, DCEP, or AC. Because a GTAW welding process requires a constant current power source, it can also be used to weld with the SMAW process. **TABLE 6-3** shows recommended polarities.

WHY IS DCEN THE POLARITY NEEDED TO WELD STEEL?

DCEN is the polarity used to weld the majority of metals with the GTAW process. When welding starts, the welding current or electrons flow from the negative electrode to the positive base metal. When the electrons flow through the base metal, the base metal's resistance produces heat, which causes the base metal to melt. Seventy percent of the heat is at the base metal, and only 30% is at the tungsten.

Because the purpose of GTAW is to melt the base metal and not the electrode, DCEN is the polarity of choice for almost all metals. Another benefit of this polarity is that the heat on the tungsten electrode is low and the tungsten is able to maintain its prepared shape.

IS DCEP USED FOR GTAW?

DCEP is the opposite of DCEN. Seventy percent of the heat is now on the tungsten and 30% at the base metal. Thus, the useful current flow that melts the base metal now produces heat at the tip of the tungsten electrode. Because the majority of heat is now on the tungsten, there is excessive melting of the tungsten tip and limited melting of the base metal.

Because of these two negative effects, the use of the DCEP polarity for the GTAW process is very limited and not useful for standard welding. The one area that it is used during the GTAW process is to prepare (ball) pure and zirconiated tungsten for welding in AC.

BECAUSE DCEN IS USED FOR MOST METALS, WHY IS AC CURRENT USED?

AC combines the characteristics of DCEN and DCEP. To understand the need for AC, you first need to understand two characteristics of aluminum,

FIGURE 6-27 DCEN melts aluminum but not the oxide, resulting in a blackish-coated weld.

which is the main metal welded with AC. First, aluminum has an average melting temperature of 1,120°F (604°C). Second, aluminum oxide—the coating that covers all aluminum—has a melting temperature of 3,600°F (1,980°C).

When GTAW aluminum with DCEN, the aluminum will melt, but the aluminum oxide does not. The unmelted oxide floats on the surface of the liquid aluminum puddle, producing a blackish coating of aluminum oxide. **FIGURE 6-27** shows the results of welding aluminum with DCEN only.

This uncompromising coating prohibits the addition of aluminum filler rod to the puddle. Unless the aluminum oxide coating is removed, acceptable welds will not be produced. Aluminum oxide can be removed by means of grinding/machining, chemical etching, or **cathotic etching**.

- **Grinding/Machining:** Removes the oxide but will reduce the aluminum's thickness and possibly scar the surface of the aluminum. The oxide layer begins to develop immediately after the grinding/machining is done; thus, welding needs to be done relatively soon afterward.
- **Chemical Etching:** The acids used for chemical etching are very caustic and highly toxic and must be completely removed afterward. If not completely removed, the acids will damage the aluminum and toxic vapors can be released during welding.
- **Cathotic Etching:** This is a direct result of DCEP. Positive ions are released from the electrode and strike the oxides on the surface of the aluminum. The loose oxides are then lifted away by the current flow from the aluminum to tungsten. The removal of oxides is done while welding, so there is no problem with oxides redeveloping or the need to use and remove acids. **FIGURE 6-28** shows the etching along the side of GTAW aluminum. The white area on both sides of the weld puddle is the area in which the oxides have been removed. It is generally called etching. The etching zone is a permanent scar left on the aluminum.

WHAT DOES THE BALANCE CONTROL DO? The balance control is used to change the ratio of DCEN versus DCEP welding current when using AC. DCEN current provides the melting and penetration of the base metal essential to GTAW. Because of the aluminum oxide layer, DCEP current is needed to produce acceptable welds.

FIGURE 6-28 The DCEP side of AC allows chaotic etching (cathotic cleaning), which shows up as a white area around the weld. Changing the balance by using more DCEP increases the cathotic etching white area, while using less DCEP decreases the cathotic etching white area.

The same DCEP current flow that removes oxides also causes excessive melting and deforming of the tungsten. When tungsten deforms, the welding arc will begin to wander. This wandering causes a loss of control over the size and shape of the weld puddle.

To obtain best results, the balance control should be set at the highest percentage of DCEN that the power source and aluminum will allow. Remember that all aluminum has an oxide layer, but the thickness of the layer can vary from one piece to another. The thicker the layer, the greater the amount of cathotic etching required.

To determine the correct setting, perform the following steps. Set the balance control at the highest setting available. Begin welding; if floating black spots appear in the puddle and/or the addition of filler is difficult, the balance setting is too high. Turn the control down. Repeat these steps until the puddle becomes clear and the addition of filler is not hampered. This will allow the aluminum to be welded at the highest balance setting possible and will keep the deforming of the tungsten to the minimum.

V I D E O

WHY DO DIFFERENT POWER SOURCES HAVE DIFFERENT BALANCE RANGES?

When power sources were first developed to weld aluminum, AC current had a balanced output of 50% DCEN and 50% DCEP. Because of excessive heat, the tungsten was normally melted into a ball shape. The arc would wander around the ball on the end of the tungsten. This wandering caused the weld puddle to be harder to control and typically wider than needed. At that time, the welds produced with these power sources were very acceptable and the preferred choice for more than 30 years. **FIGURE 6-29** shows a drawing of a 50% DCEN/50% DCEP balanced AC welding sinewave.

In the 1970s, solid-state power sources were developed that allowed the ratio of DCEN to DCEP current to be changed. The balance control on these power sources allow the welder to weld with a maximum of 68% DCEN versus 32% DCEP. These changes resulted in decreased melting of the tungsten and increased penetration. The width of the etching was also decreased. A narrower etching zone has a cosmetic benefit.

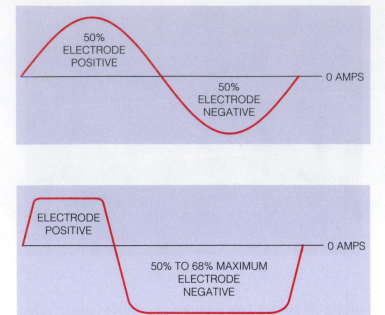

FIGURE 6-29 A 50% DCEN/50% DCEP balanced AC welding sinewave

FIGURE 6-30 A 68% DCEN/32% DCEP balanced AC welding sinewave

Thirty-two percent DCEP still causes some melting of the tungsten. The ball that is formed is noticeably smaller, resulting in decreased arc wander. **FIGURE 6-30** shows a drawing of a 68% DCEN/32% DCEP balanced AC welding sinewave.

Unfortunately, the DCEP side of the arc was called the cleaning side or cleaning action. It is important to remember that the only thing that DCEP does is to remove oxides. It does not clean oil, grease, dirt, etc. from the surface of the aluminum. In the mid 1990s, inverter power sources where developed that provided an even wider range of balance control. Inverters can have a maximum of 99% DCEN. This wider range provided increased penetration and decreased tungsten melting. A welder has more control over the welding arc and more control over the resulting weld. **FIGURE 6-31** shows a drawing of a 99% DCEN/1% DCEP balanced AC welding sinewave.

WHAT IS HIGH FREQUENCY? **High-frequency** current uses a high voltage (3,000 volts) oscillated at a high frequency (1 MHz). The voltage level is high enough that it has the potential for the arc to jump across the tungsten to base metal gap. As the high-frequency current flows across the gap, it ionizes the argon shielding gas, producing a low-resistance electrical path.

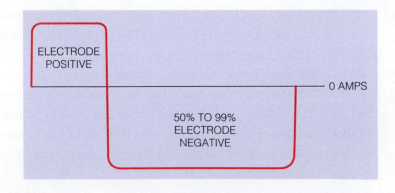

FIGURE 6-31 A 99% DCEN/1% DCEP balanced AC welding sinewave

FIGURE 6-32 The high-frequency arc jumps across the arc gap.

VIDEO

With a low-resistance electrical path now available, the welding current can now arc across the gap. **FIGURE 6-32** shows how the high-frequency arc jumps across the arc gap.

THERE ARE THREE HIGH-FREQUENCY CURRENT START SETTINGS. WHEN SHOULD EACH ONE BE USED?

High-frequency current is one of several different ways that is used to start a GTAW arc. Most magnetic amplifier and solid-state power sources have three different settings. These settings are off, start only, and continuous. They operate as follows:

- **Off:** Disables the high frequency
- **Start Only:** Allows the high frequency to turn on when the welder triggers the remote amperage control. The high frequency will automatically turn off when the arc is started. Used when welding in the DCEN polarity. Start only is used for welding in the DCEN and AC when using an inverter power source. This greatly reduces the use of high frequency. There is just one button or switch position that enables the start-only high frequency.
- **Continuous:** Allows the high frequency to turn on when the welder triggers the remote amperage control. The high frequency will remain on until the welder turns off the remote amperage control. Only used when welding with AC on noninverter power sources.

WHEN WOULD I USE LIFT ARC?

Lift arc allows a welder to start an arc without the use of high frequency. High frequency uses a very high voltage oscillated at a high frequency to start the arc without touching the tungsten to the base metal. The use of high frequency may not be desirable when welding is done around sensitive electronic equipment. Another example is when welding is done without a remote control. Most power sources will not allow the high frequency to operate without the minimum of a remote start switch.

Lift arc allows the welder to start the arc by touching the tungsten to the base metal and then lifting it up to start. Without the lift arc option, touching the tungsten to the base metal will cause the arc to start but will also cause contamination of the tungsten and base metal. The contamination is caused by current flow that melts the tungsten to the base metal.

(B)
LIFT TORCH

(A)
TOUCH TORCH TO BASE METAL

FIGURE 6-33 The lift arc technique

VIDEO

When lift arc is used, the power source limits the current flow to an amount that is enough to heat the tungsten but not melt it. Hot tungsten is a better emitter of electrons than cold tungsten. When the hot tungsten is lifted from the base metal, the arc is started. **FIGURE 6-33** shows how the lift arc functions.

WHAT BENEFIT DOES AC FREQUENCY CONTROL PROVIDE?

Magnetic amplifier and solid-state AC/DC power sources have always had an AC welding frequency equal to the frequency of the primary current used to run the power source. In the United States and Canada, the primary electrical frequency is 60 Hz.

A 60 Hz welding arc will come off the tungsten at a wide angle. This wide angle produces a wide and somewhat swallow penetrating weld pool.

AC/DC inverter power sources have the ability to change the AC welding frequency. Frequency range is generally from 20 to 250 Hz. The benefit of having the ability to change the welding frequency is the control of the angle at which the arc leaves the tungsten. As the AC welding frequency is increased, the angle of the arc decreases and the width of the puddle decreases. **FIGURE 6-34** shows the effect of increasing the AC welding frequency.

The correct frequency to use is determined by the geometry of the joint and the required weld penetration and weld puddle width. For example, a lap joint on 1/8-inch (3.2 mm) aluminum may have the best results welded with a higher frequency of 215 Hz. The finished weld benefits from the narrow and deep weld pool. The opposite may be true for the fill pass on a groove weld on 12-inch schedule 40 aluminum pipe. The frequency may be set at 40 Hz; the resulting puddle would be wider, with shallower penetration. When choosing a frequency, a welder needs to make a decision based on personal welding style and the width and penetration needed in the resulting weld.

LOW AC FREQUENCY

HIGH AC FREQUENCY

WIDER PUDDLE

NARROWER PUDDLE

WIDER AND SHALLOWER PENETRATION

NARROWER AND DEEPER PENETRATION

FIGURE 6-34 Effects of increasing the AC welding frequency

V I D E O

WHAT AC WAVE FORM SHOULD BE CHOSEN? The ability to select the AC wave form is available on a limited number of power sources. The first power sources developed had a sinewave type output. In the 1970s solid-state power sources were developed with a soft squarewave-type output. Inverters then came along that typically have advanced squarewave. Each of these waveforms had strengths and weaknesses, and each had applications in which they excelled.

With the development of microprocessor-based inverters, the welder now has the ability to choose the wave form best-suited for his or her present application. There are four different wave forms available at this time. **TABLE 6-4** below describes the wave forms and benefits.

WHAT DOES THE INDEPENDENT AMPERAGE CONTROL DO? For AC welding, when current is changed on most AC/DC power supplies, the DCEN and DCEP currents change equally. If more penetration is needed, the welding current is increased, but in turn, the heat on the tungsten increases, causing the tungsten

Wave Form	Benefit
Advanced Squarewave	Fast-freezing puddle Deep penetration Increased travel speed
Soft Squarewave	Maximum puddle control Good wetting action
Advanced Sinewave	Arc and response similar to original sinewave power sources; eliminates need for continuous high frequency
Triangular	Lower average amperage

TABLE 6-4

to melt. The balance control can help to reduce the melting, but its effect is limited. This control is only available with a limited number of AC/DC inverter power supplies.

A power source with independent amperage control has the ability to control the DCEN and DCEP current levels separately. An increase in DCEN produces greater penetration without the increased melting of tungsten from the DCEP. For example, the DCEN current can be set at 120 amps, and the DCEP current can be set at 40 amps. With these settings and the balance control set at the correct setting, the resulting arc is at its most productive for welding aluminum.

VIDEO

OPTIONAL CONTROLS AND CHARACTERISTICS

Beyond the standard controls, there are several options available with GTAW power sources:

- Pulser
- Sequencer
- Spot Timer

Options may be available as part of the power source or as kits that can be installed after the purchase. Not all manufactures make kits available. Inquire before making a purchase.

Often, older power source options were only available as a separate control box. The box sat on top or was attached to the side of the power source. It was connected to the power source through external connections.

HOW DOES A PULSER WORK?   A GTAW pulser simply operates by switching the welding amperage from high current (peak current) to low current (background current) at a set rate (pulses per second). **FIGURE 6-35** shows a pulser's current waveform.

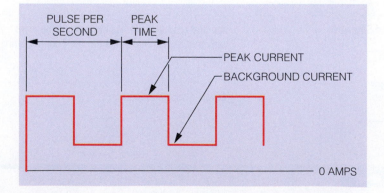

FIGURE 6-35 A pulser's current waveform

There are generally four controls that affect the operation of a pulser.

- **Peak Current:** Maximum current level. It can be controlled by the remote control or the amperage control on the power source. Please note that when pulsing, the peak current may need to be set higher to compensate for the lower average current.
- **Peak Time (Pulse Time):** Controls the percentage of time spent in the peak current mode of each pulse cycle.
- **Background Current:** Sets the minimum current level. Is often set as a percentage of the peak current. Using percentage versus a set current level provides the advantage of aligning the background current level to any change of the peak level. For example, when a welder increases the peak current, the background current will increase. The same is true when the peak amperage is decreased. This is important in that it maintains a constant ratio of peak and background currents.
- **Pulses per Second (Pulsing Frequency):** The number of times per second the current will change from peak to background. Finding the correct frequency setting is often a personal choice—selected because of the resulting bead appearance and/or penetration level.

WHAT ARE THE BENEFITS OF PULSING? The benefits of using GTAW pulsing are lower-average welding current, increased control over the weld pool size, weld pool agitation, and welder training.

The pulsing from peak to background current results in a lower average welding current. An example of this is when a welder is welding a lap weld on 16-gauge stainless steel. With no pulsing, the welding current will average out to be around 45 amps. This amperage is enough to melt the two base metal pieces and the filler metal being added. The problem is that thin stainless steel will warp and suffer from carbide precipitation if the heat input is too high. When pulsing is used, the average heat input is reduced; thus, warping and carbide precipitation can be reduced if not eliminated.

At the peak current level, the molten weld pool will be a certain width based on the amount of welding current. The current then switches to the background level, and the width of the weld pool decreases. This increasing and decreasing of weld pool size helps to produce a consistent appearance and width to the finished weld. **FIGURE 6-36** shows the effects of the pulser on puddle width.

The increasing and decreasing of weld pool size also has an agitating effect. This effect helps gas to escape from the molten weld pool before it solidifies. An example of this would be when doing buildup or a repair to cast

PEAK CURRENT

HIGHER CURRENT LEVEL CAUSES
PUDDLE TO BECOME WIDER

BACKGROUND CURRENT

LOWER CURRENT LEVEL CAUSES
PUDDLE TO SHRINK

FIGURE 6-36 Effects of the pulser on puddle width

aluminum. The aluminum cast is porous, with gas and moisture trapped within the pores. When the aluminum is melted, the gases attempt to escape before the weld pool solidifies. Often, the gases will get caught in the puddle, producing porosity. When pulsing is used, the agitation allows more gas to escape before solidification occurs.

Finally, pulsing can be helpful for training new welders. When pulsing at frequencies from 0.1 to 2 Hz, an inexperienced welder can use the pulsing to coordinate torch movement with filler wire addition. When the welding current is at the peak level, filler metal is added to the puddle. The filler will be withdrawn and the torch advanced forward when the welding current drops to the background level.

VIDEO

WHAT DOES A SEQUENCER DO?

A GTAW sequencer is most often used for welding repetitive production welds. One of the advantages of GTAW welding is the ability to change the welding current while welding, to adapt to changes in shape, fit-up, and base metal thicknesses. This is often done with a remote control.

When a weldment is produced in a production environment, the shape, fit-up, and base metal thickness become very consistent. Changes in welding current are the same for each weldment. The sequencer can be programmed, eliminating the need for a remote control and allowing the process to be controlled by the movement of a switch or by a timer.

WHEN WOULD THE SPOT TIMER OPTION BE USED?

The function of a spot timer is to stop a GTAW weld after a programmed period of time. It is often used to control the welding time of a GTAW spot weld. A GTAW spot weld is an alternative to a Resistance Spot Weld (RSW).

An RSW uses two copper electrodes to squeeze separate pieces of sheet metal together. Welding current is then conducted between the electrodes and through the sheet metal. This current flow causes the sheet metal to melt, and the melted metal blends together forming a weld nugget.

The length of the tongs determines the maximum distance of the weld from the edge of the sheet. A GTAW spot weld is done from one side of the sheet metal, allowing a weld to be performed anywhere on the surface of the sheet. A limitation of GTAW spot welding is that the sheets have to be very tight together. To ensure that the sheets are tight, the welder presses against the outside sheet with the cup of the torch. If the one or both of the sheets are not rigid enough and a gap forms between the sheets, the weld will only be on the outside sheet.

Prior to welding, the welder selects the following settings on the power source:

- Polarity
- Welding Current
- Start Method
- Postflow Time
- Preflow Time
- Balance

- Welding Frequency
- Tungsten Type
- Filler Wire
- Shielding Gas

POLARITY **TABLE 6-5** shows the welding polarities and their uses.

DCEN	DCEP	AC
Steel and stainless steel	Used to ball pure and zirconiated tungsten for AC welding	Aluminum

TABLE 6-5

AMPERAGE SETTINGS **TABLE 6-6** shows recommended welding amperages and tungsten electrode diameters for mild steel, stainless steel, and aluminum when using a remote current control.

These settings are a general reference. The amperage required when using a remote amperage control does not need to be exact. The correct current is determined by the welder's ability to look at the weld pool and adjust the remote

Mild Steel Thickness	DCEN Welding Amperage	Tungsten Diameter
1/32 (0.79 mm)	25–40	.040 (1 mm)
1/16 (1.6 mm)	50–90	1/16 (1.6 mm)
1/8 (3.2 mm)	75–110	3/32 (2.38 mm)
3/16 (4.8 mm)	115–145	3/32 (2.38 mm)
1/4 (6.4 mm)	155–200	1/8 (3.2 mm)
Stainless Steel Thickness	**DCEN Welding Amperage**	**Tungsten Diameter**
1/32 (0.79 mm)	20–30	.040 (1 mm)
1/16 (1.6 mm)	40–80	1/16 (1.6 mm)
1/8 (3.2 mm)	65–100	1/16 (1.6 mm)
3/16 (4.8 mm)	105–135	3/32 (2.38 mm)
1/4 (6.4 mm)	140–180	1/8 (3.2 mm)
Aluminum Thickness	**AC Welding Amperage**	**Tungsten Diameter**
1/32 (0.79 mm)	25–40*	.040 (1 mm)**
1/16 (1.6 mm)	60–95*	1/16 (1.6 mm)**
1/8 (3.2 mm)	100–150*	3/32 (2.38 mm)**
3/16 (4.8 mm)	140–210*	1/8 (3.2 mm)**
1/4 (6.4 mm)	200–300*	5/32 (4 mm)**

TABLE 6-6

*Welding amperage may vary when using a power source with an adjustable welding frequency.
**Tungsten diameter depends on the power source balance setting

	Start Only High Frequency		Continuous High Frequency	Lift Arc or Touch Start	
	Mag Amp Solid State	Inverter	Mag Amp Solid State	Mag Amp Solid State	Inverter
	DC	AC/DC	AC	DC	AC/DC

TABLE 6-7

control accordingly. The maximum current may be set higher than the current required to melt the base metal; this allows the welder to have a hot start. A hot start is exactly as the name suggests: The welder starts the weld at a high current, which facilitates a fast start.

When a remote current control is not used, the welding amperage has to be more accurate.

START METHOD   **TABLE 6-7** shows recommended start settings for DCEN and AC welding.

POSTFLOW   Postflow time should be set at 1 second for every 10 amps of welding current. If the tungsten turns blue or black after welding, postflow time is insufficient.

PREFLOW   A setting of 0.2 seconds of preflow time is sufficient in most cases. Increased time may be needed when increased purging of weld area is needed.

BALANCE   Balance (AC only) should be set at its highest setting available. If floating black spots appear in the puddle and the addition of filler is difficult, turn the control down until the spots disappear and the addition of filler metal becomes easier. A balance setting that is set too low causes excessive melting of the tungsten tip. The melting is apparent as balling or crowning and will cause arc wandering.

WELDING FREQUENCY   Welding frequency (AC only) is available only on inverter power sources. As the frequency is increased, the width of the welding arc is narrowed. There is no set rule for operation. Required puddle size, depth of penetration, and personal preference are often the deciding factors.

TUNGSTEN TYPE   **TABLE 6-8** shows the recommended tungsten for DCEN and AC used with each power source type and tip preparation.

	Magnetic Amplifier		Solid State		Inverter
Polarity	AC	DC	AC	DC	AC/DC
Tungsten	Pure Zirconiated	Ceriated Lanthaniated Thoriated	Pure Zirconiated	Ceriated Lanthaniated Thoriated	Ceriated Lanthaniated Thoriated
Tip Preparation	Balled	Sharpened	Balled	Sharpened	Sharpened (DC) Sharpened and blunted (AC)

TABLE 6-8

1 Select a power source (AC/DC inverter).

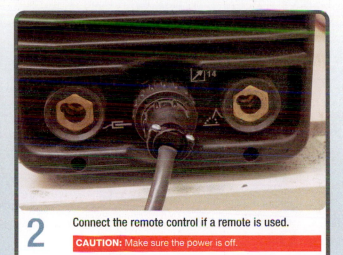

2 Connect the remote control if a remote is used.

CAUTION: Make sure the power is off.

3a Connect the work lead to the power source.

3b **CAUTION:** Inspect electrical cables (leads). Never use damaged electrical cables.

4 Clamp the work lead to the table.

5 Connect the torch lead to the power source.

6 Connect the torch gas line to the power source.

7 Prepare the tungsten (see the section on tungsten to determine the best shape for the polarity and type of power source being used).

8 Select the filler for welding.

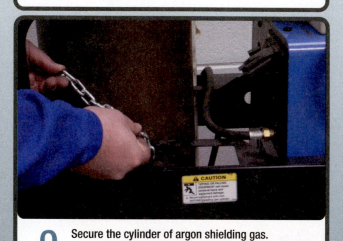

9 Secure the cylinder of argon shielding gas.

CAUTION: Never remove the cylinder cap until the cylinder has been secured with a chain. The cylinder must be chained to a cart or structure that is big enough or designed to support the weight of the cylinder.

10 Crack the cylinder valve.

CAUTION: Always crack or momentarily open the valve of a compressed gas cylinder before attaching the regulator. Make sure the valve is not pointed toward yourself or other individuals. Cracking the cylinder valve clears foreign material form the valve seat. Foreign material may cause damage to the valve or regulator connections, which can cause leaking.

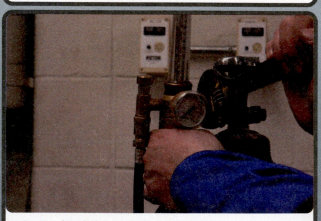

11 Attach the regulator and gas hose.

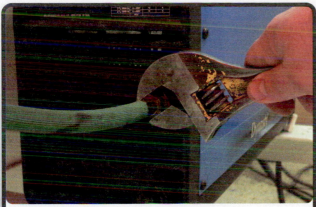

12 Attach the gas hose to the power source.

13 Install tungsten, collet, and collet body into the front of the torch.

> **CAUTION:** Do not overtighten the collet body. Overtightening can cause damage to the collet body or the torch body.

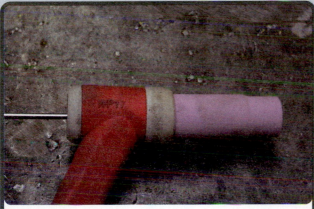

14 Install the cup.

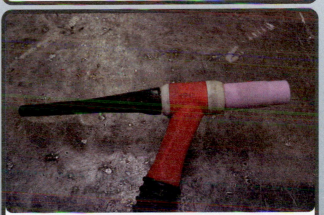

15a Install the back cap. Before tightening down, set the tungsten stick-out. If the tungsten falls out, the wrong size collet or collet body is being used or the collet body is not tight (do not install the cup on the collet body before installing it on the torch body).

15b Check for correct tungsten stick-out and to ensure the tungsten is secure.

16a Connect the power source to the primary power supply.

16b Turn on the power source.

17 Open the argon cylinder valve and set the flow rate. On most power sources, remote control has to be triggered to start the gas flow.

CAUTION: Stand to one side of the regulator gauge. The tungsten electrode will be electrically live. Do not trisser unless torch is secure and safe.

18 Determine which polarity is needed. Remember that DCEN is used for all metals except aluminum or magnesium. Aluminum and magnesium require AC. Consult the owner's manual if unsure how to set the correct polarity. For DCEN welding, set the welding amperage, gas postflow time, start method, and pulser setting (if used).

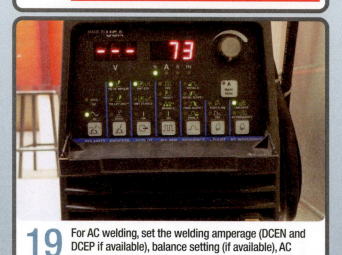

19 For AC welding, set the welding amperage (DCEN and DCEP if available), balance setting (if available), AC frequency (if available), gas postflow time, start method, and pulser setting (if used).

V I D E O

20 Put on your personal protective equipment before welding.

V I D E O

There are a number of variables over which the welder has control after the arc is initiated. Each of these variables has a direct effect on the final appearance and strength of the finished weld:

- Torch Angle
- Tungsten to Work Distance (Arc length)
- Travel Speed
- Filler Addition
- Electrode Oscillation

TORCH ANGLE   The torch angle used generally ranges from 90° to 45°, with the torch pointed toward the direction of travel (forehand or push technique). A backhand or pull technique in which the tungsten points away from the direction of travel is not recommended. At 90°, the resulting puddle is round and directly underneath the cup. As the angle is decreased the puddle begins to become oblong-shaped, with the puddle spreading out in front of the cup. Extreme torch angle will allow atmosphere into the shielding gas, causing contamination of the weld. **FIGURE 6-37** shows the recommended torch angle and **TABLE 6-9** shows results of a 90° and 45° torch angle.

VIDEO

TRAVEL ANGLE 5° TO 15°

FIGURE 6-37 The forehand technique is the correct way to weld with the GTAW process.

TUNGSTEN TO WORK DISTANCE (ARC LENGTH)   Short tungsten to work distance (arc length) of 1/16 inch to 1/8 inch (1.5 mm to 3 mm) allows the welder to maintain more control over the weld puddle and the addition of filler metal. When the arc length increases, the puddle begins to widen and becomes harder to control. This loss of control is caused by the increase of arc voltage as the arc length increases. When arc voltage is allowed to increase, the puddle begins to widen.

Having a short tungsten to work distance is a problem when an inexperienced welder is learning the GTAW process. Maintaining a short arc length without touching the tungsten to the base metal or puddle is difficult.

VIDEO

TRAVEL SPEED   Unlike the other common arc welding processes, travel speed is not controlled by the melting rate of the electrode. The welder

	90°	45°
Penetration	Deeper	Shallower
Width of Puddle	Wide	Narrow
Shielding Gas Coverage	Round pattern directly over puddle. Diameter of shielding pattern depends on diameter of cup.	Oblong pattern providing shielding ahead and directly over puddle; poor post-shielding

TABLE 6-9

controls when and how much filler is added to the weld pool—if filler is added at all. Often, travel speed is determined by the specifications of the finished weld. For example, on a lap joint on 1/8-inch (3 mm) aluminum, the customer requires a 1/8-inch (3 mm) fillet weld with a finished profile of flat to convex. Using 3/32-inch (2.4 mm) filler rod, an experienced welder will travel at the speed required to produce a finished weld meeting the customer's requirements. The welder may even change travel speed while welding in order to compensate for the changing temperature of aluminum, traveling faster as the aluminum heats up.

The same welder needs to weld a T-joint on 16-gauge to 16-gauge stainless steel. The customer requires that the finished product be free of warpage and have a straw-colored weld bead finish. The welder understands that too high of a heat input will cause the product to warp and turn blue or black in color. In order to meet the customer's requirements, the welder needs to reduce the heat input by welding at a lower current level and increasing travel speed. Travel speed can be faster than the travel speed used to weld the aluminum.

VIDEO

In both examples, the welder uses different travel speeds based on the type of metal being welded, the metal thicknesses, and the customer's wishes.

FILLER ADDITION Similar to travel speed, the amount of filler added to the weld puddle depends on the requirements of the desired weld. Unlike the other common arc welding processes, the filler is added manually, and at times, no filler is added at all. For example, consider the previous 16-gauge stainless steel T-joint. In order to increase travel speed and reduce heat input, the welder would choose to use a smaller diameter filler rod, such as a 1/16-inch (1.6 mm) or 0.40-inch (1 mm). Because the smaller diameter filler rods melt at lower temperatures, the welding current can be kept at a lower amount and thus heat input is reduced.

When a GTAW weld is examined, there are two sources of heat: the welding arc and the molten puddle. When the filler rod is melted by the heat of the welding arc, the melted wire will form into a drop on the end of the filler rod. When the drop becomes big enough, it will fall from the end of the rod and drop into the molten puddle below. The combination of the large molten drop and the effect of it landing into the puddle will cause a loss of control over the finished weld. The end result is a finished weld that may be too wide or irregular in shape, and have spatter around the edges.

VIDEO

TRAVEL ANGLE 5° TO 15°

FIGURE 6-38 Adding aluminum filler while welding with the GTAW process

The preferred technique for adding filler rod is by adding the rod to the leading edge of the molten puddle. With this technique, the filler simply melts and flows into the puddle. This melting eliminates the negative effects of the filler dropping into the puddle. The recommended placement of the filler rod is shown in **FIGURE 6-38**.

Selecting the correct filler metal size is important. For example, a customer requires the finished fillet weld size of 3/32-inch (2.4 mm). The welder has to choose between a 1/16-inch (1.6 mm), 3/3-inch (2.4 mm) or 1/8-inch (3.2 mm) filler metal. The 1/8-inch rod would be too large and would cause the finished weld to be too big. The 3/32-inch rod would seem to be the correct choice, but here is where personal choice comes into play. One welder may feel comfortable using the 3/32-inch rod and be able to control addition so that the weld fits the required dimensions. Another welder may feel that the 3/32-inch rod adds too much fill and thus chooses the 1/16-inch filler metal. As long as the end result is the same, both welders are correct in their choice.

In another example, a fabricator needs to make a control box out of 1/8-inch (3.2 mm) mild steel. The box will be used to display controls and to house wiring. Because of the lack of stresses, the welder determines that the outside corner joints that make up a majority of the weld will be done without the use of filler. An **autogenous** or fusion weld is a weld finished by melting a portion of metal from one or each of the base metals. The welder then manipulates the molten puddle to cause the melted metal to blend together. In this example, as the two corners are melted, the welder oscillates the torch back and forth, blending the two independent puddles together.

Again both examples show that the welder makes a decision regarding the diameter and the use of filler based on the intended end result.

ELECTRODE OSCILLATION

Choosing to oscillate or weave the torch is another example of understanding what is needed to meet the finished weld requirements. The majority of GTAW welds are done with no weaving, and the torch is moved in the

VIDEO

VIDEO

VIDEO

FIGURE 6-39 "Walking the cup" is a welding technique used in GTAW to provide a uniformed and consistent weld pattern.

V I D E O

forward direction only. But there are situations in which the use of weaving can produce better results. For example, a welder welding stainless steel pipe used in the manufacture of heat exchangers is using the "walking the cup" technique **FIGURE 6-39**. This technique is done by rolling from side to side using the edge of the cup. This technique produces a very consistent weaved weld. It can be done with or without the use of filler rod.

Another use for torch oscillation is to fill gaps. For example, a maintenance welder needs to repair an aluminum shroud that is damaged. After reshaping the aluminum, the fit-up has gaps of 0.040 inches to 0.098 inches. In order to successfully repair the shroud, the welder will need to weave the torch from one side to the other. With patience and the addition of filler rod, the repair is done.

FILLER METAL

HOW IS FILLER WIRE FOR GTAW CATEGORIZED?   Filler metal is categorized in a number of different ways. They include:

- AWS Specification
- AWS Classification

Specifications for GTAW filler wire falls into different categories based on filler compositions, such as steel, aluminum, stainless steel, etc. The AWS specifications for GTAW filler metal are as follows:

Steel	=	5.18
Stainless Steel	=	5.9
Aluminum	=	5.10
Copper Alloys	=	5.7
Magnesium	=	5.19

AWS Classification of Carbon Steel Filler Metal		
ER70S-X		
ER	=	electrode rod
70	=	minimum tensile strength x 1000
S	=	solid electrode wire
X	=	chemical composition and shielding

FIGURE 6-40 The AWS classification of steel filler metals

Classifications of filler metals differ in their compositions and are generally similar to that of the base metal. Alloying and deoxidizing elements are added to the filler metal to minimize porosity in the weld and to assure satisfactory weld metal mechanical properties. The most common deoxidizing elements used in carbon steel filler meta ls are manganese, silicon, and aluminum.

Filler metals for carbon steel GTAW fall into the AWS A5.18 specification. The AWS classification is ERXXS-X. A typical example would be ER70S-3 or ER70S-6.

Each letter and number of a specific filler metal classification has significant meaning. **FIGURE 6-40** conveys these meanings.

Low-alloy carbon steel filler metal has an additional suffix. Another common filler metal classification may have an additional suffix. The filler metal ER80S-D2 contains additional alloying elements nickel and molybdenum. **TABLE 6-10** shows the major alloying elements.

WHAT ARE THE INDIVIDUAL FILLER METAL CHARACTERISTICS OF CARBON STEEL FILLER METAL FOR GTAW?

ER70S-2 This classification covers filler metals that contain small amounts of titanium, zirconium, and aluminum in addition to the normal deoxidizing elements of manganese and silicon. These wires are commonly referred to as triple-deoxidized wires. They produce sound welds in all types of carbon or mild steels. They are especially suited for welding carbon steels that are rusty or have mill scale on the surface. The ERXXS-2 filler metals work well when avoiding porosity and when encountering high sulfur contents in base metals. Weld integrity will vary with the amount of oxides on the surface of the steel. An application for the ER80S-D2 is in the use of filler metal on chrome-moly steels, such as 4130, when preheat and postheat treatments are not feasible.

AWS Classification	Carbon	Manganese	Silicon	Titanium	Zirconium	Aluminum
E70S-2	0.07	0.90–1.40	0.40–0.70	0.50–0.15	0.02–0.12	0.05–0.15
E70S-6	0.07–0.15	1.40–1.85	0.80–1.15	—	—	—

TABLE 6-10

Major Alloying Elements–% By Weight (AWS Specifications)

ER308L		
ER	=	electrode rod
308	=	electrode composition
L	=	additional requirements
		change in original alloy

FIGURE 6-41 The AWS classification of stainless steel filler metals

ER70S-6 Wires in this classification contain the highest combination deoxidizers in the form of silicon and manganese. They produce smooth, well-shaped beads and are particularly well-suited for welding sheet metal. It is the best choice for when welding on rusted scaled steels or even oily surfaces. The weld quality depends on the degree of surface impurities.

WHAT IS THE SPECIFICATION/CLASSIFICATION FOR STAINLESS STEEL?

The AWS specification for solid stainless steel electrode wire is AWS A5.9. An example of the AWS classification is ER308L. **FIGURE 6-41** conveys this meaning. **TABLE 6-11** shows AWS chemical composition requirements of solid stainless steel electrodes.

Additional requirements that may be added to the classification are:

- L = Lower carbon content; generally no more than 0.03% carbon content
- H = Limited to the upper range on the carbon content
- Mo = Molybdenum added; has pitting resistance; creep strength increased
- Ni = Nickel added; high temperature strength; corrosion resistance and added ductility

WHAT ARE THE INDIVIDUAL FILLER METAL CHARACTERISTICS FOR STAINLESS STEEL?

ER308 This electrode filler wire is a general welding filler metal and is used for welding stainless steels of similar composition. It contains 19%

AWS Classification	C	Cr	Ni	Mo	Mn	Si	P	Cu
ER308	0.08 Max	19.5–22.0	9.0–11.0	0.75 Max	1.0–2.5	0.30–0.65	0.03 Max	0.75 Max
ER308H	0.04–0.08	19.5–22.0	9.0–11.0	0.50 Max	1.0–2.5	0.20–0.65	0.03 Max	0.75 Max
ER308L	0.03 Max	19.5–22.0	9.0–11.0	0.75 Max	1.0–2.5	0.30–0.65	0.03 Max	0.75 Max
ER308LSi	0.03 Max	19.5–22.0	9.0–11.0	0.75 Max	1.0–2.5	0.65–1.00	0.03 Max	0.75 Max
ER309	0.12 Max	23.0–25.0	12.0–14.0	0.75 Max	1.0–2.5	0.30–0.65	0.03 Max	0.75 Max
ER309L	0.03 Max	23.0–25.0	12.0–14.0	0.75 Max	1.0–2.5	0.20–0.65	0.03 Max	0.75 Max
ER309LSi	0.03 Max	23.0–25.0	12.0–14.0	0.75 Max	1.0–2.5	0.65–1.00	0.03 Max	0.75 Max
ER316	0.08 Max	18.0–20.0	11.0–14.0	2.00–3.00	1.0–2.5	0.30–0.65	0.03 Max	0.75 Max
ER316H	0.04–0.08	18.0–20.0	11.0–14.0	2.00–3.00	1.0–2.5	0.30–0.65	0.03 Max	0.75 Max
ER316L	0.03 Max	18.0–20.0	11.0–14.0	2.00–3.00	1.0–2.5	0.30–0.60	0.03 Max	0.75 Max
ER316LSi	0.08 Max	18.0–20.0	11.0–14.0	2.00–3.00	1.0–2.5	0.65–1.00	0.03 Max	0.75 Max
ER347	0.08 Max	19.0–21.5	9.0–11.0	0.75 Max	1.0–2.5	0.30–0.65	0.03 Max	0.75 Max
ER410	0.12 Max	11.5–13.5	0.60 Max	0.75 Max	0.60 Max	0.50 Max	0.03 Max	0.75 Max

TABLE 6-11

AWS Chemical Composition Requirements of Solid Stainless Steel Electrode

chromium and 9% nickel. It is one of the most common stainless filler metals used for a variety of applications.

ER308L   This filler metal is preferred for use on the same materials as the ER308. The chromium and nickel content is the same as the ER308 but with a lower carbon content, as designated by the L (0.03% versus 0.08%). Lower carbon content reduces the possibility of carbide precipitation and the intergranular corrosion that can occur.

ER308LS   This filler metal not only has low carbon content but also has higher silicon content than the previous two designated by the letter S (0.65–1.00% versus 0.25–0.60%). A higher silicon level improves the wetting characteristics of the weld metal. However, if the dilution of the base metal is extensive, the higher silicon can cause greater crack sensitivity.

ER309   This filler metal is used for welding a comparable base metal (309). This is one of the most versatile stainless filler metals and is recommended for welding stainless steel to mild steel and stainless steel to low-alloy steels and to join heat-treatable stainless steels when heat treatment is not possible.

ER309L   Similar to the above filler metal but has only 0.03% maximum carbon content. The lower carbon content increases resistance to inter-granular corrosion.

ER310   Most frequently used to weld base metals of similar composition. It is also recommended in many applications where ER309 is also recommended.

ER316   This filler metal is used for welding 316 and 319 stainless steels. It contains additional amounts of molybdenum for high temperature service and for increased pitting corrosion resistance. It is also a general welding filler metal used similar to ER308 but is more costly.

ER316L   This filler metal has lower carbon content than the previous example, as designated by the letter L (0.03% versus 0.08%). It is less susceptible to inter-granular corrosion caused by carbide precipitation.

ER316LHS   This filler metal not only has a lower carbon content than the ER316 filler metal but also has a high silicon content, as designated by the HS. This filler metal produces a more fluid puddle but is more susceptible to cracking.

ER347   This filler metal has less problems with carbide precipitation because niobium (columbium) or niobium along with traces of tantalum is added as a stabilizer. It is used for welding similar metals where high temperature strength is required.

ER410   Used for welding alloys of similar composition. Also used for overlays on carbon steels to resist corrosion, erosion, or abrasion. Usually requires preheat and postheat treatments.

WHAT IS SPECIAL ABOUT ALUMINUM FILLER METAL WIRE?

Unlike steel and stainless steel filler metal, composition of the electrode for aluminum is generally *never* similar to that of the base metal. Aluminum welding is basically an alloying welding process. Using the same filler metal as the base metal often will lead to cracks. Using a filler metal specifically formulated to match the base metal is essential to sound, quality welds.

Solid filler metal for aluminum GTAW fall into the AWS A5.10 specification. The AWS classification is ERXXXX. There are many different classifications of aluminum filler metal.

WHAT ARE THE MORE COMMON FILLER METALS USED TO WELD ALUMINUM AND THEIR CHARACTERISTICS?

The two most common aluminum filler metals are ER4043 and ER5356. Some companies even prefer the ER4047, which is very similar to the ER4043 but with higher silicon content that improves puddle fluidity. If you study the complete base metal/filler metal comparison chart found in Chapter 10, you will find that the vast majority of recommended filler metal is either the ER4043 or ER5356 filler metal.

ER4043 This filler metal is used for general-purpose welding. Its main alloying element is silicon, ranging from 4.5–6%. The addition of silicon lowers the melting temperature of the filler and promotes a free-flowing puddle. ER4043 wires have good ductility and high resistance to cracking during welding. This filler metal is recommended when welding the cast aluminums and base metals.

ER5356 This filler metal has improved tensile strength due to the major alloying element of magnesium, 4.5–5.5%. The ER5356 has good ductility but only average resistance to cracking during welding. This filler metal is often used as a substitute for ER4043 on 6061T-X.

Welding metals prior to a secondary process, such as welding aluminum prior to anodizing, bears the consideration of color matching. Here, the filler metal selected may anodize poorly or anodize a different color than the base metal. Matching the filler metal becomes critical. Using a filler metal that results in good coloring but results in weld metal cracking is almost never acceptable. Using the example above of 6061, we see that the recommended filler rod would be ER4043 for best cracking resistance. However, for the best color match, the ER5654 may be the best choice, or the ER5356, ER5183, or ER5556 electrodes, depending on the color match desired.

SHIELDING GASES

Shielding gas choices for GTAW are few compared to GMAW. Argon is most commonly used, followed by helium and hydrogen additions to argon.

One of the greatest strengths of the GTAW process is that a majority of welding of all metals can be performed with argon as the only shielding gas.

Argon is a heavier than air inert gas that does a great job of protecting the molten weld pool, filler metal tip, and electrode from the oxidizing atmosphere. The typical flow rate is from 15–35 cubic feet per hour (cfh) (7–16 liters per minute (L/min)). Argon is often used as the backing gas on full penetration welds.

When greater heat input is required, such as when welding metals with higher thermal conductivity (aluminum and magnesium) or when higher travel speeds are required, pure helium or a mixture of argon/helium may be used. Helium has a higher ionization point than argon; thus, the welding voltage of a helium weld is higher, resulting in greater heat input when compared to the same welding parameters of an argon-shielded weld. The drawbacks of using helium are:

- Higher cost
- Greater flow rates required: 30–50 cubic feet per hour (cfh) (14–23 liters per minute (L/min))
- Inconsistent arc starting and arc instability when welding with alternating current

In special circumstances, a small percentage of hydrogen will be added to the argon. The addition of hydrogen results in greater heat input and a cleaner weld than an argon-shielded weld. The percentage of hydrogen mixed to argon may range from 1–25%, with the lower percentages being the most common. Hydrogen in high percentages may result in porosity.

ELECTRODES

Tungsten electrodes are categorized under the AWS A5.12 specification for tungsten and tungsten alloy electrodes for arc welding and cutting. Two examples of the AWS classification are shown in **FIGURE 6-42**.

TABLE 6-12 shows several different types of tungsten available and information regarding the selection of the correct tungsten.

EWP		
EW	=	electrode welding
P	=	pure tungsten*
EWTh-2		
EW	=	electrode welding
Th	=	type of metal oxide alloy
2	=	percentage of alloy

*P designates a tungsten electrode consisting of pure tungsten only. This designation for all other tungsten electrodes will indicate the type of metal oxide alloyed with pure tungsten.

FIGURE 6-42 The AWS classification of pure and 2% thoriated tungsten electrode

Composition	Color Code	AWS Classification	Polarity	Type Of Power Source	Tungsten Preparation
Pure	Green	EWP	AC	Mag Amp Solid State	Balled
Zirconiated	Brown	EWZr-1	AC	Mag Amp Solid State	Balled
1% Thoriated	Yellow	EWTh-1	DC	Mag Amp Solid State	Sharpened
2% Thoriated	Red	EWTh-2	AC/DC	Inverter	Sharpened DC Sharpened/Blunt AC
			DC	Mag Amp Solid State	Sharpened
3% Thoriated	Blue	EWTh-3	AC/DC	Inverter	Sharpened DC Sharpened/Blunt AC
			DC	Mag Amp Solid State	Sharpened
2% Ceriated	Orange	EWCe-2	AC/DC	Inverter	Sharpened DC Sharpened/Blunt AC
			DC	Mag Amp Solid State	Sharpened
1% Lanthanated	Black	EWLa-1	AC/DC	Inverter	Sharpened DC Sharpened/Blunt AC
			DC	Mag Amp Solid State	Sharpened
			AC/DC	Inverter	Sharpened DC Sharpened/Blunt AC
Unspecified	Gray	EWG	Proprietary tungsten oxide not specified		

TABLE 6-12

CHAPTER QUESTIONS/ASSIGNMENTS

MULTIPLE CHOICE

1. What is another common name for GTAW?
   a. Gas Arc Welding (GAW)
   b. Manual Filler Welding (MFW)
   c. Tungsten Inert Gas (TIG)
   d. Tungsten Electrode Welding (TEW)

2. What is the suggested lens shade for 50 amps?
   a. 8
   b. 13
   c. 10
   d. 11

3. Which of the other common arc welding processes can a GTAW power source be used to weld?
   a. Gas Metal Arc Welding (GMAW)
   b. Flux Cored Arc Welding (FCAW)
   c. Resistance Arc Welding (RAW)
   d. Shielding Metal Arc Welding (SMAW)

4. The air-cooled welding torch uses what to remove heat?
   a. Carbon dioxide shielding gas
   b. Cooled water
   c. Argon shielding gas
   d. 75% argon/25% carbon dioxide shielding gas mixture

5. Which is not a component of a GTAW torch?
   a. Cup
   b. Liner
   c. Torch body
   d. Collet body

6. Why are there three different standard-sized back caps available?
   a. Balances out the weight of the torch
   b. Provides a more streamlined look
   c. Fits different-sized tungsten electrodes
   d. Large cap allows a full-length tungsten to be used, and smaller caps allow the torch to fit into restricted spaces

7. Why is a remote hand control used instead of a foot control?
   a. Does not restrict the welder's movement and position
   b. Allows welder to use torch for cutting
   c. Heavier than a foot control
   d. Allows the welder to have finer control amperage control than the foot control

8. How can you tell what type of tungsten electrode you are using?
   a. The color of the welding arc when being used
   b. The diameter
   c. The color of the painted band on the end
   d. The smell while welding

9. Argon shielding gas is an asphyxiant. Why is that a cause for concern?
   a. Higher cost than other gases
   b. Greater flow rates are required to use

c. Argon is heavier than air; thus, a welder needs to be careful when welding in confined spaces with limited amounts of fresh air.
d. Cylinder has to be used at temperatures above freezing

10. Filler metal used for GTAW is often the same filler used for GMAW except that GTAW filler metal is generally available in what form?
   a. 36-inch (94.1 cm) length
   b. 1-pound spools
   c. Stackable wafers
   d. Perforated sheets

11. What is the electrode positive portion of an AC GTAW welding arc used for?
   a. Melting the base metal
   b. Sharpening the tungsten
   c. Removing aluminum oxide from the aluminum base metal only
   d. Cleaning oil and grease off the aluminum base metal

FILL IN THE BLANK

12. The tungsten electrode is a _____ electrode that is used to conduct welding current.

13. A constant _____ power source is the type used for GTAW.

14. The _____ of a water-cooled torch is used to conduct welding current and is cooled by the flow of coolant.

15. The collet and collet body size must be the same as the tungsten's _____.

16. GTAW is different from the other common arc welding processes in that it allows the welder to control _____ while welding.

17. A rule of thumb for setting _____ time is 1 second for every 10 amps of welding current.

18. Adjustable balance control allows the welder to decrease the heat on the tungsten, thus reducing the amount the tungsten tip _____.

19. _____ is the process in which positive ions are released from the electrode and strike the oxides on the surface of the aluminum. The loose oxides are then lifted away by the current flow from the aluminum to tungsten.

20. A welder may choose to use _____ to eliminate the possible damage to sensitive electronic equipment caused by high frequency.

21. Adjustable AC frequency gives the welder more control over the _____ of the molten puddle and _____ penetration.

22. One of the benefits of pulsing is an average_____.

23. A _____ technique is recommended for the angle of travel.

24. The _____ technique is done by rolling from side to side using the edge of the cup.

SHORT ANSWER

25. What does GTAW stand for?

26. What are the two most common types of GTAW torches?

27. What are the two functions of a remote foot pedal?

28. What color of band is found on the end of a pure 2% thoriated and 2% ceriated tungsten?

29. When is pure and zirconiated tungsten used?

30. Which variable does the power source control with the GTAW process?

31. What polarity is used to weld steel and stainless steel?

32. When would reverse polarity be used?

33. What does the balance control do?

34. When would continuous high frequency be used?

35. What are the advantages of using the GTAW pulsing?

Chapter 7

OTHER WELDING PROCESSES

LEARNING OBJECTIVES

In this chapter, the reader will learn how to:

1. Incorporate safe practices when using any welding process.
2. Understand the limitations, benefits, and economics of each welding process.
3. Understand how the Oxygen-Fuel Welding processes function.
4. Determine when Oxygen-Fuel Welding processes would be best used in a welding environment.
5. Understand how the Plasma Arc Welding process functions.
6. Determine when the Plasma Arc Welding process would be best used in a welding environment.
7. Understand how the Resistance Welding process functions.
8. Determine when the Resistance Welding process would be best used in a welding environment.
9. Understand how the Stud Welding process functions.
10. Determine when the Stud Welding process would be best used in a welding environment.
11. Understand how the Submerged Arc Welding process functions.
12. Determine when the Submerged Arc Welding process would be best used in a welding environment.
13. Understand how the Electroslag and Electrogas Welding processes function.
14. Determine when the Electroslag and Electrogas Welding processes would be best used in a welding environment.
15. Understand how each of the Solid State Welding processes function.
16. Determine when each of the Solid State Welding processes would be best used in a welding environment.

KEY TERMS

Flashback 220	Oxidizing Flame 227
Neutral Flame 223	Diffusion 259
Carburizing Flame 223	Asperities 259

INTRODUCTION

The majority of the welding done in today's welding and metal fabrication shops is done with the four main arc welding processes: Gas Metal Arc Welding, Flux Cored Arc Welding, Shielded Metal Arc Welding, and Gas Tungsten Arc Welding. These processes and the equipment that is used to perform them are continuously being improved, pushing the limits of quality, craftsmanship, and production. Even with all their improvements, these processes are not always the best ones for every application. Other welding processes exist that have unique benefits that allow them to stand out and to fill niches that the four main welding processes cannot fill.

These processes can range from inexpensive to expensive, simple to complex, and manual to fully automated. A majority of the processes rely on high temperature melting to produce a weld, while others are able to join metal at room temperature. Each may be different, but what they have in common is that in the right application, each excels. When applied in the right situation and used correctly, each can prove to be productive and profitable.

The purpose of this chapter is to present an introduction and a description of each process. Some processes, such as Oxygen-Fuel, Plasma Arc, Resistance, Stud, and Submerged Arc, are covered in greater detail. This is based on their more common usage in industry, welding schools, and training centers. The other processes are important but not as common and are covered by a general overview.

OXYGEN-FUEL WELDING

Oxygen-Fuel Welding (OFW) (also known as oxy-fuel or oxy-acetylene) uses the heat produced by the burning of a fuel gas, enhanced by the addition of pure oxygen, to melt the base metal. The weld can then be finished by blending base metal pools together or by manually adding filler metal. The size of the puddle and the depth of penetration are determined by the diameter of the welding tip, the volume of gas flow, the welder's manipulation of the puddle if filler metal is added, and the type and thickness of metal.

PROCESS **FIGURE 7-1** shows an oxy-acetylene torch being used to weld mild steel.

As the name implies, the Oxygen-Fuel Welding flame consists of two main components: oxygen and fuel gas. Because acetylene is the most popular fuel gas used, it will be the fuel gas discussed in this portion of the chapter.

The oxygen and acetylene are stored in two separate cylinders. Each gas is supplied to the torch through individual regulators, flashback arrestors, and hoses. At the torch handle, there is an oxygen and acetylene valve. When the valves are opened, the acetylene and the oxygen are mixed together in the mixing chamber. When the system is operating correctly, the flame exists at the outside of the welding tip only.

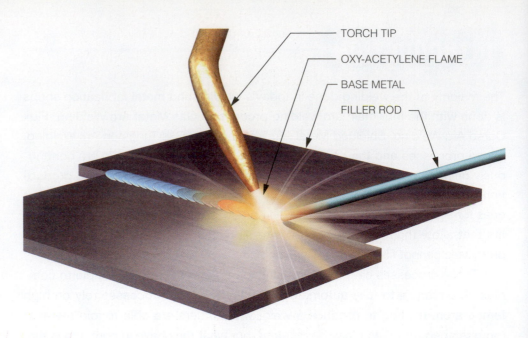

TORCH TIP

OXY-ACETYLENE FLAME

BASE METAL

FILLER ROD

FIGURE 7-1 Oxy-acetylene being used to weld mild steel

A flame is started by opening the acetylene valve and igniting the acetylene gas at the tip. The welder then opens the oxygen valve, and through a series of adjustments to both valves, a neutral flame is established.

The temperature of an oxy-acetylene neutral flame can reach 6,000°F (3,315°C), which is high enough to melt most common metals, but the best results for this process are achieved when welding carbon steels. Because carbon steel is the most common metal welded with this process, it will be the metal used for this discussion.

With a neutral flame established, the welder moves the torch tip down so the tip of the neutral flame is about 1/8 inch (3.18 mm) away from the base metal. The heat produced by the flame will cause the base metal to begin to melt. At this point, the welder can begin to manipulate the torch in order to spread out the heat and control the size of the puddle and the depth of penetration.

How the welder moves or what patterns are used to manipulate the torch are usually determined by the training the welder received and personal experience. When the correct size puddle is reached, the weld can be finished with or without the addition of a filler metal. Filler metal comes in a 36-inch (86.4 mm) rod similar to GTAW. Just like any welding process, the ability of a welder to look at a molten puddle and to correctly adjust travel speed, flame to puddle distance, degree of manipulation, and amount of filler addition is crucial to producing a sound weld. The benefits of this process are:

- **Simple to use:** Training time needed to learn the correct flame adjustment and welding technique is short.
- **Lower cost:** Equipment is a fraction of the cost of other common welding processes.
- **Portable:** There is no need for electrical power to operate the system. The system can be stored on the back of a utility vehicle and can be exposed to the elements. This makes this system perfect for repair and fabrication work in the field.
- **Weld in all positions:** The only position limitation for this process is due to the skill level of the operator.

- **Versatile:** It can be used for welding, cutting, and heat treating.
- **Thickness range:** The process is most efficient on metal ranging from 0.059 inch to 3/16 inch (1.51 to 4.76 mm). It works well on thin sheet, often providing better results than other common welding processes.
- **Visibility of molten weld puddle:** The molten weld puddle is very visible and is not obscured by a protective coating of flux or slag. When welding carbon steel, the color of puddle can be used to judge the temperature of the steel.
- **Introduction for beginner welders:** It is often used to introduce welding to new welders. It can be used to teach welders how to correctly judge a weld puddle for the correct size, shape, and temperature. Also used to develop the hand and eye coordination required for GTAW, at a lower cost (equipment, tungsten, and argon shielding gas).

The drawbacks of the process are:

- **Wider heat-affected zone (HAZ):** The extra time needed to form a weld puddle and the slower travel speed cause a larger area to be heated. This causes problems when welding medium- and high-carbon steel and other metals, negatively affecting the metallurgical and mechanical characteristics of a finished weldment.
- **No inert shielding:** When the mixture of oxygen and acetylene is correct, carbon dioxide is produced, which shields the molten weld puddle from the atmosphere. CO_2 shielding works well for carbon steels but causes weld defects in most other metals.
- **Thickness range:** This process is most efficient on metal ranging from 0.059 inch to 3/16 inch (1.51 mm to 4.76 mm). Even with a larger tip diameter and increased gas volume, performance on thicknesses greater than 3/16 inch (4.76 mm) becomes too slow to be economical.
- **Visibility of molten weld puddle:** Aluminum does not change color when melting. When trying to weld aluminum, welders find it very difficult to see and judge the molten aluminum puddle.
- **Increased safety risk:** Extra precautions need to be taken to ensure that the fuel gases and oxygen are handled and stored properly. Because the system can be very portable, extra maintenance and frequent inspection of equipment are recommended.

COMPONENTS   A typical oxygen-acetylene outfit set-up, (**FIGURE 7-2**) is made up of different components.

OXYGEN CYLINDER   Oxygen is stored under high pressure, generally around 2,200 psi (15.2 MPa). Safety precautions similar to other high-pressure welding cylinders should be followed. Unless in use, oxygen cylinders should never be stored closer than 20 feet (0.76 m) to any fuel gas. **FIGURE 7-3** shows a picture of an oxygen cylinder.

OXYGEN REGULATOR   Oxygen regulators (**FIGURE 7-4**) are designed to regulate oxygen only. The cylinder and regulator have fittings that will not fit other cylinders or regulators. Never try to adapt any other regulator to use on an oxygen cylinder.

ACETYLENE CYLINDER   Acetylene is an unstable gas when stored or used at pressures above 15 psi (0.10 MPa). The pressure of a full acetylene cylinder is 250 psi (1.72 MPa). In order to make the higher cylinder pressure safe, the

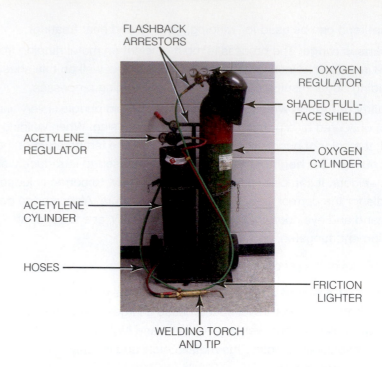

FLASHBACK
ARRESTORS

OXYGEN
REGULATOR

SHADED FULL-
FACE SHIELD

ACETYLENE
REGULATOR

OXYGEN
CYLINDER

ACETYLENE
CYLINDER

HOSES

FRICTION
LIGHTER

WELDING TORCH
AND TIP

FIGURE 7-2 A typical oxygen-acetylene outfit setup

FIGURE 7-3 Oxygen cylinder

FIGURE 7-4 Oxygen regulator

FIGURE 7-5 Acetylene cylinder

cylinder contains a fibrous filler and liquid acetone. The unstable acetylene is then stabilized by being contained in the acetone and filler. **FIGURE 7-5** shows a picture of an acetylene cylinder.

Because liquid acetone is used, two special precautions must be followed when using these cylinders. First, an acetylene cylinder should always be stored and used in an upright position. If a cylinder is laid down, the liquid acetone will flow into the valve. When the cylinder is used, the liquid acetone will flow into the regulator, hose, and torch. This will cause the flame to be erratic and negatively affect the torch's performance. Second, there is a limit to the amount of acetylene that can be drawn out of a cylinder in a given amount of time. If the withdrawal rate is too high, liquid acetone can be pulled into the valve. The maximum withdrawal rate is determined by the size of the cylinder (see the torch manufacturer's specifications). If higher withdrawal rates are required, a larger cylinder must be used or several cylinders will need to be used on a manifold system.

The manifold system allows multiple cylinders to be used at one time. This is done by attaching a regulator to each cylinder. The output gas from each regulator is then coupled into a larger diameter gas line. This allows each cylinder to supply a safe amount of gas while not surpassing its withdrawal limitations. When done correctly, each cylinder should supply an equal amount of gas with total volume equaling the required amount.

Acetylene cylinders should never be exposed to high temperatures or flames. Acetylene cylinders are designed with a fusible plug. The plug is intended to melt when external temperatures are too high, thus preventing internal pressures from reaching dangerous levels. Never use hot or boiling water to thaw out a frozen or ice-covered cylinder; this may cause the plug to melt. It is best to thaw them by placing them in a heated room.

ACETYLENE REGULATOR   Most but not all acetylene regulators and cylinder valves have left-hand threads, which are indicated by notches. The left-handed threads are intended to only allow the correct regulator to be used on a specific fuel gas cylinder. Every fuel gas regulator is designed for use with a specific fuel gas and size of tank (see the torch manufacturer's specifications). Never adapt a regulator to be used on another fuel gas or different size cylinder. **FIGURE 7-6** shows a picture of an acetylene regulator (note the notches).

NOTCHES

FIGURE 7-6 Acetylene regulator

CHECK VALVES AND FLASHBACK ARRESTORS   When operating correctly, the flame of a cutting torch will be located only on the outside of the cutting tip. There are problems that can occur that will cause the flame to move into the tip, torch, hoses, or, worst case, the cylinders. This is called a **flashback**. The extremely high risk of personal injury or property damage demands that steps must be taken and devices must be used to prevent a flashback from occurring.

The devices designed to help prevent a flashback are the check valve and the flashback arrestor. Both devices are used as a pair, with one used on the oxygen line and the other used on the fuel gas line. They can be attached between the torch and hose or between the hose and regulator. Some cutting torches manufactured today have a single flashback arrestor designed into the cutting tip.

Check valves (**FIGURE 7-7**) are designed to only allow gas to flow in one direction. Gas is allowed to flow from the regulator to the torch but cannot flow back in the other direction. A check valve is not designed to prevent a flame from traveling backward through the hose.

Foreign material can become lodged in the check valve, causing it to malfunction. It is recommended that the valves be checked periodically to ensure that they still operate properly.

Flashback arrestors (**FIGURE 7-8**) are similar to check valves in that they prevent the reverse flow of gas. As the name implies, a flashback arrestor has the added feature of being able to stop a flame from traveling backward down the hose

OXYGEN CHECK VALVE RIGHT-HAND FITTINGS

ACETYLENE CHECK VALVE NOTCHES INDICATE LEFT-HAND FITTINGS

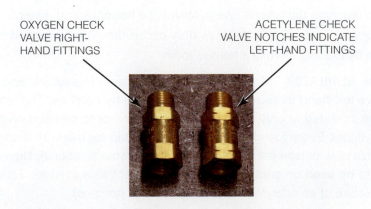

FIGURE 7-7 Oxygen and acetylene check valves

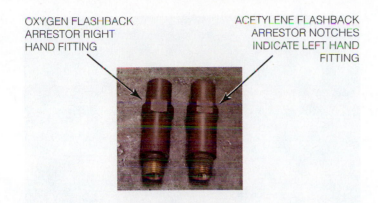

OXYGEN FLASHBACK ARRESTOR RIGHT HAND FITTING

ACETYLENE FLASHBACK ARRESTOR NOTCHES INDICATE LEFT HAND FITTING

FIGURE 7-8 Oxygen and acetylene flashback arrestors

toward the cylinder. This is done by placing a stainless steel mesh screen in the arrestor's body. The mesh screen will stop a flame that tries going through it.

The mesh screen does have the negative effect of reducing gas flow. Flow rates may need to be increased to compensate. The flashback arrestors should also be checked periodically to ensure that they are still functioning correctly.

HOSES In most cases, the hoses used consist of two individual hoses bonded together. The oxygen hose is generally green in color, with right-hand threads. The fuel gas hose is always red in color and will have left-hand threads. A hose with an R grade is used with acetylene only. All other fuel gases must use a T-grade hose. **FIGURE 7-9** shows a picture of hoses.

Inside diameter size of hoses range from 3/16 inch (4.75 mm) to 3/8 inch (9.52 mm). The smaller diameter hose has the benefit of weighing less and being more flexible than larger diameter hoses. When short lengths of hose or less volume of gas are required, the smaller hose is the best choice for welder comfort. When longer lengths or increased gas volume are required, the hose diameter must be increased. Too small of a hose diameter can result in pressure drops that negatively affect the performance of the torch.

Often, welders will combine smaller and larger diameters hoses. This is done by installing larger diameter hoses from the regulator and then coupling smaller diameter hoses to the torch. This gives the welder the benefit of less weight and more flexibility at the torch and less pressure drop throughout the line, compared to only using a smaller hose diameter. The total pressure drop of the line using this technique is still higher than if larger diameter hoses were used. If larger volumes of gas are required, the larger hose may need to be used throughout.

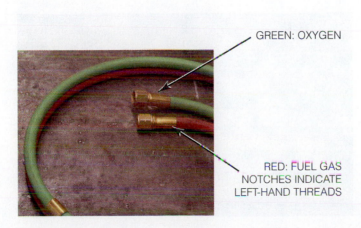

GREEN: OXYGEN

RED: FUEL GAS NOTCHES INDICATE LEFT-HAND THREADS

FIGURE 7-9 Oxygen and acetylene hoses

FIGURE 7-10 Oxy-Fuel Welding torch

TORCH HANDLE   The universal or combination handle has a connection and valve for the oxygen and the fuel at one end. On the other end, there is a threaded female fitting that allows the use of different electrode tips and torches. Oxy-Fuel systems can be used for cutting, welding, and heating. The universal torch handle allows the welder to switch between these different operations without having to disconnect the gas lines. This is a huge benefit in a shop where all three processes are used. The drawback of the universal handle is that there are more connections; thus, there is always an increased chance of gas leaking, compared to a standard torch. **FIGURE 7-10** shows a picture of a torch handle.

WELDING TIP   The size and type of welding tip **FIGURE 7-11** are determined by the thickness of the base metal. As the base metal thickness increases, the flow rate of the oxygen and fuel gas must increase to provide greater heat input in order to raise the increased mass to its melting temperature. To achieve the increased flow rate, a larger diameter tip is required.

FRICTION LIGHTER   Friction lighters **FIGURE 7-12** are the safest way to light an Oxy-Fuel torch. It is never safe to light a torch with a match, cigarette lighter, or another torch.

FILLER METAL   Filler metal for Oxy-Fuel Welding is sold in a bare rod form, similar to GTAW. The rods are available in 1/16 inch (1.58 mm) to 1/4 inch (6.35 mm) diameters and 36 inches (914.4 mm) in length.

FIGURE 7-11 Oxy-Fuel Welding tip

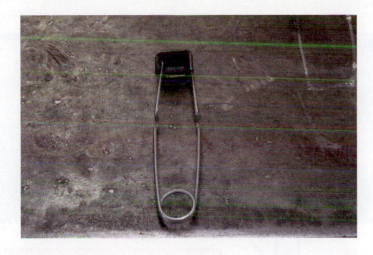

FIGURE 7-12 Friction lighter

The American Welding Society classification is based on the basis of strength. For example, a common rod used for welding mild steel is a RG45. The R stands for rod. The G represents the chemical composition of the filler metal for gas welding. The 45 equals the minimum welded tensile strength of 45,000 psi (310 MPa). Filler metal used to weld carbon steels is listed under AWS specification A5.2. **TABLE 7-1** shows the common carbon steel filler rods and the minimum tensile strengths.

FLAME TYPES There are three types of flames involved in Oxy-Fuel Welding: neutral, carburizing, and oxidizing.

Neutral flame: The neutral flame **FIGURE 7-13** is produced when an equal balance is found between the oxygen and acetylene. It provides the best results when Oxy-Fuel Welding. The operator can recognize a neutral flame by:

- The lengths of the inner cone and the outer cone are almost equal, with the inner cone recessed by 1/16 inch to 1/8 inch from the outer cone.
- The outer cone should have a relatively smooth shape.
- A slight hissing sound should be heard.

Carburizing flame: The carbonizing flame (**FIGURE 7-14**) consists of an unbalanced mix of oxygen and acetylene, with an excess amount of acetylene. The operator can recognize a carburizing flame by:

V I D E O

- The inner cone is considerably smaller than the outer cone.
- The outer cone has a rough, feathery appearance.

AWS Classification	Minimum Tensile Strength
RG45	45,000 psi (310 MPa)
RG60	60,000 psi (413.7 MPa)
RG65	65,000 psi (448.2 MPa)
RG100	100,000 psi (689.5 MPa)

TABLE 7-1

1 Secure the cylinders.

CAUTION: Only use chain to secure cylinders.

2 Crack the oxygen valve.

CAUTION: Always "crack" or briefly open the valve of a compressed gas cylinder before attaching the regulator. Make sure the valve is not pointed toward yourself or other individuals. Cracking the cylinder valve clears foreign material from the valve seat. Foreign material may cause damage to valve or regulator connections, which can cause leaking.

3 Attach the oxygen regulator.

CAUTION: Never use any petroleum-based lubricant on any pure oxygen fitting, hose, or regulator. Pure oxygen will reduce the kindling temperature of these lubricants to room temperature.

4 Crack the acetylene valve and then attach the acetylene regulator.

5a Attach a flashback arrestor onto the oxygen regulator.

5b Attach a flashback arrestor onto the acetylene regulator

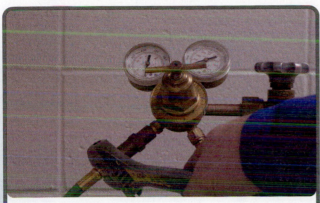

6a Attach the oxygen hose.

6b Attach the acetylene hose.

7 Attach the torch.

8 Turn out the **flow pressure adjustment screw** and then open the oxygen valve.

> **CAUTION:** When first opening an oxygen cylinder valve, always turn the valve 1/4 turn, wait several seconds, and then open it all the way. This step allows the initial in-rush of pressure to the regulator to stabilize. It is also important to stand to the side of the regulator behind the valve. If the valve were to explode, the valve may block some of the debris.

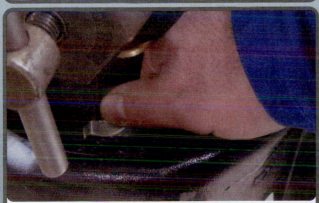

9 Turn out the flow pressure adjustment screw and then open the acetylene valve.

> **CAUTION:** Only turn the valve 1/2 to 3/4 turn. This will allow the quick shutting of the valve in case of an emergency. If a wrench or T handle is needed to operate a valve, leave the tool in place to also allow the quick shutting of the valve.

10a Set the oxygen and acetylene operating pressures (always consult the manufacturer's recommended pressure settings).

> **CAUTION:** Never use acetylene at pressures greater then 15 psi (103 kPa). Never exceed the withdrawal limits of the acetylene cylinder.

10b Set Acetylene operating pressure (always consult manufacturer's recommended pressure settings)

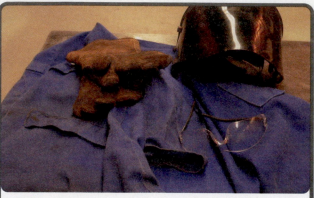

11 Put on your personal protective equipment.

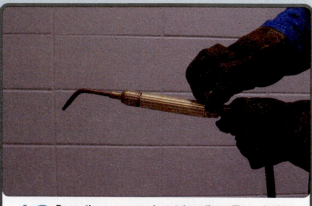

12 Purge the oxygen and acetylene lines. Then, double-check the pressure settings.

> **CAUTION:** Purging both lines will eliminate any mixed gases from the torch. Gases can become mixed if the torch was incorrectly shut down the last time it was used.

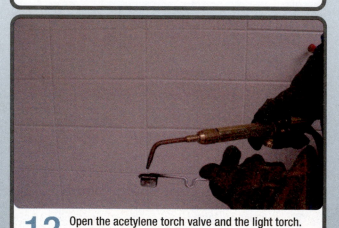

13 Open the acetylene torch valve and the light torch.

> **CAUTION:** Only use friction-style lighters to light a torch. Never use a lighter, matches, or another torch to light the torch.

14 Open the torch oxygen valve. Adjust the acetylene and oxygen valve until a neutral flame is achieved.

V I D E O

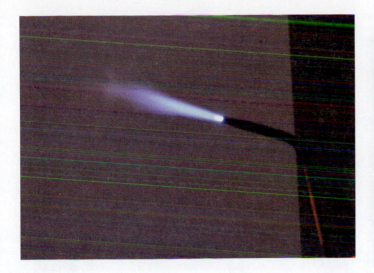

FIGURE 7-13 Oxy-Fuel neutral welding flame

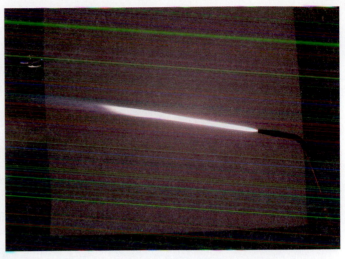

FIGURE 7-14 Oxy-Fuel carburizing welding flame

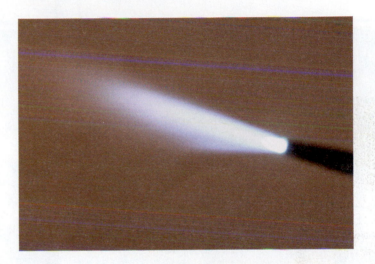

FIGURE 7-15 Oxy-Fuel oxidizing welding flame

Oxidizing flame: An oxidizing flame (**FIGURE 7-15**) consists of an unbalanced mix of oxygen and acetylene, with an excess amount of oxygen. The operator can recognize an oxidizing flame by:

- A loud hissing sound should be heard.
- It should have a sharp inner cone.

TORCH SHUTDOWN SETUP

As mentioned earlier, the Oxy-Fuel system is a very versatile and simple to use tool. Unfortunately, its simplicity often causes the dangers of its use to be taken for granted and forgotten. Too often, welders and operators will finish using the torch, shut off the torch valves, and walk away.

It is very important to properly shut down an oxy-fuel system. Failure to do so can result in damage to the system and poses a safety hazard. The following steps should be taken when finished using any oxy-fuel system. Consult the torch manufacturer's literature for specific torch shutdown procedures.

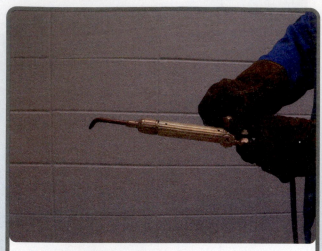

1 Turn off the torch oxygen valve first (consult the manufacturer's recommendations).

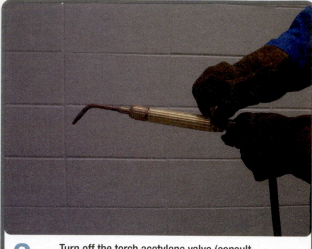

2 Turn off the torch acetylene valve (consult manufactures recommendations).

3a Close the oxygen cylinder valve.

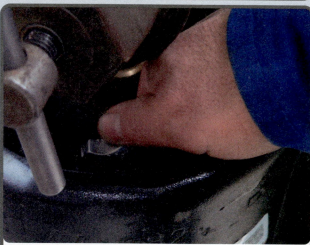

3b Close the acetylene cylinder valve.

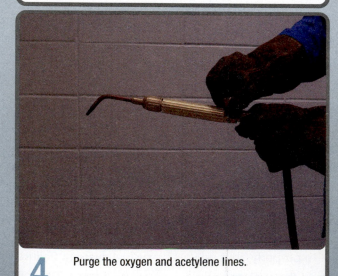

4 Purge the oxygen and acetylene lines.

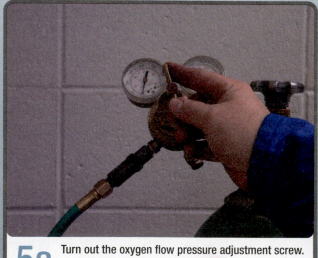

5a Turn out the oxygen flow pressure adjustment screw.

VIDEO

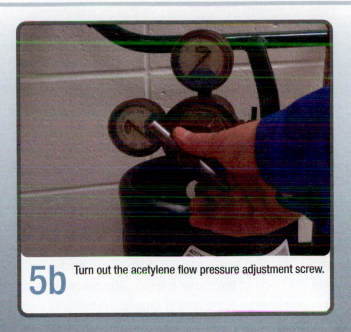

5b Turn out the acetylene flow pressure adjustment screw.

SAFETY PRECAUTIONS

- Wear eye and ear protection. Light intensity of the cutting flame requires the use of a welding helmet with a lens shade of 5 to 6.
- Use ventilation: Fumes and vapors produced by the process should always be removed from the area. Extra care should be taken when working with metal covered with paint, coatings, or cladding. Individual tolerances and safety regulations may also warrant the use of breathing protection.
- Always know where sparks are landing. Always remove combustible and flammable materials from the welding area.
- Never weld on piping or vessels that contain flammable materials.
- Never weld a vessel that has held flammable materials. All vessels that have held flammable materials contain vapors from that material. If welding must be done, proper procedures must be followed.
- Never use any petroleum oil or grease to lubricate any part of an oxy-fuel system. Pure oxygen lowers the kindling temperature of any petroleum product, causing a fire.
- Never use pure oxygen to blow off cloths and other combustibles. Pure oxygen will lower the material's kindling temperature and cause the material to ignite easier.
- Never operate acetylene at pressures greater then 15 psi (103 kPa), and never exceed the withdrawal limitations of the acetylene cylinder.

PLASMA ARC WELDING

Plasma Arc Welding (PAW) is a process that uses electrical current flow to cause a localized area of the base metal to melt. Shielding from atmospheric gases is provided by an inert shielding gas. The weld can then be finished by blending base metal pools together or by adding filler metal. The size of the puddle and the depth of penetration are determined by the welding current level, manipulation of the puddle, filler metal addition (if used), and type and thickness of metal.

PROCESS   The Plasma Arc Welding process (**FIGURE 7-16**) is a variation of the GTAW process. Both processes use an electrical arc to conduct current through a base metal. The heat of the plasma column and the electrical resistance of the base metal produces heat, which in turn causes the base metal to melt. The welding arc and molten puddle are shielded from the atmosphere by a column of inert shielding gas. When the correct size puddle is reached, the weld can be finished with or without the addition of a filler metal.

The two processes differ in the design of their torches and how the electrical arc is conducted. GTAW conducts electrical current from the end of an exposed tungsten electrode. Plasma Arc Welding also uses a tungsten electrode, but the electrode is concealed inside the body of the plasma torch. The tungsten is covered by a cylinder-shaped constricting nozzle and is closed on the end except for a small opening or orifice.

When the welder closes the on switch (in a hand control, foot pedal, or control panel), an inert shielding gas flows through a space between the tungsten and nozzle and then out of the orifice. Next, an electrical arc jumps from the negative electrode to the positive tip. As the arc travels through the inert gas stream, the gas is heated to around 30,000°F (16,000°C) and is ionized. Matter in this hot ionized state is called plasma. Plasma is the fourth state of matter, along with solid, liquid, and gas.

When gas is turned into plasma, its volume expands. This expansion of gas volume forces the plasma stream out of the small orifice of the tip. This narrow, high velocity stream of hot ionized gas shoots out of the orifice. The stream has an electrical charge, so it will bend back toward the oppositely charged tip. This arc is called the pilot arc.

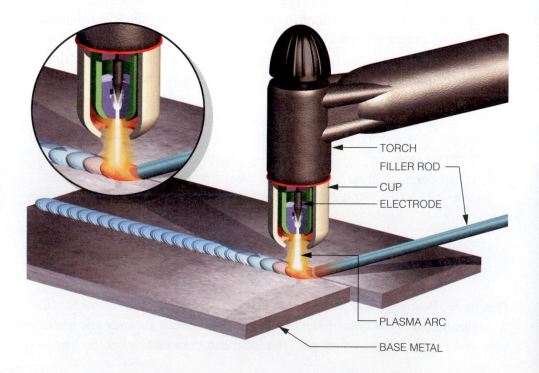

TORCH
FILLER ROD
CUP
ELECTRODE
PLASMA ARC
BASE METAL

FIGURE 7-16 Plasma Arc Welding

When the welder brings the torch tip down close to the positive base metal, the pilot arc will transfer to the base metal. At this point, the power source senses that current flow is being conducted through the base metal and will increase the amperage.

Another nozzle is located over the constricting nozzle. This nozzle is designed to direct another column of inert shielding gas. This column is used to provide protection from atmospheric gases.

The process is most often used in mechanized or automatic systems but is used by welders for handheld applications. When used manually, the process and welding techniques are very similar to GTAW. Control of welding current is done by using a remote hand control or foot pedal.

The benefits of this process are:

- **Concealed tungsten:** Concealing the tungsten inside the torch body eliminates damage to the tungsten by touching the base metal and by the filler metal. Can be used to build confidence and teach torch movement and filler addition to new welders before they move on to GTAW.

- **Reduced down time:** Less damage to tungsten reduces the amount of time needed to change damaged tungsten. Also results in a cost savings from replacing tungsten.

- **Very tight control of puddle size and location:** The constricting action of the nozzle orifice allows welders to precisely control the weld. It is often used for microscopic and low-amperage welding applications.

- **Reduced heat-affected zone (HAZ):** The area affected by heat is minimized by the degree of control.

- **Weld in all positions:** The only position limitation for this process is due to the skill level of the operator.

- **Versatile:** Can be used to weld any metal welded with the GTAW process. DCEN is used for the majority of all metals. It can even be used to weld aluminum but does require the aluminum to be extremely clean.

- **Stand off:** The distance from the puddle to the torch can vary greatly without adversely affecting the puddle shape.

- **Visibility of molten weld puddle:** The molten weld puddle is very visible and is not obscured by a protective coating of flux or slag.

- **Requires only one type of shielding gas:** Inert argon gas can be used when welding any base metal. When increased heat input is desired, helium can be mixed with argon or used alone. The use of helium is more expensive and will require increased flow rates.

- **High-frequency start:** This allows an arc to be started without touching the base metal.

- **Pulsing:** Welding output can be pulsed. It further enhances the control of the weld puddle and the HAZ.

The drawbacks of the process are:

- **Cost of system:** Plasma arc power sources and torches are more expensive than the comparable GTAW power source and torch.

- **Torch:** Not as robust as the lower-cost GTAW torch. Extra care must be taken to protect the torch in a welding environment.

- **Requires a water circulator:** All Plasma Arc torches are water-cooled. Failure to keep cooled water flowing through the torch will damage the torch.
- **Thickness range:** Process is most efficient on metal ranging from edges and surfaces requiring a microscope to welds up to 1/8 inch (3.18 mm).
- **AC welding output:** The process can be used to weld aluminum by using an AC welding output, but the increased heat of the polarity causes the tip of the tungsten to ball or crown. This change to the tungsten negatively affects the arc and may damage the constricting nozzle.
- **Must use high frequency:** Unlike newer GTAW power sources that allow the use of touch start or lift arc to start the arc, Plasma Arc must use high frequency. This can cause problems when welding around sensitive electronic equipment.

COMPONENTS   A Plasma Arc Welding system (**FIGURE 7-17**) consists of a power source, controller, torch, water circulator, remote control (optional), work cable/clamp, shielding gas, tungsten electrode, and filler wire.

POWER SOURCE   A constant current DC power source is most commonly used. A constant current AC/DC power source can be used but is not as common, with the majority of aluminum and magnesium being welded with the GTAW process. **FIGURE 7-18** shows a Plasma Arc Welding power source.

Welding current range and duty cycle should always be considered when choosing a power source. If the power source is to be used for low-current welding, extra consideration should be taken to ensure that the required welding current is within the power source's low range. Another consideration is the ability to control welding current in increments smaller than 1 amp. What this means is that the power source can make adjustments to the welding output in fractions of an amp.

CONTROLLER   The controller is either an independent component or part of the power source. When independent, the controller is the interface between the remote control and power source. The work cable, torch, and water lines from the water circulator are attached to the controller.

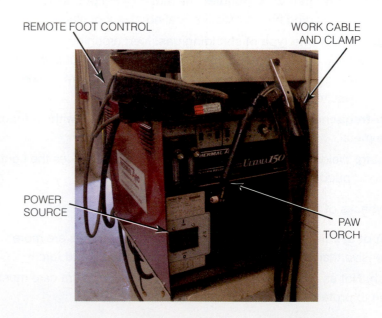

REMOTE FOOT CONTROL

WORK CABLE AND CLAMP

POWER SOURCE

PAW TORCH

FIGURE 7-17 Plasma Arc Welding power source and components

FIGURE 7-18 Plasma Arc Welding power source

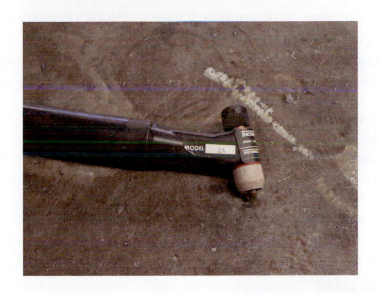

FIGURE 7-19 Plasma Arc Welding torch

TORCH The torch is very similar to GTAW torches, with the main difference being the tungsten concealed within the torch head. All torches need to have water circulating through the head and around the weld cable in order to remove heat. **FIGURE 7-19** shows a cutout of a manual Plasma Arc Welding torch.

The torch is by far the weakest link in a plasma welding system. Extra care should be taken to protect the torch from wear and tear found in a welding environment. Current range and duty cycle should always be considered when purchasing a torch.

WATER CIRCULATOR Conducting welding current produces heat within the head of the welding torch and welding cable. If allowed to overheat, the torch will be damaged. To prevent the torch and cable from being damaged, water is circulated through the weld cable and torch head.

The pumping of the water is done by the water circulator (**FIGURE 7-20**). Water circulators vary in size, with reservoirs ranging from 1 to 5 gallons (3.79 to 18.95 liters). The circulator pump is used to circulate water from the reservoir through the welding cable line to the welding torch. Water is then returned to the circulator. The hot water returns to the circulator and flows through a radiator,

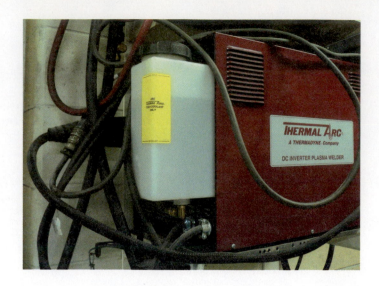

FIGURE 7-20 Plasma Arc Welding water circulator

which cools the water. Most circulators operate on 115 VAC, but there are models that operate on higher primary voltages.

REMOTE CONTROL   Another similarity between plasma welding and GTAW is the ability to remotely control the welding current. When manually welding, the remote can be in the form of a foot or hand control. For mechanized or automated systems, the remote may be a panel-mounted potentiometer or an analog signal from an automated controller. **FIGURE 7-21** shows a remote foot control.

Remote current controls are a slave to the master power source. What this means is that the power source controls the range in which the remote control operates. For example, if the current control on the power source was set to 100 amps and it had a low end range of 5 amps, the remote control would be able to control welding current from 5 to 100 amps. This type of control allows the welder to have finer control of the welding output.

WORK LEAD/CLAMP   Similar to any other arc welding process, the work lead's diameter must be large enough to handle the required current for that application. When the application is working on a larger structure, length of work lead must be taken into consideration. To compensate for the increased resistance of a longer lead, the diameter of the lead must be increased. An undersized work lead will

FIGURE 7-21 A Plasma Arc Welding foot remote control is identical in use and function to a GTAW root remote control.

negatively affect the performance of the weld. **FIGURE 7-22** shows a work cable and clamp. The work clamp is a very important part of the welding circuit. Its welding current rating must always be considered when setting up for a welding application.

SHIELDING GAS   The majority of all welding done uses pure argon as the shielding gas. Argon is inert, which means it does not react to the arc or molten weld puddle. Argon is the perfect shielding gas for the PAW process because it is inert and heavier than air. The benefit of the argon being heavier than air is that, when the argon flows out of the torch, it will fall down over the weld and push out all other atmospheric gases.

The drawback of a gas being heavier than air is that it is an asphyxiant. If inhaled an asphyxiant gas will cause oxygen to be pushed out of the lungs, causing a person to suffocate, resulting in injury or death. Care must always be taken when using argon in a confined space in which there is a limited amount of fresh air.

RESISTANCE WELDING

Resistance Welding (RW) is a group of processes that use a combination of electrical current flow and pressure to cause a localized area of base metal to melt. The degree of melting depends on the amount of current flow and pressure, time of weld, total thickness and type of base metal, and condition of base metal surfaces.

PROCESS   Resistance Welding is a group of welding processes that consists of but is not limited to Resistance Spot Welding (RSW), Projection Welding (PW), Resistance Seam Welding (RSEW), and Flash Welding (FW). All the processes differ in some ways but are the same in that they use a combination of electrical current flow and pressure to weld metals together. **FIGURE 7-23** shows a Resistance Spot Weld of mild steel.

The most commonly used process is the Resistance Spot Weld (RSW) and is the main focus of this section. Information regarding Projection, Seam, and Flash Welding are provided later in the section. **FIGURE 7-24** shows the tongs of a spot welder squeezing the sections of a mild steel lap joint together.

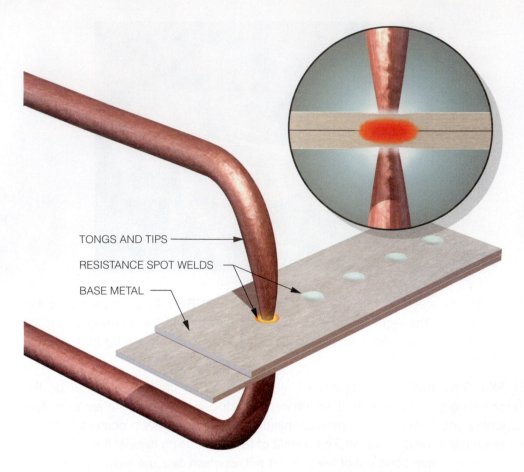

FIGURE 7-23 Resistance Spot Weld of mild steel

TONGS AND TIPS

RESISTANCE SPOT WELDS

BASE METAL

FIGURE 7-24 The amount of tong pressure is another important variable in Resistance Spot Welding.

Resistance Spot Welds are produced by squeezing the weld metal together by using a set of two electrodes or tongs. Because the metal to be welded must have a tight fit-up between surfaces, a lap joint is the most common joint to be welded. After the metals have been tightly squeezed together, AC welding current is conducted between the two tongs through the entire thickness of the pieces being welded. The interface between the metal pieces has the greatest resistance and begins to melt first. This makes a Resistance Spot Weld unique in that it truly melts from the inside out. When sufficient melting has been produced, the flow of welding current is stopped. The area of melted metal then solidifies, producing a

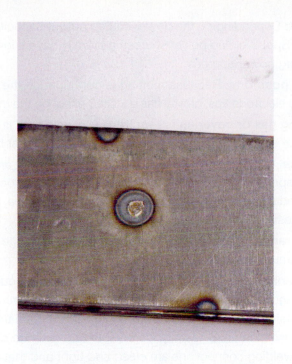

FIGURE 7-25 A spot weld nugget on mild steel

"nugget" of blended base metal. Pressure from the electrodes must remain constant before, during, and after the weld to ensure that no gap develops between the welded surfaces. Because no filler metal is added, metallurgical properties of the weld are determined by the combined properties of the weld admixture.

The process can be manual, with pressure and weld cycle controlled by the welder, or semiautomatic, with the start of the cycle controlled by the operator and the rest of the cycle controlled by the welding controller. Systems range from small handheld units used for sheet metal to fully automated robotic systems used in the automotive industry. The process is used in a broad range of industries such as HVAC, construction, fabricating, and automotive manufacturing. The size of the nugget (**FIGURE 7-25** shows a spot weld nugget) is determined by a combination of:

- **Amount of welding current:** The higher the current level is, the greater the amount of melting that is produced.
- **Pressure of tongs and fit-up of pieces:** A high-quality spot weld depends on a very tight fit between mating surfaces of the pieces to be welded. A small degree of poor fit-up can be corrected by the pressure of the tongs. Generally, tongs will be bent and deformed when trying to correct larger fit-up issues.
- **Diameter of electrode tips:** Even though the welding current is conducted through the entire area of the electrode tip diameter, the weld nugget's diameter is slightly smaller than the electrode's diameter.
- **Time of current flow:** The longer the current flows, the greater the amount of melting.

The benefits of this process are:

- **Simple to use:** Training time needed to learn the correct current, tong pressure, and welding time is short.
- **Speed:** Depending on the type and thickness of metal and power source output capacity, welds are relatively quick.

- **Lower cost:** Depending on the power source output capacity and options, Resistance Spot Welding power sources generally have a lower cost than other welding power sources.

- **Weld in all positions:** Because melting is contained in the area directly between the electrode tips, gravity has a limited effect on the molten weld pool.

- **No shielding gas or covering is required:** The weld pool is completely covered by the electrode tips; thus, exposure to the atmosphere is limited.

- **No electrical arc:** Eye and skin protection from ultraviolet arc rays is not required. Depending on the metal type and coating, the process can result in sparks and material being discharged from the weld area. Eye protection and flame-resistant clothing should always be worn.

- **Thickness range:** The maximum combined thickness range is determined by the maximum output capacity of the power source.

- **Low maintenance:** Power sources are very simple and require little maintenance. Any increase in resistance of the welding output circuit (secondary of welding transformer, tongs, and electrode tips) will reduce the welding current and cause increased heating. Care should be taken to ensure that all secondary welding connections are clean and tight and that electrode tips are in good condition.

The drawbacks of the process are:

- **Metal limitations:** Resistance Welding is most effective when welding metals that have a high resistance and low thermal conductivity, such as carbon steels. Metals such as aluminum and copper require higher current levels than for the same thickness of carbon steel with limited results.

- **Visibility of molten weld nugget:** The weld zone is completely covered by the electrode tips, and melting is at the interface, found between the pieces being welded. The operator cannot visually watch the development of the weld to determine if the welding parameters are correct. Developing the correct welding parameters is done by a sequence of producing and then destructively testing the weld.

- **Surface condition:** Any coating on the surface of the metal will have a negative effect on the weld and reduce the operating life of the electrode tips. Dip coatings or plating, such as zinc, can also produce fumes that must be vented.

- **Weld location:** The location of welds are limited by the length and shape of the tongs. The length of the tongs is limited by the capacity of the power source.

COMPONENTS

POWER SOURCE   The power source used for Resistance Spot Welding consists of several different circuits, the welding transformer, tongs, electrode tips, timer(s) and actuator. **FIGURES 7-26** and **7-27** show two different sizes of Resistance spot welders.

Power sources can range from simple handheld models to fully automated systems. All power sources are rated by their KVA (kilovolt amp) welding output capacity. They also have maximum welding amperage, OCV (open circuit voltage), and duty cycle ratings.

TRANSFORMER   Like other welding power sources, a Resistance Welding power source uses a transformer to change the incoming primary power from a high

FIGURE 7-26 Light weight, hand-held RSW models offer portability.
Source: Photo courtesy of Miller Electric Manufacturing Co.

voltage/low current to a low voltage/high current welding secondary. The transformer for a resistance welder differs in that the secondary voltage is very low, 1 to 5 volts open circuit voltage (OCV), and output current is very high. Even a small hand held 115V primary welder has a welding current of 5,000 amps.

Increases in resistance of any part of the secondary welding circuit (transformer, tongs, electrode tips, and base metal) will have a negative effect on the welding current. For example, a small spot welder that has an OCV of 2.5 volts will have a welding current of around 6,550 amps. This high welding current is possible because the total secondary resistance is 0.0004 ohms. If the secondary circuits resistance increases—because of a loose connection or an arc-damaged tip—to 0.005 ohms, the welding current will drop to 500 amps.

TIMER(S) Depending of the power source's technology level, the weld and squeeze timing are controlled by the welder or timer. An welder using a simple handheld spot welder will rely on counting or an external timer (**FIGURE 7-28**). This of course introduces a degree of error, which will show in the repeatability of welds.

As the technology level of power sources increase, so does the use of timers. Timers raise the cost of the power source, but they also increase the odds of repeating welds.

FIGURE 7-27 Heavy duty stationary RSW models offer durability and increased options such as an air filter/regulator and timer, amperage, and remote foot controls.
Source: Photo courtesy of Miller Electric Manufacturing Co.

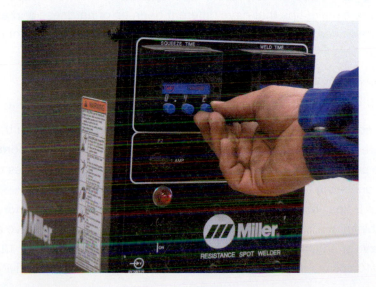

FIGURE 7-28 Setting a timer on a resistance spot welder

FIGURE 7-29 Tongs of the resistance spot welder

TONGS In order for a resistance spot weld nugget to be formed, the welding current must flow through the entire thickness of the base metal. To do this, the electrode tips must be squeezed against both sides of the base metal. Tongs serve the purpose of squeezing the tips and conducting the current. **FIGURE 7-29** shows welding tongs.

Tongs are generally made of solid copper alloy round stock. Being made of copper allows them to be low-resistance conductors of welding current. The diameter of the tongs, shaft has to be larger in order to apply pressure to both sides of the weld.

When spot welding, it is beneficial to place welds in the interior of the weldment. Their ensure their reach, manufactures have designed tongs of different lengths and shapes (to reach over obstacles and in corners). This extended reach and versatility comes at a price. The extra copper needed for the tongs is expensive, and with every increase in length, the resistance of the tongs increases. Longer and larger tongs lower the welding output of the spot welder.

ELECTRODE TIPS The electrode tip (**FIGURE 7-30**) used for spot welding must be able to withstand heat and pressure. The amount of heat and pressure can vary depending on the base metal thickness, type and surface condition, and amount of welding to be done. Replacing tips and downtime are expensive, so manufacturers make tips that have different degrees of hardness and annealing temperature values. This allows an operator to choose the correct tip for his or her application. Tips are divided into two different groups:

- Group A: Made of copper-based alloys
- Group B: Refractory metal

The most common and least expensive type is Group A. Within the group, the tips are further classified into I, II, III, IV, and V. Class I has the composition closest to pure copper. As the classification number increases, so does the hardness and annealing temperature.

Group B tips are a mixture of copper and tungsten or other refractory metals. This combination allows the tips to be more wear-resistant and to have a greater compressive strength. The group is further classified into Class 10, 11, 12, 13, and 14. Class 10 tips are the least resistant and have the greatest conductivity.

FIGURE 7-30 An electrode tip of a resistance spot welder

ACTUATOR   The actuator (**FIGURE 7-31**) moves and holds the tongs in the squeezing position. Depending on the technology level, the actuator action can be as simple as the operator manually squeezing a lever or a pneumatic or a hydraulic ram moving the tong. In most systems one tong is fixed and the other is hinged to allow movement.

Triggering the pneumatic or hydraulic ram is often done with a foot trigger or hand switch. After the initial trigger, the control of squeezing is done by the squeeze timer.

OTHER RESISTANCE WELDING PROCESSES

PROJECTION WELDING (PW)   Projection Welding is similar to Resistance Spot Welding in that the resistance at the weld interface causes localized heating to form one or more weld nuggets. Projection Welding differs from Resistance Spot Welding in that one or both pieces of the base metal are designed to have a projection protruding from its surface. When the base metals are squeezed together, the surface of the protrusion is pressed against the mating surface. Designing protrusions into the base metal eliminates weld location errors because the protrusion is the weld interface.

Another benefit of Projection Welding is that the protrusions can be different sizes and shapes. For example, a protrusion could be elongated or ring-shaped. This gives engineers and designers more flexibility and options. Tongs and electrode tips can also be specially designed to fit the different shape or size. With weld locations being set, systems can be designed that produce more than one

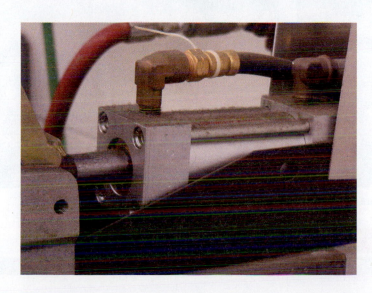

FIGURE 7-31 The actuator brings the tongs of the resistance spot welder together.

projection weld at a time. Power sources used for this process must have a higher KVA rating in order to meet the demands of larger tongs/electrode tips or performing several welds at one time.

RESISTANCE SEAM WELDING (RSEW)   Another version of the resistance spot weld is the seam weld. A seam can be produced with spot welding but is very impractical and time-consuming. The seam welding process allows a constant weld seam to be produced between base metals without stopping during the entire weld.

To allow a nonstop seam weld to be produced, electrode tips are modified into disc shapes. The disc or discs (systems may have a flat edge electrode or another disc on the opposite side) are squeezed against the base metal. As welding current is conducted through the disc edge, the metal is pulled, causing the disc to turn. As long as the pressure is constant, a seam will be produced along the entire length of travel. Power sources used for this process must have a higher KVA rating in order to meet the demands of the higher duty cycle of a constant weld.

FLASH WELDING (FW)   Flash Welding uses the interface of two metals to cause localized heating like other resistant welding processes. How it differs is that the use of tongs and electrode tips is eliminated. To produce a flash weld, the edges of two pieces of metal are pressed against each other and welding current is conducted through the joint interface. The interface has the highest resistance and will begin to melt. The variables of current, pressure, and time are crucial to producing a sound weld.

A common application is a band saw welder. Band saw blades are manufactured in long strips that are then cut to length. The ends of the band are then placed in a fixture that very tightly presses them together. Welding current is then conducted through the interface, causing it to melt and form a weld. Weld buildup is then ground off, and the band is ready for use. The power source KVA rating can vary based on the dimensions of bands.

RSW Setup

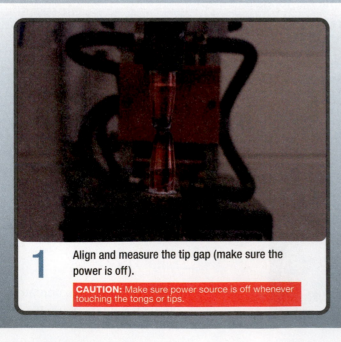

1 Align and measure the tip gap (make sure the power is off).

CAUTION: Make sure power source is off whenever touching the tongs or tips.

2 Clean the tips and/or recondition the tips (make sure the power is off).

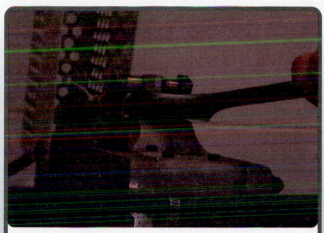

3 Measure the thickness of the base metal and then set the gap.

4 Check the pressure (make sure the power is off).

CAUTION: Place metal between closed tongs. The tongs should bend just slightly.

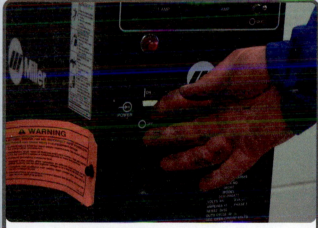

5 Turn on the power.

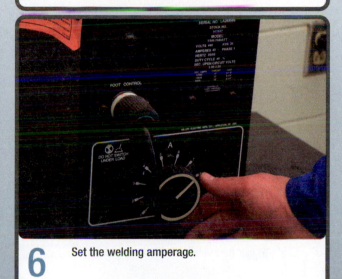

6 Set the welding amperage.

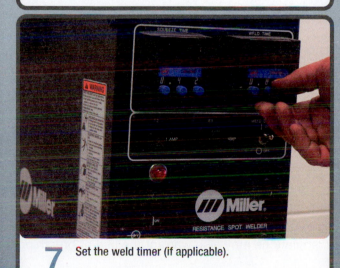

7 Set the weld timer (if applicable).

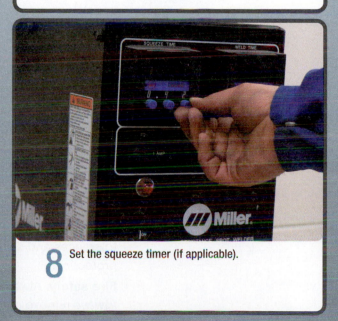

8 Set the squeeze timer (if applicable).

9 Turn on the water circulator (if applicable).

10 Put on your personal protective equipment before welding.

11 Trigger the switch.

SAFETY PRECAUTIONS

- **Wear eye protection:** Shaded eye protection is generally not required for this process because there is not an open electrical arc. It is very important that either safety glasses or a face shield is always used. It is common for molten metal or sparks to be expelled from the weld.

- **Use ventilation:** Fumes and vapors produced by the process should always be removed from the area. Extra care should be taken when working with metal covered with paint, coatings, or cladding. Individual tolerances and safety regulations may also warrant the use of breathing protection.

- **Fire safety:** Always know where sparks and molten metal are landing. Always remove combustible and flammable materials from the welding area.

- **Electrical safety:** Always be aware that the tongs and electrode tips are noninsulated electrical conductors. Never come in contact with any part of the welding circuit during the weld cycle without the proper personal protection and safety training. Never under any circumstance use the process when standing in water or when wet.
- **Maintenance:** When performing maintenance on the secondary circuit, always ensure that insulated spacers and covers are in place. Manufacturers of resistance welding power sources use the insulators to isolate the secondary circuit from the frame and case of the power source. Failure to reinstall insulators can make the frame or case of the power source electrically live during the welding cycle.

STUD WELDING

Stud Welding (SW) is a process that uses a combination of electrical current flow and pressure to cause a localized area of the base metal and a metal fastener (electrode) to melt. The degree of melting depends on the amount of current flow and pressure, time of weld, total thickness and type of base metal, diameter and shape of metal fastener, and condition of metal surfaces.

PROCESS   A stud weld is produced when electrical current is conducted through the metal fastener (electrode) and the base metal (**FIGURE 7-32** shows a stud being welded). Resistance at the interface causes a localized portion of each to melt. Pressure is then applied to the fastener, forcing it into the molten weld puddle. When the flow of welding current is stopped, pressure remains on

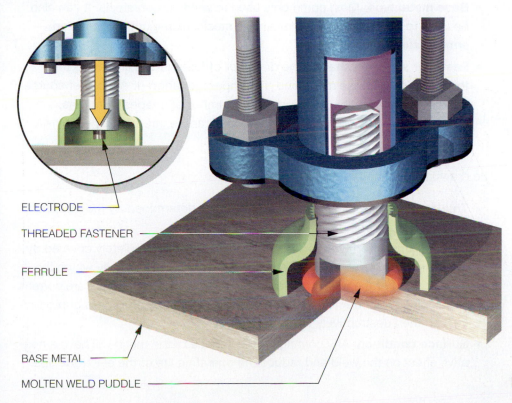

ELECTRODE

THREADED FASTENER

FERRULE

BASE METAL

MOLTEN WELD PUDDLE

FIGURE 7-32 Stud Welding

the fastener, keeping it in the solidifying puddle. Because no filler metal is added, metallurgical properties of the weld are determined by the combined properties of the weld admixture.

The molten puddle may or may not be shielded from the atmosphere. When shielding is required, it is done by the use of shielding gas, flux, or disposable ferrules. Ferrules are also used to contain and shape the molten puddle.

Stud Welding is generally semiautomatic, with the start of the cycle controlled by the operator and the rest of the cycle controlled by the welding controller. Systems range from small low-amperage units used for sheet metal to high-amperage units made for attaching bolts to larger structures. Control can also be fully automated and controlled by robotic systems used in the automotive industry. The process is used in industries including construction ranging from construction, fabricating, automotive, ship building, and military manufacturing.

The benefits of this process are:

- **Simple to use:** Training time needed to learn the correct current, pressure, and welding time is short.
- **Speed:** Depending on the type and thickness of metal and power source output capacity, welds average 1 second. The loading of fasteners can be automated by increasing the cycle time.
- **Weld in all positions:** Because melting is localized and quick, gravity has a limited effect on the molten weld pool. Ferrules can be use to hold the molten puddle in place.
- **No electrical arc:** Eye and skin protection from ultraviolet arc rays is generally not required. The metal type and coating process can result in sparks and material being discharged from the weld area. Eye protection and flame-resistant clothing should always be worn.
- **Base metal type:** Most commonly used to weld carbon steels. It can also be used to weld low-alloy and stainless steels, aluminum, copper, and armor plating.
- **Fastener size and shape:** The diameter of fasteners is limited to the thickness of the base metal and the amperage range of the power source. Fasteners are available in an unlimited number of shapes from pins, hooks, eye bolts, and brackets. Fasteners can be internally or externally threaded.

The drawbacks of the process are:

- **Thickness range:** The maximum thickness is determined by the maximum output capacity of the power source.
- **Visibility of molten weld puddle:** The weld zone is completely covered by the fastener or ferrule when used. The operator cannot visually watch the development of the weld to determine if the welding parameters are correct. Developing the correct welding parameters is done by a process of producing and then destructively testing the weld.
- **Surface condition:** Any coating on the surface of the metal will have a negative effect on the weld and reduce the operating life of the electrode tips.

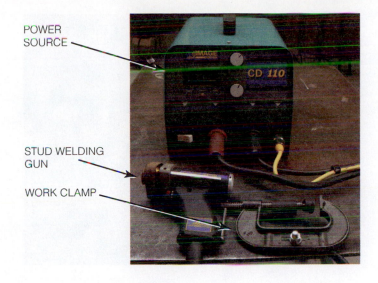

POWER SOURCE

STUD WELDING GUN

WORK CLAMP

FIGURE 7-33 Power source and components of a stud welder

Dip coatings or plating, such as galvanized and zinc, can also produce fumes that must be vented.

- **Cost:** All systems must consist of a stud weld gun, controller, and power source. The duty cycle and amperage range of the welding gun and power source must be high enough for the required fastener diameter and base metal thickness.

COMPONENTS   A Stud Welding System (**FIGURE 7-33**) consists of a power source, stud welding gun, controller, work cable/clamp, metal fastener, and ferrule.

POWER SOURCE   There are two types of power sources: the direct current (DC) constant current (CC) and the capacitor discharge. The polarity of the weld is DCEN. The power sources' main function is to provide the welding current. Power sources can be operated off of primary voltages from 120 VAC to 575 VAC single or three-phase. There are even engine-driven generator models available for construction at remote locations. Each power source is rated by its welding output range and duty cycle. **FIGURE 7-34** shows a Stud Welding power source.

FIGURE 7-34 The stud welder power source

FIGURE 7-35 Stud welding gun

STUD WELDING GUN The stud welding gun (**FIGURE 7-35**) is the heart of a Stud Welding System. When the trigger is squeezed, a solenoid slightly retracts the fastener away from the base metal. Welding current then flows through the gun, with the fastener holding the chuck and fastener. After melting of the base metal and the fastener has started, a spring or air piston pushes the fastener into the weld pool.

The chuck is used to hold the fastener during the entire welding process. The size and shape of the chuck can limit the shape and size of fasteners the gun can handle. Gun size can also have a limiting effect on the length of fastener that can be used.

CONTROLLER Stud Welding is a semiautomatic welding process. The welder has to move and hold the welding gun in place and start the process by squeezing the trigger; after that, the rest of the process is regulated by the controller. The controller can be a separate device that interfaces between the gun and power source or it may be integrated into the power source.

WORK LEAD/CLAMP Similar to any other arc welding process, the work lead's diameter must be large enough to handle the required current for that application. When an application is working on a larger structure, the length of work lead must be taken into consideration. To compensate for the increased resistance of a longer lead, the diameter of the lead must be increased. An undersized work lead will negatively affect the performance of the weld. **FIGURE 7-36** shows a work cable and clamp.

FIGURE 7-36 Work lead/clamp for a stud welder

The work clamp is a very important part of the welding circuit. Its welding current rating must always be considered when setting up for a welding application. Another important aspect is how the clamp mechanically attaches to the base metal. In applications such as high-current stud welding, the clamp should be the bolt on type that ensures a low-resistance connection.

METAL FASTENERS The wide array of shapes, sizes, and metallurgical properties of fasteners available today is what makes Stud Welding so versatile. Manufacturers of fasteners have developed product lines designed to fit the needs of the industry. Before choosing a product, review the manufacturer's literature, consult sales representatives, and ask for samples. **FIGURE 7-37** shows several types of welding fasteners.

Fasteners are available in an unlimited number of shapes, such as pins, hooks, eye bolts, and brackets. Fasteners can be internally or externally threaded. Standard mild carbon steel fastener diameters range from 0.125 inch (3.175 mm) to 1 inch (25.4 mm). The main limitations of shape, diameter, and length are the welding range of the power source and gun as well as the gun's limitations.

The fastener is used as the electrode, and a portion of it is melted, affecting the mechanical properties of the finished weld. It is important that the fastener's metallurgical properties be considered whenever choosing fasteners for an application. For example, suppose the base metal is 1022 steel (low-carbon steel), and the attached fastener will be used to hold heavy load. The chosen fastener's metallurgical properties must have a positive effect on the mechanical properties of the finished weld.

The shape and coating of the weld end of fasteners are other important considerations. The shape of the end greatly affects how much of the fastener melts during the weld. The amount of fastener that melts changes the admixture of base and fastener metals, affecting the mechanical properties of the finished weld. The size of the weld is also determined by the amount of the fastener melting. Fasteners can be purchased that have weld ends coated with flux. The flux provides the shielding of the weld and can contain scavengers that purify the weld metal.

FIGURE 7-37 Various fasteners/studs

FERRULES   A ferrule is a ceramic device that is placed over the weld end of the fastener. After the weld is completed, ferrules are removed with a specially designed tool or broken away with a hammer and discarded. During the weld, a ferrule performs several functions:

- Contains the molten weld pool
- Limits the amount of oxidation of the molten metal
- Concentrates the heat
- Reduces heat and smoke damage to the area surrounding the weld

SAFETY PRECAUTIONS

- **Wear eye protection:** Shaded eye protection is generally not required for this process because there is not a open electrical arc. It is very important that safety glasses or a face shield is always used. It is very common for molten metal or sparks to be expelled from the weld.
- **Use ventilation.** Fumes and vapors produced by the process should always be removed from the area. Extra care should be taken when working with metal covered with paint, coatings, or cladding. Individual tolerances and safety regulations may also warrant the use of breathing protection.
- Always know where sparks and molten metal are landing. Always remove combustible and flammable materials from the welding area.

SUBMERGED ARC WELDING

Submerged Arc Welding (SAW) is a process that uses electrical current flow conducted through an electrode filler wire. Resistance of the wire, arc gap, and base metal causes the filler wire and a localized area of the base metal to melt. The weld and arc are shielded from the atmosphere by a covering of granular flux. The degree of melting depends on the amount of welding voltage and current (filler wire diameter and wire feed speed), thickness and type of base metal, and condition of metal surfaces.

PROCESS   The Submerged Arc Welding process (**FIGURE 7-38** shows a submerged arc weld) is a variation of the GMAW process. Both processes use a wire electrode to conduct welding current and provide filler metal. Welding voltage is controlled by the welding power source, and welding current is determined by the wire feed speed (WFS) and the diameter of the filler wire electrode.

One of the differences between GMAW and SAW is how the arc and weld pool are shielded from the atmosphere. GMAW use a column of shielding gas to temporaily push atmospheric gas away from the weld. Submerged Arc Welding uses no external shielding gas. Instead, it uses a layer of granular fusible flux to cover and protect the weld.

The flux is designed to completely cover the arc and weld pool, isolating it from the atmospheric gases. The layer of flux is thick enough to prevent sparks, spatter, and light radiation from escaping. Thus, the name Submerged Arc Welding or its slang name of sub arc. Because no light radiation can escape, welders only need to wear safety glasses with no shade protection.

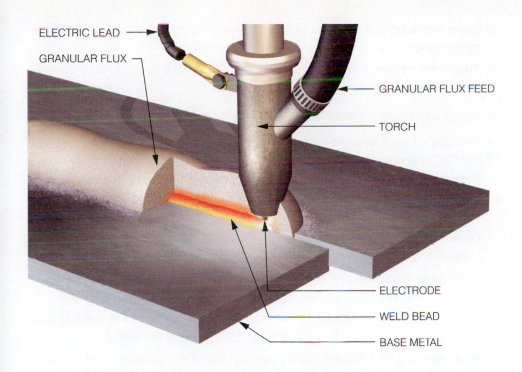

ELECTRIC LEAD

GRANULAR FLUX

GRANULAR FLUX FEED

TORCH

ELECTRODE

WELD BEAD

BASE METAL

FIGURE 7-38 Submerged Arc Welding

Another difference between SAW and GMAW is the size and number of electrodes. Electrode filler wire for Submerged Arc Welding ranges from 1/16 inch to 1/4 inch (1.6 mm to 6.4 mm) in diameter, which is larger than the typical GMAW filler wire. The ability to use a larger diameter filler wire gives Submerged Arc Welding higher deposition rates, higher welding currents, and the ability to fill larger grooves in one pass and even eliminate the need to prep weld joints. For even higher deposition rates, two or more electrodes can be used. Additional electrodes can be melted by the molten puddle, which increases the deposition rate. Another version is to add an additional power source for each electrode, increasing heat input and deposition.

Because of the effects of gravity on the flux, the majority of Submerged Arc Welding is done on grooves and fillets in the flat position. Flux dams can be used to hold the flux in place when welding in the horizontal position. Submerged Arc Welding can also be used for hard facing and resurfacing. The majority of Submerged Arc Welding systems will be mechanized by using a side beam system or weld tractor, with control of the system being semiautomatic or fully automatic. Handheld versions are also available.

The benefits of this process are:

- **Limited welding training needed:** Operating Submerged Arc Welding systems requires special training but not the extensive welding training required for the manual welding processes.
- **Deposition rate:** It has one of the highest depositions rates.
- **Reduced joint prep:** Time spent prepping joints for welding can be greatly reduced if not totally eliminated.
- **Flux:** The choice of flux can be used to affect the finished weldment's metallurgical and mechanical characteristics.
- **No electrical arc:** Eye and skin protection from ultraviolet arc rays is generally not required. Eye protection and flame-resistant clothing should always be worn.

- **Base metal type:** It can also be used to weld carbon, low-alloy, chromium-molybdenum, and stainless steels, and nickel alloys.
- **Thickness range:** It can weld a thickness beyond the range of GMAW and FCAW.

The drawbacks of the process are:

- **Visibility of molten weld puddle:** The weld zone is completely covered by the flux. The operator cannot visually watch the development of the weld to determine if the welding parameters are correct. Developing the correct welding parameters is done by a process of producing and then destructively testing the welds.
- **Base metal:** It cannot be used to weld some metals, such as aluminum, copper, and titanium.
- **Storage and handling of fluxes:** Some fluxes require special treatment.
- **Cost:** All systems must consist of a minimum of a wire feeder, flux, and power source. The welding current range and duty cycle of the welding gun and power source must be high enough for the metal thickness.
- **Maintenance:** The system has a large number of components that require preventative maintenance.

COMPONENTS   A Submerged Arc Welding system consists of a power source, wire feeder, travel system, flux delivery system, welding torch, and work cable/clamp.

POWER SOURCE   DC constant voltage and CC power sources are used for this process. No matter what type of power source, the welding current range and duty cycle must be considered. Welding currents for Submerged Arc Welding can range from 300 to 1,500 amps. A 100% duty cycle is required because of the extended welding duration. **FIGURE 7-39** shows the Submerged Arc Welding system.

Constant voltage power sources monitor and adjust the welding voltage, allowing a constant speed feeder to control the wire feed speed. When using a CC power source, a more costly voltage sensing feeder may need to be used.

FIGURE 7-39 Submerged Arc Welding power source
Source: Photo courtesy of Miller Electric Manufacturing Co.

This type of feeder monitors the arc voltage and changes the wire feed speed to manipulate the arc gap. Increasing the arc gap increases the arc voltage, and decreasing the arc gap decreases the arc voltage.

AC constant voltage or CC power sources are also used for Submerged Arc Welding. DC welding of ferrite metals can cause the development of magnetic fields, leading to arc blow. Arc blow is the term used to describe how a magnetic field can push an electrical arc. In a tight groove, the arc blow can cause the arc to be pushed to one side of the groove, causing incomplete fusion on the other side. AC does not cause the buildup of magnetic fields; in fact, it will break down existing fields. AC power sources are used for multiple wires, high welding current, and narrow-groove applications.

The original AC output waveform is a sinewave. Sinewave welding outputs are not the most efficient and experience arc outages—which at times are slow to restart—as the current flow changes direction. In order to correct these deficits, power source manufacturers have developed power sources that use a more efficient squarewave-type output. A squarewave still has arc outages, but because of its shape, hesitation or failure to restart is eliminated. Manufacturers have also developed power sources that allow the welder to manipulate the squarewave's output frequency and current level of each polarity independently. Independent current level and frequency control allow the welder to fine-tune the shape, size, and penetration levels.

WIRE FEEDER   Depending on the type of welding power source, wire feeders are either a constant speed or voltage-sensing feeder. No matter what type, the feeder needs to be able to consistently push larger diameter wires for extended periods of time. Wire drive motors and systems need to be more robust than standard feeders.

Two options that are useful for Submerged Arc Welding are run-in speed and burn-back. Run-in speed allows the welder to have a different feeding speed for the initial start of the welding. Often, the run-in speed is slower than the welding speed. This allows the start of the arc to be smoother, with less stumbling. When the arc is ended, the electrode can be left fused into the solidified weld. Burn-back stops this by momentarily keeping the welding output on after the wire feeding has stopped. With no feeding and the output on, a small portion of the electrode wire melts, producing a space between the electrode wire tip and weld.

WELDING TORCH   With the majority of systems being mechanized by using a side beam system or weld tractor, torches are fixed in position and moved along the weld. Straight-line torches are best suited for this application. The use of high welding currents, high duty cycles, and large electrode diameters must be considered when purchasing a torch. **FIGURE 7-40** shows a straight-line fixed torch.

TRAVEL SYSTEM   Movement along the weld joint is generally done by moving the welding torch and flux delivery system. Because weld joints are in a straight line, side beams and weld tractors are used (**FIGURE 7-41**).

Submerged Arc Welding is also used for hard facing and resurfacing. An example is a paper roll whose surface is damaged or reduced in thickness. Submerged Arc Welding is used to apply a new layer, which is then machined down

FIGURE 7-40 Straight-lined fixed torch
Source: Photo courtesy of Miller Electric Manufacturing Co.

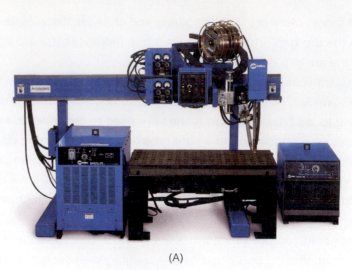

(A)

FIGURE 7-41 (a) Side-beam
SAW welding equipment
(b) Self-propelled SAW tractor
Source: Photos courtesy of Miller Electric
Manufacturing Co.

(B)

to specification. The roll is turned under the torch during welding, with the torch and flux system only indexing forward with each revolution.

FLUX DELIVERY SYSTEM   Flux is stored in a hopper and then deposited to the weld area by a tube. **FIGURE 7-42** shows a flux hopper and a delivery tube. The tube is generally attached to the welding torch and spaced to allow the flux to be deposited right ahead of the weld. The method of flux delivery can be by gravity or pressure. If the flux is gravity-fed, the hopper is attached at the top of the travel system. When pressure-feeding is used, the hopper can be located anywhere, and air pressure is used to deliver the flux.

WORK LEAD/CLAMP   Similar to any other arc welding process, the work lead's diameter must be large enough to handle the required current for that application. An undersized work lead will negatively affect the performance of the weld.

The work clamp is a very important part of the welding circuit. Its welding current rating must always be considered when setting up for a welding application. Another important aspect is how the clamp mechanically attaches to the base metal. In applications such as high-current Submerged Arc Welding, the clamp should be the bolt-on type that will ensure a low-resistance connection.

ELECTRODE   One of the strengths of Submerged Arc Welding is the variety of different electrodes that can be used. Electrodes can be solid, tubular, or even flat strips. Having a greater variety of electrodes allows engineers and welders more control over the finished weldments' metallurgical and mechanical characteristics.

The AWS classification for Submerged Arc Welding electrodes is E followed by three to four numbers or letters. For example, an EH11K is an electrode used for mild steel:

- E stands for electrode.
- H represents the level of manganese:
  - L = low maximum of .6%
  - M = medium maximum of 1.25%
  - H = high maximum of 2.25%
- The letter C may be found after the second digit. The C stands for composite wire, which means the wire is a cored wire.
- 11 represents the percentage of carbon.
- K indicates that the electrode is made from steel that was silicon-killed. The silicon helps to eliminated porosity. This digit is not always used and can be left blank.

AWS classifications for stainless, nickel, and nickel alloys are the same as electrodes used for GMAW, GTAW, and PAW. For example, ER308L is used for all four processes.

Electrode diameters range from 1/16 to 1/4 inch (1.6 to 6.4 mm). **TABLE 7-2** shows the welding current range for each diameter of mild steel slid electrodes.

FLUX   Another advantage of the Submerged Arc Welding process is the use of granular flux. The flux covers the weld and arc, shielding them from atmospheric gases. It also serves several other important functions:

- Shapes the weld bead
- Supports the weld bead and hold it in position
- Along with the filler metal, it helps to produce a finished weld with desired metallurgical and mechanical characteristics.

FIGURE 7-42 Flux hopper delivery system
Source: Photo courtesy of Miller Electric Manufacturing Co.

Electrode Diameter	Welding Current Range
1/16 inches (1.59 mm)	150–400 amps
5/64 inches (1.98 mm)	200–500 amps
3/32 inches (2.38 mm)	300–600 amps
1/8 inches (3.18 mm)	300–800 amps
5/32 inches (3.97 mm)	400–900 amps
3/16 inches (4.76 mm)	500–1,100 amps
7/32 inches (5.56 mm)	600–1,200 amps
1/4 inches (6.35 mm)	700–1,400 amps

TABLE 7-2

Normally, the filler wire electrode (solid or cored) alone is used to control the metallurgical and mechanical characteristics of a finished weldment. Metallurgical and mechanical characteristics can be controlled by the addition of deoxidizers or ionizing and alloying elements to the electrode or the flux core. The Submerged Arc Welding process not only takes advantage of the electrode filler wire (solid or cored), but the metallurgical and mechanical characteristics of the finished weldment can be enhanced by the type of granular flux chosen.

There are three basic type of fluxes: fused, bonded, and agglomerated. The names indicate how each of the fluxes are made. How they are made determines how the flux affects the finished weld as well as how the flux is stored and possibly recovered and reused.

Fused fluxes are made by melting raw materials together in a furnace. The heated flux is cooled and then crushed and screened to the desired size. These fluxes have the advantage of not picking up moisture from the atmosphere, and they can be recycled and used for multiple welds. Their disadvantage is that it is difficult to add deoxidizers and alloys to the flux mix. Fused fluxes have a neutral affect on the finished welds' metallurgical and mechanical characteristics.

Bonded fluxes are made by mixing raw materials together with a binder. The wet mixture is then baked at a low temperature, crushed, and then screened. The manufacturing process allows the addition of deoxidizers and alloys that affect the finished welds' metallurgical and mechanical characteristics. The flux also has a thicker protective layer than the fused flux. This flux will absorb moisture from the atmosphere, and special storage and handling procedures need to be followed.

Agglomerated fluxes are similar to the bonded flux except that the binder is different and the flux is baked at a higher temperature. The effect over the finished weld is also similar except that the flux has less deoxidizers and alloys.

Choosing the correct flux for a weld is determined by the type of base metal composition, electrode filler metal composition, desired metallurgical and mechanical characteristics, and the flux delivery/storage/recovery capabilities of the location. Fluxes can also be mixed to give the welder more options. Another variable that can affect the finished weld is the size of the flux particles. Larger powders can allow gases to escape with greater ease.

Manufacturers have developed a huge array of different fluxes. In order to choose the correct one for an application, it is important to know the type of base metal and desired finished weld characteristics. After this information is found, consult the manufacturer's data sheets or ask for assistance from a welding distributor or a manufacturer representative.

The AWS classification for submerged arc welding flux is F followed by two or three numbers. For example, an F6A4:

- F stands for flux.
- 6 represents the 60,000 psi tensile strength.
- A stands for "as welded." If a P is present, it means "postweld heat treatment."
- 4 represents the low-temperature impact value.

ELECTROSLAG WELDING

Electroslag Welding is a process that uses electrical current flow conducted through an electrode filler wire. Resistance of the wire, molten puddle, slag, and base metal causes the filler wire and a localized area of the base metal to melt.

PROCESS   Electroslag Welding (**FIGURE 7-43**) was designed to be a more cost-effective way to weld thick sections of steel while in the vertical position. For example, the in-position welding of a thick-walled vessel would require manual vertical-up SMAW or FCAW. Even when utilizing the higher deposition rate of the FCAW process, the weld would take a very long time to complete and would require multiple passes. Manual welding in the vertical-up position can lead to welder fatigue, which could lead to weld discontinuities. Removal of slag would also be required between each pass, further extending the time required to finish the weldment. Electroslag eliminates the need for a manual welding operation at a much more cost-effective rate.

The start of the weld is located at the bottom of the groove weld. Generally, there is no edge preparation done to the joint, other than wire wheel cleaning of the surface. Copper water-cooled shoes are placed on both sides of the weldment. The shoes are designed to contain the molten weld until it has solidified. A nonconsumable guide tube is positioned so the electrode filler wire can be fed into the area between the shoes. Before a weld is started, a granular flux is added between the shoes.

The arc is started when the electrified electrode wire comes in contact with the base metal, producing an electrical arc. As base and filler metals melt, the area between the shoes begins to fill up with molten metal and slag. The molten puddle eventually rises to the level of the electrical arc; at this point, the process

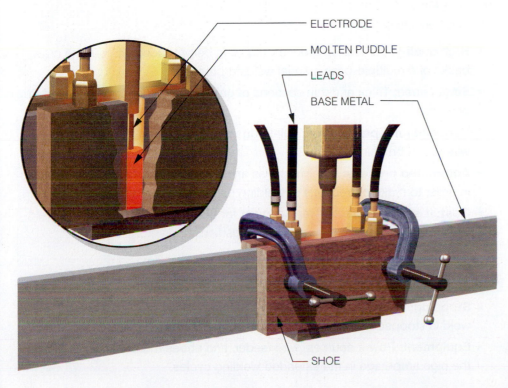

ELECTRODE

MOLTEN PUDDLE

LEADS

BASE METAL

SHOE

FIGURE 7-43 Electroslag Welding, shown here, is very similar to Electrogas Welding.

is no longer an arc welding process. Electrical current is conducted directly through the conductive slag, while the resistance of the wire, molten puddle, slag, and base metal causes the filler wire and a localized area of the base metal to melt. The shoes and electrode guide are moved along with the rising level of the weld. This continues until the entire weld is completed.

Shielding from the atmosphere is done by the slag. Depending on the thickness and width of the weld joint, multiple electrodes can be used. Electrodes can also be oscillated to provide even melting on both base metal sides.

Typically, a constant voltage DC power source is used. Welding amperages range from 500 to 1,000 amps. Because the duration of the welds can last for several hours, the power source must have a duty cycle of 100% at the required welding amperage. A separate power source is required for each electrode when using multiple electrodes.

ELECTROSLAG CONSUMABLE GUIDE WELDING (ESW-CG) This process is a variation of Electroslag Welding. It differs in that the non-electrode filler guide is replaced with a consumable guide. The guide is as long as the entire length of the weld. As the weld progresses upward, the guide is melted along with the electrode filler wire. The benefit of this system is that the welding head and wire feeder can be placed at the top of the weldment and do not need to be moved along with the weld.

ELECTROGAS WELDING (EGW) Electrogas Welding is a very similar process to Electroslag Welding in that the welding is done in the vertical-up position, with nonconsumable shoes used to contain the molten weld puddle. How it differs is that Electrogas Welding is an arc welding process that is used for thinner applications than the Electroslag process.

Electrogas is very similar to GMAW and FCAW in that the welding current is conducted across an arc gap. Filler wire can be solid or flux-cored. When solid or certain cored wires are used, shielding is provided by a shielding gas of a mixture of argon and carbon dioxide.

The benefits of this process are:

- **High quality:** Vertical-up groove welds can be produced without the drawbacks of a multiple-pass manual welding process.
- **Single pass:** Thick and thin sections of groove welds can be completed in one pass.
- **Reduced joint preparation:** Joint edge preparation can be a minimum of wire wheel brushing.
- **Automated process:** It only requires an operator to watch over the process in order to make corrections to welding parameters and travel.
- **Cost:** It is a cost-effective replacement for the labor-intensive manual welding processes done in the vertical-up position.

The drawbacks of the process are:

- **Skill level:** It requires extensive training to produce a high operator skill level.
- **Stopping:** Time-consuming repair work must be performed whenever a weld is stopped before the end of the weld.
- **Equipment:** Power sources, wire feeder, and shoes must be able to handle the high amperage in the extended welding cycles.

SOLID STATE WELDING PROCESSES

The four main arc welding processes (GMAW, FCAW, SMAW, and GTAW) and the nine found in this chapter all use high temperatures to produce a weld. The high temperature causes base and filler metals, when used, to melt and the individual molten puddles to fuse together. When the source of heat is removed, the molten mixture solidifies, and a weld bead or nugget is formed. This technique of joining metal works great and is by far the most common way to weld. But producing a weld by using high-temperature melting is not always the best way to join metal and has several drawbacks:

- Some dissimilar metals cannot be joined together.
- The resulting heat-affected zone (HAZ) can negatively affect the metallurgical and mechanical characteristics of a finished weldment.
- It will always cause some degree of distortion or warpage of the base metal.
- There is an increased risk of fire, and personal protective equipment must be used to protect welders from welding arc radiation and sparks, spatter, and hot metal.

The Solid State Welding processes provide an alternative to the high-temperature welding processes. The Solid State Welding processes covered in this section are: Forge, Cold, Diffusion, Explosion, Ultrasonic, and Friction.

Each of these processes would appear to be very different from the others. They range from simple to complex to high skill manual to fully automated. Even though they appear different, they are similar in that the joining of metal together is done by using the variables of pressure, time, and heat individually or in combination to produce a weld at a temperature below the melting temperature of the metal or in a solid state. Please note that some of the processes may produce limited melting of the base metal.

The question can be asked that without melting, how can two individual pieces of metal be permanently welded together? The Solid State Welding processes take advantage of metals' ability to form a permanent bond when two surfaces are held together. The process that allows this bond to form is called **diffusion**. Diffusion is the transfer of atoms between the individual pieces of metals' outer surface layers, resulting in a permanent bond.

If two pieces of metal are laid one on top of the other, atoms from the surface of each piece would diffuse between each. The problem with this scenario is that the diffusion process would take a very long time to occur. Producing welds in this manner would be way too slow to make it a viable process. There are several reasons why the process is so slow:

- All metals have a naturally occurring oxide layer. This layer acts as a barrier to the movement of atoms.
- Surface contaminations, such as oil and dirt, restrict atom movement.
- All metal surfaces, even those with very smooth finishes, have surface imperfections. To the naked eye, these imperfections may not be visible, but when viewed under magnification, they appear as high and low spots (**asperities**). When two metal surfaces are placed together, these imperfections result in

gaps between the surfaces, which inhibit atoms' movement. Also, the total contact area between the pieces is limited to just the areas where the high spots are in contact; thus, any resulting diffusion would be limited.

To increase the speed of the diffusion process, several steps can be taken. First is the removal of the oxide layer and any surface contamination. This speeds up the process. Even with an oxide-free, very clean surface, the speed of diffusion would be too slow.

In order to greatly increase the speed of the process, the mating surfaces must be held in very close contact. The greater the contact, the faster the diffusion process can occur and the greater the strength of the resulting welds. The most effective way to produce the intimate contact needed is by applying pressure. When pressure is applied, the gaps, which are restricting diffusion, are reduced or possibly eliminated, resulting in a faster process.

Pressure can take the form of simple compression that squeezes and holds the surfaces together. For example, pressure is used to press the two pieces of aluminum sheet together, forming a mechanical bond that holds the aluminum very tightly together. In time, a welded bond will form between the two pieces. The degree of success is limited by the metal's thickness and hardness. This method is effective, but there is still some gap between the pieces, which causes diffusion to be slowed.

To make the diffusion process even faster and more economical, the solid-state processes take advantage of metals' ability to yield when placed under stress. An example of this ability is when a tensile test is performed on a weld sample. The sample is placed under tensile stress (stress pulling in opposite directions); as the stress increases, the sample begins to stretch. Depending on the metals' elasticity, the sample will stretch to a point where it yields to the stress. Before this point is reached, the metal will return to its original shape and size when the stress is removed. After the yield point has been reached, the sample will stretch, and when the stress is removed, the sample will never return to its original shape. Even though the sample yielded, it did not tear apart.

This ability to yield to stress allows metal to be reshaped at a temperature below the metals' melting temperature. An example of this is when incorrectly drilling aluminum. If the drilling pressure and/or bit speed is incorrect, aluminum will stick to the tip of the drill bit. The temperature of the process was never high enough to melt the aluminum, yet the aluminum is formed around the bit. The downward pressure (compression stress) applied by the drill bit to the aluminum causes the aluminum to yield and change shape around the drill bit. The yielding was aided by the increase in temperature caused by the friction of drilling.

The yield point of each metal can be reduced by increasing the temperature of the metal. An example of this is Forge Welding. Hammering (compression stress) is used to press two pieces of steel together. At room temperature, the amount of pressure needed would be beyond the range of a blacksmith's arm. Raising the temperature of the steel and the pressure of the blacksmith's arm is sufficient to cause yielding, forcing the surfaces of the steel together. A majority of the Solid State Welding processes take advantage of this ability. Compression pressure is used by Forge, Explosive, and Ultrasonic Welding. Torsion pressure is used by Friction Welding.

The following text is intended to provide an overview of each of the Solid State Welding processes. Like any welding process, each has its advantages and disadvantages. What each of the processes has in common is that they use pressure, temperature, and time—alone or in combination—to form a weld.

COLD WELDING (CW)   Cold Welding, also called Cold Forging, is produced by the use of pressure on surfaces at room temperature. Pressure is applied to the individual pieces, forcing their surfaces together. Often, dies or rollers are used to press the metal together, forming a mechanical bond. Other times, welds can be produced by simply hammering the surfaces of metal sheet together. The thickness of the metal determines if pressure is applied manually by hand tools or by using mechanical, pneumatic, or hydraulic presses.

Soft ductile metals, such as aluminum, magnesium, and copper, are most often joined. In order for the diffusion process to occur, all surface contamination must be removed by using degreasers or cleaners. The oxide layer is then mechanically removed by using a rotary stainless steel brush.

The benefits of this process are:

- In its simplest form, it can be done with a hammer and a punch.
- It can be done with simple hand tools.
- The position of metal has no effect on the process.
- Mechanical joints can be produced in a matter of seconds, which will hold metal together while the diffusion process is occurring. It can be used in high-speed applications, such as the canning industry.
- Welds can be produced between dissimilar metals.

The drawbacks of the process are:

- Mechanical joints will hold pieces together, but the diffusion process takes time. Ultimate strength is not reached quickly.
- Maximum thickness is determined by how much pressure can be applied. Without mechanical, pneumatic, or hydraulic presses, the process is very limited.
- It works best with soft ductile metals.

FORGE WELDING   Forge Welds are produced by a combination of pressure and temperature. Individual pieces are heated and then pressure is applied. This is the oldest know welding process.

In its simplest form, metal (iron) pieces are heated between 2,100°F and 2,300°F (1,150°C to 1,260°C). Then, hammering pressure is applied to the outside surfaces, forcing the mating surfaces together.

A flux—often in the form of borax—is applied to the surfaces to aid in the removal of oxides. The physical action of the hammering or pressing also helps expel oxides from between pieces.

Rollers or power-assisted hammering are also used to press the metal together. The thickness of the metal determines if pressure is applied manually or by using mechanical, pneumatic, or hydraulic presses.

The benefits of this process are:

- It requires a source of heat and, in its simplest form, an anvil and a hammer.
- It can be done with simple hand tools.

The drawbacks of the process are:

- It requires a high skill level that is not developed quickly.
- The process often requires multiple cycles of heating and hammering before the desired weld and shape are reached.
- Maximum thickness is determined by the how much pressure can be applied. Without mechanical, pneumatic, or hydraulic presses, the process is very limited.
- It is limited to mild and low carbon steels, although other metals—such as chromium and nickel—may be layered within carbon steels.

DIFFUSION WELDING   Diffusion welds are produced by a combination of pressure and temperature. Pieces are placed under pressure and then heated. Complete diffusion occurs over a period of time. Filler metal can be used, generally in the form of an intermediate layer between base metals.

This process can be slower than other Solid State Welding processes, which often limits its use to high-quality, single, or limited production runs. This process has found its niche in the nuclear and aerospace industries.

All surface contamination has to be removed before the process is started. Any contamination left on the metal surface will produce a barrier to the diffusion process. Welding is accomplished by placing prepared pieces in a heated atmosphere and subjecting them to continuous pressure. This combination allows the diffusion process to occur. Diffusion Welding can be performed in an inert atmosphere or vacuum chamber when welding refractory metals.

The benefits of this process are:

- It is used to produce welded bonds that are used by the most demanding industries.
- It can produce welded bonds between most metals, including refractory metals.

The drawbacks of the process are:

- It requires a high operator skill level.
- Pressure is applied in the form of static compression; thus, there is no movement that will quicken the process.
- It requires expensive equipment.

EXPLOSION WELDING (EXW)   Explosion welds are produced by the use of pressure on surfaces at room temperatures. Individual base metal pieces are held apart and then an explosive charge is used to force the surfaces of both pieces together. The force of charge produces enough pressure to cause the surface of both pieces to yield and thus produce diffusion.

All surface contamination has to be removed before the process is started. Any contamination that is left on the metal surface will produce a barrier to the completion of the diffusion process.

The benefits of this process are:

- It can be used to produce welds on larger weldments.
- It can be performed in any welding position.
- It can produce welded bonds between most metals.

- A limited amount of equipment is required.
- A welded bond is produced in a matter of seconds.

The drawbacks of the process are:

- It requires a high operator skill level. Operators must be trained to safely perform the process.
- It requires specialized equipment to protect operators.

ULTRASONIC WELDING (USW) Ultrasonic Welds are produced by a combination of pressure at room temperature. Pieces are placed under pressure, and ultrasonic vibrations are used to cause the mating surfaces to yield. Friction produced by the vibration produces heat, which aids in the process. Diffusion is completed over a relatively short period of time.

All surface contamination must be removed before the process is started. Any contamination that is left on the metal surface will produce a barrier to the complete diffusion process. Ultrasonic Welding can be performed in an inert atmosphere or vacuum chamber when welding refractory metals.

The benefits of this process are:

- It is used to produce welded bonds that are used by the most demanding industries.
- It can produce welded bonds between most metals, including refractory metals.
- The combination of pressure and high-frequency vibration produces a very fast Diffusion Weld.

The drawbacks of the process are:

- Metal thickness is limited by the strength of the transducer.
- It requires expensive equipment.

FRICTION WELDING (FRW) Friction welds are produced by a combination of pressure at room temperature. One or both pieces are rotated while being pressed against each other. Friction causes the interface of the two pieces to heat up, producing yielding. Diffusion is completed when movement has stopped.

All surface contamination should be removed before the process is started. Any contamination that is left on the metal surface may be forced out of the weld by the force of rotation, but if not forced out, it may cause incomplete diffusion. This process can be performed in an inert atmosphere or vacuum chamber when welding refractory metals.

The benefits of this process are:

- It is used to produce welded bonds that are used by the most demanding industries.
- It can produce welded bonds between most metals, including refractory metals.
- The combination of pressure and friction produces a very fast Diffusion Weld.

The drawback of the process is:

- It requires specialized equipment.

MULTIPLE CHOICE

1. Which of the following is not an advantage of an Oxy-Fuel Welding system?
   a. Portability
   b. Does not require electricity
   c. Requires an inert shielding gas
   d. Can be used for welding, brazing, cutting, and heating

2. What type of flame is recommended for Oxy-Fuel Welding?
   a. Neutral
   b. Carburizing
   c. Liquidizing
   d. Oxidizing

3. What is the operating pressure that acetylene should never be operated above?
   a. 20 psi (138 kPa)
   b. 15 psi (103 kPa)
   c. 30 psi (207 kPa)
   d. 10 psi (69 kPa)

4. Why is argon the shielding gas of choice for Plasma Arc Welding?
   a. Has a higher transfer temperature than other gases
   b. Is cheaper in cost than other gases
   c. Smells better when used
   d. Is an inert gas

5. What is used to melt the base metal that is located between the tips when Resistance Welding?
   a. Vibrating energy
   b. High-temperature flame
   c. Electrical current flow
   d. DCEN welding arc

6. Which one of the following is an advantage of Submerged Arc Welding?
   a. Can be used to weld in all positions
   b. Is very portable
   c. A layer of granular flux buries the electrical arc
   d. Can be used to weld any metal

7. Which is a disadvantage of the Electroslag and Electrogas processes?
   a. Welds can be done in any position.
   b. It is very difficult to stop and restart the process once it has started without causing some kind of weld discontinuity.
   c. It can be used to weld a thin sheet.
   d. They are both manual welding processes.

8. Which is the oldest form of welding?
   a. Explosion Welding
   b. Forge Welding
   c. Diffusion Welding
   d. Ultrasonic Welding

FILL IN THE BLANK

9. The four main arc welding processes are Gas Metal Arc Welding, _____, Shielded Metal Arc Welding, and Gas Tungsten Arc Welding.

10. Oxygen and _____ are the most popular combination for Oxy-Fuel Welding.

11. _____-based lubricants should never be used on any oxygen fitting, hose, or regulator.

12. The _____ is the maximum amount of acetylene that can be pulled from an acetylene cylinder.

13. Plasma Arc Welding is very similar to _____.

14. Welding aluminum cannot be done with the Plasma Arc Welding process by using the _____ polarity because plasma welding power sources only provide a DC welding output.

15. A benefit of Plasma Arc Welding is that the electrode is _____ inside the torch.

16. _____ is the electrode of choice for Plasma Arc Welding.

17. The Resistance Welding processes all use a combination of electrical current flow and _____ to produce a weld.

18. The location of the resistance weld is limited by the length and shape of the _____.

19. The electrode that is used for Stud Welding is also called the _____.

20. An advantage of Stud Welding is that the weld can be done in _____ position.

21. The Solid State Welding processes produce welds at temperatures generally below the _____ temperature of the base metal.

22. Metal being welded by the _____ process is placed under pressure, and ultrasonic vibrations are used to cause the mating surfaces to yield. Friction produced by the vibration produces heat, which aides in the process.

SHORT ANSWER

23. What is the recommended shade range for Oxy-Fuel Welding?

24. What steps should be taken when shutting down by using an Oxy-Fuel system?

25. Any gap between metal surfaces being welded will cause what type of problem when Resistance Spot Welding?

26. Why are fasteners used in Stud Welding one of the process's biggest advantages?

27. What are three different types of granular flux used for Submerged Arc Welding?

28. Which welding process was developed to weld thick sections of base metal in the vertical position?

29. What are the disadvantages of using high-temperature melting to produce a weld?

30. In its simplest form, this type of Solid State Welding can be done with a hammer.

31. Friction Welding uses what to produce a weld?

Chapter 8

CUTTING PROCESSES

● LEARNING OBJECTIVES

In this chapter, the reader will learn how to:

1. Incorporate safe practices when using any cutting process.

2. Understand how the Oxygen-Fuel Cutting processes function.

3. Determine when the Oxygen-Fuel Cutting processes would best be used in a welding environment.

4. Understand how the Air Carbon Arc Cutting process functions.

5. Determine when the Air Carbon Arc Cutting process would best be used in a welding environment.

6. Understand how the Plasma Arc Cutting process functions.

7. Determine when the Plasma Arc Cutting process would best be used in a welding environment.

8. Understand how the Laser Beam Cutting process functions.

9. Determine when the Laser Beam Cutting process would best be used in a welding environment.

10. Understand how the Abrasive Water Jet Cutting process functions.

11. Determine when the Abrasive Water Jet Cutting process would best be used in a welding environment.

● KEY TERMS

Kerf 267

Taper (bevel angle) 268

Drag Lines 268

Dross 269

Kindling Temperature 273

Plasma 290

Laser (light amplification by stimulated emission of radiation) 300

Lasing Material 301

Without the ability to efficiently cut and shape metal, the modern welding shop would not exist today. There are several ways in which you can cut and shape metal.

- Mechanically, using a blade or bit. The blade or bit is moved across the surface of the metal and removes small amounts of the metal.
- Abrasive, using abrasive materials. The abrasive material can be applied to the metal in the form of grinding discs and belts or carried in a high-pressure abrasive stream. Metal is removed as this abrasive moves across it.
- Melting, using a source of heat. The metal is first melted by using a source of heat and then the molten metal is blown away by a flow of gas.
- Chemical reaction, using oxygen to oxidize metal at an elevated temperature.

This chapter discusses the melting processes of Air Carbon Arc Cutting, Plasma Arc Cutting, and Laser Beam Cutting, the abrasive process of Abrasive Water Jet Cutting, and the chemical reaction process of Oxygen-Fuel Cutting.

Each of these processes has unique advantages that allow them to be very beneficial; however, each also has unique disadvantages that keep it from being the only solution for each situation. An example is the use of Plasma Arc Cutting by a farm implement's repair shop. A Plasma Arc Cutting system is an extremely efficient method for cutting metal and works great in the shop. Part of the implement repair business is the ability to repair equipment in the field, and in this situation, the Oxy-Fuel Cutting system may be the better choice.

It is important for an operator to understand each process and its advantages and disadvantages in order to choose the best process for his or her needs.

CUTTING TERMINOLOGY

FIGURE 8-1 is a view of a typical cut and the terminology that describes it.

All the processes that are covered in this chapter use the same terms to describe the cut. It is important to know what each of these terms mean and to understand what cutting variable controls them.

Kerf is the width of the cut. Material within the kerf is essentially waste; thus, the smaller the kerf width, the better. Each process has a minimum kerf width. Variables that can cause the kerf width to increase are:

- Incorrect Settings
- Excessive Gas Flow Rates
- Incorrect Size Tips or Electrodes
- Slow Travel Speed
- Worn or Damaged Tips or Nozzles
- Incorrect Settings

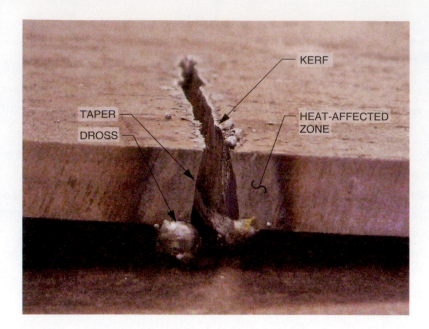

FIGURE 8-1 A view of a typical cut from an oxy-fuel torch

Taper (bevel angle) is when the kerf of the cut is wider at the top or bottom. Typically, the taper has a wider width at the bottom of the cut (**FIGURE 8-2**), but this is not always the case. Variables that can cause taper are:

- Incorrect Travel Speed (can result from too fast or too slow of travel speed)
- Incorrect Cutting Angle
- Worn or Damaged Tips or Nozzles
- Incorrect Settings

The heat-affected zone (HAZ) is the area located on each side of the cut. Heat caused by the cutting process causes the metal within the HAZ to be physically changed. Other than the Abrasive Water Jet Cutting process, there will always be some degree of HAZ. With the correct technique and settings, the amount of HAZ can be reduced. Variables that affect the HAZ are:

- Too Slow of a Travel Speed
- Incorrect Settings
- Excessive Gas Flow Rates
- Incorrect Tips or Electrode Sizes
- Worn or Damaged Tips or Nozzles

Drag lines are ripples that form along the cut edge. Generally, the number and profile of the lines increase as the thickness of the metal increases.

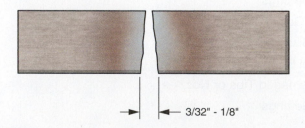

FIGURE 8-2 Bevel angle

3/32" - 1/8"

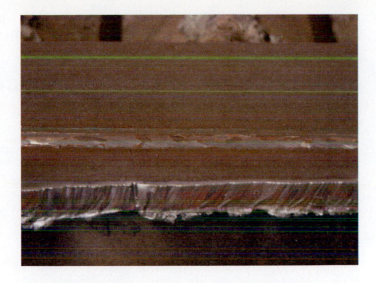

FIGURE 8-3 Drag lines formed when Plasma Arc Cutting

FIGURE 8-3 shows drag lines formed when using Plasma Arc Cutting. Variables that affect the drag lines are:

- Incorrect Travel Speed
- Incorrect Amperage or Gas Flow Settings
- Incorrect Tip Size

Dross is a rough-textured buildup of resolidified molten metal waste or oxide on the backside of the cut. When any of the processes are operating correctly, the molten or oxidized metal is blown away from the cut area. When operating incorrectly, some of the metal fails to be blown clearly away and builds up on the backside of the cut. Buildup generally needs to be ground off, causing a loss in productivity. Variables that affect the dross are:

- Incorrect Settings
- Excessive Gas Flow Rates
- Incorrect Tips or Electrode Size
- Incorrect Travel Speed
- Worn or Damaged Tips or Nozzles

TYPES OF OPERATIONS

COMPLETE CUT   This cutting process penetrates completely through the entire thickness of the metal being cut. Metal removed from the cut area is eliminated from the cut by the flow of gas (Oxygen-Fuel, Air Carbon Arc, Plasma Arc, and Laser Beam) or water (Abrasive Water Jet).

Complete cuts can be broken down into two types: the edge cut and the piercing cut. Each differs in where the cut is started.

EDGE CUT   Cutting is started on the edge of the metal and then progresses into the metal. This type of start has several advantages. **FIGURE 8-4** shows a plasma arc cut starting on the edge of a plate of mild steel.

FIGURE 8-4 A plasma arc cut starting on the edge of a plate of mild steel

- The start of the cut is much faster because the removed metal is quickly blown away from the cut.
- When cutting a thicker plate, the torch can be angled to start the cut on the edge of a corner.
- Removed metal is blown away from the cutting torch or head, extending the cutting life of the consumable components.

PIERCING CUT The cut is started on the interior of the metal being cut. As the name implies, the cutting cannot start until the arc, stream, or beam has gone through the entire thickness of the metal. **FIGURE 8-5** shows a plasma arc piercing cut starting in the middle of a plate.

Caution must be taken when piercing because molten metal and sparks can be directed back at the cutting torch or head; this blowback reduces the cutting life of the consumable components and/or damages the torch or head. To reduce the negative effects of the blowback angle, the torch should be positioned so that molten metal and sparks are blown away. As soon as the metal is pierced,

FIGURE 8-5 When piercing a plate, the torch is started at an angle to prevent molten metal blowing back toward the operator and torch.

FIGURE 8-6 Cutting at an angle with an Oxy-Fuel Cutting torch

rotate the torch angle back and then proceed with the cut. Whenever possible, try to start at the edge of the metal you are cutting.

BEVEL CUT The cutting torch or head is angled in order to form an angled cut face on the metal. It is often used to produce a bevel for fillet and groove welds (**FIGURE 8-6**).

GOUGING This cutting process is used to remove a specific amount of a metal thickness but does not cut entirely through the metal. It is used to remove base or weld metal to prepare metal for repairs, modifications, or backwelds.

The technique for gouging is similar to a piercing cut except that the torch head must be angled in order that molten metal and sparks are blown away. **FIGURE 8-7** shows a plasma arc cut gouging the weld metal from a groove weld.

SCARFING This is similar to gouging in that the cut does not go all the way through the metal. It is often used to remove bolt or rivet heads. **FIGURE 8-8** shows an oxy-fuel cut scarfing of the head of a bolt.

ETCHING Cutting is limited to the surface layer of the base metal. Generally, etching is limited to the Laser Beam Cutting process in which the power level of the laser beam can be reduced. As the reduced power laser beam melt's the metal's surface layer, a stream of compressed gas is used to blow molten metal away from the cut.

FIGURE 8-7 Gouging with a plasma arc

FIGURE 8-8 Scarfing with oxy-fuel

FIGURE 8-9 A sheet metal gauge etched with a laser

Etching is often used to print on the surface of the base metal. The color of the lettering can be changed by using a non-inert shielding gas, thus causing a slight oxidation of the cut area. **FIGURE 8-9** shows a metal thickness gauge cut and etched with a laser beam.

OXYGEN-FUEL CUTTING

Oxy-Fuel Cutting (OFC) is performed by use of a chemical reaction (oxidation). The heat produced by the burning of a fuel gas, enhanced by the addition of pure oxygen, is used to raise the temperature of the base metal. When the metal has reached its burning temperature, an additional stream of pure oxygen is added, causing the metal to be burned and blown away. **FIGURE 8-10** shows a typical oxygen-fuel flame and cut.

Combining oxygen and a fuel gas to produce a very hot flame is one of the oldest methods used to weld and cut metal, mainly steel. The use of Oxy-Fuel Welding today has greatly diminished, but Oxy-Fuel Cutting is still very popular. Oxy-Fuel Cutting is a simple and inexpensive method of cutting, which makes it

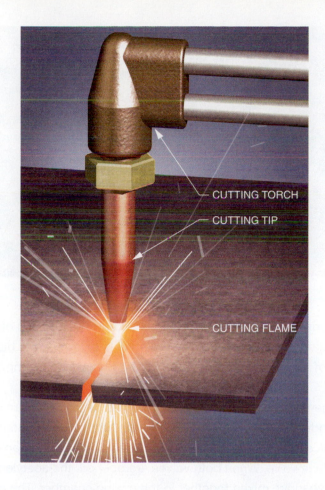

CUTTING TORCH
CUTTING TIP
CUTTING FLAME

FIGURE 8-10 Part of Oxy-Fuel Cutting

a very versatile and cost-effective tool. The oxy-fuel system can also be used as a source of heat for heat treating, heating nuts and bolts that will not loosen, and heating metal to be formed.

PROCESS   As the name implies, the Oxy-Fuel Cutting flame consists of two main components: oxygen and fuel gas. Because acetylene is the most popular fuel gas used, it is the fuel gas discussed in this portion of the chapter. The alternative fuel gases are discussed later in this chapter.

It is often assumed that the cutting of metal with the Oxy-Acetylene Cutting process is accomplished by melting the metal with the extremely hot flame, but the metal is actually burned, not cut. The temperature of an oxy-acetylene flame can reach 6,000°F (3,315°C). This temperature is high enough to melt carbon steel, but the resulting cut is not the quality demanded by industry. Carbon steel is the most common metal cut with this process; thus, it is the metal used for this discussion.

The flame is used to heat the steel to its **kindling temperature**. Kindling temperature is the temperature at which a material will begin to burn or oxidize when exposed to oxygen. In order for a material to burn, it requires three components: heat, fuel (steel), and oxygen. If you remove any of the three components, oxidization stops. If you reduce the amount of any of the three components, oxidization becomes slow or insufficient.

The flame is a mixture of oxygen and acetylene that when applied to steel causes the steel to heat up. When the steel reaches its kindling temperature, a

FIGURE 8-11 An Oxy-Fuel
Cutting torch

narrow pressurized stream of pure oxygen is added. The addition of pure oxygen causes the heated steel to oxidize. As the steel is oxidized, it is blown away by the pressure of the oxygen stream. **FIGURE 8-11** shows an example of a cutting torch and tip.

The oxygen and acetylene, which are the two gases used, are stored in two separate cylinders. Each gas is supplied to the torch through individual regulators, flashback arrestors, and hoses. At the torch handle, there is an oxygen valve and an acetylene valve. When both valves are opened, the acetylene and a portion of the oxygen are mixed together in the mixing chamber. The mixing chamber is either located in the torch handle or in the head of the torch. The remainder of the oxygen is channeled to the cutting jet.

The mixed gases from the mixing chamber then flow out of the cutting tip, where they are used to support the cutting flame. The flame will be located only outside the cutting tip when the torch is operating correctly.

The heat of the flame is then used to raise the temperature of the steel to its kindling temperature. Once the kindling temperature is reached, the operator squeezes the cutting lever, which allows a stream of oxygen to flow out of the center of the cutting tip. The combination of the steel at its kindling temperature and the stream of pure oxygen causes the steel to oxidize. The oxidized steel is then blown away by the cutting jet, exposing more steel to the process.

The benefits of this process are:

- **Simple to use:** Training time needed to learn the correct flame adjustment and cutting technique is short (see the Technique section).
- **Lower cost:** Equipment is a fraction of the cost of other common arc cutting processes.
- **Versatile:** It can be used for welding, brazing, cutting, and heat treating.
- **Portable:** There is no need for electrical power to operate the system. The system can be stored on the back of a utility vehicle and can be exposed to the elements. This makes this system perfect for repair and fabrication work in the field.

- **Cutting range:** It can be used to cut carbon steel, from thin to thick. Generally, changing tip sizes, gas flow rates, and travel speed is all that is needed to switch between different thicknesses. Maximum cutting thickness is determined by the system's gas withdrawal limitations (see the Gas Withdrawal Limitations section).

The drawbacks of the process are:

- **Slower cutting speeds:** This is true in comparison to Plasma Arc Cutting on thinner sections.
- **Wider heat-affected zone (HAZ):** The extra time needed to get to the kindling temperature and the slower cutting speed cause a larger area to be heated. Heat treatment may be required when cutting higher carbon or alloy steels.
- **Metal limitations:** Does not work well on metals that have a high oxide melting temperature. Iron oxide—the oxide that forms on steel—melts around the same temperature of the steel. This causes the oxide to be burned away and exposes more high-temperature steel to the pure oxygen.

Aluminum oxide melts around three times the temperature of the aluminum it covers. Thus, when aluminum is heated with this process, the aluminum under the outside layer of oxide melts before the oxide layer can be burned and removed. This causes the aluminum to be melted, producing an inaccurate cut. **FIGURE 8-12** shows a picture of aluminum melted with an oxy-acetylene flame.

- **Larger cut width (kerf):** This is true in comparison to Plasma Arc Cutting.
- **Increased safety risk:** Extra precautions need to be taken to ensure that the fuel gases and oxygen are handled and stored properly. Because the system can be very portable, extra maintenance and frequent inspection of equipment are recommended.

EQUIPMENT   Other than the cutting torch, equipment used for Oxygen-Fuel Cutting is the same as Oxygen-Fuel Welding. Refer to Chapter 7 for details. **FIGURE 8-13** shows a picture of a typical oxygen-acetylene outfit set-up.

FIGURE 8-12 While not impossible, cutting aluminum with oxy-fuel is not a very good method to use

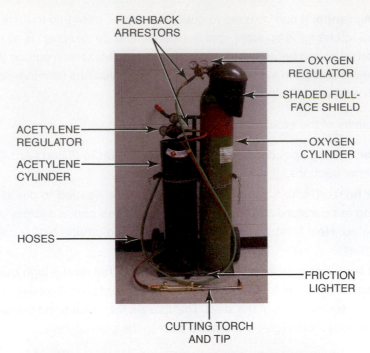

FLASHBACK ARRESTORS

OXYGEN REGULATOR

SHADED FULL-FACE SHIELD

ACETYLENE REGULATOR

OXYGEN CYLINDER

ACETYLENE CYLINDER

HOSES

FRICTION LIGHTER

CUTTING TORCH AND TIP

FIGURE 8-13 The Oxy-Fuel Cutting system

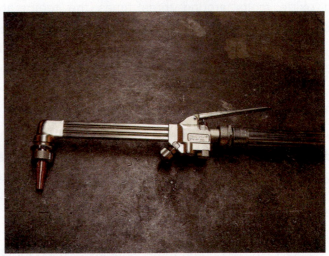

FIGURE 8-14 A typical cutting torch

TORCH Most cutting torches are very similar in design, with small differences in shape, size, and the placement of the cutting lever. **FIGURE 8-14** shows a typical cutting torch.

CUTTING TIP The size and type of cutting tip is determined by several factors:

- **Type of fuel gas:** Each fuel gas burns at different temperatures and operates at specific flow rates and pressures. These differences make it necessary to have special tips for each fuel gas. **FIGURE 8-15** shows different examples of tips for common fuel gases.

- **Thickness of steel:** As the steel thickness increases, the flow rate of the oxygen and fuel gas must also increase to provide greater heat input in order to raise the increased mass to its kindling temperature.

- **Type of operation (scrapping, demolition, general cutting, or heavy construction):** A tip that is designed for general cutting will not last very long in a scrapping or demolition operation. Tips used for these operations are exposed to more spatter, heat, and physical damage; thus, the tips must be designed to withstand the abuse.

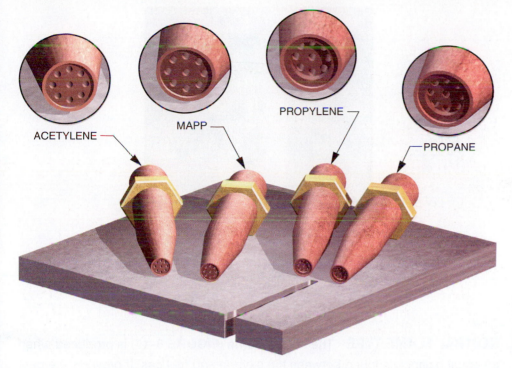

ACETYLENE

MAPP

PROPYLENE

PROPANE

FIGURE 8-15 Cutting tips have been specifically designed to work with different types of fuel gases.

- **Type of cutting process:** The versatility of the process allows it to be used for gouging and scarving. The tips for each of these processes are different from the standard cutting tips.

- **Condition of the steel's surface (clean, rusted, painted, or coated):** The cleaner the steel, the easier and faster it is to cut. Iron oxide (rust) does not transfer heat as well as steel; thus, it takes longer to reach the kindling temperature. Paint and coatings (galvanized or aluminized) also slow the rise in temperature. As a safety note, coatings can also produce gases and vapors that can be harmful to the operator.

FRICTION LIGHTER Friction lighters (**FIGURE 8-16**) are the safest way to light an oxy-fuel torch. It is never safe to light a torch with a match, cigarette lighter, or another torch.

FIGURE 8-16 A friction lighter is the correct tool to light an oxy-fuel torch

FIGURE 8-17 A neutral flame has the correct mixture of oxygen and fuel gas.

NEUTRAL FLAME TYPE The neutral flame (**FIGURE 8-17**) is produced when an equal balance is found between the oxygen and fuel gas. It provides the best results when Oxy-Fuel Cutting. The operator can recognize a neutral flame by:

- The lengths of the inner cone and the outer cone are almost equal, with the inner cone recessed by 1/16 inch to 1/8 inch from the outer cone.
- The outer cone should have a relatively smooth shape.
- A slight hissing sound should be heard.

ALTERNATIVE FUEL GASES While acetylene is needed for Oxy-fuel Welding, there are many gas choices for Oxy-Fuel Cutting. Along with acetylene, there is MAPP, propylene, propane, and natural gas. What they all have in common is that they are all hydrocarbons.

There are several variables that determine which fuel gas should be used:

- **Flame temperature:** Each fuel gas, when combined with oxygen, has a different flame temperature. **TABLE 8-1** shows the flame temperatures of different fuel gases mixed with pure oxygen.
- **Cost:** The actual cost of a cut made with each fuel gas is determined by the combined values of gas price, volume of gas required, and travel speed.
- **Safety:** The use of acetylene requires special and more stringent safety precautions. Some situations may warrant the use of a safer fuel gas.

Fuel Gas	Flame Temperature
Acetylene	6,000°F (3,316°C)
MAPP	5,300°F (2,927°C)
Propylene	5,000°F (2,760°C)
Propane	4,900°F (2,704°C)
Natural Gas	4,600°F (2,538°C)

TABLE 8-1

1 Secure the cylinders.

2 Crack the oxygen and acetylene valves.

CAUTION: Always "crack" or briefly open the valve of a compressed gas cylinder before attaching the regulator. Make sure the valve is not pointed toward yourself or other individuals. Cracking the cylinder valve clears foreign material out of the valve seat. Foreign material may cause damage to the valve or regulator connections, which can cause leaking and damage to regulator. Only use chain to secure cylinders.

3 Attach the oxygen regulator.

CAUTION: Never use any petroleum-based lubricant on any pure oxygen fitting, hose, or regulator. Pure oxygen will reduce the kindling temperature of these lubricants to room temperature.

4 Attach the acetylene regulator.

5a Attach a flashback arrestor onto the oxygen regulator.

5b Attach a flashback arrestor onto the acetylene regulator.

Lighting Torch Setup

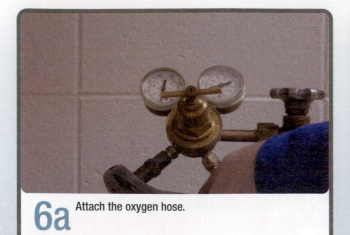

6a Attach the oxygen hose.

6b Attach the acetylene hose.

7 Attach the torch.

8 Turn out the flow pressure adjustment screw and then open the oxygen valve.

CAUTION: When first opening an oxygen cylinder valve, always turn the valve 1/4 turn, wait several seconds, and then open all the way. This step allows the initial in-rush of pressure to the regulator to stabilize. It is also important to stand to the side of the regulator behind the valve. If the valve were to explode, the valve may block some of the debris.

9 Turn out the pressure adjustment screw and then open the acetylene valve.

CAUTION: Only turn the valve a 1/2 to 3/4 turn. This will allow the quick shutting of the valve in case of a emergency. If a wrench or T handle is needed to operate the valve, leave the tool in place to also allow the quick shutting of the valve.

10a Set Oxygen operating pressure (always consult manufactures recommended pressure settings).

10b Set Acetylene operating pressure (always consult manufactures recommended pressure settings).

CAUTION: Never use acetylene at pressures greater than 15 psi (103 kPa). Never exceed the withdrawal limits of the acetylene cylinder.

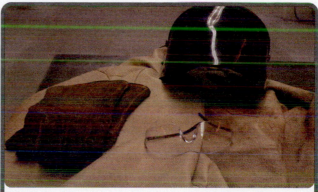

11 Put on your personal protective equipment.

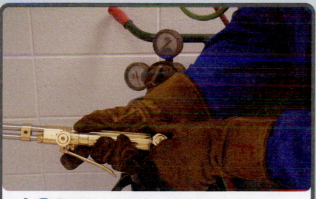

12 Purge the oxygen and acetylene lines. Then, double-check the pressure settings.

CAUTION: Purging both lines will eliminate any mixed gases from the torch. Gases can become mixed if the torch was incorrectly shut down the last time it was used.

13 Open the acetylene torch valve and then light the torch.

CAUTION: Only use friction, style lighters to light a torch. Never use a lighter, matches, or and other torch to light a torch.

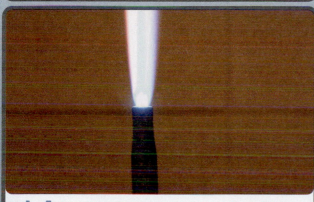

14 Open the torch oxygen valve. Adjust the acetylene and oxygen valves until a neutral flame is achieved.

As mentioned earlier, the oxy-fuel system is a very versatile and simple to use tool. Unfortunately, its simplicity often causes the dangers of its use to be taken for granted and forgotten. Too often, welders and operators will finish using the torch, shut off the torch valves, and walk away.

It is very important to properly shut down an oxy-fuel system. Failure to do so can result in damage to the system and poses a safety hazard. The following steps should be taken when finished using any oxy-fuel system. Consult the torch manufacturer's literature for the specific torch shutdown procedure.

Torch Shutdown

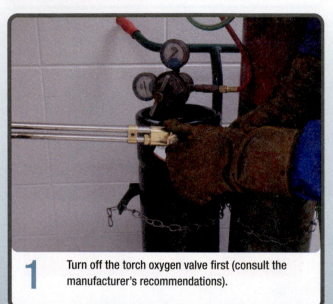

1 Turn off the torch oxygen valve first (consult the manufacturer's recommendations).

2 Turn off the torch acetylene valve (consult the manufacturer's recommendations).

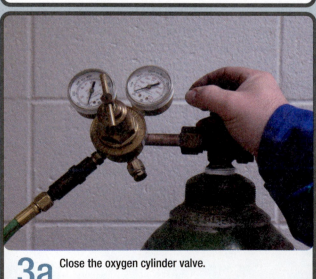

3a Close the oxygen cylinder valve.

3b Close the acetylene cylinder valve.

4 Purge the oxygen and acetylene lines.

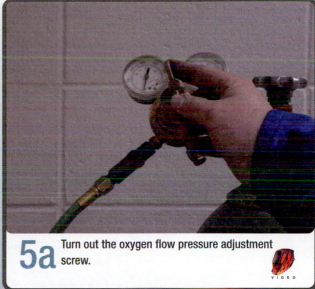

5a Turn out the oxygen flow pressure adjustment screw.

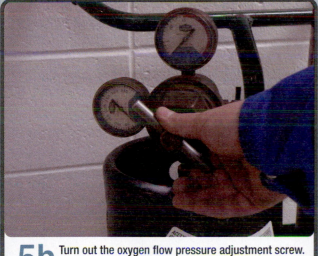

5b Turn out the oxygen flow pressure adjustment screw.

SAFETY PRECAUTIONS

- Wear eye and ear protection. The light intensity of the cutting arc requires the use of a welding helmet with a lens shade of 5 to 6. Arc and the high velocity air stream produce a high noise level.

- Use ventilation. Fumes and vapors produced by the process should always be removed from the area. Extra care should be taken when working with metal covered with paint, coatings, or cladding. Individual tolerances and safety regulations may also warrant the use of breathing protection.

- Always know where sparks and molten metal are landing: Always remove combustible and flammable materials from the cutting area. When finished cutting, always sweep and clean up the cutting area. This helps remove and extinguish hot embers.

- Never cut on concrete or rocks: Heat from the flame and slag can cause moisture in concrete or rocks to turn to steam. The steam will then cause the concrete or rock to explode.
- Never cut on piping or vessels that contain flammable materials.
- Never cut a vessel that has held flammable materials. All vessels that have held flammable materials contain vapors from that material. If cutting must be done, proper procedures must be followed.
- Never use any petroleum oil or grease to lubricate any part of an oxy-fuel system. Pure oxygen lowers the kindling temperature of any petroleum product, causing a fire.
- Never use pure oxygen to blow off cloths and other combustibles. Pure oxygen will lower the material's kindling temperature and cause the material to ignite easier.
- Never operate acetylene at pressures greater than 15 psi (103 kPa), and never exceed the withdrawal limitations of the acetylene cylinder.
- Operators should always use caution when using the system to remove sections of a structure or metal under stress. Cutting may cause the structure to shift, collapse, or spring out, causing injury.

AIR CARBON ARC CUTTING

Cutting is performed by the use of heat. The heat is produced by an electrical arc, conducted between an electrode and the base metal, to melt the base metal. A stream of compressed air is then used to blow the molten metal away.

The process is a very convenient process for welders in that the arc is produced from a welding power source and the air is from an air compressor. Both pieces of equipment can be found in most welding shops and on most jobsites.

PROCESS The Air Carbon Arc Cutting system (CAC-A) consists of three main components: an electrical arc, a stream of compressed air, and a carbon electrode. The process is very similar to SMAW in that an electrical arc is transferred across a gap between an electrode and base metal. Unlike SMAW, the electrode is not intended to be used as filler metal but instead is used only as the electrical current conductor. **FIGURE 8-18** shows an air carbon arc torch with an electrode about to gouge mild steel.

Air Carbon Arc Cutting uses the same power source as SMAW. The polarity of the process is generally set up as DCEP. Polarity can also be AC, which requires an AC electrode. When the positive electrode comes in contact with the negative base metal, electrical current is conducted. The operator then pulls the electrode back away from the base metal, producing an arc gap.

Just like the SMAW process, the heat produced from the current flow causes the melting of the base metal. As the base metal melts, a high velocity stream of air flows parallel to the electrode and blows the molten base metal away.

The process can be used to cut thicknesses from sheet metal to thick plate, with changes in welding output, electrode size, and operator manipulation. It is most commonly used on carbon steels but can be used on any metal with varying

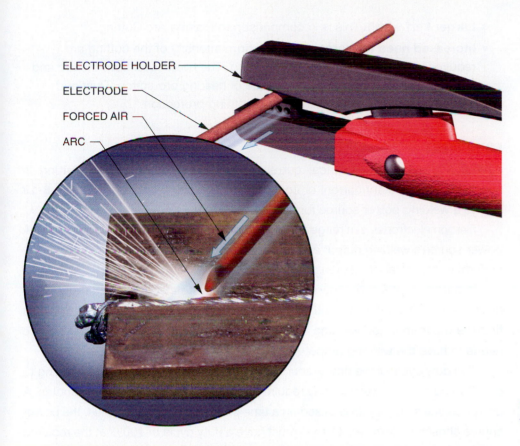

ELECTRODE HOLDER

ELECTRODE

FORCED AIR

ARC

FIGURE 8-18 An air carbon arc torch with an electrode

degrees of success. By changing welding output settings and electrode angle, the process can be used to bevel or gouge.

The benefits of this process are:

- **Simple to use:** Training time needed to learn the correct setup and cutting technique is short.
- **Low cost:** A shop that already has a large enough welding power source and compressor only needs to purchase a special electrode holder and electrodes to begin cutting.
- **Portable:** Engine-driven power sources can be used to supply cutting amperage. It is important that the operator compensates for changes in the power source duty cycle when the generator part of the power source is used to run an air compressor. This makes this system perfect for repair and fabrication work in the field.
- **Cutting range:** Can be used to cut from thin to thick base metals. Generally, changing the electrode diameter and cutting the amperage and travel speed are all that is needed to switch between different thicknesses. The maximum cutting thickness is determined by the amperage range and duty cycle of the welding power source.
- **No metal limitations:** This can be used on ferrous and nonferrous metals.

The drawbacks of the process are:

- **Requires electrical power:** This process must be operated off of the building's primary or generator power.
- **Carbon contamination in cut area:** Carbon from the electrode can build up along the surface of the cut edge. The surface may need to be removed, especially when cutting aluminum and stainless steel.

- **Larger kerf width:** This is in comparison to Plasma Arc Cutting.
- **Increased need for personal protection:** Intensity of the cutting arc requires lens shades equal to arc welding. The noise level from the arc and high-velocity air stream require the use of hearing protection. Dust and smoke may also warrant the use of breathing protection.

EQUIPMENT

POWER SOURCE   The electrical current used for Air Carbon Arc Cutting is supplied by a constant current or constant voltage power source. **FIGURE 8-19** shows a welding power source for carbon arc cutting.

Carbon electrodes can range from 0.25 inches (6.35 mm) to 1 inch (25.4 mm). A power source's welding output and duty cycle must be high enough to adequately perform within the electrode range.

It is important for the operator to understand the electrode size required to finish the job. Once the size of the electrode is determined, the operator needs to find the recommended welding amperage for the electrode. The power source needs to have the welding output range match the electrode.

The duty cycle of the power source must also be considered before starting to cut. Cutting a longer section may require a cutting time longer than 10 minutes. A power source's duty cycle is based on a time period of 10 minutes; thus, the power source chosen to complete this job must have a duty cycle of 100% at the required welding output. Trying to force an undersized power source to complete a job will cause excessive downtime and possible damage to the power source.

ELECTRODE AND WORK CABLE   The electrode and work cable must have an amperage rating high enough to conduct the proper amount of amperage for the amount of time required to finish the job. Using a cable that is too small causes voltage drops, which negatively affect the cutting performance.

Cable length must also be considered. When the total length of the electrode and work cable increase, so must the amperage rating of the cables. A common

FIGURE 8-19 An engine-driven power source

Source: Photo courtesy of Miller Electric Manufacturing Co.

mistake made by operators is to use several smaller lengths of cable connected together to make a long enough cable. Each time a connection is made, there is an increase in the amount of resistance in the cutting circuit. Resistance causes heat and voltage drops, which again can cause a decrease in cutting performance. Connections should be kept to a minimum. If connections must be made, the mating surfaces of the connections must be clean and held together as tightly as possible.

The work clamp must also be considered. Often, the cutting process is performed on base metal that is corroded or painted. The covering on the base metal should be removed where the work clamp is connected. If the coating is not removed, the clamp must be able to provide enough jaw pressure to break through the coating.

ELECTRODE HOLDER   The Air Carbon Arc Cutting electrode holder is similar to a SMAW electrode holder except for the connection and jet hole for the compressed air. **FIGURE 8-20** shows an air carbon arc torch. The holder is designed to hold an electrode at several different angles.

The holder must have a high enough amperage rating for the required amperage. Using a underrated holder will cause overheating and voltage drops.

CARBON ELECTRODE   The role of the electrode in the Air Carbon Arc Cutting process is to carry the cutting amperage used to melt the base metal. **FIGURE 8-21** shows an air carbon electrode.

The larger the diameter, the greater the current-carrying range of the electrode. **TABLE 8-2** shows the required amperage for electrodes, ranging from 5/32 inches to 5/8 inches (4 mm to 15.9 mm) for DC electrodes with DCEP.

Electrodes are made of carbon or graphite. The graphite version is less expensive and cannot carry as much current. Copper-coated electrodes are also available. Their advantages are that they have a higher current-carrying capability and melt at a slower rate.

AIR SUPPLY   Air that is used to blow away molten metal is generally supplied from an air compressor. Air pressure ranges from 80–100 psi (552–690 kPa). The diameter of the electrode determines the flow rate of gas, with 5 cfm (2.5 L/min)

FIGURE 8-20 Air Carbon Arc Cutting electrode holder

FIGURE 8-21 Electrodes for Air Carbon Arc Cutting can be uncladded or copper-cladded.

Electrode Diameter	Cutting Amperage
5/32 inches (4 mm)	90–150
3/16 inches (4.8 mm)	150–200
1/4 inches (6.4 mm)	200–400
5/16 inches (7.9 mm)	250–450
3/8 inches (9.5 mm)	350–600
1/2 inches (12.7 mm)	600–1,000
5/8 inches (15.9 mm)	800–1,200

TABLE 8-2

Source: American Welding Society (AWS) Welding Handbook Committee, 2004, *Welding Handbook*, 9th ed., Vol. 2, *Welding Processes*, Miami: American Welding Society.

for the smallest diameter electrode to 50 cfm (24 L/min) for the largest. An air compressor that is too small will negatively affect the cut quality.

TECHNIQUE

1. Select the electrode size. The diameter of the electrode is determined by the thickness of the metal to cut.
2. Connect the cables. The process polarity is either DCEP or AC.
3. Connect the work clamp. The clamp should be located as close to the work area as possible. Care should be taken to ensure that the cable and clamp are not in the way of the operator while cutting.
4. Set the power source cutting amperage. The larger the electrode diameter, the higher the cutting amperage.
5. Turn the power source output on. Unless a remote on/off switch is used, output will be turned on at the control panel of the power source.
6. Open the compressed air valve. The valve can be located at the electrode holder or at the air supply.
7. Insert the electrode into the holder.
8. Start the arc. Tap or scratch the tip of the electrode on the base metal and then pull the tip away. This process is very similar to SMAW; if the operator pulls the tip too far away, the arc will go out.

9. Start the travel. The travel should be in a push (forehand) direction, with a torch angle of around 45°.
10. Stop the arc. When finished, pull the electrode tip farther away from the base metal, and the arc will stop.

VIDEO

SAFETY PRECAUTIONS

- Wear eye and ear protection. The light intensity of the cutting arc requires the use of a welding helmet with lens shade of 10 to 14. The arc and high-velocity air stream produce a high noise level.
- Use ventilation. Fumes and vapors produced by the process should always be removed from the area. Extra care should be taken when working with metal covered with paint, coatings, or cladding. Individual tolerances and safety regulations may also warrant the use of breathing protection.
- Always know where sparks and molten metal are landing. Always remove combustible and flammable materials from the cutting area. When finished cutting, always sweep and clean up the cutting area. This helps remove and extinguish hot ambers.
- Never cut on concrete or rocks. Heat from the arc and molten slag can cause moisture in concrete or rocks to turn to steam. The steam will then cause the concrete or rock to explode.
- Never cut on piping or vessels that contain flammable materials.
- Never cut a vessel that has held flammable materials. All vessels that have held flammable materials contain vapors from that material. If cutting must be done, proper procedures must be followed.
- Operators should always use caution when using the system to remove sections of a structure or metal under stress. Cutting or gouging may cause the structure to shift, collapse, or spring out, causing injury.
- Never cut while standing in water. Caution should be used when cutting in damp environments. Always wear dry clothes, when cutting.

PLASMA ARC CUTTING

Plasma Arc Cutting (PAC) is performed by the use of heat. The heat is produced by a combination of an electrical arc traveling through a constricted column of supplied gas, usually compressed air, and the cutting current being conducted through the base metal. The column of air flow then blows away the molten metal.

Plasma Arc Cutting provides a very accurate and fast method of cutting most conductive metals.

PROCESS The plasma arc system consists of a power source and a plasma torch. The torch is a unique and critical part of the system in that it is designed to facilitate the production of the plasma arc. **FIGURE 8-22** shows a plasma arc torch cutting mild steel.

When the operator squeezes the torch's trigger, compressed air begins to flow through the torch and out the tip's orifice. Next, an electrical arc jumps from the negative electrode to the positive tip. **Plasma** is the state of matter that is

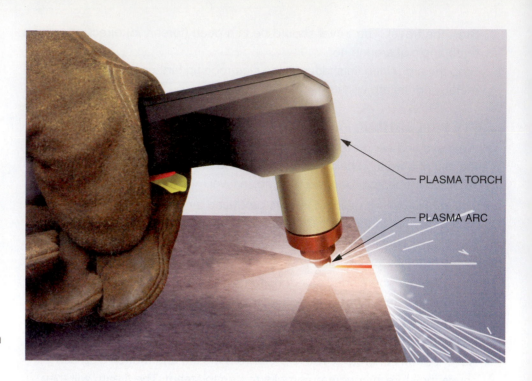

FIGURE 8-22 Plasma arc torch cutting mild steel

PLASMA TORCH

PLASMA ARC

produced when an electrical arc travels through the supplied gas stream. The gas is ionized, which generates heat to around 30,000°F (16,000°C).

Plasma is the fourth state of matter, along with solid, liquid, and gas.

When gas is turned into plasma, its volume expands. This expansion of gas volume forces the plasma stream out of the small orifice of the tip. This narrow, high-velocity stream of hot ionized gas is propelled out of the orifice. The stream has an electrical charge, so it will bend back toward the oppositely charged tip. This arc is called the pilot arc.

When the operator brings the torch tip down close to the positive base metal, the pilot arc transfers to the base metal. At this point, the power source senses that current flow is being conducted through the base metal and will increase the amperage. This arc is called the cutting arc. The base metal is melted by the plasma arc, and the molten metal is blown away by the gas column.

The plasma cutting arc of very hot ionized gas flows through a restricted opening. This combination produces a very accurate, fast cutting process.

The process can be used to cut thicknesses from sheet metal to plate, with changes in cutting amperage, tip size, and travel speed. This process will cut almost all conductive metals. By changing cutting amperage settings and torch angle, the process can be used to bevel or gouge.

The benefits of this process are:

- **Simple to use:** Training time needed to learn the correct setup and cutting technique is short.
- **Increased cutting speed:** Within the cutting range of the plasma cutter, the cutting speed—compared to the Oxygen-Fuel Cutting process—is much faster.
- **High-quality cut finish:** The finish of the cut can be controlled by correctly setting the cutting amperage and controlling travel speed within the cutting range of the plasma cutter.

- **Portable:** Inverter plasma cutters available today allow systems to be carried by a single welder. Some models are designed or used with 110 VAC or single-phase 220 VAC. Low primary current draw also allows units to be operated off an engine-driven power source or generator. Plasma cutters are available that have integrated air compressors.

- **Can be used for gouging:** The gouging of base metal or welds can be done with a change in the cutting angle. Some power sources have a gouging setting.

- **Few metal limitations:** It can be used on most conductive metals.

- **Limited or no contamination of cut edge:** Surface oxidation caused when molten metal is exposed to the compressed air stream is expected. When other shielding gases, such as nitrogen and argon, are used, oxidation is greatly reduced if not eliminated.

The drawbacks of the process are:

- **Requires compressed air or gas:** The process will not function without compressed air or gas. Moisture that is very common in a compressed air system can reduce the life of the electrode and tip. Air line dryers are highly recommended. Air line oilers must never be installed in the line when compressed air is used as the cutting gas.

- **Limited consumables life:** Consumables such as the tip and electrode need to be periodically replaced. An improper amperage setting, poor technique, and moisture in compressed air supply can greatly affect consumable life.

- **Increased need for personal protection:** The intensity of the cutting arc requires lens shades. The noise level from the arc and the high-velocity air stream require the use of hearing protection. Dust and smoke may also warrant the use of breathing protection.

EQUIPMENT   Equipment, or Plasma Arc Systems (**FIGURE 8-23**), consists of a Plasma Arc Power Source (**FIGURE 8-24**), a work cable and lamp, and a Plasma Arc Cutting Torch.

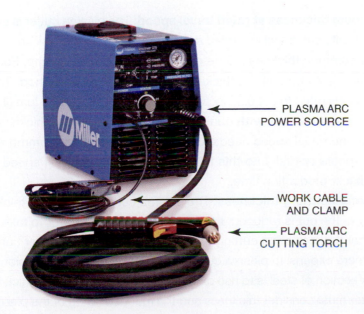

PLASMA ARC POWER SOURCE

WORK CABLE AND CLAMP

PLASMA ARC CUTTING TORCH

FIGURE 8-23 A Plasma Arc Cutting system
Source: Photo courtesy of Miller Electric Manufacturing Co.

FIGURE 8-24 Plasma Arc
Cutting power source
Source: Photo courtesy of Miller Electric
Manufacturing Co.

POWER SOURCE (PLASMA CUTTER) The power source used for Plasma Arc Cutting is a DC constant-current power source. The plasma cutting power source differs from a welding power source in that its open-circuit and load voltages are much higher. The higher voltages allow the plasma arc to be conducted at greater arc gap lengths than a welding arc. A plasma cutter's amperage range determines its cutting thickness range. Manufacturers do not rate their plasma cutters on cutting thickness alone. Speed of travel and quality of cut are also very important variables. When purchasing a plasma cutter, an operator should know the most common thickness of steel that will be cut. Ideally, the plasma cutter purchased should be able to provide a high-quality cut at this thickness. Because time is money, the faster the travel speed, the better. Manufacturers generally use three ratings to help purchasers determine the plasma cutter they need. Each rating is based on cutting carbon steel. Ratings are reduced when cutting highly conductive metals, such as aluminum.

Maximum thickness at rated travel speed: The plasma cutter is designed to provide a quality cut, travel at a rated speed, and cut a specific thickness. A operator's most common thickness to be cut should fall into this rating. For example, 0.187 inches (4.75 mm) is the thickness most often cut in a shop. The chosen plasma cutter is rated to cut 0.25 inches (6.35 mm) traveling at 10 ipm (4.23 mm/s).

Maximum thickness with quality cut: Cuts made on this thickness are high quality, but the travel speed needs to be reduced. The most common thickness to be cut should not fall into this thickness. A reduced travel speed costs the operator in lost production time.

Maximum cutting thickness: The plasma cutter will cut this thickness, but the quality of the cut is reduced. It is important for an operator to take a realistic look at what the maximum cutting thickness needs to be. A larger cutting range equals a more expensive plasma cutter. An alternative to the occasional need to cut a thick section of steel is to use oxy-fuel instead. When cutting thick aluminum, the operator must consider thickness and the reduced rating of the plasma cutter.

Along with the three ratings, manufacturers also provide a duty cycle for their plasma cutters. The duty cycle is based on a 10-minute cycle. Cutting a longer section may require a cutting time longer than 10 minutes. A plasma cutter chosen to complete this job must have a duty cycle of 100% at the required cutting output. Trying to force an undersized plasma cutter to complete a job will cause excessive downtime and possible damage to the power source.

Some plasma cutters have integrated air compressors. These models are designed for use in the HVAC industry. Technicians can use the plasma cutter throughout the jobsite without the need of a separate compressor. Generally, these plasma cutters are smaller and have limited cutting ranges.

TORCH Plasma torches are designed only for the purpose of cutting metal with a plasma arc. **FIGURE 8-25** shows a plasma arc torch. The torch not only has to produce and control the plasma arc, but it must do so while protecting the operator from the very high temperature and high voltage. The torch also must be able to continue to hold up in the tough environment of the welding industry.

Most torches consist of the torch body, power cable, swirl ring, electrode, tip, cup, and shield. Each individual component is designed to perform a specific task in the production of a plasma arc. If any of the components are damaged, worn out, or the wrong type or size, the plasma arc may operate incorrectly or not be produced at all.

Torch body: A well-designed plasma torch body allows the operator to move the torch across the metal without stressing the wrist. Just like welding, if the operator's wrist and arm are in an uncomfortable position, the quality of the plasma cut will be reduced. Before purchasing a system, an operator should try a few cuts. If cutting is not possible, at least hold on to the torch body. Try moving it forward, backward, vertical up and down, and overhead. If it does not feel comfortable, then try a different model or brand.

Every torch has a trigger that when squeezed starts the flow of air and the pilot arc. Most torches will have a trigger safety that needs to be moved before the trigger can be closed. This safety is used to prevent accidental triggering of the system.

FIGURE 8-25 Plasma arc cutting torch

The torch body is the weakest link in a plasma-cutting system. If a torch does not look like it will stand up to the abuse of a shop, then it may not last. Trying to cut costs, on the torch may end up costing more in the end.

Power cable: The larger the cutting range of the plasma cutter, the larger the diameter of the power cable needs to be. Using an undersized torch to cut thicker metal can cause overheating and damage to the torch. Power cables are generally available in 10-feet (3.048 m), 25-feet (7.62 m) and 50-feet (15.24 m) lengths. Home hobbyist or light industrial plasma cutters use the shorter torches. Industrial systems often allow the purchaser to choose between 25-feet (7.62 m) and 50-feet (15.24 m) lengths.

Combined together within the power cable are the conductor, air line, trigger leads, cup detect leads (in some models), and, in some larger systems, water lines. The covering over the cable is extremely important and must be able to withstand sparks and heat. Leather and synthetic covers can be placed over the power cable, if additional protection is needed.

Swirl ring: The swirl ring is designed to cause the supplied gas that is flowing through the torch body to swirl or spin. This spinning action helps to produce a tighter plasma column as it passes through the hole in the tip. **FIGURE 8-26** shows two different-sized swirl rings.

Electrode: The electrode is the electrical path for the cutting current in the torch body. It is made of copper, with an insert of hafnium or tungsten. The insert is designed to prolong the life of the consumable electrode. Electrode size is determined by the maximum cutting amperage of the plasma cutter. **FIGURE 8-27** shows two different electrodes.

Tip: The tip serves two purposes in the torch. First, it has the opposite polarity of the electrode. The pilot arc jumps from the electrode to the tip. Second, at the top of the tip is the orifice in which the plasma arc leaves the torch. **FIGURE 8-28** shows two different-sized tips.

The size of the tip orifice is determined by the amount of cutting current. As the current increases, so must the size of the hole. Each has a current rating. Using a tip beyond its rating will cause the orifice to be enlarged. Enlargement of

FIGURE 8-26 Swirl rings used in the plasma arc torch come in different sizes.

FIGURE 8-27 Electrodes used in the plasma arc torch come in different sizes.

FIGURE 8-28 Tips used in the plasma arc torch come in different sizes.

the orifice causes the plasma column to become larger, thus causing the kerf to become larger and a loss of some of the control of the plasma arc.

Both the electrode and the tip are considered consumables and need to be changed over time. Replacement consumables are expensive, and replacing them causes downtime. The operating life of the consumables can be extended by following these tips:

- **Limit cutting amperage:** The thicker the metal being cut, the higher the amount of current needed to perform the job. Higher current levels reduce the life of the consumables. If the plasma cutter has an amperage control, it is a good practice to turn the amperage setting down when cutting thinner sections of metal. This will increase consumable life.
- **Use air dryers for air-supplied cutting gas:** Moisture trapped in air lines enters the torch body and greatly reduces consumable life. Moisture can also negatively affect the performance of the plasma arc. Always install a dryer in the compressed air line. It is also important to frequently drain the compressor and inline trap. Do not use an air line that has an oil install in the line. **FIGURE 8-29** shows an air dryer.

FIGURE 8-29 Air used for Plasma Arc Cutting must be clean and dry.

- **Limit pilot arc use:** The pilot arc is the arc that jumps from the electrode to the tip and forms the plasma arc. The pilot arc also causes arcing on the tip. This arcing causes damage to the orifice and increases resistance. Increased resistance results in overheating of the tip. It is good practice to limit the use of the pilot arc by reducing the number of starts and stops. Most plasma cutters turn off the pilot arc when the cutting arc has been established. Trying to use longer cut lengths and preplanning cut paths can limit the number of stops and restarts. Some plasma cutter models have an additional setting that allows the pilot arc to stay on when cutting expanded metal. Use this setting only when necessary.

- **Use the correct technique when pierce-cutting and gouging:** The angle for pierce cuts and gouging should be around 45°. This angle allows the molten metal to be pushed out away from the tip. Use edge starts instead of the piercing starts whenever possible.

Retaining cup: This holds the consumable parts in place and insulates them from the base metal. Some torches have a safety switch that detects if the cup is in place. The cup must be placed on tight enough to close the switch or else the plasma cutter will not operate. **FIGURE 8-30** shows two different-sized retaining cups.

Shield: The shield protects the front of the torch from the molten metal. There are two types of shields: standard and drag. **FIGURE 8-31** shows drag and standard shields.

The standard shield has a hole in the middle from which the tip sticks out. This shield is designed to deflect molten metal away from the cup and front of the torch body. The tip is not protected. Correct standoff—around 0.125 inches (3.175 mm) from the tip to the base metal—must be maintained, not allowing the tip to touch the base metal. Allowing the tip to touch the base metal causes damage to the tip. Too long of a standoff can cause the arc to go out and or inconsistent cutting performance.

FIGURE 8-30 Nozzles used in the plasma arc torch come in different sizes.

DRAG SHIELD

STANDARD SHIELD

FIGURE 8-31 The drag shield may be dragged on the surface of the metal while cutting.

The drag shield differs from a standard shield in that it covers the tip and insulates it from the base metal. As the name implies, the drag shield allows the operator to place the shield directly on the base metal while cutting. The shield is designed to maintain the correct standoff when in contact with the base metal.

The ability to drag the shield is an advantage to the novice operator just learning to cut and the operator that is struggling to maintain the correct stand-off. The disadvantage of the drag shield is that it is impossible to see the leading edge of the cut. This makes following a line very difficult and can negatively affect the accuracy of the cut.

AIR SUPPLY Typically, most plasma cutting is done with compressed air supplied from an air compressor. Air pressure ranges from 80–100 psi (552–690 kPa). The diameter of the electrode determines the flow rate of gas, with 5 cfm (2.5 L/min) for the smallest electrode diameter to 50 cfm (25 L/min) for the largest. An air compressor that is too small negatively affects the cut quality.

Compressed air is the most common gas used, but it is not the only one. Nitrogen, argon, and a mix of nitrogen/argon are used. When to use them and what types of gas used are determined by required production rates, types of metal, and thickness of metal. Benefits of each gas need to be weighed against the extra cost when contemplating their use.

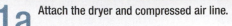

1a
Attach the dryer and compressed air line.

1b
Attach air line to compressed air supply.

2
Connect the plasma power source to the primary supply.

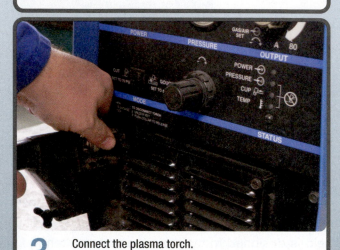

3
Connect the plasma torch.

4
Install the components in the torch head.

CAUTION: If components are installed incorrectly or if the retaining cup is not tight enough, the system will not operate. There is a small micro switch that has to be closed before cutting can start.

5
Turn on the plasma power source.

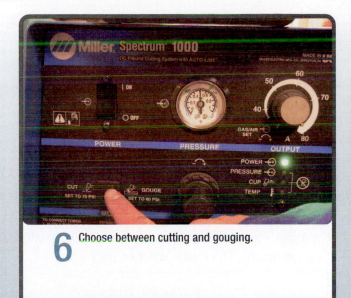

6 Choose between cutting and gouging.

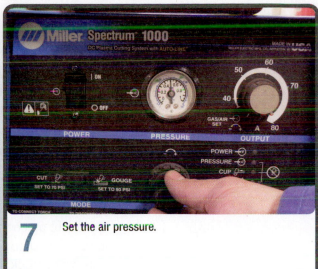

7 Set the air pressure.

8 Set the cutting amperage (always set as low as possible).

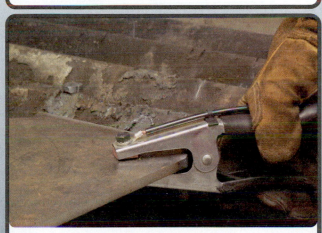

9 Attach the work clamp as close to the cut as possible and as convenient.

10 Disengage the trigger safety and then squeeze the trigger.

SAFETY PRECAUTIONS

- Wear eye and ear protection. The light intensity of the cutting arc requires the use of a welding helmet with lens shade of 4 to 8. The arc and high velocity air stream produce a high noise level.

- Use ventilation: Fumes and vapors produced by the process should always be removed from the area. Extra care should be taken when working with metal covered with paint, coatings, or cladding. Individual tolerances and safety regulations may also warrant the use of breathing protection.

- Always know where sparks and molten metal are landing. Always remove combustible and flammable materials from the cutting area. When finished cutting, always sweep and clean up the cutting area. This helps remove and extinguish hot embers.

- Never cut on concrete or rocks. Heat from arc and slag can cause moisture in concrete or rocks to turn to steam. The steam will then cause the concrete or rock to explode.

- Never cut on piping or vessels that contain flammable materials.

- Never cut a vessel that has held flammable materials. All vessels that have held flammable materials contain vapors from that material. If cutting must be done, proper procedures must be followed.

- Operators should always use caution when using the system to remove sections of a structure or metal under stress. Cutting may cause the structure to shift, collapse, or spring out, causing injury.

- Never cut white standing in water. Caution should be used when cutting in damp environments. Always wear dry cloths when cutting.

LASER BEAM CUTTING

Laser Beam Cutting (LBC) is performed by the use of a narrow, high-intensity electromagnetic energy in the form of a light beam. The energy of the light is focused into a very constricted beam, 0.004 to 0.02 inches (0.1 to 0.5 mm) in diameter. When the laser beam is focused on a base metal, the energy of the beam will cause the metal to quickly melt.

The process is extremely versatile in that there are very few materials it cannot cut. Cutting is generally done on the workpiece in the horizontal position over a cutting bed. Most systems use two axes, with a control system moving the nozzle or metals with an X and Y axis. Some systems utilize a third axis, allowing cutting in the horizontal and vertical positions.

PROCESS   There are many types of lasers produced today, from low-power pointers used in the classroom to high-power versions that are used to measure the distance between Earth and the moon. In cutting, there are two main types of lasers: the gas CO_2 and the solid-state Nd:YAG. **FIGURE 8-32** shows a laser beam cutting head.

Laser is an acronym for Light Amplification by Stimulated Emission of Radiation. What that means in short is that the laser's light beam is produced by raising the energy level of specific matter to the point at which that matter releases

FIGURE 8-32 A head of a Laser
Beam Cutting machine

photons of light (stimulated emission of radiation). Once the photons of light are released, they are focused and refocused until the energy level is raised even further, producing the laser beam (light amplification). The best way to understand this is to break it down into sections.

The first part of the process is the stimulated emission of radiation. All matter is composed of atoms. Every atom has a natural energy level. If the atom is stimulated—by intense light or electrical current—the energy level of the atom will rise. This higher energy level is unnatural and causes the atom to release the extra energy in the form of photons of light. A common example of this process is the fluorescent bulb.

A fluorescent bulb contains mercury vapor. When electrical current is applied to the mercury vapor, the atoms of the vapor are raised to a higher energy level. In order to get back to the atoms' normal energy level, the atoms release photons of ultraviolet light energy. These photons travel through the space of the bulb when released. As the photons travel through the bulb, they run into the atoms of a phosphorus-type material that is also contained in the blub. The energy level of these atoms is now raised, resulting in the release of photons. The photons of the phosphorus materials are at a wavelength that is visible to the human eye, thus usable light is produced. The difference between a laser and the fluorescent bulb is that the photons of light emitted from the bulb are not focused and travel randomly in any direction.

The laser produces light in a similar fashion. The matter that is used to produce the high-energy photons is called the **lasing material**. The lasing material can be in the form of gas, liquid, or solid. In the case of cutting and welding lasers, the most common material is the gaseous CO_2 and the solid neodymium (Nd). The neodymium is stored as an impurity in a crystal structure consisting of yttrium, aluminum, and garnet (YAG). Thus, the name Nd:YAG laser.

To produce the ultraviolet photons, the lasing material needs to be stimulated. The CO_2 gas laser uses electrodes to conduct electrical current through the CO_2. The Nd:YAG solid-state laser uses high-intensity strobe lights. The result of the stimulation is that the lasing material's atoms release photons.

So far, the functioning of the laser and the fluorescent bulb are not that different. Both are releasing photons, but what makes the laser different is that the photons are not allowed to just randomly escape in any direction. The stimulation of the lasing material is done within a vessel. The vessel is opaque and will not allow the photons to escape. On the back end of the vessel, there is a reflecting mirror. On the front end is another mirror. As photons are released, they begin to bounce around inside the vessel. When a photon strikes another atom, the atom is stimulated again, thus releasing another photon. The process of photons being released, then striking another atom, and then another photon being released is the light amplification part of the laser's name.

While the amplification process is going on, the electrical or optical stimulation is either operating continuously or pulsing on and off. The continuous operation produces a constant level of laser beam energy.

To produce the laser beam, the built-up energy within the vessel needs a way of being released. This is done by allowing some of the light energy to escape through the front of the vessel. Both the front and back of the vessel are blocked by mirrors. The front mirror is different in that it is not 100% reflective; it allows a certain percentage of light to pass through it.

The light beam that leaves the front mirror is unique in that every single photon is traveling in the same direction at the same speed and wavelength. The beam also has a low divergence rate. This means that each photon travels in a very straight line, with each one traveling parallel to the next. This allows a laser beam to travel great distances without the loss of intensity from the spread of the beam's diameter and photons knocking each other out of the beam.

A laser is coherent light and has a wavelength that is invisible. Industrial lasers often have a laser light that travels at the wavelength, which produces a red to infrared light. Special precautions must be taken to ensure that operators and other individuals are protected from the laser beam. Both visible and invisible beams can cause damage to the skin and eyes.

The beam that leaves the front mirror has to be directed toward the cutting nozzle. This is done with a series of precisely aligned mirrors that reflect the beam at angles that direct the beam toward the cutting nozzle. Before leaving the cutting nozzle, the beam's diameter is reduced by the focusing lens. The smaller diameter beam that leaves the lens is now ready to be used for cutting.

Beam power level control is done by the laser system's controller. Depending on the type and thickness of the metal to be cut, the controller will vary the stimulation of the laser material. For example, if more cutting power is required, the controller will increase the duration of the peak stimulation pulse. This in turn will cause a greater production of photons, resulting in a higher cutting power.

The type of metal being cut is critical to the success of the cut. Every metal has a certain amount of thermal conductivity and reflectivity. Metals such as aluminum that have a higher thermal conductivity require a higher cutting power than stainless steel, which has a lower thermal conductivity. Thermal conductivity is always a factor in any cutting operation but is less of an issue because of the laser's intensity.

Reflectivity is the measure of percentage of light that is reflected away from the base metal and not used for cutting. A highly reflective metal such as

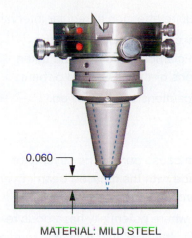

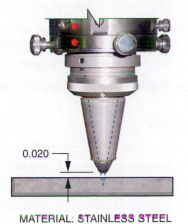

0.060

MATERIAL: MILD STEEL

0.020

MATERIAL: STAINLESS STEEL

FIGURE 8-33 For materials such as mild steel, the focus of the beam is on the top surface of the metal, whereas for materials such as stainless and galvanized steel, the focus of the beam is below the top surface.

stainless steel requires a higher cutting power to compensate for losses. The percentage of light that is reflected away from the metal is also a safety concern. Light beams that bounce off the base metal can damage equipment and injure individuals. Care should always be taken when cutting highly reflective materials.

The type of metal being cut also determines the focus distance between the end of the nozzle and the base metal. **FIGURE 8-33** shows a common focus distance for mild steel and stainless steel.

With the correct wattage setting, the heat produced by the high-intensity focused laser beam causes the material being cut to quickly heat up. For most materials, the heat level is so high that the area directly under the beam is vaporized. Removal of the vaporized material is done by a high-velocity jet of gas that flows parallel to the laser beam. Inert gases may be used to protect molten metal along the sides of the cut from being oxidized.

In the case of cutting carbon steel, the assist gas is pure oxygen. The oxygen in this case is used to burn the carbon steel away, similar to the Oxy-Fuel Cutting process.

The benefits of this process are:

- Operators with basic computer skills can be trained to operate a system.
- The kerf width is typically 0.008 inches (0.2032 mm).
- It has a narrow heat-affected zone.
- Tolerances of ±0.005 inches (±0.1 mm) can be achieved.
- The part can quickly go from design to completion. The cut surface can often be coated without any further processing.
- The drawing of the part can be saved to allow the part to be cut whenever needed.
- It has reduced scrape.

The drawbacks of the process are:

- Systems are a major financial investment.
- The system will take up valuable square footage. Systems are not designed to be moved easily.
- Maximum thickness is determined by the wattage output of the laser.
- Ventilation is required.

- Different components of the system will require attention at regular intervals.
- The nozzle of the torch can be damaged by crashing it down into the cutting material or traveling into hold-down clamps. Care must be taken when cutting material that has raised sections or that is warped or bent.
- The use of a laser requires extra safety precautions to be followed.

SAFETY

- The direct, unfocused (CO_2) laser beam can cause burns. The severity of burns can be much greater if contact is made with the focused beam. Even a stray reflected beam can permanently damage your eyesight.
- The nominal hazard zone (NHZ) is the area where potentially hazardous reflections can occur. For example, assume for the NHZ that a laser is running at 1000 watts with a 5-inch focus lens in place and that the entire beam is reflected from a workpiece surface. This region (where the reflected energy can exceed safe limits) would extend up to 14 inches from the nozzle horizontally and up to 22 inches vertically. If the beam is reflected from a tilted mirror-like surface, the danger zone can be up to 20 feet from the point of discharge.
- Some materials can emit ultraviolet radiation when struck by the laser beam. This invisible UV light is called secondary radiation. It can pass directly through clear safety glasses, causing the same kind of discomfort and possible damage as from watching electric arc welding.
- Never look directly at the point at which the laser beam comes into contact with the material.
- If viewing is required for troubleshooting or service, wear welding protective eyewear (AWS) grade 5 or better.
- Always ensure that the work area is well-ventilated.
- If the laser beam contacts your skin, it can cause serious cuts or burns. If the laser beam contacts your eyes, it can cause temporary or permanent blindness.
- Do not touch the surface of the laser lens and oscillator inside mirrors with your bare hands; they are made of the toxic material zinc selenide.
- Do not inhale the yellow vapors generated in the unlikely event the toxic coating on the lens is burned. This yellow vapor is as dangerous as mustard gas.

ABRASIVE WATER JET CUTTING

Abrasive cutting is performed by the use of a jet of high-pressure water and abrasive. Material is cut by the abrasive action of the abrasive-laden water jet stream flowing through the thickness of the material.

The process is extremely versatile in that there are very few materials it cannot cut. Cutting is generally done on the workpiece in the horizontal position over a cutting bed filled with water. Most systems have two axes, with a control system moving the nozzle with an X and Y axis. There are other systems that utilize multiple axes, allowing cutting in the horizontal and vertical positions.

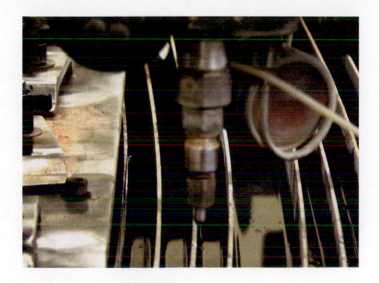

FIGURE 8-34 The nozzle of an Abrasive Water Jet Cutting machine

PROCESS There are actually two different types of water jet systems. The original system used a high-pressure water jet to cut. This type of system had thickness and material hardness limitations. Later, an abrasive jet system was invented. The abrasive jet uses an abrasive mixed into the high-pressure water jet. The addition of an abrasive allows the abrasive jet system to cut thicker and harder materials, such as metals. The name "water jet" is often used to describe both types of systems. The limitations of the water-only jet has made the abrasive water jet the most popular system for the welding and metal fabrication industries. **FIGURE 8-34** shows an abrasive water jet head.

An example of how an abrasive mixed with water can erode is the Grand Canyon. Over millions of years, the sand carried by the water of the Colorado River has slowly eroded the surface of the Colorado Plateau. After the sand erodes away the surface, the water carries away the material that has just been removed. The abrasive-laden water jet cuts by using the same abrasive action—but intensified. The jet typically has an operating pressure of 55,000 psi (379,000 kPA), moving across the material at 1,000 feet/sec (305 m/sec). This extremely fast high-pressure stream quickly cuts through the metal.

The abrasive jet uses water that is pumped by a high-volume pump. The water is then funneled though an orifice jewel. The orifice diameter of the jewel ranges from 0.010 inches to 0.014 inches (0.25 mm to 0.35 mm) and is used to focus the water, producing a very narrow, high-pressure speed jet. Orifice jewels (**FIGURE 8-35**) are typically made of ruby, sapphire, or diamond, with the diamond jewel having the longest lifespan but the highest cost.

The jet then enters a mixing chamber. As the water enters the chamber, it begins to spin, causing a spinning effect that pulls the abrasive into the water. The abrasive most commonly used for abrasive water jet systems is garnet. Garnet is a mineral that has long been used as a gemstone and abrasive. It is delivered to the mixing chamber dry and about the same consistency of sandblasting medium.

The amount of garnet mixed into the water is the same no matter what type, hardness, or thickness of material is being cut. Garnet is simply gravity-fed into the mixing chamber, with the size of the feeding tube and mixing opening

FIGURE 8-35 The jewel in the abrasive water jet nozzle may be made of a material such as sapphire or industrial diamond.

determining how much garnet is used. The tube that delivers the garnet to the chamber goes up to a gravity hopper. Filling the hopper can be done manually or by a pressurized system. **FIGURE 8-36** shows a garnet hopper.

The abrasive-laden water in the mixing chamber is then funneled into a tapered mixing tube. The tube is designed to refocus the energy of the water into the cutting jet. The high-pressure stream leaves the tip of the tube and flows through the thickness of the material being cut.

As the abrasive stream cuts through the material, it does not travel in a straight line but instead cuts in an arc. Because the stream travels in an arc, centrifugal force causes the garnet to flow across the surface of the cut, thus producing erosion. The water pressure and amount of garnet stays the same no matter what is being cut. The abrasive jet system compensates for different materials and thicknesses by changing the travel speed of the nozzle. The travel speed also needs to be adjusted for changes in the cut. For example, the speed needed for a straight cut is different from the speed around a curve. **FIGURE 8-37** shows the arc created by the abrasive stream through a section of mild steel.

The travel speed of the first system was determined by the system operators and their experience with different material hardness and thickness. These system

GARNET HOPPER

FIGURE 8-36 An abrasive—garnet—is fed through the nozzle and out the mixing tube under high pressure to cut any material without heat.

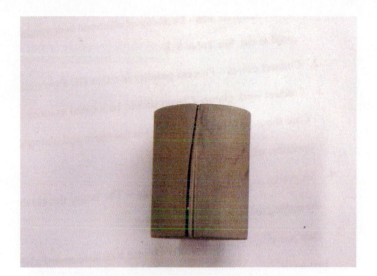

FIGURE 8-37 A slug cut out of mild steel with the abrasive water jet machine

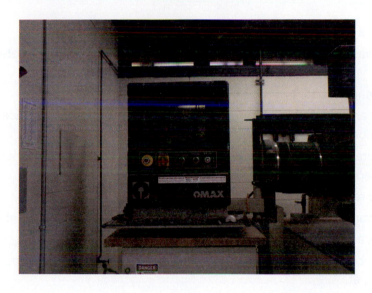

FIGURE 8-38 Abrasive water jet machines are computer-controlled, typically for tolerances of 0.005 inches.

operators needed to use a system for years and made lots of scrap in order to become proficient cutting just one type of material. Today, the system controller's database contains the information used to determine the correct travel speed for unlimited combinations of material that can be cut. **FIGURE 8-38** shows a controller.

Typical systems today use a DXF drawing to determine cutting paths. An operator simply downloads a drawing or uses the system's drawing software. The operator then selects the quality of the cut, ranging from high quality to coarse. The operator then enters the thickness and the type of material being cut. With this information, the controller its database to determine travel speeds. **FIGURE 8-39** shows a cutting bed.

Material to be cut is placed on the horizontal slats of the cutting bed. The tank of the bed is filled with water. The level of the water can be controlled, thus allowing the material to be submerged when cutting. Cutting underwater helps to reduce noise and dust from the abrasive. **FIGURE 8-40** shows an X-Y gantry system. The benefits of this process are:

- Operators with basic computer skills can be trained to operate a system.
- It uses filtered water and garnet, which are both readily available.

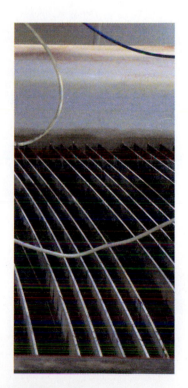

FIGURE 8-39 The cutting bed of an Abrasive Water Jet Cutting machine

FIGURE 8-40 Cutting underwater helps to reduce noise and prevents the spread of garnet used in cutting metals and other hard materials.

- It has a small kerf width.
- The water temperature does not get hot enough to produce a heat-affected zone.
- Tolerances of ±0.005 inches (±0.1 mm) can be achieved.
- The part can quickly go from design to completion. The cut surface can often be coated without any further processing.
- The drawing of a part can be saved to allow the part to be cut whenever needed.
- It has reduced scrap.
- Software can adjust the travel speed to compensate for stacked sheets.
- It has an extremely wide range of cutting thicknesses.
- It can be used to cut metals, composites, plastics, glass, ceramics, wood, stone, etc.

The drawbacks of the process are:

- Systems are a major financial investment.
- The system will take up valuable square footage. Systems are not designed to be moved easily.
- The jewel, mixing chamber, and mixing tube have a limited lifespan range. The quality of components, the condition of the water, and proper maintenance will determine where in the range the components will fail.
- Unfiltered tap water can be too hard. Dissolved deposits or suspended solids (hard water) can negatively affect the nozzle components, seals, valves, and pump.
- Different components of the system will require attention at regular intervals.
- The nozzle of the torch can be damaged by crashing it down into the cutting material or traveling into hold-down clamps. Care must be taken when cutting material that has raised sections or that is warped or bent.
- The upward pressure produced by the jet hitting the bottom of the tank and bouncing back is high enough to move materials. Hold-down clamps should be used to keep material in place.

MULTIPLE CHOICE

1. What is the kerf of the cut?
   a. The depth of the cut
   b. The width of the cut
   c. The length of the cut
   d. The waste on the bottom of the cut

2. Which process typically has the largest heat-affected zone (HAZ)?
   a. Oxygen-Fuel Cutting
   b. Plasma Arc Cutting
   c. Laser Beam Cutting
   d. Abrasive Water Jet Cutting

3. Why is the kindling temperature important for the Oxygen-Fuel Cutting process?
   a. The temperature in which the oxygen burns
   b. The temperature in which the fuel gas ignites
   c. The temperature in which the metal being cut will begin to burn (oxidize) in the presence of pure oxygen
   d. The temperatures in which pure oxygen will mix with the fuel gas to produce a flame hot enough to melt the metal being cut

4. What is the operating pressure that acetylene should never be operated above?
   a. 20 psi (138 kPa)
   b. 15 psi (103 kPa)
   c. 30 psi (207 kPa)
   d. 10 psi (69 kPa)

5. What is the most common polarity used for Air Carbon Arc Cutting?
   a. DCEP
   b. AC
   c. DCEN

6. The electrode holder used for Air Carbon Arc Cutting is similar to the electrode holder used for what other process?
   a. GMAW
   b. PAC
   c. SMAW
   d. SAW

7. The gas most often used for Plasma Arc Cutting is compressed air. What is the most common source of this gas?
   a. High-pressure cylinder
   b. Low-pressure cylinder
   c. Air compressor

8. What step should an operator follow in order to extend the operating life of Plasma Arc Cutting tips and electrodes?
   a. Reduce moisture in the air supply
   b. Reduce cutting amperage
   c. Limit the number of piercing cuts
   d. Limit the use of the pilot arc
   e. All of the above

9. What types of metal can be cut with Plasma Arc Cutting?
   a. Ferrous metals only
   b. Reactive metals only
   c. Alloys but no pure metals
   d. Almost all metals

10. Laser is an acronym for what?
    a. Loud Air Scorching Equals Radians
    b. Light Amplification by Sound Emissions of Radiation
    c. Light Amplification by Stimulated Emission of Radiation
    d. Laser Amplification of Semi-Emitters of Rays

11. What method does an abrasive water jet use to cut metal?
    a. Melting caused by heat
    b. Heat produced by the abrasive action of the water
    c. The abrasive action of high-pressure water and an abrasive
    d. The chemical reaction between the metal's surface and the water

12. What is added to the high-pressure water stream of an abrasive water jet to cause erosion of the material being cut?
    a. Pulverized diamonds
    b. Flakes of stainless steel
    c. Garnet
    d. Aluminum oxide

FILL IN THE BLANK

13. Plasma Arc Cutting, Air Carbon Arc Cutting, and Laser Beam Cutting all use _____ of the base metal to produce a cut.

14. Kerf is the name for the _____ of a cut.

15. When performing a _____ cut, care should be taken to make sure that the cutting torch is angled so that molten metal and sparks are ejected away from the torch.

16. The Oxygen-Fuel Cutting process does not cut _____ very well because of the metal's high melting temperature.

17. The use of check valves or _____ are recommended with any oxygen-fuel systems.

18. _____-coated electrodes are the most common electrodes used for Air Carbon Arc Cutting.

19. One of the ratings of plasma cutting power sources is the maximum thickness at rated _____ speed.

20. Plasma Arc Cutting uses a _____ arc that jumps from the negative electrode to the positive tip.

21. Caution should always be used when using a cutting process that removes sections of a structure or metal under stress. Cutting may cause the structure to shift, collapse, or _____, causing injury.

22. The matter that will be used to produce the high-energy photons is called the _____.

23. The water used for Abrasive Water Jet Cutting is pressurized to _____ of pressure.

24. Unlike the other cutting processes in this chapter, the Abrasive Water Jet and Laser Beom Cutting processes are not limited to the cutting of _____ only.

SHORT ANSWER

25. What are variables that affect the size and shape of cutting drag lines are?

26. List two of the benefits of Oxy-Fuel Cutting.

27. Never use petroleum-based lubricants on what part of an oxy-fuel system?

28. Lighting an oxy-fuel torch should never be done with what?

29. Air Carbon Arc Cutting is very similar to Plasma Arc Cutting in that they both most commonly use what gas to blow molten metal away from the cut?

30. Using an air supply that has moisture or oil in it will cause what to wear out sooner?

31. One advantage of laser cutting is that the power level can be reduced, producing a cut that does not pierce completely though the metal. What is this type of cut called?

32. How can a person get burned from a laser?

33. What welding protection eyewear is recommended for viewing a laser beam?

34. Why is ventilation necessary for lasers?

35. With what is the laser lens coated?

36. Why should material being cut with an abrasive water jet system be held in place by hold-down clamps?

Chapter 9

METALS AND WELDING METALLURGY

WELDING METALLURGY

Metallurgy is a broad scientific field that studies metals, their extraction from ores, purification and alloying, heat treatment, and working. The purpose here is to relate metallurgy to the selection of materials for welding and to provide an understanding of heat effects from welding on base metals. In addition, the effects of cold and hot working created during the manufacturing processes are examined.

In addition, heat treatment processes will be studied for two purposes. The first purpose is to gain an understanding of time and temperature effects on steels. Considering the nonuniform temperature gradients created by heat generated during welding, a thorough understanding of time-temperature effects is critical. Heat treatments occur at varying temperatures, and structural effects vary accordingly. These effects are similar to those distributed throughout the weld and the **heat-affected zone (HAZ)**. The HAZ is the area of base metal adjacent to the weld where grain structure and properties have been altered due to heat from welding. Knowledge of heat effects is critical to developing adequate welding procedures.

The second reason to understand heat-treating processes is also related to successful welding procedures. Stresses from welding and from hot and cold working processes—such as drawing, rolling, and bending metals—can be reduced or eliminated through heat treatment.

The first step in understanding metals is the study of metal properties. Understanding metal properties aids in the selection of materials for job specifications and service conditions under which they are used. Discussion will follow on metal classifications and identification. Metals fall into two basic groupings: ferrous or iron alloys and nonferrous alloys that do not contain iron. Ferrous metals include carbon steels, alloy steels—including stainless steels—and the cast irons. Nonferrous metals include aluminum, magnesium, copper, titanium, and nickel alloys. After a basic understanding of what metals are available is achieved, the focus shifts to more basic structures: the building blocks of metals and metal alloys. Considerable focus is then made on phase changes in steels and isothermal transformation or heat effects on steels based on temperature changes and cooling speeds. Finally, heat treatments used to condition metals to prepare them for use in service are discussed.

PHYSICAL AND MECHANICAL PROPERTIES OF METALS

Welding and fabricating metal products cannot be accomplished successfully without design considerations based on two categories of metallurgical properties. Physical and mechanical properties displayed by specific metal alloys help to determine their suitability for service conditions. Strength, electrical or thermal conductivity, weight, and formability are just a few factors to consider.

HOW DO PHYSICAL AND MECHANICAL PROPERTIES DIFFER? **Physical properties** can be measured without changing the chemical nature, structure, behavior, or composition of the metal. These are characteristics inherent in metals, such as magnetic properties, electrical properties, thermal properties, and the metal's atomic weight.

Mechanical properties measure the ease with which the metal can be shaped. Mechanical properties include strength, hardness, ductility, brittleness, elasticity, plasticity, malleability, and toughness. Measuring properties helps to determine the usefulness of a metal for a particular job or design. One of the most important properties is **strength**. Strength is the measure of a material's resistance to external forces applied before it fails. Several notable strengths of particular interest to the welder are shown in **TABLE 9-1**, which also indicates the types of stress applied and measures that result from testing.

One of the most broadly used tests for determining strength is the tensile test. Data obtained from a tensile test can be used to measure tensile strength, yield strength, and breaking strength. Two other important measurements derived from tensile testing are percent elongation and percent reduction in area—both measure a material's ductility.

HOW IS A TENSILE TEST PERFORMED? **Tensile strength** is the maximum load-carrying capability of a material based on a tensile test. The tensile test pulls a metal specimen apart and is used to obtain information about the mechanical properties of a material. These properties include ductility, tensile strength, proportional limit, elastic limit, yield point, yield strength, ultimate tensile strength, and breaking strength.

The tensile strength of a metal, measured in pounds per square inch (psi) or Mega pascal (Mpa), is determined by dividing the maximum load applied during the test by the original cross-sectional area of the specimen before testing occurred. **FIGURE 9-1** portrays a reduced section tensile specimen used for measuring strength of materials and also used as part of welding procedure qualifications.

Types of Strength		
	Method of Stress	**Test Results**
Tensile strength	Pulling apart	Reduction of area (%)
		Elongation (% change in length)
		Tensile strength (psi or Mpa)
Impact strength	Impact load	Energy absorbed during fracture
		Foot-pounds of force
Fatigue strength	Withstand repeated loading	Number of cycles to failure
Compressive strength	Pushing together	Change in length
Shear strength	Sliding action	Force/area in a slip-slide action
		in opposite directions
Torsion strength	Twisting action	Ultimate strength (psi or Mpa)

TABLE 9-1

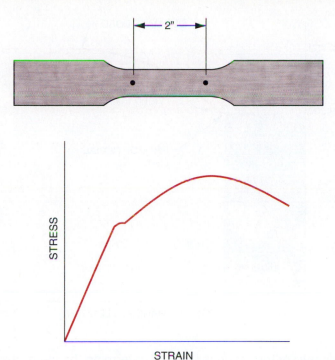

FIGURE 9-1 For determining the percentage of elongation, a 2-inch mark is made on the tensile specimen.

FIGURE 9-2 A stress strain curve

The wider end portions of the specimen are held in the jaws of a tensile-pulling machine and pulled in opposite directions. The reduced middle section with two markings (gage length) set at a 2-inch (5.08 cm) interval is pulled apart until the specimen fractures and breaks.

Two forms of data taken at proportional intervals during a tensile test can be plotted on a graph to help understand and to show relationships among several physical properties and from which to derive strengths important to designers and fabricators (see **FIGURE 9-2**).

The vertical axis on the left shows the load force or stress applied during testing. As the load is applied, the material stretches and the quantity of change in length is plotted on the horizontal axis and is called strain. The resulting line plotted across the range of the test is called a stress-strain curve.

Elasticity of a metal refers to its resilience or ability to resist deformation by stretching. It is the characteristic of a material to return to its original shape and dimensions after a deforming load has been removed. Elasticity is determined by the ratio of stress to strain below the proportional limit. As stress increases, strain increases by direct proportion within what is called the elastic range (see **FIGURE 9-3**), starting at 0 and ending at the **proportional limit (elastic limit)**. The proportional limit is the greatest load a material can withstand and still return to its original length when the load is removed. Within the elastic range and up to the proportional limit, a material stretches, and if the load is removed, it returns to its original length.

Given load increases beyond the proportional limit, the material yields to the stress force and permanent deformation occurs. It has reached its **yield point**, which is the point at which a marked increase in deformation occurs without an increase in load stress. This phenomenon is seen in low-carbon or medium-carbon steel but not in nonferrous metals or alloy steels.

The yield point is often represented on a graph as a 2% offset from the curve at the proportional limit. The significance of the yield point is that the stress at the

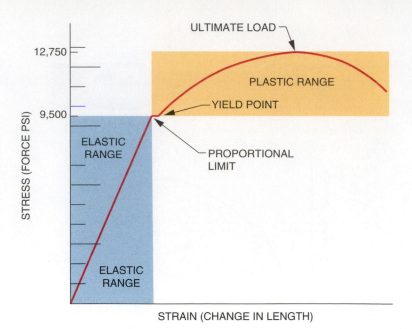

FIGURE 9-3 Beyond the elastic limit, metal will yield and plastically flow

point of yielding to deformation is used to determine the **yield strength** of the material. Yield strength is the amount of measured stress in pounds per square inch or Mega pascals required to cause a material to plastically deform.

Plasticity refers to the ability of a metal to sustain permanent deformation or to yield without rupture. Plastic deformation occurs when enough stress is applied to a solid that it does not return to its original shape. The plastic range is the area of the stress-strain curve beyond the elastic limit. The importance of the plastic range is that this is the area where plasticity occurs while a metal is under stress and helps determine its ability to be formed without failure. Metals that display good plasticity are readily formed.

The property of a material to deform permanently or to exhibit plasticity without rupture while under tension is called **ductility**. Iron, aluminum, gold, silver, and nickel are ductile materials. Percent elongation and percent reduction in area are two measures of ductility obtained from a tensile test. **Elongation** (see **FIGURE 9-4**) is the increase in length that occurs before a material is

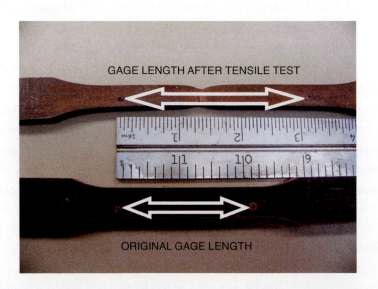

FIGURE 9-4 Measuring a tensile specimen for calculating the elastic limit

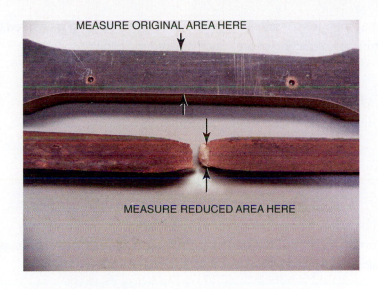

MEASURE ORIGINAL AREA HERE

MEASURE REDUCED AREA HERE

FIGURE 9-5 Measuring a tensile specimen to calculate reduction of area

fractured when subjected to tensile stress. This is expressed as a percentage of the original gage length. Typically, a 2-inch (5.08 cm) or 8-inch (20.32 cm) gage length is marked on the reduced section of a tensile specimen before the load stress is applied. After the specimen fractures, percent elongation is calculated, dividing the change in length by the original length. Similarly, reduction in area (see **FIGURE 9-5**) is the decrease in cross-sectional area at the narrowest point where a material is fractured. This is expressed as a percentage of the original area by dividing the change in area by the original area.

Calculations for tensile strength, percent elongation, and reduction of area follow:

$$\text{Tensile Strength} = \frac{\text{maximum load}}{\text{original area}}$$

$$\% \text{ Elongation} = \frac{\text{gage length after pull} - \text{original gage length}}{\text{original gage length}} \times 100$$

$$\% \text{ Reduction of Area} = \frac{\text{original area} - \text{area after tensile pull}}{\text{original area}} \times 100$$

Figure 9-2 shows the plot of the stress/strain curve for a tensile test on ASTM A36 steel with near-median values. The tensile specimen has a [reduced section reduced area] of 0.188 in^2 (121.3 mm^2). Given the 12,750 pounds (56,715 N) force required to break the material apart and a yield point of 9,500 pounds (42,258 N), yield and tensile strengths can be calculated as follows:

$$\text{Tensile Strength} = 12{,}750 \text{ lbs} \div 0.188 \text{ in}^2 = 67{,}819 \text{ psi}$$

or

$$56{,}715 \text{ N} \div 121.3 \text{ mm}^2 = 468 \text{ N/mm}^2 = 468 \text{ Mpa}$$

$$\text{Yield Strength} = 9{,}500 \text{ lbs} \div 0.188 \text{ in}^2 = 50{,}532 \text{ PSI}$$

or

$$\text{Yield Strength} = 42{,}258 \text{ N} \div 121.3 \text{ mm}^2 = 348 \text{ N/mm}^2 = 348 \text{ Mpa}$$

For the same tensile specimen, if the area after breaking measured 0.176 in^2 (113.5 mm^2), the reduction of area is calculated as follows:

The original area was 0.188 in^2 (121.3 mm^2).

Reduced Area After Pull = 0.176 in^2 (113.5 mm^2)

$$\% \text{ Reduction in Area} = \frac{0.188 - 0.176}{0.188} \times 100 = 6.38\%$$

or

$$\% \text{ Reduction in Area} = \frac{121.3 - 113.5}{121.3} \times 100 = 6.38\%$$

Percent elongation if the gage length after pulling is 2.62 inches (6.65 cm) can be calculated:

$$\% \text{ Elongation} = \frac{2.625 \text{ in} - 2 \text{ in}}{2 \text{ in}} \times 100 = 31\%$$

or

$$\% \text{ Elongation} = \frac{6.65 \text{ cm} - 5.08 \text{ cm}}{5.08 \text{ cm}} \times 100 = 31\%$$

WHAT DOES A TENSILE TEST WITH LOW DUCTILITY INDICATE? **Brittleness** is the property of materials that exhibit low-fracture toughness and do not deform under load but shatter or break suddenly. Brittleness is the opposite of ductility. Brittle metals tend to be harder than ductile metals, displaying little plasticity, and are not suitable for forming and bending operations. Cast irons tend to be brittle, as do hardened steels and cold-formed materials. Notch-type flaws in welds, such as undercut, can cause brittle failures—even on metals considered to be ductile.

HOW ARE OTHER STRENGTHS TESTED? Impact strength is an expression of resistance of a material to fracture resulting from impact loading. An impact test applies an impact load to a small notched specimen. The test is performed with specimens at various temperatures, where test results indicate the notch toughness or ability to absorb energy in the presence of a flaw.

The two main types of impact testing are Charpy and Izod (see **FIGURE 9-6**). Historically, testing has been performed by hitting a notched test specimen with a weighted hammer swung on a pendulum. More modern methods of notch testing employ a drop hammer to impact a specimen placed on the shoulder of two anvil-like blocks.

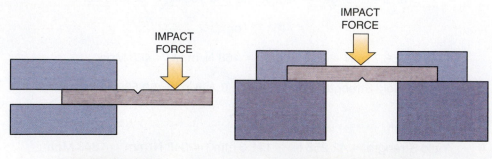

FIGURE 9-6 The Izod and Charpy impact tests

IZOD IMPACT TEST   CHARPY IMPACT TEST

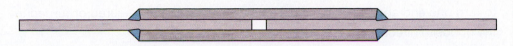

Grain size and structure have considerable effect on impact strength at low temperatures. As the temperature decreases, strength and hardness increase for almost all metals, and this often results in loss of ductility.

Shear strength is a measure exhibited by a load that tends to force materials apart by the application of forces pushing in opposite directions, pushing grains or crystals out of position by a slip-slide action. Shearing and punching operations exhibit a yielding action caused by the slippage of crystal planes along the maximum shear stress surface between opposing forces.

In welding applications, a fillet weld shear test is employed for establishing welding procedures for fabrication purposes. **FIGURE 9-7** shows an example of a fillet weld shear test.

The test specimen is placed in a tensile testing machine and pulled apart until failure or breakage occurs. Test results are obtained by dividing the maximum load by the effective weld area, which is the theoretical throat times the weld length.

Compressive strength is the maximum stress that can be applied to a material in compression without permanent deformation or failure. Compression forces are the opposite of tensile forces. While tensile forces pull a metal apart, compression loads push in on a material from opposite sides. Cast irons display considerable compressive strength.

On the other hand, malleable materials do deform under compression. **Malleability** is the ability of a metal to deform permanently when loaded in compression. Some highly malleable pure metals are lead, gold, silver, iron, and copper. Alloys tend to be less malleable and usually require annealing to soften the material.

Fatigue strength measures the load-carrying ability of a material subjected to loading, which is repeated a definite number of cycles. For part designs that are subject to cyclical loading, fatigue strength is often more important than ultimate tensile strength. Fatigue life can be enhanced by a smooth design, no undercut in welds, and no sharp corners. Endurance limit is the stress below which the test specimen fails, regardless of the number of load cycles. An example of a useful fatigue or endurance test application is to test a vehicle frame by simulating vibration conditions matching those real stresses the frame would undergo on the road.

Creep strength refers to a measure of a material's tendency to relieve applied stresses by gradual deformation over time. These stresses and deformations occur differently from most stress forms discussed thus far in that deformation occurs below the yield point rather than above it. Room temperature creep is a form of stress relaxation where plastic deformation is caused by sustained loading below the measured yield point. High temperature creep strength is a measure of a material's ability to withstand creep at elevated temperature. This is an important design consideration because creep generally occurs more rapidly at elevated temperatures than at room temperature. Creep strength of a material is given in terms of an allowable amount of plastic flow (creep) per 1,000-hour period.

Torsion strength is a measure of a material subjected to a twisting-type stress. The twisting or wrenching force is exerted on a material by forces tending to turn one end about a longitudinal axis while the other is secured or turned in the opposite direction.

HOW IS HARDNESS MEASURED?   There are many methods of measuring hardness. The most common methods of measuring hardness by measuring a metal's resistance to penetration. The Rockwell test uses either a small hardened steel ball or an industrial diamond-point penetrator to indent the metal and then measure the depth of the indentation that is made. Hardness values are read from a scale on the Rockwell testing machine.

Two common scales shown in **FIGURE 9-8** are used for Rockwell testing of soft metals or hardened steels. The Rockwell "C" (Rc) scale (black numbers) is used for testing hard steels and employs a 150 kg load to push the diamond (brale) penetrator into the metal surface. The Rockwell "B" (Rb) scale (red numbers) is used with a ball penetrator for testing soft metals, such as coppers, brasses, aluminum, and soft steels, and uses a 100 kg load.

It is important to use the correct penetrator (see **FIGURE 9-9**) and scale for hard and soft metals. A diamond penetrator has no problem penetrating a hardened steel surface. However, using the 1/16-inch (1.6 mm) hardened steel ball on very hard steel will damage the penetrator.

Another hardness test, Brinell testing, is similar to Rockwell testing in that a hardened steel ball is used to penetrate the metal surface. Rather than measuring the depth of penetration, it measures the major diameter of the mark left by the penetrator. The Brinell test is used for measuring the hardness of soft steels and soft metals.

WHAT PHYSICAL PROPERTIES ARE IMPORTANT IN WELDING AND FABRICATION?   Thermal expansion and thermal conductivity are two significant properties that affect welding and fabrication. These properties are considered

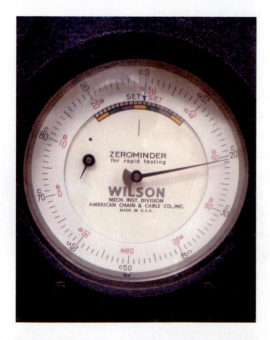

FIGURE 9-8 For the Rockwell "C" (Rc) scale, read the black numbers and for the Rockwell "B" (Rb) scale, read the red numbers.

1/16" HARDENED STEEL BALL PENETRATOR   DIAMOND (BRALE) PENETRATOR

FIGURE 9-9 The ball penetrator (left) is used for Rockwell "B" testing, and the diamond penetrator is (right) used for Rockwell "C" hardness testing.

FIGURE 9-10 Distortion

in product design to prevent **distortion** and in writing welding procedures (**FIGURE 9-10**). Distortion is the warping or displacement of base metal parts due to the nonuniform expansion and contraction forces caused by heat effects of welding and heating operations and the subsequent cooling cycles that follow.

Thermal expansion is the tendency of a metal to grow in volume when heated. When heated, metals expand or grow larger, and when they cool down, they shrink. This tendency to change size when heating and cooling impacts the heated weld metal and structure being welded. If a piece of metal is heated without restraints and is then allowed to cool down, it grows outward uniformly and then shrinks uniformly to its original size. However, if the part is under restraint—as in welding two base metals together—the welded assembly cools the weld unevenly and the parts pull out of alignment. This is called distortion.

The amount a metal expands is based on the metal's coefficient of thermal expansion (**TABLE 9-2**), which is the amount of expansion per degree temperature per unit of length. The greater the amount of thermal expansion, the more a heated part expands and therefore the more distortion can be expected.

Thermal conductivity is a measure of the rate at which a quantity of heat travels through a metal per unit time, per unit cross-section, and per temperature gradient. All materials in some measure conduct heat. Thermal conductivity rates for several welded metals are given in **TABLE 9-3**.

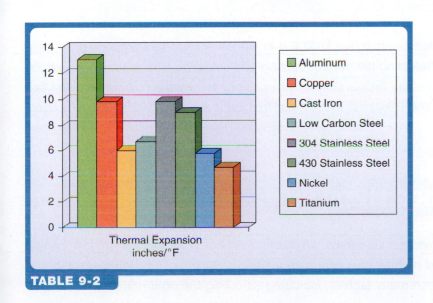

Aluminum
Copper
Cast Iron
Low Carbon Steel
304 Stainless Steel
430 Stainless Steel
Nickel
Titanium

Thermal Expansion
inches/°F

TABLE 9-2

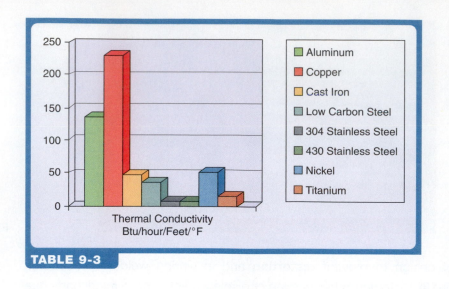

TABLE 9-3

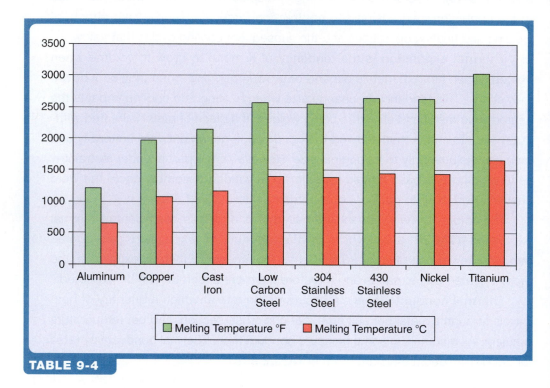

TABLE 9-4

Melting temperature (see **TABLE 9-4**) is another important property of concern to welders. Selecting the correct brazing and soldering filler metals depends on base metal melting temperatures. For example, to braze an aluminum alloy that melts near 1,260°F (682°C), a filler metal that melts around 1,150°F (621°C) is used.

CLASSIFICATION AND SELECTION OF FERROUS METALS

Identifying base metals can be difficult for the novice. Metals fall into groups and are classified by their chemical compositions. Successful welding and fabrication depend on many factors, including selecting the right material for the job.

Studying metals and comparing known materials can simplify material identification and help ensure that quality procedures are employed.

Steel is an alloy of iron and carbon. Carbon is added to iron to strengthen a material that would be quite weak without it. Carbon steels come in many shapes and are processed in many ways. In addition to carbon, they contain other elements. Manganese and silicon are common deoxidizing elements; these and other elements, such as phosphorus and sulfur, are found as impurities left behind by the steel manufacturing process. Or they may be added intentionally to enhance other properties. Sulfur is added to make steel free machining, but it makes it virtually unweldable.

Steels may be purchased in wrought forms as structural shapes, such as angle iron, channel iron, wide flange beams, and pipes. Structural shapes are commonly used for welding applications. Furthermore, carbon steels can be purchased in different forms based on the temperature conditions under which they are rolled, drawn, or extruded into shapes. Hot-rolled steels (HRS) are produced at high temperatures at or near the forging range. They are coated with a slight to heavy mill scale created through oxidation of the surface metal at elevated temperature. Cold-rolled steels (CRS) are produced by dipping hot-rolled materials into an acid bath to remove the scale (known as pickling) and then rolling them again at lower temperatures, producing steel with a light metallic shine and considerably more accurate dimensional size tolerances when compared to hot-rolled steels. Cold-rolled steels are sometimes dipped in oil to prevent oxidation; these are called pickled and oiled steels. The oil requires chemical removal before welding.

On occasion, hot-rolled steels are pickled to remove the mill scale. They are not subsequently rolled cold but are instead coated with an oil to prevent corrosion. These hot-rolled steels are referred to as hot-rolled pickled and oiled (HRPO) and labeled accordingly.

WHAT ARE CARBON STEELS? Carbon steels are discussed in terms of carbon content, which is the principle alloying element. From welding, fabricating, and metallurgical viewpoints, carbon content is critical to the categorization of steels. Low-carbon steels contain 0.05%–0.29% carbon and are soft and ductile. They are readily welded with fairly simple procedures and are generally easy to bend, shape, and form. In addition, they can be surface-hardened by carburizing in a carbon-rich environment.

Medium-carbon steels contain 0.30%–0.59% carbon and require more complicated welding procedures. Increased carbon can lead to hardening and possible cracking along the fusion line and in the heat-affected zone (HAZ) if cooling rates are not controlled. These materials can be heat-treated, resulting in enhanced strength.

High-carbon steels contain 0.60%–1.0% carbon. Increased carbon gives them greater response to heat treatment and increased wear resistance. Again, increased carbon content requires strict control of procedures for welding.

Ultra-high-carbon steels contain 1.0%–2.0% carbon. These steels are used for their high hardness properties and are suitable for use in axles and hardened industrial knives.

Most welding applications, procedures, and codes require that ferrous materials conform to American Society for Testing Materials (ASTM) standards. ASTM A36 is a common structural steel with a range of chemical compositions and carbon content of 0.15%–0.29%. Using a range of chemical composition allows for making adjustments of alloying elements so the base metal meets physical and mechanical properties required by the ASTM standard without increasing costs like exact specifications do. Because base metal properties depend on cooling rates, chemical composition can be used to compensate for changes in cooling rates based on the mass of the steel structure produced.

Mill Certification Test Reports can be requested upon purchase and used to ensure that material properties meet those of standards used by welding and fabrication professionals. **TABLE 9-5** shows an example of mill test results for an A36 steel plate. The heat number listed identifies a specific batch of steel for which the mill certification test is completed.

Tracking materials through ordering, mill certification, and marking can eliminate use of the wrong material for any given application. When ordering materials, a good practice is to require a Certified Material Test Report (CMTR) and to have the material labeled with the metal type and heat or batch number. When materials arrive, match the label with numbers from the test record. Once a material is acquired, transfer the label information to any drop material not used. This identifies the material for future use.

HOW ARE STEELS CLASSIFIED?

The American Iron and Steel Institute (AISI) and the Society of Automotive Engineers (SAE) have developed a system for classifying plain carbon and alloy steels. Alloy steels have significant additions of elements to improve properties. In the SAE and AISI classification, there are four digits. The only difference between the AISI and the SAE systems is that the AISI system adds a prefix to identify the method used to process the steel. Some AISI prefixes are:

- **C:** Basic Open Hearth Furnace
- **D:** Acid Open Hearth Furnace
- **E:** Electric Furnace

The first digit from the left is a number representing the major alloying element. The next digit is the percentage of that major alloying element. The last two digits, shown as xx below, denote the percentage of carbon. You will note that in carbon steel the 1 represents the major alloying element as carbon. Plain carbon steel is less than 1.00% carbon, so a 0 is the second-place digit.

Material Type	Heat Number	Yield Strength psi (Mpa)	Tensile Strength psi (Mpa)	% Elongation
ASTM A36	16432	43,068 (297)	68,156 (470)	31
Chemical Analysis %				

Carbon	Manganese	Phosphorus	Sulfur	Silicon	Copper
0.21	0.51	0.81	0.91	0.03	0.05

TABLE 9-5

AISI-SAE Designations		
Carbon Steels		
Plain Carbon	10xx	1006–1095
Free Cutting (resulfurized)	11xx	1108–1151
Resulfurized-Rephosphorized	12xx	1211–1214
Manganese Steels		
Manganese	13xx	1320–1340
Nickel Steels		
0.50% nickel	20xx	
1.50% nickel	21xx	
3.50% nickel	23xx	
5.00% nickel	25xx	
Nickel-Chromium Steels		
1.25% nickel, 0.65% chromium	31xx	
1.75% nickel, 1.00% chromium	32xx	
3.50% nickel, 1.57% chromium	33xx	
3.00% nickel, 0.80% chromium	34xx	
Corrosion and Heat Resisting Steels		
	303xx	
Molybdenum Steels		
Carbon-Molybdenum	40xx	4024–4068
Chromium-Molybdenum	41xx	4130–4150
Nickel-Chromium-Molybdenum	43xx	4317–4340
Nickel-Molybdenum	46xx	4608–4640
Nickel-Chromium	47xx	
Nickel-Molybdenum	48xx	4812–4820
Chromium Steels		
Low Chromium, 0.50%	50xx	
Medium Chromium, 1.0%	51xx	5120–5152
High Chromium, 1.5%	52xxx	52095–52101
Chromium-Vanadium	6xxx	AISI 400 series stainless steels
–	61xx	6120–6152
Tungsten	7xxx	
Triple Alloy	8xxx	
Silicon-Manganese	9xxx	9255–9262

TABLE 9-6

The AISI and SAE classifications for carbon steels and alloyed steels are shown in **TABLE 9-6**.

Alloying elements are added to steels to enhance metallurgical properties. **TABLE 9-7** groups alloying elements according to their effects. Adding alloys impacts a metal's weldability and must be taken into consideration for developing welding purposes.

Effects	Alloying Elements
Increase hardness	carbon
Increase hardenability	boron, chromium, manganese, molybdenum, nickel, silicon, tungsten
Increase wear resistance	carbon, cobalt, tungsten
Increase strength	carbon, cobalt, copper, manganese, nickel, phosphorus
Improve toughness	nickel, vanadium
Improve corrosion Resistance	chromium, copper
Increase machinability	lead, sulfur, tellurium
Deoxidizer	aluminum, manganese, silicon
Increase high temperature strength	molybdenum, tungsten
Eliminate carbide precipitation	columbium, titanium, tantalum

TABLE 9-7

WHAT ARE STAINLESS STEELS?   The stainless steel numbering system is universal for AISI, SAE, and ASTM. These numberings system group stainless steels into several categories by alloy types listed in **TABLE 9-8**. An L for low carbon (not to exceed 0.03%) and an S for high silicon is sometimes used. Stainless steels, used for their corrosion-resistant properties, are iron alloys with chromium additions ranging from 11%–30%. Nickel, molybdenum, copper, titanium, aluminum, silicon, niobium, and nitrogen are added to improve characteristics, corrosion resistance, and mechanical properties.

The 300 series stainless steels cover the greatest tonnage of stainless alloys specified and produced. Along with the 200 series, they are known as austenitic stainless steel. Austenitic stainless steels are nonmagnetic, nonheat-treatable, and easily work-hardened. They primarily contain iron, carbon, chromium, and nickel. They are used in cooking pots and pans, restaurant kitchens, and butcher shops, as well as process pipe, tanks and marine applications.

Alloy Type	Classification	Characteristics/Useability
Austenitic chromium-nickel	304, 308, 308L, 309, 309S, 316, 316L, 321	formable, good weldability
Austenitic chromium-manganese-nickel	201, 202, 205, 216	formable, good weldability
Ferritic chromium	405, 430, 442, 446	very soft, limited corrosion resistance toughness, formability, and weldability not heat treatable
Duplex (ferritic/austenitic)	2205, 2304, 2507	improved stress corrosion cracking resistance, formability, weldability, and high strength
Martensitic chromium	403, 410, 416, 420, 431, 440	increased hardness and strength, decreased ductility and toughness
Precipitation hardening	15-5 PH, 17-4 PH, 17-7 PH, PH 13-8Mo	high strength, fair weldability, aerospace applications

TABLE 9-8

The 200 series alloys became popular during time periods when nickel was in short supply during WWII and in the 1950s due to civil wars in Africa and Asia, the major suppliers. These alloys have additions of manganese to offset lower nickel contents. Although raw materials are cheaper to produce, processing costs and other factors, such as poorer deep drawing when compared to more common 300 series alloys, have limited the use of these alloys.

The 400 series stainless steels fall into two categories: ferritic or martensitic. All 400 series alloys are magnetic and they can develop a light layer of rust. Considered straight chromium alloys because they lack the nickel of austenitic types, weld quality is reduced especially with the martensitic alloys.

Ferritic alloys have a structure similar to low-carbon steels but with improved corrosion resistance. They are not heat-treatable. Chemical compositions contained in ferritic stainless steels are iron, carbon, chromium, and manganese.

Martensitic stainless steels are heat-treatable alloys that contain primarily iron, carbon, and chromium. They are used in such areas as surgical knives, high-quality knives, and the like. Closely-monitored welding procedures are required because many martensitic stainless steels are air-hardening alloys, meaning they quench-harden in air.

Duplex stainless steels retain a balance of characteristics borrowing from the austenitic and ferritic alloys. While not hardenable by heat treatment, they are stronger than the ferritic and austenitic types, with good weldability and formability. They primarily contain chromium, nickel, molybdenum, and nitrogen.

Precipitation-hardening (PH) alloys are hardenable by aging heat treatment. They are more readily welded than the hardenable martensitic alloys and retain corrosion-resistant properties comparable to austenitic alloys. Precipitation-hardening alloys are sometimes referred to as semi-austenitic.

WHAT ARE CAST IRONS? Cast irons are ferrous alloys containing 2%–4% carbon and for that reason tend to be harder and more brittle than steels. Alloying elements nickel, chromium, and molybdenum are sometimes added to improve hardness, strength, toughness, ductility, wear resistance, and corrosion resistance.

Cast iron classifications fall into five basic groupings: gray, ductile, malleable, compacted graphite, and white cast irons. The most common metal poured for castings is gray cast iron. Gray cast iron contains 2%–3% of carbon and is sometimes referred to as gray iron. High carbon contents make cast iron brittle. Ductile cast iron, sometimes referred to as ductile iron, contains the least amount of carbon, and white cast iron contains the most carbon, up to 4% or more.

The difference between ductile and malleable cast iron can be explained in the definition of ductility and malleability. Ductility is the ability to deform permanently without rupture while under tension. Malleability is the ability to deform permanently without rupture while under compression. Both malleable and ductile cast irons are tougher than the common gray cast irons and stronger in their own ways. Malleable irons resist shock better, while ductile irons have improved tensile strength.

Cast iron is inherently stable. This stability is desirable for machine tools when accuracy and repeatability are required. Resistance to vibration makes

Carbon Tool Steels		AISI Designation
W	Water-hardening	W1–W7
Cold Work Tool Steels		
A	Air-hardening, chromium die steels	A2–A10
0	Oil-hardening, low alloy	01–07
D	High carbon, high chromium	S1–S7
Hot Work Tool Steels		
H	Chromium, tungsten, molybdenum	H10–H43
High-Speed Tool Steels		
T	Tungsten	D2–D7
M	Molybdenum	T1–T15
Special Purpose		
S	Tungsten, shock resisting, finishing	M1–M47
L	Low alloy tungsten, finishing	L1–L7
F	Carbon-tungsten, finishing	F1–F3
P	Low-carbon steel, die casting mold	P1–P21

TABLE 9-9

it suitable for use in machine bases, such as mills and lathes. Increased carbon in cast irons makes repairing and welding parts, when necessary, more difficult.

HOW ARE TOOL STEELS CLASSIFIED?   Tool steels are carbon steel alloys with strength, toughness, and hardenability characteristics that render them appropriate for use in making tools of many types. The AISI tool steel classifications fall into numerous categories based on heat treatment methods, types of alloys, properties, and uses (see **TABLE 9-9**).

Care must be taken when welding, heating, and cooling tool steels. Each category has its own sensitivity to heat treatment and to heat effects of welding. They are used for their ability to be heat-treated to obtain specific properties, such as deep hardening and wear resistance. The degree of hardening relies on the respective heat treatment procedures for each alloy. Air-hardening alloys harden by quenching in air, oil-hardening alloys in oil, and water-hardening alloys in water. Commercially available tool steels must be heat-treated, welded, and fabricated in accordance with its manufacturers specified procedures.

CLASSIFICATION AND SELECTION OF NONFERROUS METALS

Aluminum is a lightweight metal alloyed to provide desirable strength, formability, and corrosion resistance. The Aluminum Association classifies aluminum alloys by using a four-digit numbering system (see **TABLE 9-10**). The first digit denotes the alloy group based on the major alloying element.

Series Designation	Major Alloy Element	Hardening Method
1xxx	>99% pure	Strain
2xxx	Copper	Heat-treatable
3xxx	Manganese	Strain
4xxx	Silicon	Heat-treatable
5xxx	Magnesium	Strain
6xxx	Magnesium and silicon	Heat-treatable
7xxx	Zinc	Heat-treatable
8xxx	Other alloying elements	Varies
9xxx	Future alloy	Varies

TABLE 9-10

The last three digits, represented here by xxx, indicate different criteria for 1xxx series alloys than for all the remaining alloys, 2xxx through 9xxx. For the nearly pure 1xxx series aluminums, the second digit indicates the number of impurities, while the last two digits represent the minimum aluminum content in hundredths of a percentage. For the remaining alloys, the second digit signifies modifications to the alloy, while the last two digits identify a specific alloy.

In addition to the four-digit identification, a suffix may be added to indicate the delivery condition of the material, whether as rolled, cast, strain-hardened, tempered, or annealed. **TABLE 9-11** breaks down the Aluminum Association's suffix system by letter and numerical listings.

Aluminum alloys are used for their lightweight and highly corrosion-resistant properties. While aluminum strength is lower than that of steels, it is also only one-third the weight. Aluminum products can be made lighter with excellent strength-to-weight ratios. Strength can be improved by strain hardening, alloying, and through heat treatment when using the correct alloys.

Copper has high electrical and thermal conductivity. Copper is used extensively as an electrical conductor for leads, contacts, and in electrical coils, both in wire and sheet forms. It is easily formed, especially in the annealed state. Copper alloys include bronzes and brasses. Brasses are zinc-copper alloys that are usually yellow in appearance. Bronzes are copper-tin alloys that are usually reddish-brown in color. Copper alloys are shown in **TABLE 9-12**.

ALL THESE CLASSIFICATION SYSTEMS SEEM CONFUSING. HOW CAN ONE KEEP TRACK OF ALL THE METALS?

The Unified Numbering System (UNS) was developed by ASTM and SAE to combine metal alloys into series or designations based on alloy types. Eighteen UNS series designations (see **TABLE 9-13**) are used to group both ferrous and nonferrous metals and give them a universally used number to eliminate confusion encountered when different associations assign a different number to the same material. Furthermore, UNS incorporates a system that includes international specifications from countries including Australia, Germany, and Japan.

Aluminum Suffix Denotation
Basic Suffix Applied

F	As fabricated	No special treatment
0	Annealed	Softest condition available
H	Strain-hardened	Wrought products only
W	Solution heat-treated	Unstable temper, slow natural aging occurs
T	Solution heat-treated	Stable temper, various methods apply

Level of Strain Hardening

H1	Strain-hardened only	A second digit following the H indicates the degree of hardening as follows:
H2	Strain-hardened and partially annealed	0 = full hard
		2 = 1/4 hard
H3	Strain-hardened and stabilized	4 = 1/2 hard
		6 = 3/4 hard
H4	Strain-hardened and coated (lacquer or paint)	8 = full hard
		9 = extra hard
		A third digit may be added for variations.

Level of Stable Temper

T1	Naturally aged		
T2	Annealed		
T3	Solution heat-treated	Cold-worked	
T4	Solution heat-treated	Aged-hardened naturally	
T5	Artificially aged		
T6	Solution heat-treated	Artificially aged	
T7	Solution heat-treated	Stabilized	
T8	Solution heat-treated	Cold-worked	Artificially aged
T9	Solution heat-treated	Artificially aged	Cold-worked
T10	Artificially aged	Cold-worked	

TABLE 9-11

Copper Alloys

Name	Alloying Elements
Copper	CU
Brasses	CU-ZN
Leaded brasses	CU-ZN-PB
Tin brasses	CU-ZN-SN-PB
Phosphor bronzes	CU-CN-P
Leaded phosphor bronzes	CU
Copper-phosphorus and copper-silver-phosphorus	CU
Aluminum bronzes	CU
Copper-nickels	CU
Nickel silvers	CU

TABLE 9-12

UNS Series	Metal Alloy Group
A00001–A99999	Aluminum and Aluminum Alloys
C00001–C99999	Copper and Copper Alloys
D00001–D99999	Specified Mechanical Properties Steels
E00001–E99999	Rare Earth and Rare Earthlike Metals and Alloys
F00001–F99999	Cast Irons
G00001–G99999	AISI and SAE Carbon and Alloy Steels (except tool steels)
H00001–H99999	AISI and SAE H-Steels
J00001–J99999	Cast Steels (except tool steels)
K00001–K99999	Miscellaneous Steels and Ferrous Alloys
L00001–L99999	Low-Melting Metals and Alloys
M00001–M99999	Miscellaneous Nonferrous Metals and Alloys
N00001–N99999	Nickel and Nickel Alloys
P00001–P99999	Precious Metals and Alloys
R00001–R99999	Reactive and Refractory Metals and Alloys
S00001–S99999	Heat- and Corrosion-Resistant (stainless) Steels
T00001–T99999	Tool Steels, Wrought and Cast
W00001–W99999	Welding Filler Metals
Z00001–Z99999	Zinc and Zinc Alloys

TABLE 9-13

A prefix is used to identify the metal grouping. Ferrous alloys make up the groupings: D, F, G, H, J, K, S, and T. The nonferrous alloy groups are A, C, E, L, M, N, P, R, and Z. W consists both ferrous and nonferrous welding filler metals.

METAL IDENTIFICATION METHODS

Given the wide ranges of metals used in fabrication, care must be taken to guarantee that the correct materials are used. To ensure product quality when purchasing materials, a manufacturer requests material certification documents to show that it meets the required metal classification criteria for the job. However, inventory control is not always ideal. Known materials can be misplaced or mixed with other materials. A custom repair welder may have to figure out what a broken part is made from in order to select the correct weld material.

HOW CAN UNKNOWN METALS BE IDENTIFIED?
A good practice is to have known materials marked and labeled for comparison to the unknown pieces. Known samples can be used for visual, magnetic, hardness, and spark-testing comparison.

Visual examination is the first factor in identifying a metal. Visual examination can include surface color or grain color. For example, a contrast can be seen between grain colors of broken white cast iron and gray cast iron samples.

Metal	Color	Magnetic Property	Density lb/in.^{3}
Aluminum	silvery white	nonmagnetic	0.098
Copper	red	nonmagnetic	0.322
Brass	light yellow	nonmagnetic	0.306
Bronze	reddish-brown	nonmagnetic	0.320
Carbon Steels	light gray	strong attraction	0.284
Stainless Steels (austenitic)	silvery white	slight attraction if cold-worked	0.286
Stainless Steels (ferritic)	silvery white	strong attraction	0.276
Stainless Steels (martensitic)	silvery white	strong attraction	0.286
Lead and Zinc Alloys	dark gray	nonmagnetic	0.25–0.45
Magnesium	silvery white	nonmagnetic	0.063
Titanium	silvery white	nonmagnetic	0.164

TABLE 9-14

The gray cast has a gray appearance, while the white cast has a white appearance. External colors of various metals are given in **TABLE 9-14**. To illustrate how visual examination can be used to help a welder identify a metal for repair welding, **FIGURE 9-11** shows six different metals:

- **A:** Copper
- **B:** Beryllium Copper
- **C:** Brass
- **D:** Stainless Steel
- **E:** Cold-Rolled Steel
- **F:** Hot-Rolled Steel

Metal colors and appearance vary with surface conditions. Mill scale, oxidation, and finish techniques—such as polishing, bead or sand blasting, or created brushed finishes—may complicate metal identification. In these cases, it may be helpful to use a combination of identification methods. For example, both alu-

FIGURE 9-11 Different metals can be separated by visual appearance.

minum and stainless steels can be given a polished mirror finish; however, comparing their weights will quickly distinguish one from the other. Furthermore, magnesium and aluminum look very much the same. Metal weights or densities are given in Table 9-14. Magnesium is lighter, about two-thirds that of aluminum, and this may be helpful in identification if a scale is available, but a simple acid test can distinguish between the two materials. Nitric acid will make magnesium turn gray but will not discolor aluminum.

Metals can also be separated to some degree through testing magnetic properties. Steels are magnetic at room temperature, and most nonferrous metals are nonmagnetic. Magnetic properties vary with alloy content. While testing magnetic attraction is not an accurate method of measuring differences among alloys, it can be used in the process of elimination, leaving a smaller group of possibilities. See Table 9-14 for metals and their respective magnetic properties.

Hardness testing is useful in metal identification. Hardness values of known metals can be compared with samples needing identification. Combining color identification with hardness values can be very helpful in metal identification.

Steel hardness values depend on several factors, including carbon content, heat treatment, and forming methods, such as hot- or cold-forming. Steels increase in hardness with increased carbon content even in the as rolled condition The higher the carbon content the harder a steel can become from heat effects. Steels hardened through heat-treating are harder than the as-rolled forms, while steels that have been full annealed are softer. Hardness testing is often used to determine if a heat treatment has produced the correct hardness value. Knowing the heat-treat condition is significant to metal identification.

Alternative methods for determining hardness of metals—steels in particular— are scratch or file testing. Although these methods give somewhat subjective, inaccurate results, they can be useful in separating or identifying hardened pieces. In both cases, it is helpful to have known materials to use as standards and then compare to the unknown materials. These types of simple tests can help to separate low-carbon steels from higher-carbon steels. The standard series of SAE steels (1018, 1045, 1060, and 1095), along with any common alloy steels used in the shop, will be quite helpful for comparison's sake. Remember that steels harden as carbon content increases. It may also be wise to have heat-treated samples with varying degrees of hardness available for comparison.

Always remove mill scale before hardness testing. File testing involves running a hardened steel file across the edge of the unknown metal and feeling the degree of drag, or resistance, and then running the file across the known standards until a match is found. The scratch test employs the task of scratching one metal with another. Using the known standards, start with the hardest (highest carbon) and simply drag the unknown material's edge across the standard. If it does not scratch the standard, retry the test on the next softer standard and repeat until a match is made. The whole point is that a harder metal will scratch a softer metal.

Another useful test—spark testing—requires a degree of experience and known materials to compare patterns. Spark pattern elements include the length and volume of carrier lines or spark streams, the number, size, and shape of forks

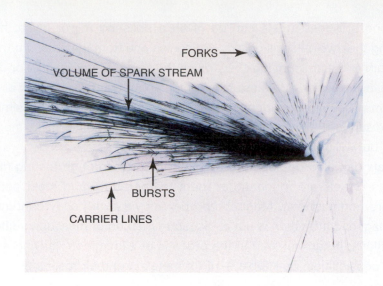

FORKS →

VOLUME OF SPARK STREAM

BURSTS

CARRIER LINES

FIGURE 9-12 When ground, different metals produce a unique spark and, when read correctly, provide metallurgical information.

and burst patterns, and the color of the streams and bursts. **FIGURE 9-12** illustrates how to read a spark pattern.

When performing a spark test, it is important to test samples with mill scale and other impurities removed because they will alter the spark pattern. Also, carbon content may be lower near the surface due to decarbonization that occurs during oxidation and the forming of mill scale. Visible light filters out color, so spark testing with lights dimmed or off is helpful. Spark patterns also vary with the amount of pressure applied with a grinding wheel or stone. When comparing metal samples, keep pressure even from one piece to another.

Spark testing is a good means for testing carbon content in steels. As carbon content increases, the spark pattern changes. Wrought iron (see **FIGURE 9-13**) with less than 0.03% carbon has a very light yellow-orange, long carrier line that tends to thicken at its midpoint and has no bursts or forks.

An SAE 1018 (see **FIGURE 9-14**) has a pattern similar to wrought iron. Carrier lines broaden wider at their ends, and the color is a lighter yellow than wrought iron. Carbon content is still low, and no significant burst patterns exist.

FIGURE 9-13 Wrought iron

FIGURE 9-14 SAE 1018, low-carbon steel

FIGURE 9-15 SAE 1045, medium-carbon steel

FIGURE 9-16 SAE, high-carbon steel

Medium-carbon steels, such as this SAE 1045 (see **FIGURE 9-15**), have more distinguished patterns. Notice the increase in stream volume and the moderate number of forks and burst patterns at the ends of carrier lines. The color is similar to that of the 1018 steel.

The increased carbon contents of SAE 1095 (see **FIGURE 9-16**) cause a marked increase in the number of bursts. Note also that the bursts are larger, with more forks than those of the 1045.

Cast irons (see **FIGURE 9-17**) have 2.0%–4.0% carbon, and this suppresses the spark pattern. The carrier lines are much shorter with many small orange-yellow bursts.

With the wide range of alloy steels available, it is difficult to become proficient or precise at spark testing unless it is used with known samples and combined with other identification methods. Alloying elements suppress the carbon-caused sparks and also affect color. A 316L stainless steel has a pattern similar to wrought iron, with a slightly orange tint (see **FIGURE 9-18**). Of course,

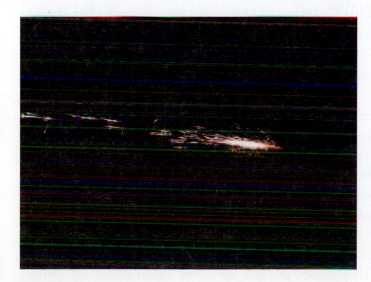

FIGURE 9-17 Cast iron

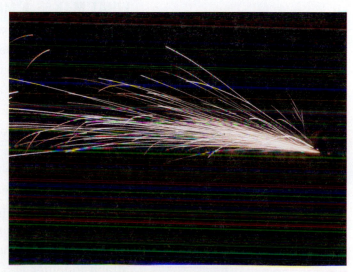

FIGURE 9-18 316L stainless steel

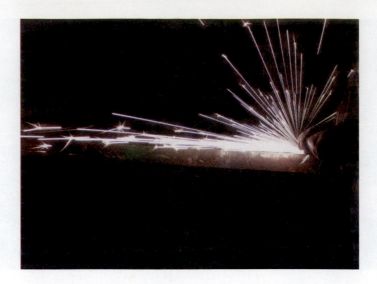

FIGURE 9-19 Titanium

VIDEO

316L is nonmagnetic, and visual identification indicates the color of base metal a shinier silver white than the gray appearance of wrought iron and carbon steels.

Some other metals have distinguishing spark patterns, such as the titanium in **FIGURE 9-19**. Titanium weighs about half as much as steels and even less than half compared to most stainless steels. The spark test easily confirms the identification as titanium.

While all of these methods are useful and practical for non-critical repairs, there are methods of positive metal identification that should be used to establish the exact identity of alloys for critical welding applications. Chemical laboratory wet testing techniques, hand-held X-ray fluorescence equipment, and even spectral analyses are possible methods for the situations that justify their high costs.

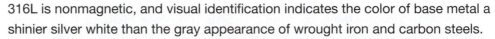

CRYSTALLINE STRUCTURE OF METALS

WHAT MAKES A METAL? Metals used for manufacturing are composed of elements that are mixed together to give them strength and properties that make them suitable for use in service. A metal may be preferred based on strength, weight, corrosion resistance, or appearance. Metals are alloys of two or more elements—of which at least one must be a metal—and generally fall into two categories: ferrous and nonferrous. Ferrous alloys are made primarily from iron, as discussed previously. Nonferrous metals are made when small quantities of alloying elements are added to metals other than iron.

Most metals are made up of crystal structures called grains. These crystal structures are formed during solidification from a molten state. The building block of matter and metals is the atom. Molten metals are composed of atoms that are moving around at a rapid rate. During solidification, these movements are slowed until some atoms become attracted to each other. Individual atoms attach together and form a recognizable structure called a unit cell. Single unit cells, called seed nuclei, form a nucleus from which metal crystalline

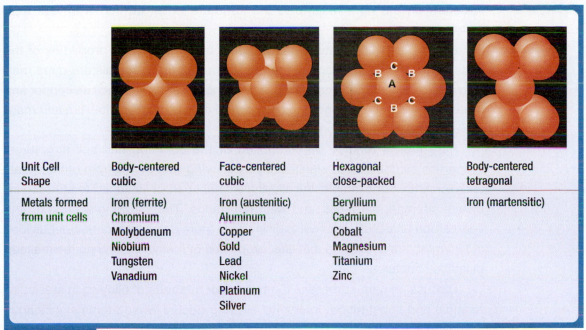

Unit Cell Shape	Body-centered cubic	Face-centered cubic	Hexagonal close-packed	Body-centered tetragonal
Metals formed from unit cells	Iron (ferrite) Chromium Molybdenum Niobium Tungsten Vanadium	Iron (austenitic) Aluminum Copper Gold Lead Nickel Platinum Silver	Beryllium Cadmium Cobalt Magnesium Titanium Zinc	Iron (martensitic)

TABLE 9-15

structures grow. Further cooling causes unit cells to bond with other unit cells, and a space lattice structure begins to form. A space lattice is the organization of atoms into an organized pattern that forms a crystalline structure or a grain in metals.

TABLE 9-15 lists four common unit cells and the metals that take their forms.

It is these unit cells that start nucleation, which leads to grain formation in metals **FIGURE 9-20**. Atoms attach to the first seed nuclei to form unit cells. Each color grouping represents the initial growth of individual grains. The first picture on the left shows how a lattice structure grows in an outward direction. The middle picture illustrates the final formation of grains where the original colonies grew outward until they came into contact with the bordering colonies. These contact points where grains collide into one another are called grain boundaries. Grain boundaries etch differently than the rest of the grain because the atoms are jammed together in a misfit and stressed condition, and this is why they are so visible under microscopic examination. The resulting grain

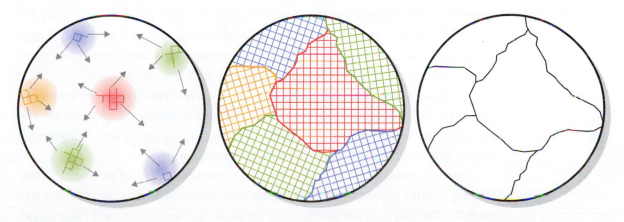

FIGURE 9-20 Unit cells

structure on the right resembles the structure of ferrite iron with less than 0.02% carbon.

Lattice structures determine to a great extent the many properties of the metals they make up. Many of the face-centered cubic metals are quite malleable, such as gold, silver and lead, while others are ductile, such as copper and nickel. Likewise, the hexagonal close-packed metals, such as titanium and magnesium, tend to be more brittle.

Pure metals tend to be composed of the common unit cell structures. Properties can be altered significantly by alloying with atoms from other metals and elements. For example, the alloying element carbon is added to steels to increase strength, hardness, and wear resistance. The following section on the iron-carbon phase diagram will explain this in more detail, but for now, let us look at two methods of alloying that alter properties of metals by altering the makeup of lattice structures.

Discussion here addresses the two major methods of alloying in steels: interstitial and substitution. Interstitial alloying occurs when atoms significantly smaller than those of the parent metal are added. These smaller atoms fit into the gaps or interstices of the parent metal's atom cubic structure. This is the case with atoms of carbon added to steels. Interstitial carbon atoms disrupt the lattice structure and therefore the grain structure in steels and give these iron alloys strength properties that can be manipulated to the manufacturer's advantage. The carbon interrupts the slip planes in the crystalline structure, making it more resistant to deformation, which means higher strength. Pure iron is very weak compared to the carbon steel alloys.

Alloying by substitution also enables iron to be used for a variety of applications that cannot be matched by pure iron or other metal alloys. Substitutional alloying occurs when atoms of similar size to those of the parent metal fill spaces that would otherwise be occupied by parent metal atoms. The elements manganese (Mn), silicon (Si), nickel (Ni), chromium (Cr), and molybdenum (MO) are used examples of substitutional alloying.

WHAT ARE THE EFFECTS OF COOLING RATES ON METAL GRAINS?
Faster cooling in steels results in smaller grains, while slower cooling results in larger grains. Slower cooling rates inhibit nucleation, which results in less seed nuclei forming during solidification and therefore more room for individual grains to form. With less seed nuclei in a given volume of metal, there is more room for each grain to grow before contact is made with other grains. Conversely, faster cooling causes more seed nuclei to form, and this means more grains in a given volume of metal are formed, therefore resulting in smaller grains.

Smaller grains are more resistant to deformation. These result in a stronger metal that is less easily bent, shaped, or formed. The smaller grains have the benefit of increasing strength. Larger grains tend to be softer and more ductile (less brittle) and are more easily bent, shaped, and formed.

Cooling rates in welds have significant impacts on weld strength. Faster cooling creates smaller, harder, and stronger grains. The cold base metal can cause rapid cooling, yet some areas close to the fusion line will have larger grains. The important thing to remember about welding is that the grain size will

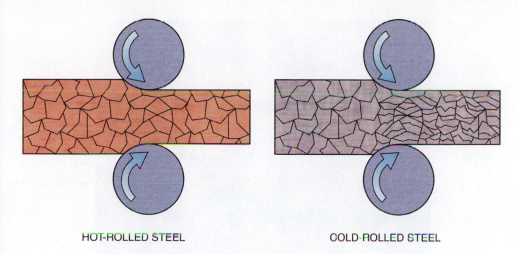

HOT-ROLLED STEEL   COLD-ROLLED STEEL

FIGURE 9-21 Cold forming (right) leaves a smaller, harder, unrefined grain structure.

vary depending on how high the weld and base metals are heated and then on how fast subsequent cooling occurs.

WHAT ARE THE EFFECTS ON METAL GRAINS FROM FORMING OPERATIONS?

All forming operations distort the metal grains to some degree, usually by squeezing the grains and flattening them. While there are many methods of forming operations, the degree of grain deformation is greatly influenced by the temperature at which the operation is done. Forming operations fall into two categories: cold forming and hot forming. Hot forming occurs at elevated temperatures, and the large high-temperature grains are broken down into smaller elongated grains. Depending on how hot the metal is after forming, the grains may then be refined to some degree. Cold forming is done without heating the metal and leaves a smaller, harder, unrefined grain (see **FIGURE 9-21**). In rolling steel plate, for eample, the grains are elongated and run lengthwise down the metal piece.

Grain direction is an important factor to consider when welding and fabricating. Grain direction affects steels in two significant ways on cold-forming operations. First, the amount of spring-back on bending operations varies when bending along the grain as opposed to across the grain. For this reason, it is important to cut multiple parts to be bent from the plate so they are all bent with the grain running in the same direction. This aids in the repeatability of parts meeting tolerances due to spring-back because parts bent across the grain tend to spring back more than those bent along the grain.

Another outcome of bending along the grain is that metals tend to fail under strain of bending more readily than those bent across the grain. **FIGURE 9-22** shows potential effects of bending and grain direction. The two pieces shown were both cut from the same piece of stock. The piece on the left was bent across the grain and withstood bending forces. The piece on the right was bent and failed along the grain. This example confirms the reason why welding codes recommend bending weld test samples across the grain.

Grain direction can also have adverse effects on welded heavy structures that cool rapidly. Rapid cooling can cause hard structuring to develop in weldments. This can lead to weld heat affected zone cracking, especially on less ductile

FIGURE 9-22 Knowing grain direction aids in the prevention of cracking when forming.

higher-carbon and alloys steels that do not give or stretch much under pressure. In these situations, welds made across the grain are less likely to crack than welds made along the grain.

IRON-CARBON PHASE DIAGRAM

An equilibrium phase diagram for metals is a graph that represents changes in structures in metals based on temperature and alloy content. The **iron-carbon phase diagram** is a graph of phase changes in steels based on carbon percent by weight at equilibrium temperatures from ambient to above the melting points of steels. These changes are plotted based on slow cooling rates, so that the phases reach equilibrium, and changes documented include liquid to solid, as well as structural changes in solid structures of steels. A study of the iron-carbon phase diagram (**FIGURE 9-23**) not only helps understand how carbon values relate to microstructure changes among ASTM and AISI-SAE plain carbon steels, but it also helps clarify important changes in atomic structure and grain size that affect metal properties.

Steels are alloys of iron with carbon contents up to 2%. Pure iron (Fe) is a weak structure, and carbon added increases strength, hardness, and wear resistance. Common structures in steels vary with carbon content and with changes in temperature. Those structures—austenite, ferrite, cementite, and pearlite—are important in any discussion of iron-carbon diagrams.

Austenite, or gamma iron, is an interstitial solid solution of carbon in iron and occurs in steels above a critical temperature in the face-centered cubic form. Austenite is stable above the upper transformation temperature (A_3) indicated by the blue line on **FIGURE 9-23**. Steels become nonmagnetic in the austenitic form.

Ferrite (**FIGURE 9-24**), or alpha iron, is an interstitial solid solution of iron containing small amounts of up to 0.02% carbon in the body-centered cubic form. Because of the near absence of carbon, ferrite iron is the softest structural form in steels.

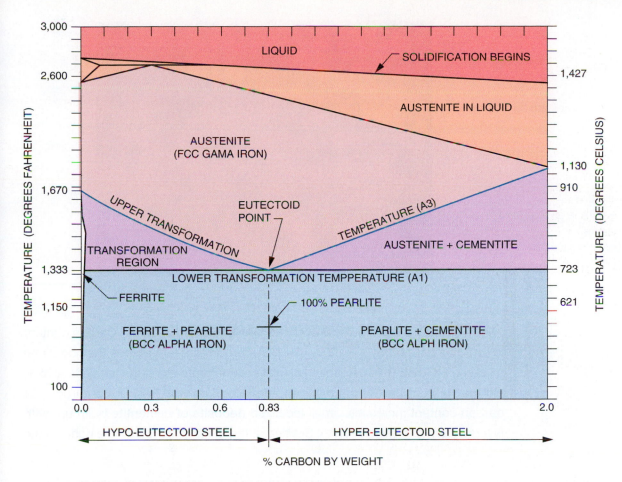

FIGURE 9-23 The iron-carbon phase diagram

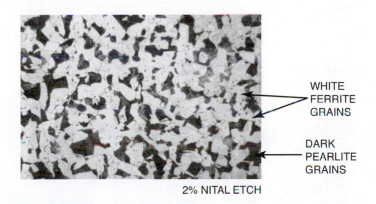

2% NITAL ETCH

WHITE
FERRITE
GRAINS

DARK
PEARLITE
GRAINS

FIGURE 9-24 Ferrite (alpha iron)
Source: Neely, John E. and Thomas J. Bertone, *Practical Metallurgy and Materials of Industry*, 5th Edition, p. 112. Reprinted by permission.

Cementite, or iron carbide, is the hardest structure represented on the iron-carbon diagram. It contains 6.67% carbon by weight and is found as small deposits in steels. The higher the carbon content, the more cementite is found in the grain structure. Low-carbon steels have small amounts; medium-carbon steels have more.

Pearlite (see **FIGURE 9-25**) is a lamellar grain structure composed of alternating layers of ferrite and cementite. Carbon steel with 0.83% carbon has a balance of ferrite and cementite and is considered to have a 100% pearlite grain structure.

FIGURE 9-25 Pearlite is a lamellar grain structure.

Source: Neely, John E. and Thomas J. Bertone, *Practical Metallurgy and Materials of Industry*, 5th Edition, p. 112. Reprinted by permission.

2% NITAL ETCH

WHAT ARE THE PARTS OF THE IRON-CARBON DIAGRAM? Carbon content is listed as percent by weight on the iron-carbon diagram across the horizontal axis (see **FIGURE 9-26**). Notice the small area of ferrite along the left vertical axis. On the very left part of the diagram, ferrite occurs with little or no carbon. As carbon content increases, small localized deposits of cementite build up, with layers of ferrite forming pearlite. Starting at the left and moving to the right, as the

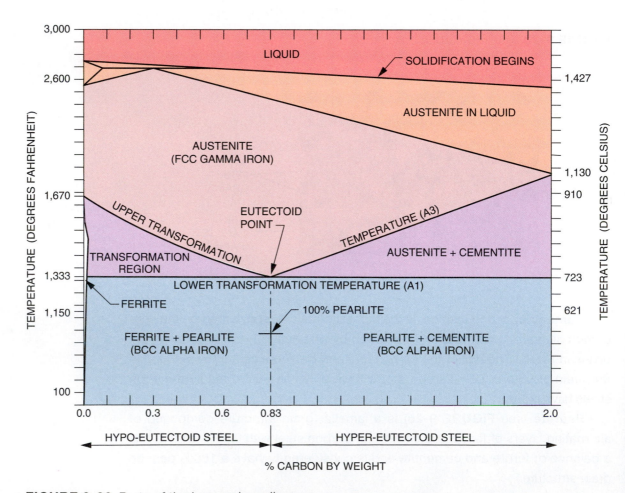

FIGURE 9-26 Parts of the iron-carbon diagram

carbon percent increases, so too does the amount of pearlite mixture, reducing the amount of ferrite grain structure. All these steels to the left of the eutectoid point (0.83% C) are composed of a mixture of ferrite and pearlite. These are called hypo-eutectoid steels. Under adequate magnification, ferrite shows up as white areas, whereas cementite appears as dark bands.

Eutectoid steels have 0.83% carbon and are considered to have a 100% pearlite structure. Adding more carbon beyond 0.83% reduces the amount of pearlite in the grain and increases the percentage of cementite. Cementite is a hard brittle structure, and steels with high concentrations of cementite are difficult to weld and form. **FIGURE 9-26** represents a portion of the iron-carbon diagram only up to 2.0% carbon. This includes the range of carbon steels up to the high-end ultra-high carbons. Beyond this point, increased carbon is not readily dissolved in iron, and the materials become very brittle. Cast irons generally contain 2.0%–4.0% carbon.

Several important lines plotted on the iron-carbon diagram help to communicate phase changes based on temperature and carbon content. The lower transformation temperature (A_1) occurs at 1,333°F (723°C) in all carbon steels. Its significance is that a transformation occurs when steel cools from above this line or when heated from below. Below the lower transformation temperature, steels' lattice structures are the body-centered cubic (BCC) form, while above, the cubic structure takes on varying amounts of both BCC and face-centered cubic (FCC) until the temperature reaches an upper transformation temperature (A_3). An understanding of these phase changes underscores the importance of preheating and postweld heat treatments. It is important to understand that these are allotropic transformations. That is, they occur in the solid state, and the steel changes crystal structure, gains or loses magnetism, and abosorbs or rejects carbon, without changing size or shape. This makes it possible to heat treat fabricated objects without changing their shape. It is the key to understanding why we live in the "Iron Age."

The upper transformation temperature varies with changes in carbon content, and transformation here is from an austenitic FCC structure above to FCC/BCC mixtures below, except in eutectoid (0.83% C) steel. The transformation temperature drops from 1,670°F (910°C) for steels with 0.02% carbon to 1,300°F (723°C) at the eutectoid point and then climbs again as carbon content increases. At the eutectoid point, transformation moves directly from austenite (FCC) to pearlite (BCC) if cooling, or from pearlite (BCC) to Austenite (FCC) upon heating.

WHAT USE IS THE IRON-CARBON DIAGRAM?

It should be apparent now that increased carbon has an impact on steel's microstructure. Furthermore, temperature changes also impact the crystalline structural forms of steels. The iron-carbon diagram (**FIGURE 9-27**) shows the effects of slow cooling on two different steels. Assume this case with an ASTM 1060 (0.60% C) steel and an ASTM 1080 (0.80% C) steel: They have just been manufactured as hot-rolled plates and cooling has just begun. Both steels—SAE 1060 at point A and SAE 1080 at point F—have an austenite grain structure with carbon atoms dissolved uniformly within. The carbon atoms settle in the interstices between the iron atoms in the FCC cubic structure, because FCC can absorb all the carbon in steel.

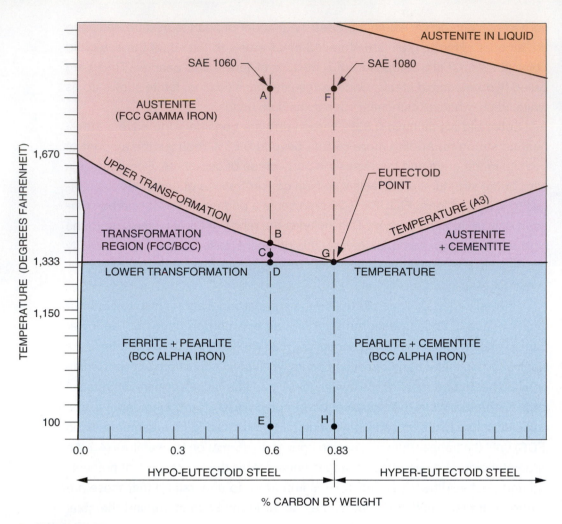

FIGURE 9-27 The iron-carbon diagram showing the effects of slow cooling on two different steels

At point A, the ASTM 1060 steel has a relatively large grain in the forging temperature range. As the steel cools to near the upper transformation temperature, no significant change takes place. At point B, the carbon atoms begin to migrate out from the FCC structure, and transformation from austenite to a ferrite/pearlite mixture begins. Nucleation of ferrite grains and alternating layers of ferrite and cementite begin to form on the grain boundary of the austenite grains. At point C, about one-half the structures have transformed, resulting in a balance of austenite (FCC) and ferrite/pearlite (BCC) grains. Finally, at point D, transformation from austenite to a ferrite/pearlie mixture is complete. Slow cooling has resulted in a large and somewhat soft grain at point E (70°F/21°C). For all carbon steels with less than 0.83% carbon, a similar narrative can be made for the cooling rate and the changes that occur. Steels with more than 0.83% carbon undergo similar changes except that the transition is from austenite to a structure composed of pearlite and cementite rather than ferrite and pearlite.

The ASTM 1080 steel undergoes a slightly different cooling transition, starting at point F with large austenite grains. Notice that at point G, the upper transformation temperature line appears to intersect with the lower transformation temperature line. This is called the eutectoid point, and the 1080 is essentially eutectoid steel. At point G, transformation from austenite to 100% pearlite structure occurs. At point H, slow cooling has resulted in a large and somewhat soft grain. The eutectoid point for steels occurs in those containing approximately 0.77%–0.83% carbon.

WHAT HAPPENS WHEN STEELS ARE HEATED?   When steels are heated, the whole scenario described for cooling of steels, such as the ASTM 1060 and ASTM 1080 discussed above, is reversed—with one exception. Follow the reverse course from point E, and the 1060 begins transformation from ferrite/pearlite at point D. Then, the BCC structures begin to break down and transform to a relatively fine-grained austenite. At point B, transformation to austenite is complete, and grain size remains small until the steel is heated to about 50–100°F(10–38°C) above the upper transformation temperature. This fine austenitic grain is an extremely important factor in heat-treating processes used for hardening steels. The 1080 steel transforms directly from 100% pearlite to austenite at the eutectoid point (G).

This heating/cooling cycle is desirable when manufacturing products that are required to have these properties. Keep in mind that this heating/cooling cycle also occurs during welding and that could create undesirable effects if these transformations are not monitored and dealt with.

HEAT-TREATING PROCESSES

WHAT IS HEAT TREATING?   Heat treating consists of heating a metal to a specific temperature and cooling at a controlled rate to achieve a predictable grain structure to produce refined metallurgical properties. Heat treating includes methods of grain refinement that make a structure either harder or softer than before heat treatment was employed.

Hardening heat treatments are used to increase strength and wear resistance. Softening processes are often used to improve machinability and formability. Stress relief treatments improve toughness, reduce stresses, and improve machinability in some steels without softening them significantly. An awarenes of how steels are hardened and softened is important in understanding how thick sections of metal tend to cool and quench welds as well as how preheat and postheat treatments deter the formation of undesirable hard transformation products, such as martensite in weldments. Furthermore, some heat treatable alloy steels are required to be welded in the annealed condition.

HOW ARE STEELS HARDENED?   The fine austenite grain is an extremely important factor in heat-treating processes used for hardening steels because it is transformed through controlled, sometimes rapid cooling to a hard structure called **martensite**. Martensite is a hard, brittle grain structure found in steels caused by cooling, usually rapidly, from austenite. Interstitial carbon atoms are trapped in the body-centered tetragon space lattice. When steels with significant carbon content are cooled from austenite at accelerated rates, carbon is not able to migrate out to stable forms of BCC structures like those formed from slow cooling. Carbon atoms are trapped in the interstices between iron atoms, and an elongated body-centered tetragon (BCT) forms into the needlelike structure martensite. Although martensite can be a desired transformation product, it is not a desired end result. It is in too brittle a state, and a follow-up softening process called **tempering** is performed to reduce stresses and prevent cracking. Tempering is a process of heating a metal to a precise temperature where a controlled cooling rate is used to soften steels slightly, reducing the stressed condition in martensite. Tempering increases toughness and makes them less brittle.

The importance of the correct heat-treat temperature and **quenching** rate cannot be overstated. In order to harden them, steels must be quenched from austenite. Quenching is a controlled cooling rate used to harden steels. Quenching is often referred to as rapid cooling; however, this may be misleading. Cooling rates vary depending on steel composition. Cooling a high-carbon steel in brine (saltwater) is a rapid process, compared to cooling an air-hardening tool steel in air under a controlled air flow. When welding thick steels that are not preheated, a similar quenching effect can occur as the heat is conducted away from the weld rapidly by the surrounding steel. This rapid cooling can create a hard, brittle weldment and heat-affected zone, which can be avoided by slowing the cooling rate by use of a preheat and/or maintaining interpass temperatures and/or postweld heat treatments.

The correct temperature for forming austenite before quenching is about 50°F (10°C) above the upper transformation temperature. This temperature allows for a stable, small-grained austenite suitable for hardening to form. Heating significantly above this temperature will lead to grain growth, and even a moderately larger grain is undesirable. If steel is heated too high and quenched after grain growth occurs, the larger, weaker grain is likely to fail by cracking. The remedy for preventing failure due to overheating before quenching is to allow the steel to cool back down slowly to room temperature and then reheat to the correct temperature.

Heating in a metallurgical heat-treat furnace allows for specific temperatures to be reached. This is the preferred method for heating parts before quenching. Other methods for gauging temperature exist. An optical or radiation laser-type pyrometer has been used for measuring high-temperature steels. A not-yet-antiquated method still used by blacksmiths today is to use a magnet to test the steel by heating it until it loses its magnetic properties. Austenitic iron is nonmagnetic. The trick is to avoid overheating in order to avoid grain growth.

The quenching media used for hardening is selected based on the steel classification. Water-cooling provides rapid quenching for high-carbon steels, which are cheaper than alloy steels, but water quenching also creates more residual stresses. Residual stresses can lead to cracking and distortion. Air, oil, and molten salts are used for steels with higher alloy contents. These steels have higher **hardenability**, and if quenched in water, they would be severely overstressed and cracking would result. Hardenability is the capacity of a metal to be hardened by heat treatment. While hardness is a measure of a metal's resistance to penetration testing, hardenability is a measure of a steel's ability to become through-hardened or hardened throughout, rather than just at the surface.

Plain carbon steels tend to harden to a shallow depth on large base metal masses. Many alloy steels used in the tool-making industry and some high-strength low-alloy steels used in fabrication will harden to an appreciably deeper depth. While hardness is a measure of a metal's resistance to deformation by scratching or indentation, hardenability is a function of its ability to be hardened. Hardenability depends on carbon and alloy content, mass of the part, austenitizing temperature, and the cooling rate of the steel. Materials with a high degree of hardenability are more readily hardened to deeper depths. A notable concern to welders is that the higher the steel's hardenability, the more difficult it is to weld successfully.

Alloys with high amounts of chromium, copper, manganese, molybdenum, nickel, silicon, and vanadium have high hardenability. If quenched too rapidly, these materials under-cool severely, likely resulting in cracking. This is why it is absolutely critical to know the correct cooling rate and quenching media for producing the desired hardness after quench and temper are complete. The *Machinery's Handbook* is a good reference for heat-treatment temperatures and quenching media for steels.

HOW ARE COOLING RATES DETERMINED FOR STEELS?
Time Temperature Transformation (TTT) diagrams are used to plot cooling curves for steels. These diagrams have been created by metallurgists through extensively testing thousands of samples from each type of steel alloy. Samples are tested by heating them to the correct austenite temperature, cooling them at varying cooling rates, and then examining the resulting microstructures and hardness values. These values are then plotted on a graph. Any change in alloy or carbon content changes the metal alloy and its subsequent cooling curve shown on a TTT diagram.

Understanding how to read a TTT diagram can be helpful in predicting how to attain specific properties for steels through heat treatment. Cooling steel over a specific controlled time period produces a specific transformation product, which determines the hardness or softness of that steel. Controlled heat treatments are used to improve hardness, toughness, wear resistance, and machinability.

FIGURE 9-28 illustrates the most basic features of TTT diagrams. The vertical axis on the left is used to plot the continuous cooling temperature. The horizontal

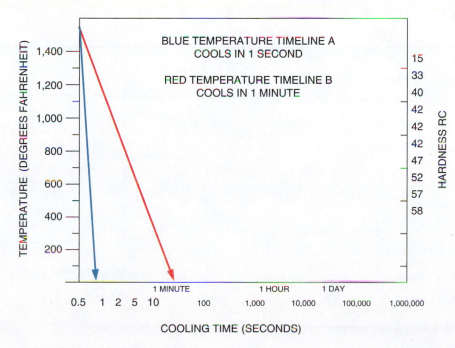

FIGURE 9-28 A Time Temperature Transformation (TTT) diagram

axis at the bottom plots the cooling time by using a logarithmic scale that condenses the time results into a useful format. Notice also a scale on the right-side vertical axis that plots the Rockwell hardness. The hardness scale is used to predict the hardness value of final transformation products for some steels and heat-treat processes.

Finally, the two colored timelines are used here to represent two possible cooling rates for specific steels. A temperature timeline is used to plot the path of the cooling curve. The blue temperature timeline A demonstrates a rapid cooling from above the upper critical temperature to room temperature in 1 second. The red timeline represents a slower cooling rate over a 1-minute time span.

WHAT IS THE TRANSFORMATION REGION? From discussion of the iron-carbon diagram, it was determined that steels transform from a BCC (ferrite + pearlite) structure to a FCC structure called austenite when heated to above the upper transformation temperature. When cooling from austenite, transformation to a room temperature structure begins at the upper transformation temperature, and if cooling is slow enough, the steel returns to its original BCC structure that existed before heating occurred. The transformation region on the iron-carbon diagram is based on plots of temperature and carbon, as shown in **FIGURE 9-29**.

The TTT diagram also has a transformation region superimposed onto it, as shown in **FIGURE 9-30**. The transformation region on the TTT diagram loosely

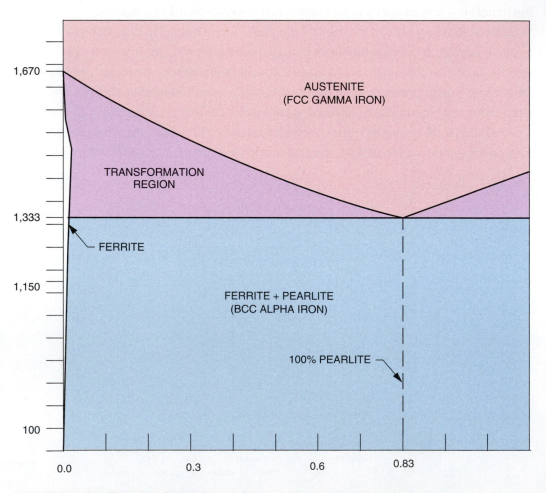

FIGURE 9-29 The transformation regions on the iron-carbon diagram

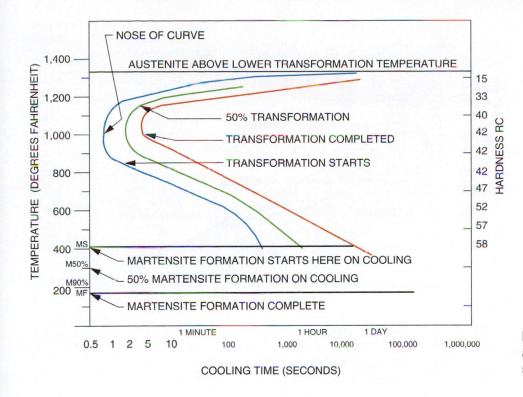

FIGURE 9-30 The TTT diagram also has transformation regions superimposed onto it.

represents the same region as the iron-carbon diagram—only in relation to temperature and time rather than temperature and carbon content. Transformation from austenite to another room temperature product takes place by cooling through the transformation region starting at the blue line and then passing through the brown line to the right or by dropping to the left of the nose of the diagram and passing through the Ms line and eventually cooling completely. Notice the two horizontal lines near the bottom of **FIGURE 9-30**. The top line labeled Ms indicates where, upon rapid cooling, austenite begins transforming to martensite. The Mf line indicates where the transformation to martensite is complete. With all heat-treat processes, the intent is to cool the steel through the appropriate portion of these transformation regions, and with welding, the intent is to avoid passing through the martensitic transformation region.

Austenite is stable above the upper transformation temperature, with carbon dissolved uniformly throughout the grains. After cooling begins and the temperature drops below the upper transformation temperature, carbon starts to precipitate out of the austenite, and if cooling is slow, it returns to the BCC structure. Austenite in the area to the left of the blue line is unstable, as carbon is trapped in the interstices of the FCC structure. Precise control of cooling will result in a grain structure ranging from a soft, slowly cooled ferrite/pearlite one to a hard, rapidly cooled structure called martensite.

WHAT ARE THE POSSIBLE TRANSFORMATION PRODUCTS FROM THE CONTROLLED COOLING OF STEELS?

To understand the relationships between time-temperature cooling rates and end-transformation products, it helps to overlay transformation regions and respective transformation products onto the TTT diagram. **FIGURE 9-31** breaks the transformation region into six areas based on grain structures that result due to varying cooling rates.

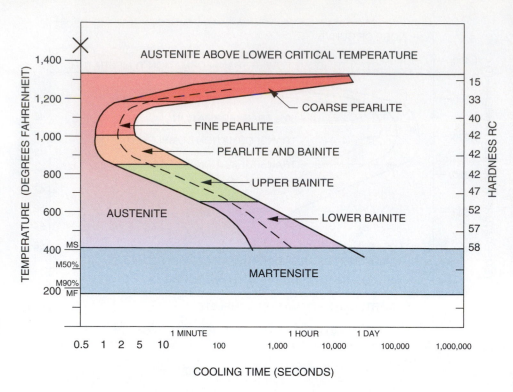

FIGURE 9-31 This TTT diagram breaks the transformation region into six areas based on grain structures.

The softest transformation product—coarse pearlite—is produced with the slowest cooling rate. Subsequent cooling rates of progressive speeds produce fine pearlite, pearlite + bainite, upper bainite, lower bainite, and martensite, respectively. The faster the cooling rate for specific steels, the harder the transformation products. Martensite—being the hardest and most brittle structure—requires an immediate follow-up heat treatment called tempering to stabilize the structure and to prevent cracking. **Bainite** is a hard transformation structure in iron alloys—harder than pearlite but softer than martensite.

HOW DO COOLING CURVES AFFECT STEEL STRUCTURES?

Where the temperature timeline passes into the transformation region determines how the grain structure begins to form. If it passes halfway through the region, half the austenite transforms to that structure, and if passes all the way through, then it completely transforms. In order to achieve a hardened structure, the steel must be cooled rapidly enough to ensure that the cooling curve passes to the left of the nose on the TTT diagram as shown in **FIGURE 9-32** (blue cooling curve). When welding, the intent is to cool the weld slower to avoid these hardened structures.

The steel in this example cools in 1 second, as indicated by the blue timeline. In this case, martensite starts to form at point A, where it crosses the Ms line. Martensite formation is complete when the material cools through the Mf line at point B. Because cooling occurred completely through the martensite region, the entire structure transforms to martensite.

The second cooling curve represents a slower cooling time, crossing into the fine pearlite zone at point C at a little over 1 second and passing halfway through transformation there. At this point, only half the austenite is transformed to fine pearlite. Because the cooling curve does not pass beyond the halfway point, no more transformation occurs until the timeline passes through the Ms line at

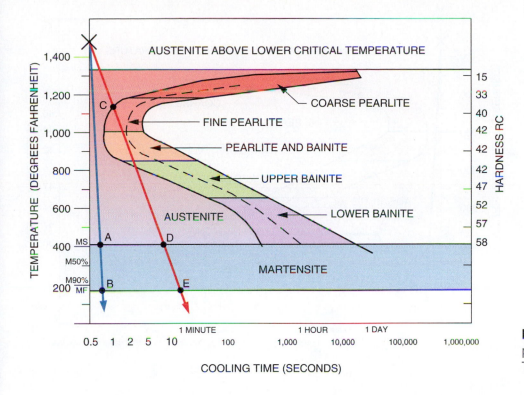

FIGURE 9-32 Rapid cooling passing through the nose of a TTT diagram

point D. At point D, the remainder of the austenite begins transforming to martensite and then completion occurs at point E. This grain structure is half fine pearlite because it first passed halfway through the fine pearlite region and half martensite because the second half of cooling passed through the martensite region. Tempering or postweld heat treatment are required to reduce stresses caused by the martensite grain structure.

WHAT ARE THE COMMON HARDENING PROCESSES, AND HOW DO COOLING CURVES DIFFER FOR HARDENING AND SOFTENING?

Several hardening processes exist. Those discussed here are: standard quench and temper; isothermal quench and temper; martempering; and austempering. The goal with hardening is typically to cool the steel through the martensite range and then temper or to cool it through the lower bainite range. The goal with welding is to prevent the steel from cooling rapidly and to eliminate the formation of martensite through postweld heat treatment.

Standard quench and tempering (**FIGURE 9-33**) is typical of steels heated in a furnace to the austenite transformation temperature A and cooled rapidly. Martensite transformation begins at point B and is complete at point C. Follow-up tempering is done immediately to reduce residual stresses caused by internal stresses on the brittle martensite grains. The steel is heated to an appropriate tempering temperature—usually between 400°F (204°C) and 600°F (316°C). In this case, the tempering temperature is about 550°F at point D. The steel is held for a suitable time at this temperature, depending on its mass. After holding at the tempering temperature, the metal is cooled completely down to E, and the final microstructure is tempered martensite. Tempering has relaxed the body-centered cubic structure martensite and eased the internal stresses to some extent.

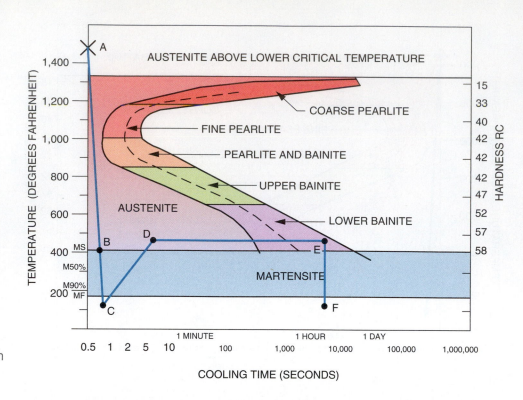

FIGURE 9-33 Standard quench and tempering

Martempering (**FIGURE 9-34**) is a method for allowing some grain refinement before quenching by allowing the unstable austenite FCC structure to relax a bit and allow more carbon to come out of solution. Notice that instead of cooling in a straight line through the Ms line, the steel is quenched and held in a molten salt bath from points B to C. This allows for austenite on the internal portions of the part to transform more substantially and reduces internal stresses. The steel is then quenched to form martensite at point D. Tempering treatment follows at about 550°F (288°C) from point E to point F and cooled to point G. A tempered

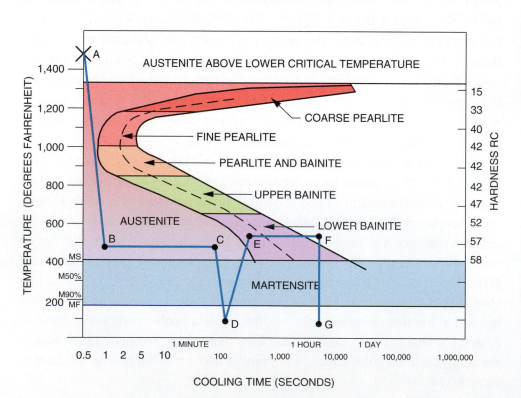

FIGURE 9-34 Martempering shown on a TTT diagram

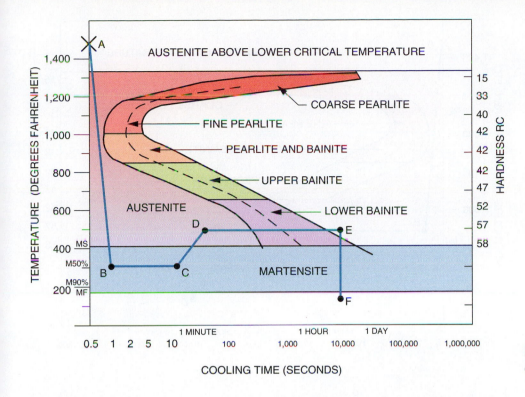

FIGURE 9-35 Isothermal quench and tempering

martensite structure results from martempering. Martempering produces a more through-hardened and stable part than standard quench and tempering.

Isothermal quench and tempering (**FIGURE 9-35**) produces a softer, less stressed grain than either standard quench and tempering or martempering by quenching the steel through only part of the martensite transformation region. In this case, the steel is quenched only halfway through the martensite region to point B and held for a length of time to point C. This allows time for half the structure to uniformly transform to martensite before transferring the material to a different quenching media at point D (500°F/260°C) and holding for a suitable time through the lower bainite transformation region to point E. A subsequent quench to point F leaves a through-hardened grain structure of roughly 50% tempered martensite and 50% lower bainite. Bainite is a softer and tougher structure than tempered martensite.

A fourth and final hardening method is called austempering (see **FIGURE 9-36**). Austempering transforms austenite to bainite by cooling the steel and holding it in a molten salt bath between 400°F (204°C) and 800°F (427°C), depending on the alloy composition. No martensite is formed, and no tempering is required. In this case, the steel is quenched from point A to point B and held there until complete transformation to lower bainite has occurred at point C. The part is then cooled completely. Austempering has improved toughness and ductility over other hardening processes while also eliminating the problems of distortion and cracking effects of more rapid quenching methods. The point where the temperature timeline passes through the transformation region can be used to predict an approximate hardness value—in this case, RC 57.

WHAT ARE THE COMMON SOFTENING PROCESSES? Full annealing and normalizing are heat treatments used to soften steel products. Annealing is the favored treatment to create a maximum softness, making metals easier to cold-form,

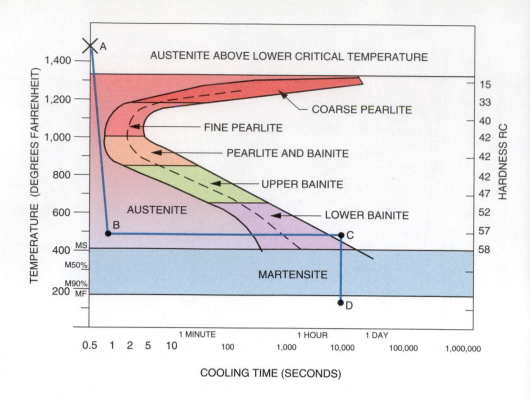

FIGURE 9-36 Austempering shown on a TTT diagram

and it is sometimes used to improve machinability of metals. Normalizing is used to refine grain structures in cold-worked, hot-worked, and sometimes welded parts to a soft uniform grain structure throughout the part. Normalized parts are not as soft as fully annealed parts. These processes are sometimes used as an important part of welding procedures when welding alloy steels.

FIGURE 9-37 shows a cooling curve representing a full anneal heat treatment of a steel. Full anneal is achieved by heating to 50°F (10°C) above the upper transformation temperature and holding for a suitable time—typically

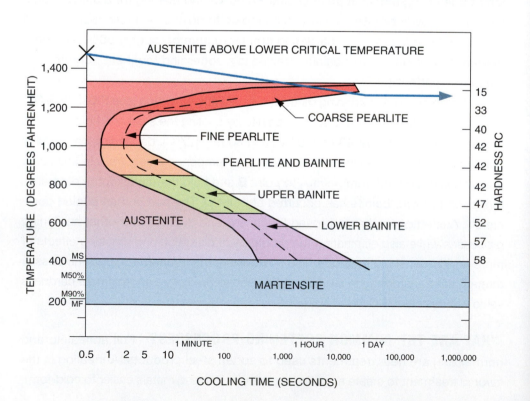

FIGURE 9-37 Cooling curve representing a full anneal heat treatment of a steel

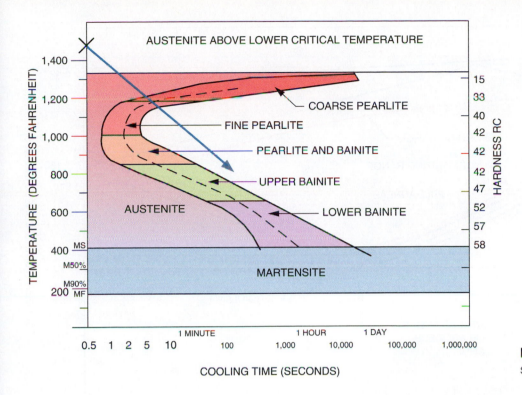

FIGURE 9-38 Normalizing shown on a TTT diagram

1 hour/inch of base metal thickness—to allow full transformation of austenite throughout. The part is then cooled extremely slowly in a heat-treat furnace or wrapped, enclosed, or buried in an insulating material. The steel represented here has a final hardness value of slightly more than Rc 15, as indicated by the blue arrow.

An example of normalizing is given in **FIGURE 9-38**. Normalizing heats steels to approximately 100°F (38°C) above the upper transformation temperature and holds them for an appropriate time to allow full transformation of austenite throughout. Then, slow cooling is achieved by cooling in still air. Parts are set on a mesh or screen to allow cooling from all sides to ensure that an even, uniform grain is attained.

FIGURE 9-39 displays softening heat treatments that occur below the lower critical temperature. Stress relief treatments do not heat steels up above the lower transformation temperature. The methods for relieving stresses and refining grains below the lower transformation temperature are stress relief annealing, bright annealing, process annealing, and spheroidizing.

Stress relief annealing is used to reduce stresses in forged, cast, and welded low-carbon steel products. Stress relief annealed parts are not nearly as soft as full annealed parts. Parts are heat-treated to temperatures below, but sometimes nearing the lower transformation temperature.

Bright annealing and process annealing are performed at similar temperatures to those for stress relief annealing on low-carbon steels. Process annealing is used on parts that have been cold-worked and softens them enough that they can undergo further forming operations.

Bright annealing is performed with parts held in a chamber with an inert gas atmosphere to prevent oxidation. The term bright anneal evolves from the brighter, shiny appearance of annealed parts.

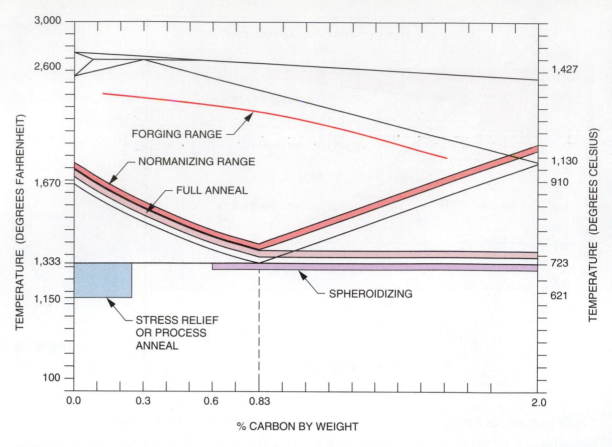

FIGURE 9-39 Heat treatments for carbon steels

Spheroidizing involves holding high-carbon steel at a temperature below and near the lower transformation temperature. In spheroidized parts, the carbon forms small spheroidal shapes. The process is used to improve machinability of high-carbon steels.

HOW DO HEAT TREATMENTS AFFECT NONFERROUS METALS? It is important here to remember that these TTT diagrams relate cooling rates for steels. For most nonferrous metals, cooling rates do not affect the final structure. For example, annealing copper is accomplished by heating to an appropriate temperature (cherry red) and then cooling. This leaves a soft, easily formed grain structure. Cooling can be done rapidly in water or slowly in air, with little variation in hardness or formability.

Some nonferrous metal alloys are hardenable to some degree. These are hardened through solution heat treatment and subsequent aging to allow precipitation of alloying elements that distort the space lattice structure. Others are hardened by the strain induced during forming, such as rolling aluminum plate.

HOW DOES THE HEAT OF WELDING AFFECT A WELD? Heat effects from welding on base metals are significant. **FIGURE 9-40** compares effects of heat from welding on hot-rolled steel and cold-rolled steel in and around the

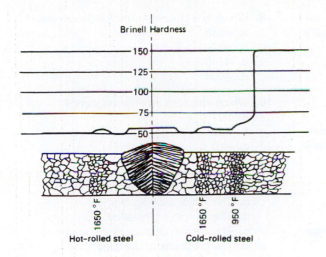

FIGURE 9-40 Heat-affected areas from welding

Source: Neely, John E. and Thomas J. Bertone, *Practical Metallurgy and Materials of Industry*, 5th Edition, p. 112. Reprinted by permission.

heat-affected zone. The molten weld metal cooled slowly, forming a columnar grain structure caused by nucleation starting on the cold base metal and growing slowly toward the center of the weld. Grain growth can be seen in the base metal outside the fusion line where the weld meets the base metal. This portion of the base metal reached temperatures high above the upper transformation temperature. Just outside the coarsened grains is an area of small grains where base metal temperatures ranged from the lower transformation temperature to near the upper transformation temperature.

It becomes apparent when examining a weld microstructure that heat from welding can greatly influence base metal structures. Varying temperature ranges and cooling rates on steel base metals—both carbon and alloy—causes a wide range of structures in steels. Medium- and high-carbon steels readily form hard martensitic structures due to rapid cooling in the heat-affected zone by the cold base metal. Welding carbon and alloy steels can easily result in weld and base metal cracks.

Welding procedures must take into account carbon and alloy content and often require preheating base metals and controlling interpass temperatures and postheat treatment. Further discussion of heat effects on base metals and welding procedures for reducing their influences are detailed in Chapter 11.

CHAPTER QUESTIONS/ASSIGNMENTS

MULTIPLE CHOICE

1. What is the measure of the maximum load-carrying capacity of a material based on a tensile test?
   a. Breaking
   b. Fatigue strength
   c. Tensile strength
   d. Yield strength

2. What property of a metal refers to its resilience or ability to resist deformation by stretching?
   a. Elasticity
   b. Hardenability
   c. Malleability
   d. Plasticity

3. What strength is the amount of stress in pounds per square inch (psi) required to cause a material to plastically deform?
   a. Compressive
   b. Tensile
   c. Torsion
   d. Yield

4. Which is the property of a material to deform permanently or to exhibit plasticity without rupture while under tension?
   a. Elasticity
   b. Ductility
   c. Plasticity
   d. Malleability

5. Which property is the ability of a metal to deform permanently when loaded in compression?
   a. Elasticity
   b. Ductility
   c. Plasticity
   d. Malleability

6. What is an expression of resistance of a material to fracture resulting from impact loading?
   a. Compressive strength
   b. Fatigue strength
   c. Impact strength
   d. Tensile strength

7. Which strength measures the load-carrying ability of a material subjected to loading, which is repeated a definite number of cycles?
   a. Compressive strength
   b. Fatigue strength
   c. Impact strength
   d. Torsion strength

8. Which steels are readily welded with fairly simple procedures and are generally easy to bend, shape, and form?
   a. Low-carbon
   b. Medium-carbon
   c. High-carbon
   d. Low-alloy

9. Which is a chromium-molybdenum steel?
   a. 4068
   b. 4130
   c. 4340
   d. 4608

10. Which stainless steel is magnetic?
    a. 201
    b. 309
    c. 410

11. For classifying aluminum alloys, the first digit denotes the alloy group based on what?
    a. The major alloying element
    b. The temper of the alloy
    c. The carbon content of the alloy
    d. The hardness of the alloy

12. Ferrite contains how much carbon?
    a. Less than 0.02%
    b. 0.3%–0.15%
    c. 0.15%–0.29%
    d. 0.30%–0.59%

13. Faster cooling in steels results in which?
    a. Smaller grains with low strength
    b. Smaller grains with high strength
    c. Larger grains with low strength
    d. Larger grains with high strength

14. Which is a lamellar grain structure composed of alternating layers of ferrite and cementite?
    a. Austenite
    b. Iron carbide
    c. Martensite
    d. Pearlite

15. Which is a hard structure found in steels caused by cooling from austenite at controlled cooling rates?
    a. Austenite
    b. Ferrite
    c. Pearlite
    d. Martensite

FILL IN THE CHART

16. Label the iron-carbon phase diagram. Include the following items:

Carbon percent by weight
100% pearlite
Ferrite + pearlite
Temperature degrees Fahrenheit
Temperature degrees Celsius
Lower transformation temperature
Upper transformation temperature
Austenite region
Ferrite
Ferrite + cementite

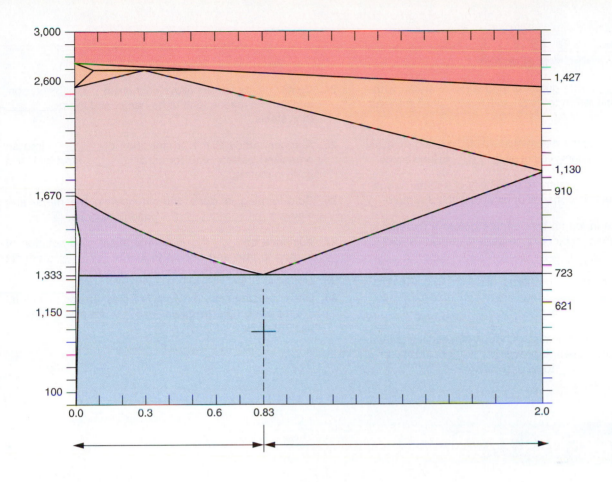

17. Correctly label the TTT diagram below:

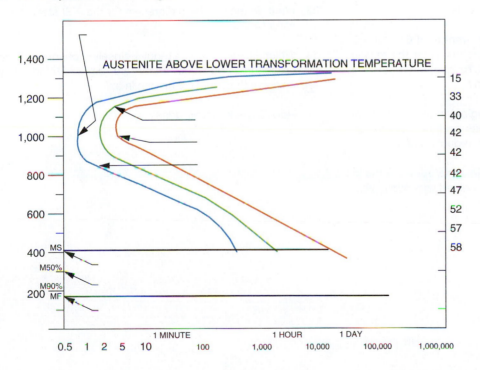

18. The _____ is the greatest load a material can withstand and still return to its original length when the load is removed.

19. The _____ point is the point at which a marked increase in deformation occurs without an _____ in load stress.

20. The term _____ refers to the ability of a metal to sustain permanent deformation or to yield without rupture.

21. The increase in length that occurs before a material is fractured when subjected to tensile stress is called _____.

22. Brittle metals tend to be harder than _____ metals, displaying little _____, and are not suitable for forming and bending operations.

23. If a piece of metal is heated without restraints and is then allowed to cool down, it grows _____ uniformly and then shrinks _____ to its original size.

24. Cast irons are ferrous alloys containing _____ carbon, and for that reason tend to be harder and more _____ than steels.

25. A space lattice is the organization of _____ into an organized pattern that forms a _____ structure or a grain in metals.

26. Cold-forming is done without heating the metal and leaves a _____, _____, unrefined grain.

27. Austenite or _____ is an interstitial solid solution of carbon in iron and occurs in steels _____ a critical temperature in the face-centered cubic form.

28. Understanding how to read a TTT diagram can be helpful in predicting how to attain _____ for steels through heat treatment.

29. What is a heat-affected zone?

30. List the 6 common strengths of metals.

31. A tensile specimen had a tensile strength of 67,819 psi. The original gage length was 2 inches. After the tensile pull, the gage length measured 2.75 inches. What is the % elongation of the test specimen?

32. A tensile specimen had an original cross-section of 0.605×0.187. The specimen cross-section after the pull was 0.502×0.093. What is the % reduction of area?

33. If a piece of metal is heated while under restraint, as in the case of two base metals in a welded assembly, what happens?

34. Which carbon steel contains 0.30%–0.59% carbon?

35. What do the numbers represent for the AISI steel classification 4330?

36. Which stainless steels require closely monitored welding procedures because many are air-hardening alloys, meaning they quench-harden in air?

37. How is spark testing done?

Chapter 10

WELDING ALLOYS

INTRODUCTION

The American Welding Society defines **weldability** as "The capacity of a material to be welded under the imposed fabrication conditions into a specific, suitably designed structure and to perform satisfactorily in the intended service."[1] Just because a metal can be joined and a weld appears visually satisfactory does not mean it is a good weld or is suitable for service. Changes in base metal structures caused by heat from welding may result in cracking or metallurgical changes that render the part useless. Heat from welding can reduce corrosion resistance in stainless steel, cause cracking in carbon steels, and result in undesirable distortion of any type of base metal.

Metal products used throughout industry are vital for building structures, bridges, aircraft, and watercraft as well as for ordinary products used to enhance everyday living and comfort. Weldability of a metal is determined by the welding procedures used. This chapter discusses welding techniques, surface preparation, preheat temperatures, interpass temperatures and postheat treatments, cracking in welds and base metals, discontinuities and defects, and procedures that help to ensure weld quality.

WELDING CARBON AND LOW-ALLOY STEELS

Carbon steels are classified based on carbon content (see **TABLE 10-1**). Weldability of carbon steels depends on following procedures that take into account the effects of carbon and other alloying that can contribute to weld and base metal cracking caused by several factors. Martensite formation, hydrogen in base metals, and external stresses imposed on the weld joint and heat-affected zone all contribute to cracking in carbon steel welds. Martensite is a hard, brittle grain structure found in steels caused by cooling, usually rapidly, from austenite. Further contributors to weld metal failures are discontinuities,

Common Name	Carbon, %	Typical Hardness	Typical Use	Weldability
Low-carbon steel	0.15 max	60 HRB	Special plate and shapes, sheet strip, welding electrodes	Excellent
Mild steel	0.15–0.30	90 HRB	Structural shapes, plate, and bar	Good
Medium-carbon steel	0.30–0.50	25 HRC	Machine parts and tools	Fair (preheat and postheat normally required; low-hydrogen welding process recommended)
High-carbon steel	0.50–1.00	40 HRC	Springs, dies, railroad rail	Poor (low-hydrogen welding process, preheat and postheat required)

TABLE 10-1

Classification and Weldability of Carbon Steels

Source: American Welding Society (AWS) Welding Handbook Committee, 1998, *Welding Handbook,* 8th ed., vol. 4, *Materials and Applications, Part 2,* Miami: American Welding Society.

---

[1] ANSI/AWS A3.0-94, Standard Welding Terms and Definitions, p. 38, 1995.

such as undercut, incomplete fusion, and incomplete joint penetration. Unlike hot cracking of aluminum welds, which are visible upon completion of the weld, cracks in steel weldments often form over a longer time period and may not even be visible on the initial completion of the weld.

Rapid cooling in the heat-affected zone can cause formation of martensite. **Carbon equivalence** is a measure of a steel alloy's chemical composition, which is used to determine its weldability. Medium-carbon, high-carbon, and alloy steels with high-carbon equivalence are likely to have hardened structures in the heat-affected zone caused by martensite formation resulting from rapid cooling. These structures must be avoided to the greatest degree possible. In most cases this can be accomplished by developing a sound welding procedure that includes ductile filler metals, eliminates moisture, reduces base metal cooling rates, and minimizes residual joint restraint.

A carbon equivalence formula can be used to predict a steel alloy's susceptibility to weld-metal cracking and its ability to be welded. Carbon equivalence can be used to help determine preheat temperatures, to improve weldability, and to reduce weld-metal cracking. The American Welding Society recommends using this formula to determine carbon equivalency:

$$CE = C + ((Mn + Si) \div 6) + ((CR + Mo + V) \div 5) + ((Ni + Cu) \div 15)\ [2]$$

FIGURE 10-1 shows the relationships between carbon content and maximum hardness in steels with martensitic structures. Steel alloys with carbon equivalence similar to the carbon contents shown in Figure 10-1 should be welded with care to avoid formation of these hardened structures.

WHAT ARE THE MAJOR PROBLEMS ENCOUNTERED WITH CARBON STEEL WELDS?
Low- and mild-carbon steels are less susceptible to cracking than medium-carbon, high-carbon, or alloy steels because of their reduced tendency to form martensite. However, low-carbon and mild steels can suffer damaging effects from hydrogen absorption as well as effects of grain direction and residual stresses. **Residual stresses** are stresses present in a welded joint caused by

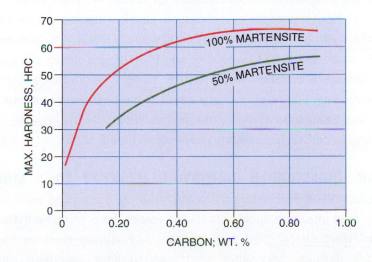

FIGURE 10-1 Relationship between carbon content and maximum hardness of steels

Source: American Welding Society (AWS) Welding Handbook Committee, 1998, *Welding Handbook*, 8th ed., vol. 4, *Materials and Applications, Part 2*, Miami: American Welding Society.

[2] *AWS Handbook*, Volume 4, 8th ed., vol. 4, p. 4.

nonuniform heating and cooling, which in turn leads to nonuniform expansion and contraction of the weld and heat-affected zone. While martensite formation is reduced in low- and mild-carbon steels, it is not entirely eliminated because diffusion of carbon in steels through the transformation region can result in concentrations susceptible to martensite formation. Due to its brittle structure, high-strength martensite tends to crack from cooling and effects of stresses.

Hydrogen cracking is a form of cold cracking in steels caused by the absorption of hydrogen in the base metal while it is at high temperatures. Several factors contribute to hydrogen cracking. The first factor is the presence of hydrogen dissolved in the heat-affected zone of the weld joint. Steel can absorb and retain hydrogen induced during welding. Moisture in electrodes and coatings, contaminated shielding gases, and hydrocarbons such as oil, grease, or paints can contribute to hydrogen entrapment in the heat-affected zone and must therefore be eliminated. One other major contributor to hydrogen-induced cracking is martensite formation in the heat-affected zone.

Martensite—formed during rapid cooling—is a low-ductility structure that cannot easily withstand the external stresses imposed on welded structures. Rather than stretch as a more ductile material would, martensite is more likely to cause cracking under these conditions. Hydrogen is readily diffused into matensitic structures, building up in the misfit lattice structure. If stresses are not relieved or if postweld heat treatments are not used to allow the hydrogen to dissipate into the local base metal, the misfit grains are likely to succumb to stresses caused by external joint restraint, and cracking occurs.

Finally, residual stresses not only cause cracking in steels, but they also cause the warping or distorting of base metal pieces. This distortion is undesirable and must be kept to a minimum.

WHAT CAN BE DONE TO ELIMINATE CRACKING IN CARBON AND ALLOY STEELS?

Several steps can be taken to eliminate cracking in carbon and alloy steels. Start by using an electrode manipulation that results in a slightly convex weld bead and fill craters to their full cross-section to reduce crater cracking. Regulate travel speeds to generate an **oval weld pool** as opposed to a rapid travel speed that produces an elongated V-shaped weld pool. Where possible, use high heat input processes or techniques. Use low-hydrogen-type electrodes to eliminate hydrogen entrapment. For Shielded Metal Arc Welding, those electrodes—such as the EXX15, EXX16, and EXX18—are stored in a rod oven to prevent moisture absorption (see **FIGURE 10-2**) in accordance with welding code requirements. Finally, control preheat and interpass weld temperatures and postweld heat treatments to reduce thermal stresses and the negative effects of rapid cooling as carbon equivalence increases in steels.

HOW CAN UNDESIRABLE DISTORTION BE KEPT TO A MINIMUM?

Distortion can be minimized by controlling fixturing, controlling weld sequence, controlling heat input, balancing shrinkage forces, and, in some instances, by use of postweld stress relief heat treatment. Fixturing involves the use of strongbacks, braces, and other structures to limit weldment movement. Controlling weld sequence involving techniques such as using staggered and

FIGURE 10-2 Electrodes being stored in a rod oven to prevent moisture absorption

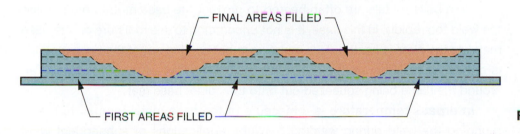

FINAL AREAS FILLED

FIRST AREAS FILLED

FIGURE 10-3 Block sequence

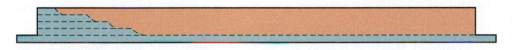

FIGURE 10-4 Cascade sequence

chain intermittent welding, which helps to balance shrinkage forces in opposite directions. Using the backstep sequence on sheet and thinner plate or using of the block sequence (see **FIGURE 10-3**) or cascade sequence (see **FIGURE 10-4**) on heavy plate assemblies can offset shrinkage forces.

Controlling heat input can be accomplished by choosing a low heat input welding process, narrowing groove angles, and reducing root openings, therefore minimizing the number of weld passes. Fewer weld passes can also be accomplished through the use of large electrode diameters or by using double-groove weld joints rather than single-groove weld joints.

WHAT ARE PREHEAT, INTERPASS, AND POSTHEATS? Preheat is the

process of heating a base metal immediately before welding is initiated and every time welding is resumed (see **FIGURE 10-5**). Preheat is used to reduce damaging effects from hydrogen absorption, residual stresses that result from nonuniform expansion and contraction of the weld, and martensite formation along the fusion line and in the heat-affected zone. Preheating the base metal reduces temperature gradients that contribute to nonuniform expansion and contraction that contribute to residual stresses in the completed weld assembly and to weld cracking. Preheat also slows the cooling rate through the critical transformation region, reducing the probability of martensite formation, which leads to weld-metal cracking.

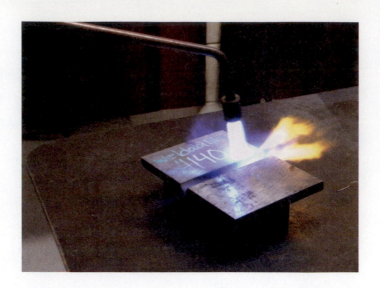

FIGURE 10-5 Preheat

Cold base metals are often heated to prevent the base metal from cooling the weld too rapidly. In this case, it is not uncommon to see moisture on the base metal as the plate is heated. This moisture in not coming out of the base metal; it is coming out of the air. It is condensation, a product of the warm moist air around the flame being squeezed out onto the colder base metal.

Interpass temperature is a measure of the base metal temperature surrounding the weld during welding between applications of subsequent weld deposits. In some cases, a maximum interpass temperature is applied to ensure that the base metal temperature does not get too high as when welding cast irons or quenched and tempered steels.

Postheat treating usually refers to a stress relief heat treatment that reduces residual stresses in the weld assembly after welding is complete. Weld assemblies are heated to an appropriate temperature, held 1 hour per inch (25.4 mm) thickness, and cooled slowly. Stress relief heat treatment temperatures and cooling rates vary with the base metal type being used. For carbon steels, the weld assembly is heated to a temperature below the lower critical temperature in the range of 900°F–1,200°F (482°C–649°C).

While preheating of small parts may be accomplished by use of a heat treating furnace or oven, a more practical method is to use an oxy-fuel torch (see **FIGURE 10-5**). Resistance heaters that come in blankets or strips are used for larger fabrication.

WHAT CONSIDERATIONS ARE GIVEN TO WELDING ALLOY STEELS? Welding of low-alloy steels requires special considerations. As with increases in carbon content, the addition of alloying elements increases the probability of hardened structures forming in the weld and heat-affected zone. Increases in carbon and other alloying elements makes the base metal more sensitive to weld-metal cracking, resulting in a greater need for control of heating and cooling cycles. The use of low-hydrogen welding procedures is recommended for welding all alloy steels. Low-hydrogen welding procedures include the use of electrodes and fluxes that do not contain moisture in their coatings. These are the low-hydrogen-type SMAW electrodes such as the EXX15, EXX16, EXX18, and EXX28 as well as

FIGURE 10-6 Temperature-indicating crayon

common electrodes used for the GMAW, GTAW, FCAW, and SAW processes. Electrode recommendations are given for each type of alloy steel discussed in the sections and tables that follow. Low-hydrogen procedures consider and eliminate moisture from shielding gas sources and the welding atmosphere surrounding the weld. With these steels, preheat, interpass, and postheat treatment must be considered.

Weldment temperatures can be monitored by using temperature crayon sets. In these sets of multiple markers, each crayon-like marker has a different and specific melting point. The crayon is rubbed against the base metal surface and the base metal temperature is determined by the melting temperature of the crayon (see **FIGURE 10-6**). More accurate temperature measurements can be made by using various types of available thermal (contact type) pyrometers and optical (infrared) thermometers (see **FIGURE 10-7**).

Matching the filler metal composition to the base metal is always a likely consideration. **TABLE 10-2** lists suffixes that are attached to AWS filler metal classifications and their corresponding chemical compositions that can be used to help match filler metals to similar base metal compositions. These suffixes can be consulted when selecting electrodes for specific alloys according to Tables 10-7, 10-8, 10-11, and 10-15.

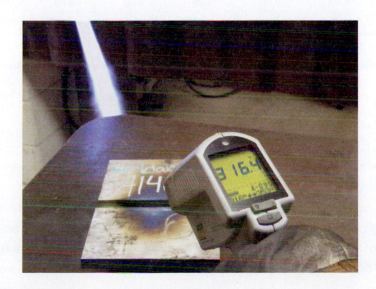

FIGURE 10-7 Optical thermometer

Suffix	C	Mn	Si	Ni	Cr	Mo	Addition
AI	0.12	0.60–1.00	0.40–0.80	—	—	0.40–0.65	—
BI	0.05–0.12	0.90	0.60–0.80	—	0.40–0.65	0.40–0.65	—
B2	0.05–0.12	0.90	0.60–0.80	—	1.00–1.50	0.40–0.65	—
B2L	0.05	0.90	1.00	—	1.00–1.50	0.40–0.65	—
B3	0.05–0.12	0.90	1.00	—	2.00–2.50	0.90–1.20	—
B3L	0.05	0.90	1.00	—	2.00–2.50	0.90–1.20	—
B4L	0.05	0.90	1.00	—	1.75–2.25	0.40–0.65	—
B5	0.07–0.15	0.40–0.70	0.30–0.60	—	0.40–0.60	1.00–1.25	V–0.05
B6	0.05–0.10	1.0	0.90	0.40	4.0–6.0	0.45–0.65	—
B6L	0.05	1.0	0.90	0.40	4.0–6.0	0.45–0.65	—
B7	0.05–0.10	1.0	0.90	0.40	6.0–8.0	0.45–0.65	—
B7L	0.05	1.0	0.90	0.40	6.0–8.0	0.45–0.65	—
B8	0.05–0.10	1.0	0.90	0.40	8.0–10.5	0.85–1.20	—
B8L	0.05	1.0	0.90	0.40	8.0–10.5	0.85–1.20	—
B9	0.08–0.13	1.25	0.30	1.0	8.0–10.5	0.85–1.20	(1)
CI	0.12	1.25	0.60–0.80	2.00–2.75	—	—	—
CIL	0.05	1.25	0.50	2.00–2.75	—	—	—
C2	0.12	1.25	0.60–0.80	3.00–3.75	—	—	—
C2L	0.05	1.25	0.50	3.00–3.75	—	—	—
C3	0.12	0.40–1.25	0.80	0.80–1.10	0.15	0.35	V–0.05
C3L	0.80	0.40–1.40	0.50	0.80–1.10	0.15	0.35	V–0.05
C4	0.10	1.25	0.60–0.80	1.10–2.00	—	—	—
C5L	0.05	0.40–1.00	0.50	6.00–7.25	—	—	—
NM1	0.10	0.80–1.25	0.60	0.80–1.10	0.10	0.40–0.65	(2)
DI	0.12	1.00–1.75	0.60–0.80	0.90	—	0.25–0.45	—
D2	0.15	1.65–2.00	0.60–0.80	0.90	—	0.25–0.45	—
D3	0.12	1.00–1.80	0.60–0.80	0.90	—	0.40–0.65	—
G	—	1.00 MIN	0.80 MIN	0.50 MIN	0.30 MIN	0.20 MIN	(3)
M	0.10	0.60–2.25	0.60–0.80	1.25–3.80	0.15–1.50	0.20–0.55	V–0.05
PI	0.20	1.20	0.60	1.00	0.30	0.50	V–0.10
WI	0.12	0.40–0.70	0.40–0.70	0.20–0.40	0.15–0.30	—	V and 0
W2	0.12	0.50–1.30	0.35–0.80	0.40–0.80	0.45–0.70	—	Cu

TABLE 10-2

Suffixes for Low-Alloy Steel Electrodes

Note (1) Various amounts of V, Cu, AL, Nb (Cb), and N
Note (2) V-0.02, CU-0.10, & AL 0.05
Note (3) V-0.10 MIN Cu-0.20 MIN; see ANS A5.5 for details
See ANS A5.5 for exact analysis; this is a summary

Source: Cary, Howard B., *Modern Welding Technology*, 4th ed., © 1998. Reproduced by permission of Pearson Education, Inc., Upper Saddle River, New Jersey.

High-strength low-alloy (HSLA) steels display high strength and toughness as well as corrosion resistance greater than those properties of carbon steels. These alloys (see **TABLE 10-3**) are grouped by ASTM specifications based on chemical compostion. Both carbon and alloy content influence the weldability

TABLE 10-3

ASTM Specifications for High-Strength Low-Alloy Structural Steels

ASTM Specification	Type or Grade	Composition, %[a]										Minimum Tensile Strength		Minimum Yield Strength	
		C	Mn	P	S	Si	Cr	Ni	Mo	V	Other	ksi	MPa	ksi	MPa
A242	1	0.15	1.00	0.15	0.05	—	—	—	—	—	0.20 min Cu	63–70	434–482	42–50	289–344
A441	—	0.22	0.85–1.25	0.04	0.05	0.30	—	—	—	—	0.20 min Cu; 0.02 min V	60–70	413–482	40–50	275–344
A572	42[b]	0.21	1.35	0.04	0.05	0.30	—	—	—	—	0.20 min Cu	60	413	42	289
	50[b]	0.23	1.35	0.04	0.05	0.30	—	—	—	—	0.20 min Cu	65	448	50	344
	60[b]	0.26	1.35	0.04	0.05	0.30	—	—	—	—	0.20 min Cu	75	517	60	413
	65[b]	0.26	1.65	0.04	0.05	0.30	—	—	—	—	0.20 min Cu	80	551	65	448
A588	A	0.10–0.19	0.90–1.25	0.04	0.05	0.15–0.30	0.40–0.65	—	—	0.02–0.10	0.25–0.40 Cu	63–70	434–482	42–50	289–344
	B	0.20	0.75–1.25	0.04	0.05	0.15–0.30	0.40–0.70	0.25–0.50	—	0.01–0.10	0.20–0.40 Cu	—	—	—	—
	C	0.15	0.80–1.35	0.04	0.05	0.15–0.30	0.30–0.50	0.25–0.50	—	0.01–0.10	0.20–0.50 Cu	—	—	—	—
	D	0.10–0.20	0.75–1.25	0.04	0.05	0.50–0.90	0.50–0.90	—	—	—	0.30 Cu, 0.05–0.15 Zr; 0.04 Nb	—	—	—	—
	E	0.15	1.20	0.04	0.05	0.15–0.30	—	0.75–1.25	0.10–0.25	0.05	0.50–0.80 Cu	—	—	—	—
	F	0.10–0.20	0.50–1.00	0.04	0.05	0.30	0.30	0.40–1.10	0.10–0.20	0.01–0.10	0.30–1.00 Cu	—	—	—	—
	G	0.20	1.20	0.04	0.05	0.25–0.70	0.50–1.00	0.80	0.10	—	0.30–0.50 Cu; 0.07 Ti	—	—	—	—
	H	0.20	1.25	0.035	0.040	0.25–0.75	0.10–0.25	0.30–0.60	0.15	0.02–0.10	0.20–0.35 Cu; 0.005–0.030 Ti	—	—	—	—
	J	0.20	0.60–1.00	0.04	0.05	0.30–0.50	—	0.50–0.70	—	—	0.30 min Cu; 0.03–0.05 Ti	—	—	—	—
A633	A	0.18	1.00–1.35	0.04	0.05	0.15–0.50	—	—	—	—	0.05 Nb	63–83	434–572	42	289
	C	0.20	1.15–1.50	0.04	0.05	0.15–0.50	—	—	—	—	0.01–0.05 Nb	65–90	448–620	46–50	317–344
	D	0.20	0.70–1.60	0.04	0.05	0.15–0.50	0.25	0.25	0.08	—	0.35 Cu	65–90	448–620	46–50	317–344
	E	0.22	1.15–1.50	0.04	0.05	0.15–0.50	—	—	—	0.04–0.11	0.01–0.03 N	75–100	517–689	55–60	379–413
A710	A	0.07	0.40–0.70	0.025	0.025	0.35	0.60–0.90	0.70–1.00	0.15–0.25	—	1.00–1.30 Cu; 0.02 min Nb	65–90	448–620	55–85	379–586
	B	0.06	0.40–0.65	0.025	0.025	0.20–0.35	—	1.20–1.50	—	—	1.00–1.30 Cu; 0.02 min Nb	88–90	606–620	75–85	517–586

[a]Single values are maximum unless otherwise noted.

[b]These grades may contain niobium, vanadium, or nitrogen.

Source: American Welding Society (AWS) Welding Handbook Committee, 1998, *Welding Handbook*, 8th ed., vol. 4, *Materials and Applications*, Part 2, Miami: American Welding Society.

and the needs for heat treatment of high-strength low-alloy steels. Alloy elements such as manganese, molybdenum, chromium, nickel, and copper must be considered when determining the need for preheat. The ANSI/AWS D1.1 Structural Welding Code—Steel, Table 3.2, lists minimum preheat and interpass temperatures for some mild steels and HSLA steels.

Welding of high-strength low-alloy steels can be accomplished by using the SMAW, FCAW, GMAW, and SAW processes. Low-hydrogen-type welding procedures are preferred for welding HSLA steels due to their sensitivity to weld cracking. **TABLES 10-4** though **10-6** make recommendations for matching

Steels						Matching Filler Metals				
Steel Specification[a, b]		Minimum Yield Point/Strength		Tensile Strength Range		Electrode Specification[c]	Minimum Yield Point/Strength		Tensile Strength Range	
		ksi	MPa	ksi	MPa		ksi	Mpa	ksi	Mpa
ASTM A203	Grade A	37	255	65–85	448–586	Shielded Metal Arc Welding (See AWS A5.1 or A5.5)[e]				
	Grade B	40	276	70–90	483–620					
	Grade D	37	255	65–85	448–586	E7015, E7016, E7018, E7028	60	414	72 min.	496 min.
	Grade E	40	276	70–90	483–620					
ASTM A204	Grade A	37	255	65–85	448–586	E7015-X, E7016-X, E7018-X	57	390	70 min.	483 min.
	Grade B	40	276	70–90	483–620					
ASTM A242	—	42–50	290–345	63–70	435–483	E7010-X[d]	60	414	70 min.	483 min.
ASTM A441	—	40–50	275–345	60–70	415–483	Submerged Arc Welding (See AWS A5.17 or A5.23)[e]				
ASTM A572	Grade 42	42	290	60 min.	415 min.	F7XX-EXXX or F7XX-EXX-XX	58	400	70–95	483-660
	Grade 50	50	345	65 min.	450 min.					
ASTM A588	4 in. (102 mm) and under	50	345	70 min.	483 min.	Gas Metal Arc Welding (See AWS A5.18)				
ASTM A633	Grade A	42	290	63–83	430–570	ER70S-X	60	414	72 min.	496 min.
	Grades C and D 2.5 in. (64 mm) and under	50	345	70–90	483–620	Flux Cored Arc Welding (See AWS A5.20)				
ASTM A710	Grade A, Class 2, under 2 in. (51 mm)	55	380	65 min.	450 min.	E7XT-X (except -2, -3, -10, -GS)	60	414	72 min.	496 min.
API 2H	Grade 42	42	290	62-80	430–550					
	Grade 50	50	345	70 min.	483 min.					
API 5L	Grade X-52	52	360	66–72	455–495					
	Grade X-56	56	386	71–75	489–517					
	Grade X-60	60	414	75–78	517–537					

TABLE 10-4

Recommended Base Metal–Filler Metal Combinations for Matching Electrode Tensile Strengths Nominally of 70 ksi (483 Mpa) Minimum

[a]In joints involving base metals of different groups, low-hydrogen filler metal requirements applicable to lower strength group may be used. The low-hydrogen processes shall be subject to the technique requirements applicable to the higher strength group.
[b]Match API Standard 2B (fabricated tubes) according to steel used.
[c]When welds are to be stress-relieved, the deposited weld metal shall not exceed 0.05% vanadium.
[d]Cellulose electrodes for field pipeline welding.
[e]Deposited weld metal shall have a minimum impact strength of 20 ft•lb (27.1 J) at 0°F (−18°C) when Charpy V-notch specimens are required.

Source: American Welding Society (AWS) Welding Handbook Committee, 1998, *Welding Handbook*, 8th ed., vol. 4, *Materials and Applications, Part 2*, Miami: American Welding Society.

Steels						Matching Filler Metals				
Steel Specification[a, b]		Minimum Yield Point/Strength		Tensile Strength Range		**Electrode Specification**[c]	Minimum Yield Point/Strength		Tensile Strength Range	
		ksi	Mpa	ksi	Mpa		ksi	Mpa	ksi	Mpa
ASTM A202	Grade A	45	310	75–95	517–655	**Shielded Metal Arc Welding (See AWS A5.5)**[e]				
	Grade B	47	324	85–110	586–758	E8015-X, E8016-X, E8018-X	67	460	80 min.	552 min.
ASTM A204	Grade C	43	296	75–95	517–655					
ASTM A302	Grade A	45	310	75–95	517–655	E8010-G[d]	67	460	80 min.	552 min.
	Grade B	50	345	80–100	552–689	E9010-G[d]	70	483	90 min.	620 min.
	Grade C	50	345	80–100	552–689	**Submerged Arc Welding (See AWS A5.23)**[e]				
	Grade D	50	345	80–100	552–689	F8XX-EXX-XX	68	470	80–100	552–690
ASTM A572	Grade 60	60	414	75 min.	515 min.					
	Grade 65	65	450	80 min.	552 min.	**Gas Metal Arc Welding (See AWS A5.28)**[e]				
ASTM A633	Grade E	55–60	380–414	75–100	517–689	ER80S-X	68	470	80 min.	552 min.
ASTM A710	Grade A, Class 2, 2 in. (51 mm) and under	60–65	414–450	72 min.	495 min.	**Flux Cored Arc Welding (See AWS A5.29)**[e]				
	Grade A, Class 3, above 2 in. (51 mm)	60–65	414–450	70 min.	483 min.	E8XTX-X	68	470	80–100	552–690
ASTM A736		55–75	379–517	72–105	496–724					
ASTM A737	Grade B	50	345	70–90	483–620					
	Grade C	60	414	80–100	552–689					
API5L	X-65	65	450	77–80 min.	530–552					
	X-70	70	483	82 min.	565					

TABLE 10-5

Recommended Base Metal–Filler Metal Combinations for Matching Electrode Tensile Strengths Nominally of 80–90 ksi (552–620 MPa) Minimum

[a]In joints involving base metals of different groups, low-hydrogen filler metal requirements applicable to lower strength group may be used. The low-hydrogen processes shall be subject to the technique requirements applicable to the higher strength group.
[b]Match API Standard 2B (fabricated tubes) according to steel used.
[c]When welds are to be stress-relieved, the deposited weld metal shall not exceed 0.05% vanadium.
[d]Cellulose electrodes for field pipeline welding.
[e]Deposited weld metal shall have a minimum impact strength of 20 ft•lb (27.1 J) at 0°F (−18°C) when Charpy V-notch specimens are required.

Source: American Welding Society (AWS) Welding Handbook Committee, 1998, *Welding Handbook*, 8th ed., vol. 4, *Materials and Applications*, Part 2, Miami: American Welding Society.

filler metals and base metal combinations for welding HSLA steels for each welding process.

The quenched and tempered steels are low-alloy steel products that have been hardened and tempered to improve mechanical properties. These alloys are supplied in the heat-treated condition for fabrication purposes. They display good strength and toughness, corrosion resistance, and weldability. **TABLE 10-7** lists quenched and tempered steels by ASTM specifications, with chemical composition and minimum required mechanical properties. Welding processes used for welding quenched and tempered steels include the SMAW, FCAW, GMAW, GTAW, and SAW processes.

Several special considerations apply to welding quench and tempered alloys. As with other alloy steels, the use of low-hydrogen welding procedures is

Steels

Steel Specification[a,b]		Minimum Yield Point/Strength		Tensile Strength Range	
		ksi	Mpa	ksi	Mpa
ASTM A225	Grade C	70	483	105–135	724–931
	Grade D	55–60	379–414	75–105	517–724
ASTM A710	Grade A, Class 1, 0.75 in. (19 mm) and under	80	552	90 min.	620 min.
	Grade A, Class 3, 2 in. (51 mm) and under	75	517	85 min.	586 min.
ASTM A735	—	65–80	448–552	80–115	552–793
API 5L[e]	X-80	80	552	90 min.	620 min.
ASTM A353[f]	—	75	517	100–120	689–827

Filler Metals

Electrode Specification[c]	Minimum Yield Point/Strength		Tensile Strength Range	
	ksi	Mpa	ksi	Mpa
Shielded Metal Arc Welding (See AWS A5.5)[d]				
E10015-X, E10016-X, E10018-X	87	600	100 min.	690 min.
Submerged Arc Welding (See AWS A5.23)[d]				
F10XX-EXX-XX	88	610	100–120	690–830
Gas Metal Arc Welding (See AWS A5.28)[d]				
ER100S-X	88–102	610–700	100 min.	690 min.
Flux Cored Arc Welding (See AWS A5.29)[g]				
E10XTX-X	88	605	100–120	690–830
Shielded Metal Arc Welding (See AWS A5.11)[g]				
E310	60	414	80	550
ENiCrFe-2	45	310	80	550
ENiCrMo-3	60	414	110	760
Gas Metal Arc Welding (See AWS A5.14)[g]				
ERNiCr-3	40	276	80	550
ERNiCrFe-6	40	276	80	550

TABLE 10-6

Recommended Base Metal–Filler Metal Combinations for Steels with Tensile Strengths Nominally of 100 ksi (689 Mpa) Minimum

[a]In joints involving base metals of different groups, low-hydrogen filler metal requirements applicable to lower strength group may be used. The low-hydrogen processes shall be subject to the technique requirements applicable to the higher strength group.
[b]Match API Standard 2B (fabricated tubes) according to steel used.
[c]When welds are to be stress-relieved, the deposited weld metal shall not exceed 0.05% vanadium.
[d]Deposited weld metal shall have a minimum impact strength of 20 ft•lb (27.1 J) at 0°F (−18°C) when Charpy V-notch specimens are required.
[e]Undermatching cellulose electrodes may be used for root and cap passes for field pipeline welding.
[f]Applicable electrodes for welding this steel are only the stainless steel and nickel-alloy electrodes (AWS A5.11 and A5.14) shown in lower right.
[g]Yield strength may not match base metal properties.

Source: American Welding Society (AWS) Welding Handbook Committee, 1998, *Welding Handbook*, 8th ed., vol. 4, *Materials and Applications*, *Part 2*, Miami: American Welding Society.

preferred for welding quench and tempered alloys. While not required on thinner base metals, preheat (see **TABLE 10-8**) may be desired on heavy weldments with significant joint restraint. Suggested filler metals are given in **TABLE 10-9** for SMAW, FCAW, GMAW, GTAW, and SAW processes.

Heat-treatable low-allow (HTLA) steels (see **TABLE 10-10**)—another group of low-alloy steels—differ from quenched and tempered steels in that they are not supplied in the heat-treated condition. Welding these steels requires use of low-hydrogen-type procedures. The fact that they are heat-treatable indicates high hardenability, which means strict control of procedures is desired to slow the cooling rate and to eliminate hardening and cracking in the heat-affected zone. To eliminate cracking problems, HTLA steels are welded in the annealed condition by using controlled preheat and interpass temperatures

TABLE 10-7

Typical Quenched and Tempered Steels

Common Designation	Grade or Class	C	Mn	P	S	Si	Cr	Ni	Mo	Cu	Others	Min Tensile ksi	Min Tensile MPa	Min Yield ksi	Min Yield MPa
A514 or A517	A	0.15–0.21	0.80–1.10	0.035	0.04	0.40–0.80	0.50–0.80	—	0.18–0.28	—	Zr, 0.05–0.15; B, 0.0025	A514	A514	A514	A514
	B	0.12–0.21	0.70–1.00	0.035	0.04	0.20–0.35	0.40–0.65	—	0.15–0.25	—	V, 0.03–0.08; Ti,0.01–0.03; B, 0.0005–0.005	100–130	689–896	90–100	620–689
	C	0.10–0.20	1.10–1.50	0.035	0.04	0.15–0.30	—	—	0.20–0.30	—	B, 0.001–0.005	A517	A517	A517	A517
	D	0.13–0.20	0.40–0.70	0.035	0.04	0.20–0.35	0.85–1.20	—	0.15–0.25	0.20–0.40	Ti, 0.004–0.10[b]; B,0.0015–0.005	105–135	723–930	90–100	620–689
	E	0.12–0.20	0.40–0.70	0.035	0.04	0.20–0.35	1.40–2.00	—	0.40–0.60	0.20–0.40	Ti, 0.04–0.10[b]; B,0.0015–0.005				
	F	0.10–0.20	0.60–1.00	0.035	0.04	0.15–0.35	0.40–0.65	0.70–1.00	0.40–0.60	0.15–0.50	V, 0.03–0.08; B, 0.0005–0.006				
	G	0.15–0.21	0.80–1.10	0.035	0.04	0.50–0.90	0.50–0.90	—	0.40–0.60	—	Zr, 0.05–0.15; B, 0.0025				
	H	0.12–0.21	0.95–1.30	0.035	0.04	0.20–0.35	0.40–0.65	0.30–0.70	0.20–0.30	—	V, 0.03–0.08; B, 0.0005–0.005				
	J	0.12–0.21	0.45–0.70	0.035	0.04	0.20–0.35	—	—	0.50–0.65	—	B, 0.001–0.005				
	K	0.10–0.20	1.10–1.50	0.035	0.04	0.15–0.30	—	—	0.45–0.55	—	B, 0.001–0.005				
	L	0.13–0.20	0.40–0.70	0.035	0.04	0.20–0.35	1.15–1.65	—	0.25–0.40	0.20–0.40	Ti, 0.04–0.10[b]; B, 0.0015–0.005				
	M	0.12–0.21	0.45–0.70	0.035	0.04	0.20–0.35	—	1.20–1.50	0.45–0.60	—	B, 0.001–0.005				
	N	0.15–0.21	0.80–1.10	0.035	0.04	0.40–0.90	0.50–0.80	—	0.25	—	Zr, 0.05–0.15; B, 0.0005–0.0025				
	P	0.12–0.21	0.45–0.70	0.035	0.04	0.20–0.35	0.85–1.20	1.20–1.50	0.45–0.60	—	B, 0.001–0.005				
	Q	0.14–0.21	0.95–1.30	0.035	0.04	0.15–0.35	1.00–1.50	1.20–1.50	0.40–0.60	—	V, 0.03–0.08				
A533	A	0.25	1.15–1.50	0.035	0.040	0.15–0.30	—	—	0.45–0.60	—	—	80–125	551–862	50–83	344–569
	B	0.25	1.15–1.50	0.035	0.040	0.15–0.30	—	0.40–0.70	0.45–0.60	—	—	80–125	551–862	50–83	344–569
	C	0.25	1.15–1.50	0.035	0.040	0.15–0.30	—	0.70–1.00	0.45–0.60	—	—	80–125	551–862	50–83	344–569
	D	0.25	1.15–1.50	0.035	0.040	0.15–0.30	—	0.20–0.40	0.45–0.60	—	—	80–125	551–862	50–83	344–569
A537	2	0.24	0.70–1.60	0.035	0.040	0.15–0.30	0.25	0.25	0.08	0.35	—	70–100	482–689	46–60	317–414
A543	B	0.23	0.40	0.020	0.020	0.20–0.40	1.50–2.00	2.60–4.00[c]	0.45–0.60	—	V, 0.03	90–135	620–930	70–100	482–689
	C	0.23	0.40	0.020	0.020	0.20–0.40	1.20–1.50	2.25–3.50	0.45–0.60	—	V, 0.03	90–135	620–930	70–100	482–689
A553	1	0.13	0.90	0.035	0.035	0.15–0.40	—	8.50–9.50	—	—	—	100–120	690–825	85	585
	2	0.13	0.90	0.035	0.035	0.15–0.40	—	7.50–8.50	—	—	—	100–120	690–825	85	585
A678	A	0.16	0.90–1.50	0.04	0.05	0.15–0.50	0.25	0.25	0.08	0.20[d]	—	70–90	482–620	50	344
	B	0.20	0.70–1.60	0.04	0.05	0.15–0.50	0.25	0.25	0.08	0.20[d]	—	80–100	551–689	60	414
	C	0.22	1.00–1.60	0.04	0.05	0.20–0.50	0.25	0.25	0.08	0.20[d]	—	85–115	586–792	65–75	448–517
HY-80		0.12–0.18	0.10–0.40	0.025	0.025	0.15–0.35	1.00–1.80	2.00–3.25	0.20–0.60	0.25	V, 0.03; Ti, 0.02	—	—	80	551
HY-100		0.12–0.20	0.10–0.40	0.025	0.025	0.15–0.35	1.00–1.80	2.25–3.50	0.20–0.60	0.25	V, 0.03;Ti,0.02	—	—	100	689
HY-130		0.12	0.60–0.90	0.010	0.015	0.15–0.35	0.40–0.70	4.75–5.25	0.30–0.65	—	V, 0.05–0.10	—	—	130	896

Composition, %[a]

[a] When a single value is shown, it is a maximum limit. [b] Vanadium may be substituted for part or all of titanium content on a one-for-one basis. [c] Limiting values vary with the plate thickness. [d] When specified.

Source: American Welding Society (AWS) Welding Handbook Committee, 1998, *Welding Handbook*, 8th ed., vol. 4, *Materials and Applications, Part 2*, Miami: American Welding Society.

373

| | Thickness Range | | Minimum Preheat and Interpass Temperature* | | | | | | | | | |
| | | | A514/A517 | | A533 | | A537 | | A543 | | A678 | |
in.	mm		°F	°C	°F	°C	°F	°C	°F	°C	°F	°C
Up to 0.50	Up to 12.7		50	10	50	10	50	10	100	40	50	10
0.56 to 0.75	14.2 to 19.1		50	10	100	40	50	10	125	50	100	40
0.81 to 1.00	20.6 to 25.4		125	50	100	40	50	10	150	65	100	40
1.1 to 1.5	27.9 to 38.1		125	50	200	95	100	40	200	95	150	65
1.6 to 2.0	40.6 to 50.8		175	80	200	95	150	65	200	95	150	65
2.1 to 2.5	53.3 to 63.5		175	80	300	150	150	65	300	150	150	65
over 2.5	over 63.5		225	105	300	150	225	105	300	150	—	—

TABLE 10-8

Suggested Minimum Preheat and Interpass Temperatures for Arc Welding Typical ASTM Quenched and Tempered Steels

*With low-hydrogen welding practices, maximum temperature should not exceed the given value by more than 150°F (66°C).

Source: American Welding Society (AWS) Welding Handbook Committee, 1998, *Welding Handbook*, 8th ed., vol. 4, *Materials and Applications*, Part 2, Miami: American Welding Society.

| ASTM Specification or Common Name | Grade or Type | Class | Welding Process | | | |
			SMAW[a]	GMAW,[b] GTAW	SAW[c]	FCAW[d]
A514 or A517	All	—	E1X01X-M	ER1X0S-1	F1XXX-EXXX-MX	E1X0TX-KX
A533	B	1, 2	E901X-D1 E901X-M	ER100S-1	F9XX-EXXX-FX	E9XTX-NiX E9XTX-K2
		3	E1101X-M	ER110S-1	F11XX-EXXX-M2	E110TX-XX
A537	—	2	E801X-CX E901X-D1 E9018-M	ER80X-NiX	F8XX-EXXX-NiX F9XX-EXXX-NiX	E8XTX-NiX E9XTX-NiX
A543	B	1, 2	E1101X-M	ER110S-1 ER120S-1	F11XX-EXXX-FX	E11XTX-KX
A678	C	—	E901X-D1 E9018-M E1001X-D2 E10018-M	ER100S-1	F9XX-EXXX-FX F10XX-EXXX-MX	E9XTX-NJX E9XTX-MX
HY-80,HY-100	—	—	E1101X-M	ER110S-1 ER120S-1	F11XX-EXXX-MX	E11XTX-KX

TABLE 10-9

Suggested Filler Metals for Arc Welding Typical Quenched and Tempered Steels

[a]Shielded Metal Arc Welding electrodes with low-hydrogen coverings in ANSI/AWS A5.5, *Specification for Low Alloy Steel Covered Arc Welding Electrodes.*
[b]Electrodes for Gas Metal Arc or Gas Tungsten Arc Welding in ANSI/AWS A5.28, *Specification for Low Alloy Steel Filler Metals for Gas Shielded Arc Welding.*
[c]Submerged Arc Welding with electrode-flux combinations in ANSI/AWS A5.23, *Specification for Low Alloy Steel Electrodes and Fluxes for Submerged Arc Welding.*
[d]Flux Cored Arc Welding electrodes in ANSI/AWS A5.29, *Specification for Low Alloy Steel Electrodes for Flux Cored Arc Welding.* Classification of EXXT1 and EXXT5 electrodes is done with CO_2 shielding gas. However, Ar-Co_2 gas mixtures may be used when recommended by the manufacturer to improve usability.
Source: American Welding Society (AWS) Welding Handbook Committee, 1998, *Welding Handbook*, 8th ed., vol. 4, *Materials and Applications*, Part 2, Miami: American Welding Society.

Common Designation	Composition, %						
	C	Mn	Si	Ni	Cr	Mo	V
4027	0.25–0.30	0.70–0.90	0.15–0.35	—	—	0.20–0.30	—
4037	0.35–0.40	0.70–0.90	0.15–0.35	—	—	0.20–0.30	—
4130	0.28–0.33	0.40–0.60	0.15–0.35	—	0.80–1.10	0.15–0.25	—
4135	0.33–0.38	0.70–0.90	0.15–0.35	—	0.80–1.10	0.15–0.25	—
4140	0.38–0.43	0.75–1.00	0.15–0.35	—	0.80–1.10	0.15–0.25	—
4320	0.17–0.22	0.45–0.65	0.15–0.35	1.65–2.00	0.40–0.60	0.20–0.30	—
4340	0.38–0.43	0.60–0.80	0.15–0.35	1.65–2.00	0.70–0.90	0.20–0.30	—
5130	0.28–0.33	0.70–0.90	0.15–0.35	—	0.80–1.10	—	—
5140	0.38–0.43	0.70–0.90	0.15–0.35	—	0.70–0.90	—	—
8630	0.28–0.33	0.70–0.90	0.15–0.35	0.40–0.70	0.40–0.60	0.15–0.25	—
8640	0.38–0.43	0.75–1.00	0.15–0.35	0.40–0.70	0.40–0.60	0.15–0.25	—
8470	0.38–0.43	0.75–1.00	0.15–0.35	0.40–0.70	0.40–0.60	0.20–0.30	—
AMS6434	0.31–0.38	0.60–0.80	0.20–0.35	1.65–2.00	0.65–0.90	0.30–0.40	0.17–0.23
300M	0.40–0.46	0.65–0.90	1.45–1.80	1.65–2.00	0.70–0.95	0.30–0.45	0.05 min
D-6a	0.42–0.48	0.60–0.80	0.15–0.30	0.40–0.70	0.90–1.20	0.90–1.10	0.05–0.10

AISI Steel	Thickness Range		Minimum Preheat and Interpass Temperature*	
	in.	mm	°F	°C
4027	Up to 0.5	Up to 13	50	10
	0.6–1.0	15–26	150	66
	1.1–2.0	27–51	250	121
4037	Up to 0.5	Up to 13	100	38
	0.6–1.0	15–26	200	93
	1.1–2.0	27–51	300	149
4130, 5140	Up to 0.5	Up to 13	300	149
	0.6–1.0	15–26	400	204
	1.1–2.0	27–51	450	232
4135, 4140	Up to 0.5	Up to 13	350	177
	0.6–1.0	15–26	450	232
	1.1–2.0	27–51	500	260
4320, 5130	Up to 0.5	Up to 13	200	93
	0.6–1.0	15–26	300	149
	1.1–2.0	27–51	400	204
4340	Up to 2.0	Up to 51	550	288
8630	Up to 0.5	Up to 13	200	93
	0.6–1.0	15–26	250	121
	1.1–2.0	27–51	300	149
8640	Up to 0.5	Up to 13	200	93
	0.6–1.0	15–26	300	149
	1.1–2.0	27–51	350	177
8740	Up to 1.0	Up to 26	300	149
	1.1–2.0	27–51	400	204

TABLE 10-11

Recommended Minimum Preheat and Interpass Temperatures for Several AISI Low-Alloy Steels

*Low-hydrogen welding processes only.

Source: American Welding Society (AWS) Welding Handbook Committee, 1998, *Welding Handbook*, 8th ed., vol. 4, *Materials and Applications, Part 2*, Miami: American Welding Society.

(see **TABLE 10-11**). Alloys can be heat-treated to desired mechanical properties after welding is completed.

Chromium-molybdenum steels are chromium-molybdenum alloys supplied in the normalized and tempered or annealed condition. These base metals are used in the petroleum industry and for elevated temperature use, such as in boilers. **TABLE 10-12** lists ASTM specifications for chromium-molybdenum alloy steels and categorizes them based on manufacturing product forms. Welding processes used for chromium-molybdenum steels include the SMAW, FCAW, GMAW, GTAW, and SAW processes.

Filler metals for chromium-molybdenum alloys should match those of the base metal except that the carbon content may be slightly lower. As with many alloys, base metal dilution of the weld pool compensates for the lower carbon content. Chromium-molybdenum low-hydrogen electrodes with the B suffix are preferred. Recommended filler metal electrode classifications for the common welding processes SMAW, FCAW, GMAW, GTAW, and SAW are given in

Type	Forgings	Tubes	Pipe	Castings	Plate
1/2Cr-1/2Mo	A182-F2	A213-T2	A335-P2 A369-FP2 A426-CP2	A356-GR5	A387-Gr2
1Cr-1/2Mo	A182-F12 A336-F12	A213-T12	A335-P12 A369-FP12 A426-CP12	—	A387-Gr12
1 1/4Cr-1/2Mo	F182-F11/F11A A336-F11/F11A	A199-T11 A200-T11 A213-T11	A335-P11 A369-FP11 A426-CP11	A217-WC6/11 A356-Gr6 A389-C23	A387-Gr11
2 1/4Cr-1Mo	A182-F22/F22a A336-F22/F22A	A199-T22 A200-T22 A213-T22	A335-P22 A369-FP22 A426-CP22	A217-WC9 A356-Gr10	A387-Gr22
3Cr-1Mo	A182-F21 A336-F21/F21A	A199-T21 A200-T21 A213-T21	A335-P21 A369-FP21 A426-CP21	—	A387-Gr21
5Cr-1/2Mo	A182-F5/F5a A336-F5/F5A	A199-T5 A200-T5 A213-T5	A335-P5 A369-FP5 A426-CP5	A217-C5	A387-Gr5
5Cr-1/2MoSi	—	A213-T5b	A335-P5b A426-CP5b	—	—
5Cr-1/2MoTi	—	A213-T5c	A335-P5c	—	—
7Cr-1/2Mo	A182-F7	A199-T7 A200-T7 A213-T7	A335-P7 A369-FP7 A426-CP7	—	A387-Gr7
9Cr-1Mo	A182-F9 A336-F9	A199-T9 A200-T9 A213-T9	A335-P9 A369-FP9 A426-CP9	A217-C12	A387-Gr9
9Cr-1Mo and V+Nb+N	A182-F91	A199-T91 A200-T91 A213-T91	A335-P91 A369-FP91	—	A387-Gr91

TABLE 10-12

Representative ASTM Specifications for Chromium-Molybdenum Steel Product Forms

Source: American Welding Society (AWS) Welding Handbook Committee, 1998, *Welding Handbook*, 8th ed., vol. 4, *Materials and Applications, Part 2*, Miami: American Welding Society.

TABLE 10-13. Low-hydrogen electrodes must be stored in dry ovens and used within code-recommended exposure times to prevent hydrogen cracking in welds. In situations where pre-heat and post-heat are not feasible, such as on 4130 tubing, use of GTAW or GMAW with an ER80S-D2 nickel-molybdenum alloy filler metal may be used.

Preheat and interpass temperatures must be closely monitored to minimize weld metal cracking and to reduce the likelihood of hydrogen-induced cracking. **TABLE 10-14** provides minimum preheat temperatures for chromium-molybdenum alloys based on chemical composition and base metal thicknesses.

The final step to putting a welded chromium-molybdenum alloy into service is postweld heat treatment. The purposes of postweld heat treatment are twofold. The first is to improve properties of the heat-affected zone. Strength and ductility can be improved through **stress relief** annealing at a temperature slightly below the base metal's lower transformation temperature. Ranges for

Steel	SMAW[a]	GTAW,[e] GMAW	FCAW[h]	SAW[i]
1/2Cr-1/2Mo	E801X-B1	Note f	E7XT5-A1 E8XT1-A1	F8XX-EXXX-B1
1Cr-1/2Mo, 1 1/4Cr-1/2Mo	E801X-B2 or E701X-B2L	ER80X-B2 or ER70X-B2L	E8XTX-B2 or E8XTX-B2L or E8XTX-B2H	F8XX-EXXX-B2 or F8XX-EXXX-B2H
2 1/4Cr-1Mo	E901X-B3 or E801X-B3L	ER90X-B3 or ER80X-B3L	E9XTX-B3 or E9XTX-B3L or E9XTX-B3H	F9XX-EXXX-B3
3Cr-1Mo	Note b	Note b	Note b	Note b
5Cr-1/2Mo	E502-1X[d] or E801X-B6 or E801X-B6L	ER502[g] or ER80X-B6	E502T-1 or 2 or E6XT5-B6	F9XX-EXXX-B6 or F9XX-EXXX-B6H
7Cr-1/2Mo	E7Cr-1X[d] or E801X-B7 or E801X-B7L	Note c	Note c	Note c
9Cr-1Mo	E505-1X[d] or E801X-B8 or E801X-B8L	ER505[g] or ER80X-B8	E505T-1 or 2 or EX15-B8 or E6XT5-B8L	F9XX-EXXX-B8
9Cr-IMoandV+Nb+N	E901X-B9	ER90X-B9	—	F9XX-EXXX-B9

TABLE 10-13

Suggested Welding Consumables for Joining Chromium-Molybdenum Steels

[a]Per ANSI/AWS A5.5, *Specification for Low-Alloy Steel Covered Arc Welding Electrodes,* unless otherwise indicated.
[b]No matching filler metal, select between 21/4Cr-1Mo and 5Cr-1/2Mo.
[c]No matching filler metal, select between 5Cr-1Mo and 9Cr-1/2Mo.
[d]Originally classified per ANSI/AWS A5.4, *Specification for Covered Corrosion-Resisting Chromium and Chromium-Nickel Steel Welding Electrodes.*
[e]Per ANSI/AWS A5.28, *Specification for Low-Alloy Steel Filler Metals for Gas Shielded Arc Welding,* unless otherwise indicated.
[f]No matching filler metal, consider higher alloy.
[g]Originally classified per ANSI/AWS A5.9, *Specification for Bare Stainless Steel Welding Electrodes and Rods.*
[h]Per ANSI/AWS A5.29, *Specification for Low-Alloy Steel Electrodes for Flux Cored Arc Welding* (use with CO_2 or Ar-CO_2 mixture).
[i]Weld metal designation per ANSI/AWS A5.23, *Specification for Low-Alloy Steel Electrodes and Fluxes for Submerged Arc Welding.*

Source: American Welding Society (AWS) Welding Handbook Committee, 1998, *Welding Handbook*, 8th ed., vol. 4, *Materials and Applications*, Part 2, Miami: American Welding Society.

stress relief are given in **TABLE 10-15**. The second purpose is to eliminate causes of weld-metal cracking by reducing residual stresses and allowing hydrogen to dissipate from concentrations in the heat-affected zone.

WELDING STAINLESS STEELS

Stainless steels are used for their corrosion resistance. Their ability to resist corrosive and chemical reactions caused by moisture, chemicals, and gases are due to alloying with specific elements for each use. Stainless steels are iron alloys with 11% or greater chromium content. The chromium forms a protective oxide coating when exposed to oxygen in the surrounding atmosphere, providing the desired corrosion resistance.

Steel*	Thickness					
	Up to 0.5 in. (13 mm)		0.5 to 1.0 in. (13 to 25 mm)		Over 1.0 in. (25 mm)	
	°F	°C	°F	°C	°F	°C
1/2Cr-1/2Mo	100	38	200	93	300	149
1Cr-1/2Mo 1 1/4Cr-1/2Mo	250	121	300	149	300	149
2Cr-1/2Mo 2 1/4Cr-1Mo 3Cr-1Mo	300	149	350	177	350	177
5Cr-1/2Mo 7Cr-1/2Mo 9Cr-1Mo 9Cr-1Mo V+Nb+N	350	177	400	204	400	204

TABLE 10-14

Recommended Minimum Preheat Temperatures for Welding Chromium-Molybdenum Steels with Low-Hydrogen Covered Electrodes

*Maximum carbon content of 0.15%. For higher carbon content, the preheat temperature should be increased 100 to 200°F (38 to 93 °C). Lower preheat temperatures may be used with Gas Tungsten Arc Welding.

Source: American Welding Society (AWS) Welding Handbook Committee, 1998, *Welding Handbook*, 8th ed., vol. 4, *Materials and Applications*, *Part 2*, Miami: American Welding Society.

Steel	Temperature Range*	
	°F	°C
1/2Cr-1/2Mo	1150–1300	621–704
1Cr-1/2Mo 1 1/4Cr-1/2Mo	1150–1325	621–718
2Cr-1/2Mo 2 1/4Cr-1Mo 3Cr-1Mo	1250–1400	677–760
5Cr-1/2Mo 7Cr-1/2Mo 9Cr-1Mo	1300–1400	704–760
9Cr-1Mo plus V+N6+N	1350–1400	732–760

TABLE 10-15

Recommended Stress-Relief Temperature Ranges for Chromium-Molybdenum Steels

*Temperature should not exceed the tempering temperature of the steel.

Source: American Welding Society (AWS) Welding Handbook Committee, 1998, *Welding Handbook*, 8th ed., vol. 4, *Materials and Applications*, *Part 2*, Miami: American Welding Society.

As indicated by **TABLE 10-16**, there are many different stainless steel alloys. The ability to weld stainless steels and procedures for joining them is determined by the differences in their chemical compositions. The most common stainless alloys fall within one of two groupings: straight chromium and chromium/nickel. The straight chromium types have only chromium as a major alloy, with small quantities of other elements. These are the 400 series alloys, which are highly magnetic. The 400 series alloys can be divided into two groups: ferritic and martensitic. The ferritic group is not hardenable and can be welded with some

Alloy Type	Classification	Characteristics/Useability
Austenitic chromium-nickel	304, 308, 308L, 309, 309S, 316, 316L, 321	formable, good weldability
Austenitic chromium-manganese-nickel	201, 202, 205, 216	formable, good weldability
Ferritic Chromium	405, 430, 442, 446	very soft, limited corrosion resistance, toughness, formability and weldability, not heat treatable
Duplex (ferritic/austenitic)	A240, A479	improved stress corrosion cracking resistance, formability, weldability, and high strength
Martensitic Chromium	403, 410, 416, 420, 431, 440	increased hardness and strength decreases ductility and toughness
Precipitation hardening	ASTM 564	high strength, fair weldability, aerospace applications

TABLE 10-16

Abridged Chart of Stainless Steel Types

minor changes in physical properties of the base metal. The martensitic group is hardenable by heat treatment, and in some cases, welding is avoided.

The alloys most common to welding and fabrication are the 300 series chromium/nickel types. Additions of nickel to these alloys improve corrosion resistance, formability (ductility), and weldability. While readily weldable, special considerations apply when welding these alloys. In chromium-nickel alloys, small amounts of carbon can lead to loss of corrosion resistance along the heat-affected zone of the weld. The first consideration when welding chromium-nickel alloys is to design procedures that shorten the time that the weld is at high temperature. If weld and base metal temperatures are sustained during welding chromium-nickel stainless steels in a range of 800°F–1,600°F (427°C–871°C) for too long, chromium combines with carbon, forming chromium carbides. The name of this phenomenon is called **carbide precipitation**. Carbide precipitation in stainless steel welds results in a condition known as inter-granular corrosion. This means that metal grains along the fusion line can corrode and fall out. An alloy in this condition is referred to as sensitized. Knowing that there is a time-/temperature relationship to sensitization, welding procedures should be established that do not allow the base metal to reach sensitization. High welding speeds help to shorten the time the weld zone is at the undesirable temperature range.

Because chromium-nickel alloys are commonly used in food industries and water treatment applications, welding procedures that eliminate carbide precipitation are absolutely necessary. When welding must be done, use of alloys with less than 0.03% carbon is preferred. These extra-low-carbon (ELC) alloys are less prone to formation of chromium carbides that rob stainless steel alloys of the chromium needed to maintain corrosion resistance. Both ELC base metals and filler metals can be used. Use ELC filler metals such ER308L and ER316L.

Finally, special stabilized grades of stainless steel base and filler metals can reduce the likelihood of chromium carbides forming. These stabilizing elements—titanium and columbium (niobium)—tie up the carbon, creating titanium carbides or columbium carbides, leaving the chromium where it belongs in the base metal heat-affected zone and maintaining the desired corrosion resistance. The E309cb and E310cb alloy electrodes contain the stabilizing elements columbium and titanium.

Another stainless steel series, the 200 series, in addition to nickel, contains manganese to increase strength. These chromium-nickel-manganese alloys are also austenitic and are sometimes used in place of the nickel-chromium base metals.

WHAT WELDING PROCESSES ARE USED FOR WELDING STAINLESS STEELS? The most common welding processes used for welding stainless steels are GTAW and GMAW. The obvious advantage of GTAW is its ability to be used on the thinner sheet materials. Often, GTAW can be used for root passes on heavier structures, while remaining groove fill passes can be completed with a higher deposition process.

FCAW and SMAW are also used extensively for welding stainless steels. Both processes are suitable for heavier welding plates and structures. Although they must be removed after welding, slag deposits from these processes help to control cooling rates. Weld bead appearance and quality are satisfactory.

CAN ALL STAINLESS STEELS BE WELDED? Although alloys in most of the alloy series—200, 300, and 400—can be welded, the procedures for welding them vary. **TABLE 10-17** provides filler metal recommendations common for stainless steel base metals.

As discussed previously, the austenitic alloys are prone to the effects of carbide precipitation. In addition, nickel alloying lowers the thermal conductivity rate and increases the thermal expansion rate. As a result, fabrications composed of these alloys are more likely to distort. For these reasons, when welding, use the steps that follow:

1. Use high travel speeds to reduce heat input.
2. Do not preheat.
3. Keep interpass temperatures low.
4. Use low-carbon (L) or extra-low carbon (ELC) base and filler metals.
5. Use stabilized grade filler metals (E309cb or E310cb) when practical.
6. Use shielding gases with low-carbon content.
7. Tack thin metals more frequently to control distortion.
8. Modify the weld sequence to control distortion.

Welding procedures for the magnetic chromium (400) series stainless steels vary with the alloy type. The ferrite alloys are readily weldable with most common arc welding processes, provided carbon content is not altered. This can be done by using shielding gases with minimal carbon dioxide content of 2% or lower. Heavy base metals may be prone to formation of martensite, and preheating at about 400°F (205°C) can prevent this from happening.

The martensitic alloys, as their name implies, are heat-treatable and require special attention to welding procedures. As carbon content increases in these alloys, the likelihood of martensite formation and weld cracking in the heat-affected zone is increased. Preheat and postheat treatments are necessary to maintain structural integrity of the weld joint.

	Recommended Filler Metal			
AISI No.	First Choice	Second Choice	Popular Name	Remarks
Cr–Ni–Mn				
201	308	308L		Substitute for 301
202	308	308L		Substitute for 302
301	308	308L		
302	308	308L		
302B	308	309		High silicon
303	—	—		Free machining, welding not recommended, 312
303Se	—	—		Free machining, welding not recommended, 312
Cr–Ni–austenitic				
304	308	308L	18/8	
304L	308L	347	18/8 ELC	Extra low carbon
305	308	—		
308	308	—	19/9	
309	309	—	25/12	
309S	309	—		Low carbon
310	310	—	25/20	
310S	310	—		Low carbon
314	310	—		
316	316	309cb	18/12 Mo	
316L	316L	309cb	18/12 Elc	Extra low carbon
317	317	309cb	19/14 Mo	
321	347	308L		
347	347	308L	19/9 Cb	Difficult to weld in heavy sections
348	347	—	19/9 CbLTa	Difficult to weld in heavy sections
Cr–martensitic				
403	410	—		
410	410	430	12 Cr	
414	410	—		
416	410	—		Use 410-15
416Se	—	—		Free machining, welding not recommended
420	410	—	12 CrHC	High carbon
431	430	—		
440A	—	—		High carbon, welding not recommended
440B	—	—		High carbon, welding not recommended
440C	—	—		High carbon, welding not recommended
Cr–ferritic				
405	410	405Cb		
430	430	309	16 Cr	
430F	—	—		Free machining, welding not recommended 1
430FSe	—	—		Free machining, welding not recommended 1
446	309	310		
501	502	—	5 Cr–1/2 Mo	Chrome–moly steel
502	502	—	5 Cr–1/2 Mo	Chrome–moly steel

TABLE 10-17

Recommended Filler Metals for Stainless Steels (Use E or R Prefix)

Source: Cary, Howard B., *Modern Welding Technology*, 4th ed., © 1998. Reproduced by permission of Pearson Education, Inc., Upper Saddle River, New Jersey.

WELDING TOOL STEELS

As the name implies, tool steels are used for making work tools. They are alloy steels with medium to high carbon contents. The high-alloying content produces those properties that are suitable for bending, forming, and cutting operations, where high hardness and wear resistance are required. As a consequence of higher carbon and alloying, when repairs are required these steels necessitate closely monitored welding procedures. Weldable tool steels include air-hardening, oil-hardening, and water-hardening die steels as well as hot-work tool steels. An abridged chart of tool steel types is represented by the lists in **TABLE 10-18**.

WHAT TECHNIQUES ARE RECOMMENDED FOR WELDING TOOL STEELS?

Identify the tool steel type and then perform a hardness test to determine whether it is in the hardened state or has been annealed. The following techniques can be used to weld tool steels:

- Preheat before welding.
- Weld in the annealed condition. Preheat at the max preheat.

AISI-SAE Type	Classification of Tool Steels	Composition (%)					
		C	Cr	V	W	Mo	Other
W1	Water hardening	0.60	—	—	—	—	—
W2		0.60	—	0.25	—	—	—
S1		0.50	1.50	—	2.50	—	—
S5	Shock resisting	0.55	—	—	—	0.40	0.80 Mn 2.00 Si
S7		0.50	3.25	—	—	1.40	—
O1	Oil hardening	0.90	0.50	—	0.50	—	—
O6		1.45	—	—	—	0.25	1.00 Si
A2	Cold work	1.00	5.00	—	—	1.00	—
A4	Medium-alloy air hardening	1.00	1.00	—	—	1.00	2.00 Mn
D2	Cold work, high carbon, high chromium	1.50	12.00	—	—	1.00	—
M1	Cold work	0.80	4.00	1.00	1.50	8.00	—
M2	Molybdenum	0.85	4.00	2.00	6.00	5.00	—
M10		0.90	4.00	2.00	—	8.00	—
H11	Hot work	0.35	5.00	0.40	—	1.50	—
H12	Chromium	0.35	5.00	0.40	1.50	1.50	—
H13		0.35	5.00	1.00	—	1.50	—
P20	Die-casting mold	0.35	1.25	—	—	0.40	—

TABLE 10-18

Abridged Chart of Tool Steel Types

Source: Cary, Howard B., *Modern Welding Technology*, 4th ed., © 1998. Reproduced by permission of Pearson Education, Inc., Upper Saddle River, New Jersey.

- Maintain preheat throughout welding, but avoid overheating to avoid grain growth.
- Weld in the hardened condition. Do not preheat above the temper/draw temperature.
- Use low current to keep heat input low.
- Use stringers to keep heat input low.
- Peen each pass.
- Harden as recommended (air, oil, water).
- Temper.

HOW ARE FILLER METALS SELECTED FOR TOOL STEELS?   While electrodes for tool steels do not need to match the base metal, the deposited weld metal must match the heat treatment of the tool for use in service. The welder has a couple of choices for electrode selection. When the weld is not required to have a high degree of hardness, low-hydrogen-type electrodes can be used. Otherwise, the welder can consult catalogs and data sheets from specialty electrode manufacturers. Because precise welding procedures must be followed, these electrode manufacturers data sheets provide the specifics of procedures for each electrode and can recommend electrodes for each of the common tool steels.

WHAT ARE THE CONCERNS WITH PREHEAT AND POSTHEAT TREATMENT WITH TOOL STEEL?   In most cases, preheat of about 400°F (204°C) is an adequate temperature. Interpass temperatures should be monitored during welding. When welding is completed, a tempering postheat treatment can be made in accordance with **TABLE 10-19**.

WELDING CAST IRONS

Difficulties in welding cast irons center on the high carbon content that makes them brittle. Carbon contents ranging from 2%–4%—in addition to the alloying elements nickel, chromium, and molybdenum—warrant consideration for carefully-monitored welding procedures. Gray cast iron containing 2%–3% carbon is the most produced alloy and likely most often repaired. The high quantities of carbon are too great to be dissolved uniformly into a solid solution as in steels with lower carbon content. Carbon amounts too great for uniform absorption in the iron parent metal precipitate into high carbon concentrations in the form of graphite flakes, leaving a hard brittle structure that presents difficulties for welding repairs.

Ductile and nodular cast irons are alloyed with aluminum, magnesium, or cerium, which alter the form of high-carbon graphite from flakes to a spherical or nodular shape. The result is higher strength properties, virtually eliminating the brittleness associated with the gray cast irons. These nodular spherical shapes give the iron high ductility.

CAN ALL CAST IRONS BE WELDED?   The degree of weldability of cast irons varies considerably with alloying and previous heat-treatment condition. The common gray cast irons have low ductility, which can easily lead to weld cracking

AISI-SAE Type	Preheat	Temperature (°F) Interpass	Draw or Tempering
W1	250/450	250/450	300/650
W2	250/450	250/450	300/650
S1	300/500	300/500	300/500
S5	300/500	300/500	500 min.
S7	300/500	300/500	425/400
01	300/400	300/400	300/450
06	300/400	300/400	300/450
A2	300/500	300/500	350/400
A4	300/500	300/500	350/400
D2	700/900	700/900	925/900
M1	950/1,100	950/1,100	1,000/1,050
M2	950/1,100	950/1,100	1,000/1,050
M10	950/1,100	950/1,100	950/1,050
H11	900/1,200	900/1,200	1,000/1,150
H12	900/1,200	900/1,200	1,000/1,150
H13	900/1,200	900/1,200	1,000/1,150
P20	400/800	400/800	1,000

TABLE 10-19

Preheat and Interpass Temperatures When Welding Tool Steels

Source: Cary, Howard B., *Modern Welding Technology*, 4th ed., © 1998. Reproduced by permission of Pearson Education, Inc., Upper Saddle River, New Jersey.

along the heat-affected zone, especially since the graphite flakes tend to form notch-type discontinuities in the structure from which cracks can propagate. Ductile, nodular, and malleable cast irons can be welded. White cast iron is generally considered not weldable.

HOW ARE CAST IRONS WELDED?

There are several welding methods that can be used to produce welds for repair of cracks or build up worn surfaces. Arc welding with SMAW, GMAW, and FCAW are possible, with SMAW being more common. Oxy-Fuel Welding with an acetylene gas, while difficult to employ, can be done successfully by an experienced welder. Furthermore, brazing with bronze filler metals, when done properly, produces strong weld joints. However, the nature of cast irons with high carbon contents requires careful procedures, and even using them does not guarantee success.

The condition of the base metal needing repair can limit success. Old rusty manifolds or fireplace grates are examples of problematic weld repairs. Often, these products have deteriorated to a level where weld repair and adherence of a filler metal are not practical. In situations like these, try cleaning the base metal and then attempt to spot a weld on the surface with a low-hydrogen electrode such as an E7018. If the weld metal adheres, proceed with the repair using a filler metal alloyed for welding cast irons. If the weld balls up or rolls off the surface, it is not usually worth the effort to attempt a repair. Note that mild steel electrodes are not the most resistant to cracking. This "spot test" is used only in determining that a filler metal will adhere to the cast iron.

Once the alloy type is identified, the welding process, filler metal, joint design, and procedures are determined. The first step is to prepare the base metal and weld joint for welding. In order to attain maximum strength and lessen the probability of cracking, complete joint penetration is preferred. At the same time, fusion welding itself should limit melting of the base metal to reduce the size of the heat-affected zone. The difficult task is to get enough melting of the base metal to provide complete penetration without overdoing it.

While not always practical, sawing and machining are preferred methods for producing weld joints. Using a grinder to gouge or create a groove or to prep the weld joint should be avoided because it smears the graphite concentrations, creating difficulty in bonding, especially for braze-welded joints. Grinding or gouging can be followed by rotary file (die grinder) or machining. Obviously, sawing produces a V-groove joint, and machining can be used to create a U-groove joint. While a V-groove joint design may work, a U-groove joint design provides easy joint root access for the welding electrode when fusion-welding with the SMAW process. This joint can easily be prepped with a cutting electrode followed by rotary filing and/or wire brushing.

For repairing cracks, special considerations apply. A hole should be drilled at each end of the crack. These holes prevent further growth of the crack in either direction. The hole should be made before further joint penetration. If a cast part has broken, the edge preparation should be done in a manner that leaves part of the original edges intact to allow for easy alignment of the mating surfaces. Clamps may be used to keep parts in alignment.

Before welding, the complete weld assembly must be prepared by removing contaminants, such as paint, rust, or oils, from the area immediately surrounding the weld. The surface layer or skin of the casting should also be removed. Many castings, such as engine blocks or gear casings, are exposed to oils and can absorb those elements into the porous surfaces of cracks needing repair. Heating these surfaces prior to welding, up to about 500°F (260°C), will drive out oil or other impurities.

Immediately before welding, the entire base metal should be preheated. In most cases, preheat ranges from 900°F–1,200°F (482°C–649°C) are suitable. Higher preheat temperatures nearing 1,200°F (649°C) lessen temperature gradients between the molten weld bead and the surrounding base metal. The transition temperature of cast iron is 1,333°F (723°C); heating higher should be avoided. Cast iron base metals begin to melt at 2,109°F (1,154°C). The high heat of the welding arc or fusion weld increases the base metal temperature immediately surrounding the weld. Without sufficient preheat, the fusion arc welding processes compound cracking problems due to nonuniform expansion and contraction of the base metal adjacent to the weld. Too little preheat will lead to base metal cracking along the weld.

Furthermore, a postheat temperature should be maintained while the weld area temperature stabilizes to a uniform range. Then, slow cooling must be achieved, preferably by cooling from postheat temperature in an oven. The advantage of cooling in an oven is that it is insulated and contains fire bricks that will hold heat for a significantly longer time than any other method, allowing the part to cool down over a maximum time span. This cannot not be overemphasized. In some cases, a small weld assembly can be heated within a chamber composed of fire bricks. When welding is completed, the chamber can be closed and covered with thermal insulation. If these methods are not possible, wrapping the work in a thermal blanket or burying it in ash or sand can slow the cooling rate.

Fusion welding with arc welding processes can be accomplished with cast iron, mild steel, and nickel-alloy electrodes. **TABLE 10-20** matches filler metals with welding processes based on color match and ability to machine the weld. As indicated, the cast iron filler provides the best color match and machinable weld deposit combination. Mild steel welds, while capable of producing a weld deposit with a fair color match, are not machinable and are more prone to cracking. Nickel-alloyed electrodes are used successfully with machinable deposits, but color match is poor at best. Copper-based alloys produce bronze welds with no color match but good crack resistance.

HOW ARE CAST IRONS JOINED WITH AN OXY-FUEL TORCH?

An oxy-fuel torch can be used for fusion welding or brazing. Fusion welding with acetylene gas produces a good color match and machinable-finished weld deposit. Brazing or braze welding results in a machinable weld deposit but with no color match. Either joining method can produce a weld with good strength if done

Welding Process and Filler Metal Type	Filler Metal Spec[a]	Filler Metal Type[a]	Color Match
MAW (stick)			
Cast iron	E-CI	Cast iron	Good
Copper-tin[b]	ECuSn A and C	Copper–5% or 8% tin	No
Copper-aluminum[b]	ECuA1–A2	Copper–10% aluminum	No
Mild steel	E-St	Mild steel	Fair
Nickel	ENi-CI	High-nickel alloy	No
Nickel-iron	ENiFe-CI	50% nickel plus iron	No
Nickel-copper	ENiCu-A and B	55% or 65% Ni + 40% or 30%	No
Oxyfuel gas			
Cast iron	RCI & A and B	Cast iron, with minor alloys	Good
Copper-Zinc[b]	RCuZn B and C	58% copper–zinc	No
Brazing[c]			
Copper-zinc	RBCuZn A & D	Copper–zinc and copper–zinc–nickel	No
GMAW (MIG)			
Mild steel	E60S-3	Mild steel	Fair
Copper base[c]	ECuZn-C	Silicon bronze	No
Nickel–copper	ENiCu-B	High-nickel alloy	No
FCAW			
Mild steel	E70T-7	Mild steel	Fair
Nickel type	No spec	50% nickel plus iron	No

TABLE 10-20

[a]See "Specification for Welding Electrodes and Rods for Cast Iron," AWS A5.15
[b]Would be considered a brass weld.
[c]Heat source, any for brazing; also carbon arc, twin carbon arc, gas tungsten arc, or plasma arc.

Source: Cary, Howard B., *Modern Welding Technology*, 4th ed., © 1998. Reproduced by permission of Pearson Education, Inc., Upper Saddle River, New Jersey.

properly. Make sure to follow the previously discussed preparation of base metals, which includes cleaning, joint preparation, and preheating. Make sure the preheat is high—near 1,200°F (649°C). Interpass temperatures should maintain the preheat level, and heat input must remain high in order for the welding flame to sufficiently melt the base metal and maintain a weld pool.

For fusion welding, select a filler metal from **TABLE 10-20**. The cast iron filler produces a good color match. Use of a flux is required. The flux is applied by slightly heating the electrode and dipping it into the flux. Follow the electrode manufacturer's guidelines. The torch is set at a neutral-to-slightly-reducing flame. Use a properly sized welding tip per the torch manufacturer's recommendations. Heat input must remain high in order for the welding flame to sufficiently melt the base metal and maintain a weld pool.

Once the weld pool is established, the welder should oscillate the torch in a circular manner in order to agitate the weld pool. This will help remove gas bubbles that could otherwise result in porosity and also aids the flux removal of impurities, floating them out the surface of the weld. Postheat the part to near 1,200°F (649°C) and then cool as slowly as possible, preferably in a heat-treating oven.

Brazing and braze welding can be used successfully with Oxy-Fuel and GTAW processes by using copper-nickel and copper-zinc-nickel electrodes. Brazing requires narrow gaps—0.001 inches to 0.003 inches (0.0254 mm to 0.762 mm)—between the parts to allow for capillary action to occur. With both brazing and braze welding, weld joint cleanliness is critical in order for bonding to occur. Grinding as a preparation should be avoided, as it smears the cast iron graphite structures, inhibiting adherence of the braze filler metals.

Because brazing uses filler metals that melt above 840°F (450°C), a logical preheat would be above that temperature but well below the melting temperature of the filler metal. Preheat and interpass temperatures of 900°F–1,200°F (482°C–649°C) should be sufficient. Check the filler metal manufacturer's data on the filler metal melting temperature. After welding is complete, maintain a postheat temperature equal to the preheat and interpass temperature long enough for the whole part, including the weld joint, to equalize and then cool as slowly as possible, preferably in a heat-treating oven.

WELDING ALUMINUM ALLOYS

While most aluminum alloys can be welded with common welding processes, weldability factors differ from those encountered with carbon steels. Formation of hardened structures in aluminum are not a consideration as it is in steels. Actually, the opposite metallurgical condition is likely to occur as a result of welding. Many alloys are strengthened by strain hardening through cold-forming operations. Recrystallization temperatures for aluminum are as low as 175°F (79.5°C) with some alloys. This means that welding will actually soften the area in the heat-affected-zone of welded aluminum structures.

Several properties affect the ease and success for welding aluminum. Pure aluminum melts at 1,220°F (660°C), and aluminum alloys melt at 900°F–1220°F

(482°C–660°C). The low melting/solidification temperature is not necessarily disadvantageous. Combined with a high thermal conductivity, the weld pool solidifies rather quickly, and these factors combined with surface tension allow welding at high heat inputs for all position welding. The high thermal conductivity of aluminum does require high heat input to offset heat loss from the weld area.

WHAT SPECIAL CONSIDERATIONS ARE MADE FOR WELDING ALUMINUM ALLOYS?

Aluminum alloys react to atmospheric oxygen, forming a hard tenacious surface oxide. This oxide melts at nearly three times the base metal melting temperature and in many cases must be removed prior to welding. Aluminum alloys melt at around 1,260°F (683°C), and the surface oxides melt at about 3,600°F (1,982°C). In some applications of GMAW and GTAW, the oxide may be removed adequately by cathodic cleaning (etching), which is the lifting of surface oxides by the influence of the DCEP welding arc under an argon gas shield. In brazing and soldering applications on clean base metal, fluxes adequately remove and protect the weld area. Chemical cleaning can remove light to moderate oxide buildup.

In some instances, mechanical means must be used to remove more dense surface oxides. Extended exposure of aluminum to moisture (rain/snow) over extended periods can result in a dense oxide layer that neither fluxes nor cathotic etching can eliminate. Under these circumstances, mechanical removal becomes necessary to prevent contamination of the weld pool. Failure to remove heavy oxides results in oxides floating in and contaminating the weld pool. Use of stainless steel brushes or Scotch-Brite pads on the weld joint immediately prior to welding removes these detrimental oxides. It is important to remember that oxide formation on aluminum begins as soon as a clean aluminum surface is exposed to atmospheric air. For this reason, it is not practical to remove the oxides hours in advance of implementing the welding procedure.

HOW ARE SURFACE OXIDES PREVENTED FROM FORMING DURING STORAGE?

Because aluminum alloys are prone to building up of dense surface oxides when exposed to moisture, special base metal storage methods apply. Stacking of aluminum sheets and structural shapes should be avoided because this can result in trapping of moisture between the pieces, resulting in oxide buildup. Correct storage of aluminum involves placement of sheets vertically and leaving a small space between sheets to prevent trapping moisture between them. This reduces the chance of excess oxide buildup.

WHAT SHIELDING GAS IS USED FOR WELDING ALUMINUM ALLOYS?

The shielding gases used for welding aluminum alloys are pure argon and mixtures of argon and helium. These two inert gases do not react negatively with the base metal, as would other reactive gases or mixtures that include oxygen and carbon dioxide. Argon is most commonly used, in part because it produces a cleaner weld bead, aiding in the removal of surface oxides. It is the gas of choice when using the GTAW process. In some instances, helium percentages of 25% or more are added to argon for welding heavier plate structures. Argon-helium mixtures generate higher heat in the welding arc, which is a benefit on heavy aluminum with its high thermal conductivity that draws heat rapidly from the weld. Helium, however, is becoming scarce and more costly; soon, it may no longer be available for welding.

HOW ARE FILLER METALS SELECTED FOR SPECIFIC ALUMINUM ALLOYS?

Aluminum series alloys can be broken down into two basic groupings: heat-treatable alloys and non-heat-treatable alloys. While non-heat-treatable alloys may at times use filler materials of similar composition, the heat-treatable alloys cannot be welded with filler metals of the same composition. The problem with using a matching filler metal to weld a heat-treatable alloy is that weld cracking is likely to occur. This is especially a concern when welding the popular 6061 alloys.

Many filler metal charts can be used to identify the correct filler metal for the job. **TABLE 10-21** recommends filler metals based on several weld characteristics, including: ease of welding (W), strength of welded joint (S), ductility (D), corrosion resistance (C), sustained temperatures (T), and color match (C). Each of the characteristics is given a letter rating for each recommended filler metal alloy listed. The letter ratings are A to D in decreasing order of usability from A to D.

WHAT ARE CONCERNS FOR COLOR MATCHING?

Welding aluminum prior to anodizing bears the consideration of color matching. Anodizing is a process of electrochemically plating the surface of aluminum to create a uniform oxide thickness. If the wrong filler metal is used, the weld and base metals will react differently to anodization and will have a poor color match. Here, the filler metal selected may anodize poorly or anodize a different color than the base metal. Matching the filler metal becomes critical. Using a filler metal that results in good coloring but results in weld metal cracking is never acceptable. On 6061 base metal, we see that the recommended filler rod is ER4043 for ease of welding, corrosion resistance in water, and sustained high-temperature service. However, using a 4xxx series filler metal would result in a dark grayish coloring after anodizing. For color match, the ER5356 is the best choice.

HOW CAN ALUMINUM BASE METALS BE IDENTIFIED?

Beyond using chemical testing or spectra-analytical testing, it may be helpful to become familiar with the alloy tree (see **FIGURE 10-8**). The base alloys are given in white numbers, while the filler alloys are colored. The 1XXX, 3XXX, 4XXX, and the 5XXX series are generally considered non-heat-treatable. The alloy tree can be used to determine the base metal type because each type of manufacturing industry uses base metal alloys based on chemical composition as a basis for usability.

ONCE THE BASE METAL AND FILLER METAL ARE DETERMINED, WHAT SPECIAL CONSIDERATIONS ARE APPLIED TO WELDING ALUMINUM ALLOYS?

The welding process is selected based on equipment availability, heat input required, base metal thickness, and joint design required. Aluminum can be welded with common arc welding processes such as GMAW, GTAW, and SMAW. The obvious choice for welding the high production on base metals 1/4 inches (6.4 mm) and greater thicknesses is GMAW. GTAW is a popular choice on thinner weld assemblies because it offers precise heat control. Pulsing with GMAW and GTAW provides superior results on repairs of aluminum castings

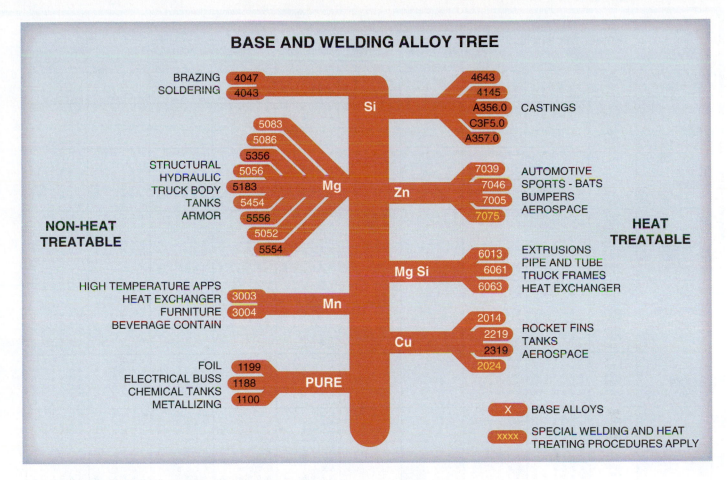

FIGURE 10-8 Aluminum base metal and alloy tree

Source: Courtesy of AlcoTec, a subsidiary of the ESAB Group, Inc.

because the agitating effects of pulsing on the weld pool aid in working impurities and gas bubbles out of the molten weld pool. Oxy-Fuel brazing has uses in repairs of thin base metals and radiators.

When welding with GMAW, a forehand technique is used. The forehand technique directs the shielding gas ahead of the weld pool and pushes atmosphere away from the arc. The backhand technique is avoided because it draws atmosphere into the arc, which contaminates the weld. Under this circumstance a contaminant—a black substance called smoot—covers the weld surface. Using the correct forehand technique produces a clean weld deposit.

Surface oxides and impurities must be removed either by chemical or mechanical means. Generally, heat input must be high enough to sufficiently melt the base metal. Although aluminum has a high thermal conductivity that draws heat away from the weld, preheat is not recommended on many alloys. Heating alters the base metal grain structure and can reduce corrosion resistance. Setting the welder at high amperages and using rapid welding speeds provides good results.

The rapid cooling rate of aluminum assemblies creates some unique problems. Hot-cracking occurs in aluminum welds, particularly in the weld crater. This is caused by two primary factors. One factor is hot shortness or reduced strength of the weld pool at elevated temperatures. The other is the nonuniform cooling rate of the weld pool and pulling forces imposed on it during solidification. After the weld is completed and during the cooling cycle, the outer weld pool in

Aluminum Filler Alloy Chart

Base Alloys	Filler Alloys	1060, 1070, 1080, 1350	1100	2014, 2036	2219	3003, ALCLAD 3003	3004	ALCLAD 3004	5005, 5050	5052, 5652
Characteristics		W S D C T M	W S D C T M	W S D C T M	W S D C T M	W S D C T M	W S D C T M	W S D C T M	W S D C T M	W S D C T M
319.0, 333.0, 354.0, 355.0, C355.0, 380.0	2319			B A A A A	B A A A A					A A A A A
	4043	B A A A A	B A A A A	C C B C A A	C C B C A A	B B A A A	B B A A A	B B A A A	B B A A A	
	4145	A A B A A	A A B A A	A B C B A A	A B C B A A	A A B A A	A A B A A	A A B A A	A A B A A	
413.0, 443.0, 444.0, 356.0, A356.0, 357.0, 359.0	4043	A A A A A	A A A A A	B B A A A	B B A A A	A A A A A	A A A A A	A A A A A	A A A A A	A B A A A A
	4145	A A B B A	A A B B A	A A B A A	A A B A A	A A B B A				
	A356.0									
	A357.0									
	5356									B A B B A
7005, 7021, 7039, 7046, 7146, 710.0, 711.0	4043	A A C A A	A A C A A	B B A A A	B B A A A	A B C A A	A D C B A	A D C B A	A B C B A	B D C B A
	4145			A A B A A	A A B A A					
	5183	B A B A A	B A B A A			B A B A A	B A B A A	B A B A A	B A B A A	A A B A A
	5356	B A A A A	B A A A A			B B A A A	B B A A A	B A A A A	A B A A A	A B A A A
	5554					C C A A A A	C C A A A A	C A A A A A	C A A A A	B C A A A A
	5556	B A B A A	B A B A A			B A B A A	B A B A A	B A B A A	B A B A A	A A B A A
	5654					C C A A B	C C A A B	C C A A A	C A A A A	B C A A A
6061, 6070	4043	A A C A A	A A C A A	B B A A A	B B A A A	A B C A A	A D C A A	A D C A A	A B C A A	A D C A A
	4145	A A D B A	A A D B A	A A B A A	A A B A A	A A D B A	B C D B A	B C D B A	A B D B A	
	4643 (1)									
	5183	B A B A	B A B A			B A B A	B A B	B A B A	B A B A	B A B C B
	5356	B A A A	B A A A			B A A A	B B A	B B A A	B A A A	B B A C A
	5554									C C A B A B
	5556	B A B A	B A B A			B A B A	B A B	B A B A	B A B A	B A B C B
	5654									C C A B A
6005, 6063, 6101, 6151, 6201, 6351, 6951	4043	A A C A A	A A C A A	B B A A A	B B A A A	A B C A A	A D C A A	A D C A A	A B C A A	A D C A A
	4145	A A D B A	A A D B A	A A B A A	A A B A A	A A D B A	B C D B A	B C D B A	A B D B A	
	4643 (1)									
	5183	B A B A	B A B A			B A B A	B A B	B A B A	B A B A	B A B C B
	5356	B A A A	B A A A			B A A A	B B A	B B A A	B A A A	B B A C A
	5554									C C A B A B
	5556	B A B A	B A B A			B A B A	B A B	B A B A	B A B A	B A B C B
	5654									C C A B A
5454	4043	A B C C A	A B C C A			A B C C A	A D C C A	A D C C A	A B C C A	A D C C A
	5183	B A B B A	B A B B A			B A B B A	B A B B A	B A B B A	B A B B A	A A A B A
	5356	B A A B A	B A A B A			B A A B A	B B A B A	B B A B A	B A A B A	A B A B A
	5554	C A A A A A	C A A A A A			C A A A A A	C C A A A A	C C A A A A	C A A A A A	C C A A A A
	5556	B A B B A	B A B B A			B A B B A	B A B B A	B A B B A	B A B B A	A A B B A
	5654									B C A B A
511.0, 512.0, 513.0, 514.0, 535.0, 5154, 5254	4043	A B C C	A B C C			A B C C	A D C C	A D C C	A B C C	A D C C
	5183	B A B B A	B A B B A			B A B B A	B A B B A	B A B B A	B A B B A	A A B B B
	5356	B A A B A	B A A B A			B A A B A	B B A B A	B B A B A	B A A B A	A B A B B
	5554	C A A A A	C A A A A			C A A A A	C C A A A	C C A A A	C A A A A	C C A A B
	5556	B A B B A	B A B B A			B A B B A	B A B B A	B A B B A	B A B B A	A A B B B
	5654	C A A A B	C A A A B			C A A A B	C C A A B	C C A A B	C A A A B	B C A A B
5086, 5056	4043	A B C B	A B C B			A B C B	A C C B	A C C B	A B C B	
	5183	A A B A A	A A B A A			A A B A A	A A B A A	A A B A A	A A B A A	A A B A A
	5356	A A A A A	A A A A A			A A A A A	A B A A A	A B A A A	A A A A A	A B A A A
	5554									C C A A A
	5556	A A B A A	A A B A A			A A B A A	A A B A A	A A B A A	A A B A A	A A B A A
	5654									B C A A B
5083, 5456	4043	A B C B	A B C B			A B C B	A C C B	A C C B	A B C B	
	5183	A A B A A	A A B A A			A A B A A	A A B A A	A A B A A	A A B A A	A A B A A
	5356	A A A A A	A A A A A			A A A A A	A B A A A	A B A A A	A A A A A	A B A A A
	5554									C C A A A
	5556	A A B A A	A A B A A			A A B A A	A A B A A	A A B A A	A A B A A	A A B A A
	5654									B C A A B
5052, 5652	4043	A B C A A	A B C A A			A B C A A	A B C A A	A C C A A	A B C A A	A D C B A
	5183	B A B A	B A B A			B A B A	B A B A	B A B A	B A B A	A A B C B
	5356	B A A A	B A A A			B A A A	B A A A	B B A A	B A A A	A B A C A
	5554									C C A A A B
	5556	B A B A	B A B A			B A B A	B A B A	B A B A	B A B A	A A B C B
	5654									B C A B A
5005, 5050	1100	C B A A A A	C B A A A A			C C A A A A			B A A A A	*1100*
	4043	A A C A A	A A C A A			A B C A A	A B C A A	A B C A A	A B D A A	*4043*
	4145	B A D B A	B A D B A			B B D B A				*4145*
	5183	C A B B	C A B B			C A B C B	B A B	B A B B	B A C B	*5183*
	5356	C A B B	C A B B			C A B C	B A A A	B A A B	B A B B	*5356*
	5556	C A B B	C A B B			C A B C B	B A B A	B A B B A	B A C B	*5556*
ALCLAD 3004	1100	D B A A A A	D B A A A A			C C A A A A			*1100*	
	4043	A A C A A	A A C A A			A B C A A	A D D A A	A D D A A	*4043*	
	4145	B A D B A	B A D B A			B B D B A			*4145*	
	5183	C A B C B	C A B C B			C A B C A	B A C C A	B A C C A	*5183*	
	5356	C A B C B	C A B C B			C A B C A	B B B C A	B B B C A	*5356*	
	5554						C C A B A A	C C A B A A	*5554*	
	5556	C A B C B	C A B C B			C A B C A	B A C C A	B A C C A	*5556*	
3004	1100	D B A A A A	D B A A A A			C C A A A A		*1100*		
	4043	A A C A A	A A C A A			A B C A A	A B D A A	*4043*		
	4145	B A D B A	B A D B A			B B D B A		*4145*		
	5183	C A B B	C A B B			C B C A	B A C C A	*5183*		
	5356	C A B B	C A B B			C A B C A	B B B C A	*5356*		
	5554						C C A B A A	*5554*		
	5556	C A B B	C A B B			C B C A	B A C C A	*5556*		
3003, ALCLAD 3003	1100	B B A A A A	B B A A A A	B A A A A	B A A A A	B B A A A A	*1100*			
	4043	A A B A A	A A B A A	A A B A A	A A B A A	A A B A A	*4043*			
	4145	A A C B A	A A C B A	A A B A A	A A B A A	A A C B A	*4145*			
2219	2319			B A A A A	A A A A A A	*2319*				
	4043	B A A A A	B A A A A	B C B C A	B C B C A	*4043*				
	4145	A A B A A	A A B A A	A B C B A	A B C B A	*4145*				
2014, 2036	2319			C A A A A	*2319*					
	4043	B A A A A	B A A A A	B C B C A	*4043*					
	4145	A A B A A	A A B A A	A B C B A	*4145*					
1100	1100	B B A A A A	B B A A A A	*1100*						
	4043	A A B A A	A A B A A	*4043*						
	5356			*5356*						
1060, 1070, 1080, 1350	1100	B B A A A B	*1100*							
	1188	C C A A A A	*1188*							
	4043	A A B A A	*4043*							

TABLE 10-21

Source: Courtesy of AlcoTec, a subsidiary of the ESAB Group, Inc.

AlcoTec — A Subsidiary of The ESAB Group, Inc.

Aluminum Filler Alloy Chart

5083, 5456 (W S D C T M)	5086, 5056 (W S D C T M)	511.0, 512.0, 513.0, 514.0, 535.0, 5154, 5254 (W S D C T M)	5454 (W S D C T M)	6005, 6063, 6101, 6151, 6201, 6351, 6951 (W S D C T M)	6061, 6070 (W S D C T M)	7005, 7021, 7039, 7046, 7146, 710.0, 711.0 (W S D C T M)	413.0, 443.0, 444.0, A356.0, 357.0, 359.0 (W S D C T M)	319.0, 333.0, 354.0, 355.0, C355.0, 380.0 (W S D C T M)	(W S D C T M)
A A A A A	A A A A A	A A A A A	A A A A A A	B B A A A A A B A A A	B B A A A A A B A A A	B B A A A A A B A A A	B B A A A A A B A A A	B A A A A A A B B B A	2319 4043 4145
A B B A A	A B B A A	A B B A A	A B B A A A	A B A A A A A A B B A	A B A A A A A A B B A	A B B A A A A A B B A	A A B A A A A A B B A B A A A A A A A A A A A A	4043 4145 A356.0 A357.0 5356	
A A A A A	A A A A A	A A A B A	A A A B A			A A A A B			
A A B A A A B A A A A A B A A	A A B A A A B A A A A A B A A	A A B A A A B A A A B C A A A A B A A B C A A A	A A B A A A B A A A B C A A A A A A B A A B C A A A	A D C B A A A B A A A B A A A B C A A A A A A B A A B C A A A	A D C B A A A B A A A B A A A B C A A A A A A B A A B C A A A	B D C B A A A B A A A B A A A B C A A A A A A B A A B C A A A	4043 4145 5183 5356 5554 5556 5654		
A D C A A A B A A A B A A A B C A A A A A B A A B C A A B	A D C A A A B A A A B A A A B C A A A A A B A A B C A A A	A D C A B A B C B B B A C A C C A B B B A B C B C C A B B	A D C B A B A B C A B B A C A C C A A A A B A B C A C C A B B	A C B A A A C B A A B A A C A B A A C A C B A B B A B A A C A C B A B B	A C B A A A C B A A B A A C B B B A C A C B A B B B B A A C B C B A B B	4043 4145 4643 (1) 5183 5356 5554 5556 5654			
A B C A A A B A A A A A A A B A A A A A A B A A B A A A B	A B C A A A B A A A A A A A B A A A A A A B A A B A A A D	A B C A B A B C A B A A C A C A A B A B A B C A C A A B B	A B C B A B A B C A B A A C A C A A A A A B A B C A C A A B B	A C D A A A C B A A B A A C A B A A C A C B A B B A B A A C A C B A B B	4043 4145 4643 (1) 5183 5356 5554 5556 5654				
A A B B A A B A B A B C A A A A A B B A	A A B B A A B A B A B C A A A A A B B A	A A B B A A B A B A B C A A A A A B B A B C A A B	A A B B A A B A B A B C A A A A A A B B A B C A A B	4043 5183 5356 5554 5556 5654					
A A B A A A B A A A B C A A A A A B A A B C A A B	A A B A A A B A A A B C A A A A A B A A B C A A B	A A B B B A B A B A B C A A B A A B B B B C A A A	4043 5183 5356 5554 5556 5654						
A A B A A A B A A A A A B A A	A A B A A A B A A A A A B A A	4043 5183 5356 5554 5556 5654							
A(2)B A A A A A A A A D A A	4043 5183 5356 5554 5556 5654								
4043 5183 5356 5554 5556 5654									

Aluminum Filler Alloy Chart

Symbol	Characteristics
W	Ease of welding (relative freedom from weld cracking).
S	Strength of welded joint (as-welded condition). (Rating applies particularly to fillet welds. All rods & electrodes rated will develop presently specified minimum strengths for butt welds.)
D	Ductility. (Rating is based upon the free bend elongation of the weld.)
C	Corrosion resistance in continuous or alternate immersion in fresh or salt water.
T	Recommended for service at sustained temperatures above 150 F (65.5°C).
M	Color match after anodizing.

A, B, C & D are relative ratings in decreasing order of merit. The ratings have relative meaning only within a given block.

NOTE: Combinations having no rating are not usually recommended. Ratings do not apply to these alloys when heat treated after welding.

(1) 4643 is a heat-treatable filler alloy and gives higher strength in thick 6xxx series weldments after postweld solution heat treatment and aging.

(2) An "A" rating for alloy 5083 to 5083. No rating for alloy 5456 to 5456.

4047 can be used in lieu of 4043 for thin section sheet due to the lower melting point of 4047.

How to Use

1. Select base alloys to be joined (one from the side blue column, the other from the top blue row).

2. Find the block where the column and row intersect.

3. This block contains horizontal rows of letters (A, B, C, or D) representative of the alloy directly across from them in the filler alloy box at the end of each row. The letters in each line give the A-to-D rating of the characteristics listed at the top of each column – W, S, D, C, T, and M (see Legend at right for explanation of each letter).

4. Analyze the weld characteristics afforded by each filler alloy. You will find that you can "trade off" one characteristic for another until you find the filler that best meets your needs.

Example

When joining base alloys 3003 and 1100, find the intersecting block. Now, note that filler alloy 1100 provides excellent ductility (D), corrosion resistance (C), performance at elevated temperatures (T), and color match after anodizing (M), with good ease of welding (W) and strength (S). However, if ease of welding and shear strength are UTMOST in importance, and ductility and color match can be sacrificed slightly, filler alloy 4043 can be used advantageously.

AlcoTec Wire Corporation

2750 Aero Park Drive · Traverse City, MI 49686 USA · Phone (231) 941-4111
Sales and Marketing Fax (231) 941-9154 · Administrative and Quality Control Fax (231) 941-1040
E-mail: alcotec@traverse.com   Web site: www.alcotec.com

contact with the base metal begins to solidify rapidly, while the center of the weld pool is still liquid. Because the weld pool is oval in shape, stress forces caused by contraction of the surrounding weld pool during cooling put a high amount of strain on the weaker center of the weld pool. The weld pool as it begins to solidify cannot withstand these forces, and a crack forms.

HOW ARE CRACKS ELIMINATED IN ALUMINUM WELDS?   Beyond selecting the correct filler metal, several welding techniques can be employed to eliminate cracking on aluminum welds. They are maintaining an appropriate travel speed, producing slightly convex weld beads, and filling the weld craters. Determining the correct travel speed can be accomplished by the welder observing the weld pool shape as the weld progresses along the weld joint. Just as an oval or elliptical weld pool is preferred for welding on other base metals, so too is it critical when welding aluminum alloys. Too fast a travel speed creates a V-shaped solidification pattern on the trailing edge of the weld pool that often leads to centerline cracking down the length of the weld. Furthermore, too small of a weld will not withstand rapid solidification forces, and the weld metal will crack.

If welding voltages are too high, the weld bead profile will tend to be flat or concave. This will contribute to weld cracking. Optimum welding parameters— including voltage, amperage (wire feed speed), electrode angles, and travel speed—result in a convex bead profile that is desirable. Adjust these parameters to attain a convex bead.

The last step in making a weld is termination of the weld pool. The molten weld pool solidifies into the crater at the end of the weld. Crater fill is important with aluminum welds. Just as a flat or convex weld bead is desirable, so too is a convex crater desired. How crater fill is accomplished depends on the welding process. GTAW is the easiest welding process to control crater fill. Using remote settings, the welder can slowly decrease current while adding filler metal to the weld pool, producing a well-filled crater. With SMAW or GMAW, the welder can straighten the electrode at the end of the weld and back the electrode into the weld pool, pause to allow it to fill, and stop.

WELDING MAGNESIUM ALLOYS

Welding magnesium alloys employs similar welding procedures to those used for welding aluminum alloys. The reasons for this are that magnesium also has a low melting temperature, forms a dense surface oxide when exposed to atmospheric conditions, and has both a high thermal conductivity and high thermal expansion rates. Welding procedures include cleaning weld joint surfaces, removing surface oxides either chemically or mechanically, selecting the appropriate filler metal, welding procedures producing appropriate sized convex welds, and filling craters fully.

HOW ARE FILLER METALS SELECTED FOR SPECIFIC MAGNESIUM ALLOYS?
The first step in selecting a magnesium filler metal is to identify the base metal that is to be welded. **TABLE 10-22** lists common ASTM base alloys. As indicated,

ASTM ALLOY	Nominal Composition (%)						
	Aluminum	Manganese	Zinc	Zirconium	Rare Earths	Thorium	Magnesium
Sand and permanent mold castings							
AX92A	9.0	0.15	2.0	—	—	—	Remainder
AZ63A	6.9	0.25	3.0	—	—	—	Remainder
AZ81A	7.6	0.13 min.	0.7	—	—	—	Remainder
AZ91C	8.7	0.20	0.7	—	—	—	Remainder
EK30A	—	—	—	0.35	3.0	—	Remainder
EK41A	—	—	—	0.6	4.0	—	Remainder
EZ33A	—	—	2.7	0.7	3.0	—	Remainder
HK31A	—	—	—	0.7	—	3.0	Remainder
HZ32A	—	—	2.1	0.7	—	3.0	Remainder
Die castings							
AZ91A	9.0	0.20	0.6	—	—		Remainder
AZ91B							
Extrusions							
AZ31B	3.0	0.45	1.0	—		—	Remainder
AZ31C							
AZ61A	6.5	0.30	1.0	—	—	—	Remainder
M1A	—	1.50	—	—	—	—	Remainder
AZ80A	8.5	0.25	0.5	—	—	—	Remainder
ZK60A	—	—	5.7	0.55	—	—	Remainder
Sheet and plate							
AZ31B	3.0	0.45	1.0	—	—	—	Remainder
HK31A	—	—	—	0.7	—	3.0	Remainder

TABLE 10-22

Composition of Magnesium Alloys. From ASTM 8275, Magnesium Alloys (abridged)

Source: Cary, Howard B., *Modern Welding Technology*, 4th ed., © 1998. Reproduced by permission of Pearson Education, Inc., Upper Saddle River, New Jersey.

AWS Classification[a]	Nominal Composition (%)										
	Mg	Al	Be	Mn	Zn	Zi	Rare Earths	Cu	Fe	Ni	Si
AZ61A	Remainder	5.8–7.2	0.0002–0.0008	0.15	0.40–1.5	—	—	0.05	0.005	0.005	0.05
AZ101A	Remainder	9.5	0.0002–0.0008	0.13	0.75	—	—	0.05	0.005	0.005	0.05
AZ92A	Remainder	8.3–9.7	0.0002–0.0008	0.15	1.75	—	—	0.05	0.005	0.005	0.05
EZ33A	Remainder	—	—	—	2.0–3.1	0.45–1.0	2.5–4.0	—	—	—	—

TABLE 10-23

Composition of Magnesium Filler Metals per AWS A5.19
[a]Use suffix letter E, electrode, or R, rod.

Source: Cary, Howard B., *Modern Welding Technology*, 4th ed., © 1998. Reproduced by permission of Pearson Education, Inc., Upper Saddle River, New Jersey.

alloys differ based on the nominal composition alloys additions of aluminum, manganese, zinc, rare earths, and thorium.

Magnesium filler metals are manufactured in accordance with filler metal specification A5.19. AWS magnesium filler classifications are listed in **TABLE 10-23**. Because more base alloys exist than do filler metal classifications, it is not possible to match each base metal to a filler metal of the same composition. **TABLE 10-24** can be used to match the desired magnesium filler metals magnesium base alloys.

Base Alloy	AM100A	AZ10A	AZ31B&C	AZ61A	AZ63A	AZ80A	AZ81A	AZ91C	AX92A	EK41A
AM100A	1. AZ92A 2. AZ101									
AZ10A	AZ92A	1. AZ61A 2. AZ32A								
AZ31B&C	AZ92A	1. AZ61A 2. AX92A	1. AZ61A 2. AZ92A							
AZ61A	AZ92A	1. AZ61A 2. AZ92A	1. AZ61A 2. AZ92A	1. AZ61A 2. AZ92A						
AZ63A	X	X	X	X	AZ92A					
AZ80A	AZ92A	1. AZ61A 2. AZ92A	1. AZ61A 2. AZ92A	1. AZ61A 2. AZ92A	X	1. AZ61A 2. AZ92A				
AZ81A	AZ92A	AZ92A	AZ92A	AZ92A	X	AZ92A	1. AZ92A 2. AZ101			
AZ91C	AZ92A	AZ92A	AZ92A	AZ92A	X	AZ92A	AZ92A	1. AZ92A 2. AZ101		
AZ92A	AZ92A	AZ92A	AZ92A	AZ92A	X	AZ92A	AZ92A	AZ92A	AZ101	
EK41A	AZ92A	AZ92A	AZ92A	AZ92A	X	AZ92A	AZ92A	AZ92A	AZ92A	EZ33A
EZ33A	AZ92A	AZ92A	AZ92A	AZ92A	X	AZ92A	AZ92A	AZ92A	AZ92A	EZ33A
HK31A	AZ92A	AZ92A	AZ92A	AZ92A	X	AZ92A	AZ92A	AZ92A	AZ92A	EZ33A
HM21A	AZ92A	AZ92A	AZ92A	AZ92A	X	AZ92A	AZ92A	AZ92A	AZ92A	EZ33A
HM31A	AZ92A	AZ92A	AZ92A	AZ92A	X	AZ92A	AZ92A	AZ92A	AZ92A	EZ33A
HZ32A	AZ92A	AZ92A	AZ92A	AZ92A	X	AZ92A	AZ92A	AZ92A	AZ92A	EZ33A
K1A	AZ92A	AZ92A	AZ92A	AZ92A	X	AZ92A	AZ92A	AZ92A	AZ92A	EZ33A
M1A MG1	AZ92A	1. AZ61A 2. AZ92A	1. AZ61A 2. AZ92A	1. AZ61A 2. AZ92A	X	1. AZ61A 2. AZ92A	AZ92A	AZ92A	AZ92A	AZ92A
ZE41A	0	0	0	0	X	0	0	0	0	EZ33A
ZX21A	AZ92A	1. AZ61A 2. AZ92A	1. AZ61A 2. AZ92A	1. AZ61A 2. AZ92A	X	1. AZ61A 2. AZ92A	AZ92A	AZ92A	AZ92A	AZ92A
ZH62A	X	X	X	X	X	X	X	X	X	X
ZK51A										
ZK60A										
ZK61A										

TABLE 10-24

Guide to the Choice of Filler Metal for Welding Magnesium (From reference 6.)

Note: 0, no data available for welding this combination; X, welding not recommended.

Source: Cary, Howard B., *Modern Welding Technology*, 4th Edition, © 1998. Reproduced by permission of Pearson Education, Inc., Upper Saddle River, New Jersey.

WELDING NICKEL ALLOYS

Nickel alloys are used for their corrosion resistance. Because of these corrosion-resistant properties, they are used extensively in the chemical and food industries, as is stainless steel. Corrosion resistance is a function of nickel's

EZ33A	HK31A	HM21A	HM31A	HZ32A	K1A	LA141A	M1A MG1	QE22A	ZE10A	ZE41A	ZX21A	ZH62A ZH51A ZK60A ZK61A
EZ33A												
EZ33A	EZ33A											
EZ33A	EZ33A	EZ33A										
EZ33A	EZ33A	EZ33A	EZ33A									
EZ33A	EZ33A	EZ33A	EZ33A	EZ33A								
EZ33A	EZ33A	EZ33A	EZ33A	EZ33A	EZ33A							
AZ92A	AZ92A	AZ92A	AZ92A	AZ92A	AZ92A	0	1. AZ61A 2. AZ92A					
EZ33A	EZ33A	EZ33A	EZ33A	EZ33A	EZ33A	0	0	EZ33A	0	EZ33A		
AZ92A	AZ92A	AZ92A	AZ92A	AZ92A	AZ92A	0	1. AZ61A 2. AZ92A	AZ92A	1. AZ61A	AZ92A	1. AZ61A 2. AZ92A	
X	X	X	X	X	X	X	X	X	X	X	X	EZ33A

surface oxide, which is caused by the base metal exposure to the atmosphere at an elevated temperature, such as base metal exposure during production of alloys. This dense surface oxide must be removed before welding, and the welding arc and electrode must be protected during welding from atmospheric contamination.

Common nickel alloys are produced by many commercial manufacturers as proprietary alloys. The terms Hastalloy, Inconel, Incoloy, and Monel are all trade names given to special nickel alloys. **TABLE 10-25** lists nominal chemical compositions of common high nickel alloys, both by each specific alloy and also its corresponding Unified Numbering System designation.

Alloy Designation	UNS Number	Nominal Composition (%) Ni	C	Mn	Fe	S	Si	Cu	Cr	Al	Ti	Cb[a]	Others
Nickel 200	N02200	99.5[b]	0.08	0.18	0.2	0.005	0.18	0.13	—	—	—	—	—
Nickel 201	N02201	99.5[b]	0.01	0.18	0.2	0.005	0.18	0.13	—	—	—	—	—
Nickel 205	N02205	99.5[b]	0.08	0.18	0.10	0.004	0.08	0.08	—	—	0.03	—	Mg 0.05
Nickel 211	—	95.0[b]	0.10	4.75	0.38	0.008	0.08	0.13	—	—	—	—	—
Nickel 220	N02220	99.5[b]	0.04	0.10	0.05	0.004	0.03	0.05	—	—	0.03	—	Mg 0.05
Nickel 230	N02230	99.5[b]	0.05	0.08	0.05	0.004	0.02	0.05	—	—	0.003	—	Mg 0.06
Nickel 270	N02270	99.98	0.01	<0.001	0.003	<0.001	<0.001	<0.001	<0.001	—	<0.001	—	Mg < 0.001, Co < 0.001
Duranickel alloy 301	—	96.5[b]	0.15	0.25	0.30	0.005	0.5	0.13	—	4.38	0.63	—	—
Permanickel alloy 300	—	98.5[b]	0.20	0.25	0.30	0.005	0.18	0.13	—	—	0.40	—	Mg 0.35
Monel alloy 400	N04400	66.5[b]	0.15	1.0	1.25	0.012	0.25	31.5	—	—	—	—	—
Monel alloy 401	—	42.5[b]	0.05	1.6	0.38	0.008	0.13	Bal.	—	—	—	—	—
Monel alloy 404	N04404	54.5[b]	0.08	0.05	0.25	0.012	0.05	44.0	—	0.03	—	—	—
Monel alloy R–405	N04405	66.5[b]	0.15	1.0	1.25	0.043	0.25	31.5	—	—	—	—	—
Monel alloy K–500	N05500	66.5[b]	0.13	0.75	1.00	0.005	0.25	29.5	—	2.73	0.60	—	—
Monel alloy 502	N05502	66.5[b]	0.05	0.75	1.00	0.005	0.25	28.0	—	3.00	0.25	—	—
Inconel alloy 600	N06600	76.0[b]	0.08	0.5	8.0	0.008	0.25	0.25	15.5	—	—	—	—
Inconel alloy 601	N06601	60.5	0.05	0.5	14.1	0.007	0.25	0.50	23.0	1.35	—	—	—
Inconel alloy 617	—	54.0	0.07	—	—	—	—	—	22.0	1.0	—	—	Co 12.5, Mo 9.0
Inconel alloy 625	N06625	61.0[b]	0.05	0.25	2.5	0.008	0.25	—	21.5	0.2	0.2	3.65	Mo 9.0
Inconel alloy 671	—	Bal.	0.05	—	—	—	—	—	48.0	—	0.35	—	—
Inconel alloy 702	N07702	79.5[b]	0.05	0.50	1.0	0.005	0.35	0.25	15.5	3.25	0.63	—	—
Inconel alloy 706	N09706	41.5	0.03	0.18	40.0	0.008	0.18	0.15	16.0	0.20	1.75	2.9	—
Inconel alloy 718	N07718	52.5	0.04	0.18	18.5	0.008	0.18	0.15	19.0	0.50	0.90	5.13	Mo 3.05
Inconel alloy 721	—	71.0[b]	0.04	2.25	6.5	0.005	0.08	0.10	16.0	—	3.05	—	—
Inconel alloy 722	N07722	75.0[b]	0.04	0.50	7.0	0.005	0.35	0.25	15.5	0.70	2.38	—	—
Inconel alloy X-750	N07750	73.0[b]	0.04	0.50	7.0	0.005	0.25	0.25	15.5	0.70	2.50	0.95	—
Inconel alloy 751	—	72.5[b]	0.05	0.5	7.0	0.005	0.25	0.25	15.5	1.20	2.30	0.95	—
Incoloy alloy 800	N08800	32.5	0.05	0.75	46.0	0.008	0.50	0.38	21.0	0.38	0.38	—	—
Incoloy alloy 801	N08801	32.0	0.05	0.75	44.5	0.008	0.50	0.25	20.5	—	1.13	—	—
Incoloy alloy 802	—	32.5	0.35	0.75	46.0	0.008	0.38	—	21.0	0.58	0.75	—	—
Incoloy alloy 804	—	41.0	0.05	0.75	25.4	0.008	0.38	0.25	29.5	0.30	0.60	—	—
Incoloy alloy 825	N08825	42.0	0.03	0.50	30.0	0.015	0.25	2.25	21.5	0.10	0.90	—	Mo 3.0
Ni–span–C alloy 902	N09902	42.25	0.03	0.40	48.5	0.02	0.50	0.05	5.33	0.55	2.58	—	—

TABLE 10-25

Composition of Nickels and Nickel Alloys

[a]Cobalt included.
[b]Not for specification purposes.

Source: Cary, Howard B., *Modern Welding Technology*, 4th ed., © 1998. Reproduced by permission of Pearson Education, Inc., Upper Saddle River, New Jersey.

WHAT SPECIAL CONSIDERATIONS ARE MADE FOR WELDING NICKEL ALLOYS?

High-nickel alloys tend to have a weld bead that is described as sluggish, meaning that the weld pool does not wet out as well as carbon steel welds. Another characteristic is that weld penetration is also reduced. In order to overcome these issues, a high degree of welder skill is required. The welder must use techniques that ensure the welding arc impinges on the leading edge of the weld pool to improve penetration. Otherwise, the weld metal is likely to roll over the cold base metal, reducing both penetration and fusion. Furthermore, the electrode must be manipulated to ensure fusion at the edges of the weld bead.

In order to increase penetration and access to the weld root and groove faces, weld joint geometry makes use of greater groove angles and root openings. Groove angles and root openings are shown in **FIGURE 10-9**. A Groove face is shown in **FIGURE 10-10**.

HOW ARE FILLER METALS SELECTED FOR SPECIFIC NICKEL ALLOYS?

Each manufacturer of these nickel alloys provides specifications and guidelines for electrode selection for specific alloys and makes recommendations for

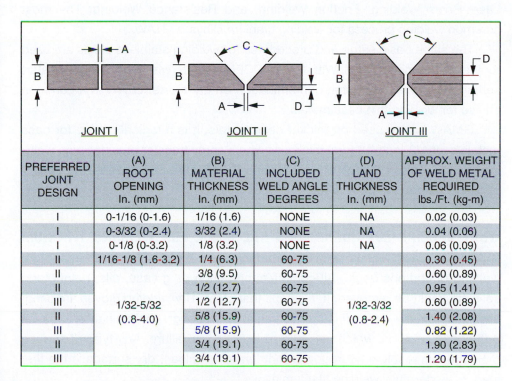

PREFERRED JOINT DESIGN	(A) ROOT OPENING In. (mm)	(B) MATERIAL THICKNESS In. (mm)	(C) INCLUDED WELD ANGLE DEGREES	(D) LAND THICKNESS In. (mm)	APPROX. WEIGHT OF WELD METAL REQUIRED lbs./Ft. (kg-m)
I	0-1/16 (0-1.6)	1/16 (1.6)	NONE	NA	0.02 (0.03)
I	0-3/32 (0-2.4)	3/32 (2.4)	NONE	NA	0.04 (0.06)
I	0-1/8 (0-3.2)	1/8 (3.2)	NONE	NA	0.06 (0.09)
II	1/16-1/8 (1.6-3.2)	1/4 (6.3)	60-75		0.30 (0.45)
II		3/8 (9.5)	60-75		0.60 (0.89)
II		1/2 (12.7)	60-75		0.95 (1.41)
III	1/32-5/32 (0.8-4.0)	1/2 (12.7)	60-75	1/32-3/32 (0.8-2.4)	0.60 (0.89)
II		5/8 (15.9)	60-75		1.40 (2.08)
III		5/8 (15.9)	60-75		0.82 (1.22)
II		3/4 (19.1)	60-75		1.90 (2.83)
III		3/4 (19.1)	60-75		1.20 (1.79)

FIGURE 10-9 Joint design for nickel alloys
Source: Courtesy of Haynes International.

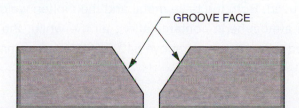

GROOVE FACE

FIGURE 10-10 Groove face

welding them. Haynes International, a leading manufacturer of high-performance heat- and corrosion-resistant nickel and base alloys, suggest two possible methods of filler metal selection. Filler metals can be selected with properties matching those of the base metal or an overalloyed filler metal may be used. Matching is easy if one base metal is being used; simply select an electrode of the same composition. When dissimilar base metals are joined together, a filler metal is preferred that matches the alloy with the higher alloying and corrosion properties. The advantage of using an overalloyed filler metal provides two benefits. The first benefit is the reduction of preferential weld metal corrosion attack. The other benefit is that of using a single filler metal on the jobsite.

WELDING TITANIUM ALLOYS

WHAT WELDING PROCESSES ARE USED FOR WELDING TITANIUM ALLOYS?
Welding processes used for welding titanium alloys are those that do not use fluxes because fluxes react with the base metal, causing brittleness that can lead to cracking and may also reduce corrosion resistance. The welding processes used may include GTAW, GMAW, PAW, Electron-Beam Welding, Laser-Beam Welding, Friction Welding, and Resistance Welding. The most common welding process for welding titanium alloys is GTAW.

The same basic setup and procedures for welding stainless steels are used for welding titanium alloys. With GTAW, DCEN is used, with high-frequency start settings to prevent tungsten contamination on arc starts. GTAW is a common choice for welding thin base metals.

GMAW is also used on thicker base metals. It is a logical choice for base metals over 1/2-inch (13 mm) thickness where spray transfer provides high welding speeds. Pulsed GMAW offers a balance of high welding speeds and control of overall heat to the base metal.

WHAT SPECIAL CONSIDERATIONS ARE MADE FOR WELDING TITANIUM ALLOYS?
Base metals and filler metals must be absolutely clean before welding operations begin. All contaminants, such as grease, oils, paints, and solvents, must be removed. It is critical that clean, dry welding gloves are used; even oils from a welder's fingerprints can supply enough contamination to cause brittleness in welds, which can lead to rejection or failure. Any type of surface moisture can result in hydrogen embrittlement. A clean, dry surface is the first step to successfully welding titanium alloys.

Proper shielding is provided by an inert shielding gas. Argon is the gas of choice. On heavy weld assemblies, a mixture of argon with additions of helium may be used. Both the base metal and the molten weld pool must be protected from atmospheric contamination; all the while, the weld or base metal is exposed to temperatures above 840°F (449°C). A common practice is to complete the welding operations with the workpiece enclosed in a container filled with a protective argon gas shield. This method provides the best results because all surfaces of the weld assembly and the weld itself are separated from any atmospheric contamination until the part has cooled sufficiently.

It is not a safe practice to protect only the weld side of the weld assembly because if the back side of the base metal reaches temperature ranges above 840°F (449°C), contamination will occur. Oxidation cracking can occur at temperatures above 1,200°F (649°C). This can be caused by carbon, nitrogen, or oxygen pick-up at elevated temperatures due to poor base metal cleaning or exposure to air.

If welding cannot be done inside an argon filled chamber, other special methods must be applied to provide adequate protection. In this case, purging the back side of the weld joint with argon gas is desired. A backing assembly made up of a channel-shaped extrusion can be designed to contain an argon shielding gas purge along the back side of the weld joint.

On the weld side of the joint, several common methods for protecting the weld pool exist. The first is the use of a gas lens when GTAW is the welding process used. A gas lens provides more streamlined and better gas coverage than a conventional GTAW gas diffuser. The second means for providing adequate protection is the use of a trailing shield. A trailing shield (argon) is a second shielding gas supplied by a hose attached to a trailing shielding cover directly behind the gas nozzle following right behind the weld pool.

HOW ARE FILLER METALS SELECTED FOR SPECIFIC TITANIUM ALLOYS?

Filler metals for welding titanium are manufactured in accordance with AWS filler metal specification A5.16-70, Titanium and Titanium-Alloy Bare Welding Rods and Electrodes. While the common practice is to select a filler metal that matches the base metal properties and composition, some occasions warrant use of a filler metal with a strength level just below that of the base metal to improve ductility. Base metal dilution to the weld pool will improve weld metal strength. **FIGURE 10-11** describes the classification system for titanium filler metals by using the example ERTi-6ELI titanium filler metal alloy.

ER	=	electrode rod
Ti	=	unalloyed titanium or base alloy
2	=	alloy type follows ASTM/ASME specifications
ELI	=	extra-low interstitial content (i.e., carbon, hydrogen, nitrogen, and oxygen)

FIGURE 10-11 Classification system for titanium filler metals

CHAPTER QUESTIONS/ASSIGNMENTS

MULTIPLE CHOICE

1. Stress relief heat treatment is considered to be which of the following?
   - a. Preheat
   - b. Interpass heat
   - c. Postheating

2. Which steel is welded in the annealed condition?
   - a. Chromium-molybdenum
   - b. Quench and tempered
   - c. High-strength low-alloy
   - d. Heat-treatable low-alloy

3. Which stainless steels are prone to carbide precipitation?
   - a. Austenitic
   - b. Ferritic
   - c. Martensitic

4. Which stainless steels require preheat?
   - a. Austenitic
   - b. Ferritic
   - c. Martensitic

5. How are surface oxides avoided in the storage of aluminum alloys?
   - a. Place sheets in a stack to prevent moisture from getting between them.
   - b. Place sheets vertically and leave a small space between sheets to prevent trapping moisture between them.
   - c. Place sheets in an oven to keep moisture out.

6. Which filler metal provides the best color match for welding the 2036 alloy?
   - a. 2319
   - b. 4043
   - c. 4047
   - d. 4145

FILL IN THE BLANK

7. Unlike hot cracking in _____ welds, which are visible upon completion of the weld, cracks in steel weldments often form over a _____ time period and may not even be visible on initial completion of the weld.

8. Hydrogen cracking is a form of _____ cracking in steels caused by_____ of hydrogen in the base metal.

9. Filler metals for chromiummolybdenum alloys should match those of the base metal with the exception that the _____ content may be slightly lower.

10. The final step to putting a welded chromium-molybdenum alloy into service is _____ heat treatment.

11. Carbide precipitation in stainless steel welds results in something called _____ corrosion.

SHORT ANSWER

12. What are some of the possible negative effects of welding on a weldment?

13. What factors contribute to cracking in carbon steel welds?

14. Which carbon steels exhibit fair to poor weldability characteristics?

15. What types of procedures are used for welding steels with high carbon equivalence?

16. What conditions contribute to hydrogen entrapment in steel welds?

17. What steps can be used to eliminate cracking in carbon and alloy steel welds.

18. What steps are taken to eliminate carbide precipitation?

19. What techniques are recommended for welding tool steels?

20. How are cast irons prepared for welding?

21. What difficulty is encountered when welding aluminum alloys?

Chapter 11
WELDING SYMBOLS

Placing welding symbols on prints, shop drawings, and welding procedures communicates precise information to the welder. If the information communicated on the welding symbol is complete, the welder knows precisely how the joint is to be designed and can perform the task as intended.

To ensure continuity from one business to another, the American Welding Society (AWS) has established guidelines for the size, shape, location, and meaning of weld symbols. For example, the angle of the other fillet weld symbol is to be drawn at a 45° angle, as shown in **FIGURE 11-1**, and not at the angles or backward, as shown in **FIGURE 11-2**. When drawn correctly, there is no doubt what symbol is being communicated.

A **welding symbol** communicates what is to be welded in a single graphic picture that may otherwise involve a paragraph of written instructions to communicate the same requirement. Welding symbols have a pattern that, once understood, enables the welder to correctly identify size, shape, length, etc., from one symbol to the next. Knowing, understanding, and producing welds in accordance to a welding symbol is as important as producing a good weld.

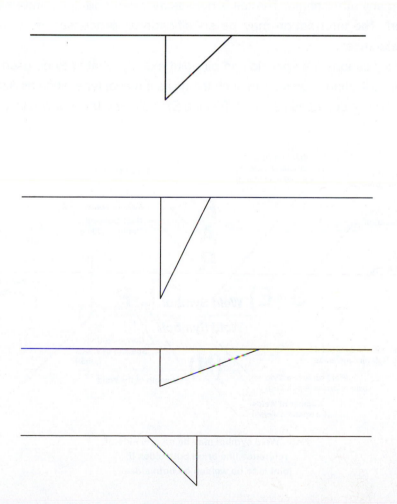

FIGURE 11-1 Correct geometry of a fillet weld symbol

FIGURE 11-2 Incorrect geometry of a fillet weld symbol

WHAT ARE THE BASIC PARTS OF A WELDING SYMBOL?   A welding symbol is shown in **FIGURE 11-3**. The basic parts of a welding symbol are the **tail**, the **reference line**, and the **arrow**. The arrow points to the joint to be welded; welds go either on the same side the arrow is pointing to or the other side of the joint. The reference line is where all the pertinent information is placed about the weld to be made. The tail may or may not be necessary depending on if any additional information required is absent from the reference line.

HOW IS A WELDING SYMBOL READ?   A welding symbol is always read the same way: from left to right. It does not matter if the arrow is pointing up or down, to the right or left, angled forward or angled backward, or even in multiple directions (see **FIGURE 11-4**). The bottom side of the reference line always indicates the **arrow side**, meaning that wherever the arrow is pointing to is where the weld is to be deposited. The information on the top of the reference line is called the **other side**; it is the opposite side of the joint.

The arrow always points at an angle—up or down or left or right. The arrow should never point straight across (horizontal) and should never point straight up and down (vertical). The reference line always lies horizontally and never at any other angle. The tail is necessary when additional information is required. The information may be specifications, processes, or notes and other references.

Specifications are specific and pinpoint exactly what is to be used on the weld. Specifications can refer to such things as material type, such as A36 steel, or specifying a consumable insert. Placing E7018 in the tail of a welding symbol

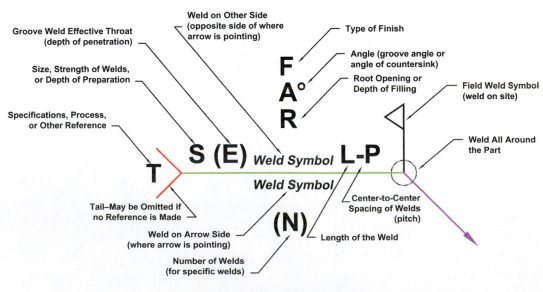

Note: Weld symbol may be on one side of
reference line or on both sides if
joint is to be welded on both sides.

FIGURE 11-3 An AWS welding symbol

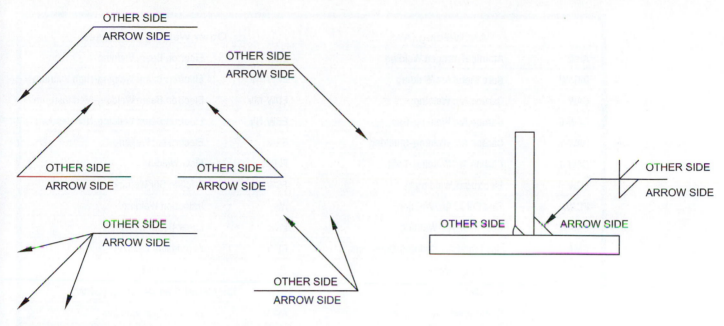

FIGURE 11-4 Weld and weld symbol location

specifies the classification of electrode to use in welding. Specifying a welding procedure provides the welder with other additional welding information, such as weld parameters, shielding gas requirements, and preheat/postheat requirements.

Process tells the welder what welding, cutting, or nondestructive weld inspection process is to be used on the weld joint. Many of the welding processes placed in the welding symbol tail are found in **TABLE 11-1**.

Notes and other references may be other designations related to the welding field. An example is the letter designations for applying welding processes (see **TABLE 11-2**). There may be a reference to "Refer to Detail A," for example, to call attention to the orientation of a slot or seam weld. A note may point out anything. A note may indicate which side of a weld joint is welded first: arrow side or other side. A TYP in the tail is an abbreviation for "typical" and indicates all joints of similar design have the same weld information. Another example is when size is not specified but the welder is to produce a weld size according to company standards (CO STD).

WELD SYMBOLS

WHAT SYMBOLS ARE PLACED ON A WELDING SYMBOL? A welding symbol, previously defined, differs from a **weld symbol**. A weld symbol refers to a particular type of weld (i.e., fillet weld, spot weld, slot weld, etc.). The weld symbols placed on the reference line are shown in **FIGURE 11-5**.

A weld symbol is a fairly accurate representation of the joint or the weld it depicts. **FIGURE 11-6** illustrates the position of a weld symbol in relationship to the weld joint and the weld bead. **FIGURE 11-7** shows various single groove welds, double groove welds, and the edge joint.

Arc Welding (AW)		Other Welding Processes	
AHW	Atomic Hydrogen Welding	EBW	Electron Beam Welding
BMAW	Bare Metal Arc Welding	EBW-HV	Electron Beam Welding-High Vacuum
CAW	Carbon Arc Welding	EBW-MV	Electron Beam Welding-Med Vacuum
CAW-G	Carbon Arc Welding-Gas	EBW-NV	Electron Beam Welding-Non Vacuum
CAW-S	Carbon Arc Welding-Shielded	ESW	Electroslag Welding
CAW-T	Carbon Arc Welding-Twin	FLOW	Flow Welding
EGW	Electrogas Welding	FSW	Friction Stir Welding
FCAW	Flux Cored Arc Welding	IW	Induction Welding
GMAW	Gas Metal Arc Welding	LWB	Laser Beam Welding
GMAW-C	Gas Metal Arc Welding-Composite	PEW	Percussion Welding
GMAW-P	Gas Metal Arc Welding-Pulsed	TW	Thermit Welding
GMAW-S	Gas Metal Arc Welding-Short Circuit	**Oxy-Fuel Gas Welding (OFW)**	
GTAW	Gas Tungsten Arc Welding	AWW	Air Acetylene Welding
GTAW-P	Gas Tungsten Arc Welding-Pulsed	OAW	Oxy-Acetylene Welding
PAW	Plasma Arc Welding	OHW	Oxy-Hydrogen Welding
SMAW	Shielded Metal Arc Welding	PGW	Pressure Gas Welding
SW	Stud Welding	**Brazing (B)**	
SAW	Submerged Arc Welding	BB	Block Brazing
SAW-S	Submerged Arc Welding-Series	DFB	Diffusion Brazing
Solid State Welding (SSW)		DP	Dip Brazing
CEW	Co-Extrusion Welding	EXB	Exothermic Brazing
CW	Cold Welding	FLB	Flow Brazing
DFW	Diffusion Welding	FB	Furnace Brazing
EXW	Explosion Welding	IB	Induction Brazing
FOW	Forge Welding	IRB	Infrared Brazing
FRW	Friction Welding	RB	Resistance Brazing
HPW	Hot Pressure Welding	TB	Torch Brazing
ROW	Roll Welding	TCAB	Twin Carbon Arc Brazing
USW	Ultrasonic Welding	**Soldering (S)**	
Resistance Welding (RW)		DS	Dip Soldering
FW	Flash Welding	FS	Furnace Soldering
PW	Projection Welding	IS	Inductions Soldering
RSEW	Resistance Seam Welding	IRS	Infrared Soldering
RSEW-HF	Resistance Seam Welding-High Freq.	INS	Iron Soldering
RSEW-I	Resistance Seam Welding-Induction	RS	Resistance Soldering
RSW	Resistance Spot Welding	TS	Torch Soldering
UW	Upset Welding	USS	Ultrasonic Soldering
UW-HF	Upset Welding-High Frequency	WS	Wave Soldering
UW-I	Upset Welding-Induction		

TABLE 11-1

Welding and Brazing Processes

Source: Reproduced from AWSA3.0:2001, "Standard Welding Terms and Definitions," with permission from the American Welding Society (AWS), Miami, Florida.

Type of Process	
AU	Automatic
MA	Manual
ME	Machine
SA	Semiautomatic

TABLE 11-2

Weld & Supplementary Symbols

	Fillet	Plug/Slot	Spot[1]	Seam[1]	Stud	Projection[2]	Edge	Surfacing
Arrow Side								
Other Side								
Both Sides								

Grooves	Square	V	Bevel	U	J	Flare-V	Flare-Bevel	Scarf[3]
Arrow Side								
Other Side								
Both Sides								

	Backing Spacer	Consumable Insert	Melt-Thru	Back or Backing Weld	Weld All Around	Field Weld	Convex Concave	Flush

[1] When the symbol is astride of the reference line, the meaning is "No Side Significant."

[2] The Projection weld symbol is *ALWAYS* accompanied by a "PW" in the tail.

[3] Scarf joints are used only when brazing.

FIGURE 11-5 Weld symbols employed on a welding symbol.

Source: AWS A2.4:2007. Figure reproduced and adapted with permission from the American Welding Society (AWS), Miami, Florida.

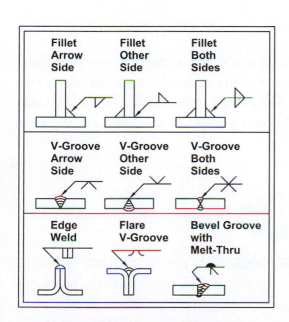

FIGURE 11-6 Typical applications of welding symbols

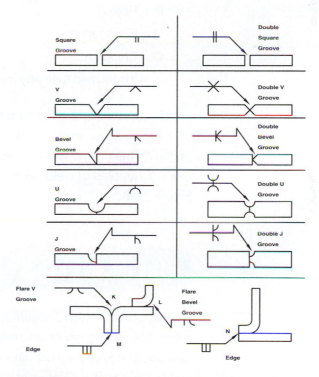

FIGURE 11-7 Applications of groove weld symbols

FINISHING METHODS AND CONTOURS Various contours and methods of finishing a weld shape may be added. The contour may be flush, concave, or convex (see **FIGURE 11-8**). The contour symbols are indicated by use of a straight or curved line. The flush or flat contour uses a straight line, while the concave and convex symbols use curved lines that cave in toward the symbol or curve out and away

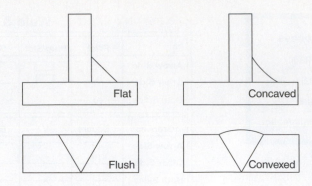

FIGURE 11-8 Weld profiles

from it, respectively. The finishing method indicates how to achieve the contour and is designated with a capital letter.

\ = Flat or Flush Contour

⌣ = Concave Contour

⌢ = Convex Contour

G = Grinding

M = Machining

R = Rolling

H = Hammering

C = Chipping

U = Unspecified (use any method to accomplish the desired contour)

There is a slight difference in meaning of flush and flat contours. Flush generally means the weld is finished flush to the surface of the base metal. Flat generally means the weld is finished in a straight plane from toe to toe of a weldment. Examples of weld contours for both groove and fillet welds are shown in Figure 11-8.

FILLET WELD SYMBOL

A weld symbol often encountered is the fillet weld symbol. **FIGURE 11-9** illustrates the wide array of information that may be used along with the fillet weld symbol. The fillet weld—T- and lap joints—are arguably the most common joints welded in industry. Fillet weld symbols are the easiest to read and form the foundation to learning other weld symbols.

The welding symbol shown in **FIGURE 11-10** indicates a fillet weld to be made on the arrow side and to be welded all around a channel iron. There is no other information given. When a size of the weld is not given, the company standard (also referred to as a user's standard or shop standard) is generally employed. This means whatever standard practice used at the place of employment is the size of weld bead to be made.

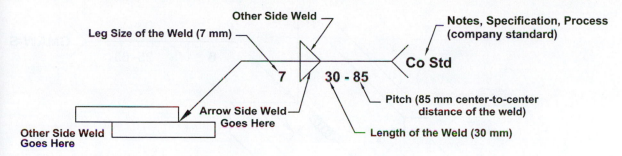

FIGURE 11-9 The fillet welding symbol

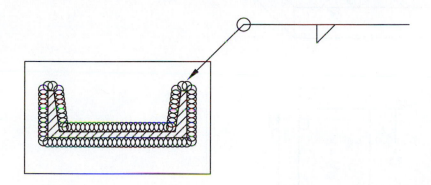

FIGURE 11-10 Application of an all-around fillet weld

The welding symbol shown in **FIGURE 11-11** indicates a fillet weld to be made on the other side. The fillet weld is to be made with 1/2-inch leg sizes. The fillet weld is to be ground (G) flush. The line below the G is the flush symbol. Leg sizes may be equal, as shown in Figure 11-11, or unequal. An unequal leg size would be indicated, for example, as (1/4 x 1/2). The parentheses are always wrapped around uneven leg sizes. A See Detail is indicated in the tail, and an accompanying drawing detailing which leg is larger or smaller is required. When no length is given on the welding symbol, as in Figure 11-11, the length of the weld is continuous, meaning it runs the full distance of the part to be welded.

The welding symbol in **FIGURE 11-12** indicates an intermittent 6 mm fillet weld made on the arrow side, 25 mm long, and is spaced 63 mm from center to center (**pitch**). **Intermittent** means a short weld is made, with space between a series of short welds. Pitch refers to the center-to-center distance between intermittent welds. When given, the length of the weld is always placed immediately to the right of the weld symbol, and pitch follows after a dash. The welding process to be used—GMAW in the short-circuit transfer mode—is given in the tail of the symbol.

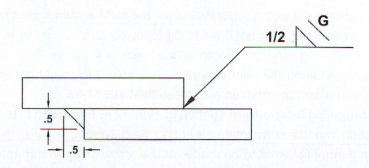

FIGURE 11-11 Fillet weld with bead contour identified

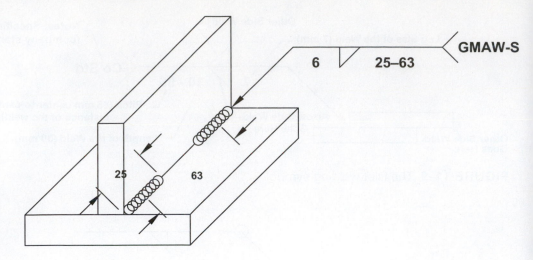

FIGURE 11-12 Fillet weld with size, length, and pitch

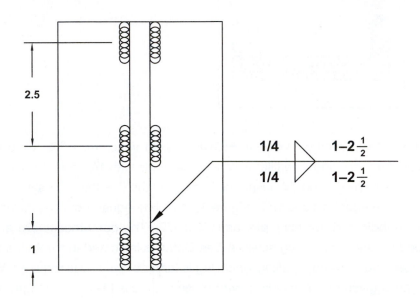

FIGURE 11-13 Chain intermittent fillet weld

This top view of **FIGURE 11-13** is a T-joint showing a chain intermittent fillet weld. The welding symbol indicates a 1/4-inch fillet weld is to be made on both the arrow side and the other side that is 1-inch long with a 2 1/2-inch pitch. **Chain intermittent** refers to the placing of intermittent fillet welds directly opposite each other on the T-joint.

FIGURE 11-14 displays a slight variation from Figure 11-13. The welding symbol shown in Figure 11-14 indicates a 6 mm fillet weld to be made on the arrow side, 25 mm long, with a 65 mm pitch and convex contour. On the other side, a 6 mm fillet weld is continuously welded throughout the entire length with a concave contour. If a length is not indicated, the weld is always continuous.

A length of a weld does not have to be followed by a pitch. For example, a part is 6 inches long, and the length of the weld is 4 inches long. A pitch is impossible. A detail on the print or a note indicates if the weld is centered or located a given dimension from an edge. See **FIGURE 11-15**.

The **staggered intermittent** fillet weld symbol in **FIGURE 11-16** shows a slight variation from the chain intermittent fillet weld symbol. The welding symbol indicates a 6 mm fillet weld to be made on the arrow side, 25 mm long, and a 65 mm pitch. On the other side, a 6 mm fillet weld is 25 mm long with a 65 mm

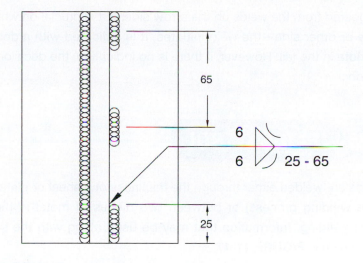

FIGURE 11-14 Continuous and intermittent fillet welds

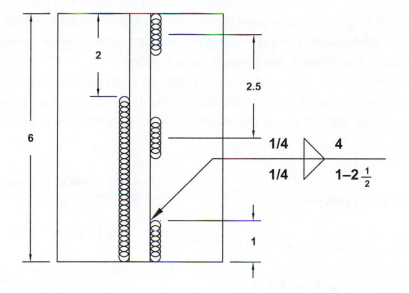

FIGURE 11-15 When a length of weld is specified and the part is longer than the weld length, the print must identify where the weld must start or end.

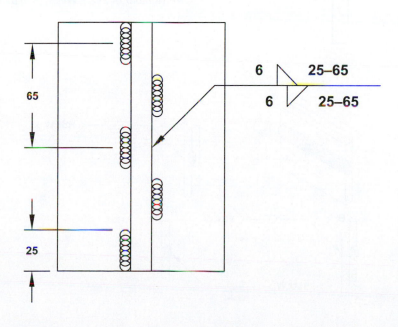

FIGURE 11-16 Staggered intermittent fillet weld

pitch, staggered from the welds on the arrow side. If it is crucial on which side—arrow side or other side—the weld initiates, it is indicated with a detail on the print or a note in the tail. However, if there is no indication, the decision is left up to the welder.

SEAM WELD SYMBOL

Seam welds are welded either through the thickness of a sheet or plate (depending on the welding process) or between two pieces of metal in the case of Resistance Welding. Information that may be used along with the seam weld symbol is shown in **FIGURE 11-17**.

Consider the drawing and the seam weld symbols in **FIGURE 11-18**. The side view A is an example of a resistance seam weld. Note that the weld symbol is astride the reference line, indicating no side significant. The welds produced by either B or C could have been made with the GTAW, GMAW, FCAW, EBW, SMAW, or SAW process or any welding process producing enough voltage and amperage to melt through one piece of metal into another.

The orientation of the seam weld is indicated on the print with centerlines, in a detail, or in a note in the welding symbol tail. In the case of centerlines, the seam weld is welded along the length of the centerline. In Figure 11-18,

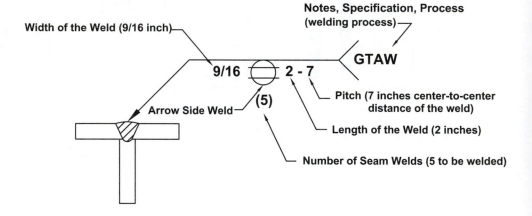

FIGURE 11-17 The seam welding symbol

FIGURE 11-18 Seam welds may be arrow side, other side, or no side significant.

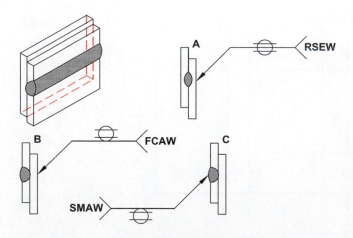

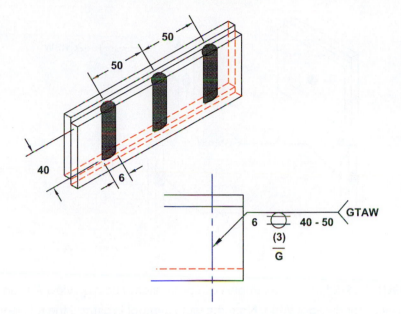

FIGURE 11-19 **FIGURE 11-19** Because a seam weld may run vertically or horizontally, the orientation of the weld must be indicated on the print.

the weld runs parallel with the lap joint, and in **FIGURE 11-19**, the weld runs perpendicular.

The example in Figure 11-19 indicates a 6 mm wide intermittent seam weld is to be made on the arrow side, 40 mm long, and spaced 50 mm from center to center (pitch). There are three seam welds to be made. The weld is to be ground flush after welding. The welding process used is GTAW.

On seam welds, the strength of the weld may be substituted for size in either pounds per square inch (U.S. standard) or newtons (metric). The symbol in Figure 11-19, for example, may be expressed as 2,000 (for metric) rather than the 6 mm as shown. Pounds is a larger unit of measurement, and the strength is usually expressed as a three-digit number, such as 400, 500, etc.

SPOT WELD SYMBOL

Spot welds are similar to seams welds except they are a single round nugget as opposed to a weld with a length. **FIGURE 11-20** exemplifies the wide range of information that may be used in conjunction with the spot weld symbol.

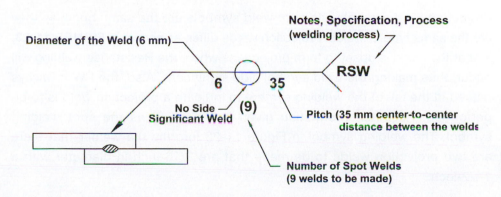

FIGURE 11-20 The spot welding symbol

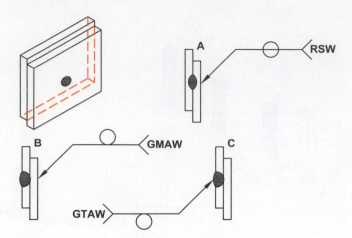

FIGURE 11-21 Spot welds may be arrow side, other side, or no side significant.

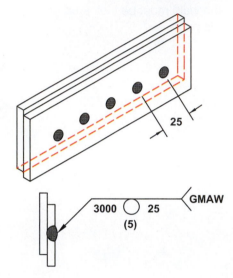

FIGURE 11-22 Strength, pitch, and the number of spot welds required

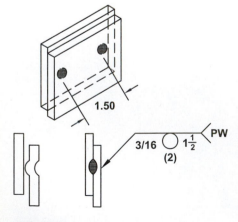

FIGURE 11-23 Projection weld with arrow side significance with size, pitch, and the number of welds required.

FIGURE 11-21 shows examples of spot welds. The side view A is an example of a Resistance Spot Weld. Note the weld symbol is astride the reference line, indicating no side significant. The welds produced by either B or C could be made by the GTAW, GMAW, FCAW, EBW, SMAW, or SAW processes. Any welding process producing enough voltage and amperage to melt through one piece of metal into another may be used to produce a spot weld.

Just as in seam welds, it is not obvious where the spot weld is to be welded, and a centerline is drawn on the print to indicate location. The spot weld is welded on the centerline location. Since a spot weld is a circular nugget, orientation (parallel or perpendicular), is not a consideration.

The welding symbol shown in FIGURE 11-22 indicates a spot weld is to be made on the arrow side with 3,000 newtons of shear strength and a pitch of 25 mm. The (5) indicates there are five spot welds to be made. The welding process to be used is GMAW. The symbol does not—but could—include a contour and/or method of finishing.

The spot weld symbol may or may not include the number of welds to be made, in parentheses, on the symbol. Because the weld has no length, only pitch is placed on the symbol. Diameter, for weld size, may be substituted for strength of the weld.

PROJECTION WELD SYMBOL

Projection weld symbols and spot weld symbols are the same because they are the same type of welds. Projection welds differ, as shown in FIGURE 11-23, in that the metal is upset to form projections where the Resistance welding will occur, thus making the weld symbol side significant. Also, the PW is *always* placed in the tail of the welding symbol to indicate a projection weld is to be performed. All other information given is the same as the spot welding symbol. The welding symbol in Figure 11-23 informs the welder that there are two projection welds to be made that are 3/16 inch in diameter with a 1 1/2-inch.

STUD WELD SYMBOL

Stud welds may be created by using a Resistance Welding process or Arc Welding. The same information found on a spot welding symbol is also contained in the stud welding symbol. Information that may be used with the stud weld symbol is shown in **FIGURE 11-24**. The symbol is always shown on the arrow side. The welding symbol in **FIGURE 11-25** indicates the plate requires a single 12 mm diameter-threaded stud.

PLUG WELD SYMBOL

Plug welds differ from spot welds in that a hole is drilled or punched through one member and then welded. **FIGURE 11-26** illustrates the wide array of information that may be used along with the plug weld symbol. The welding symbol

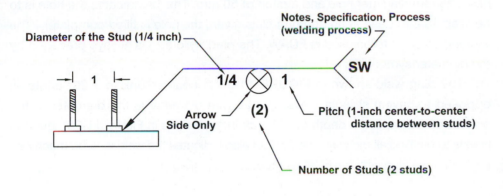

FIGURE 11-24 The Stud Welding symbol

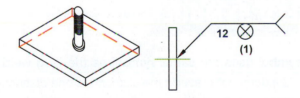

FIGURE 11-25 Stud weld showing size and number of studs required

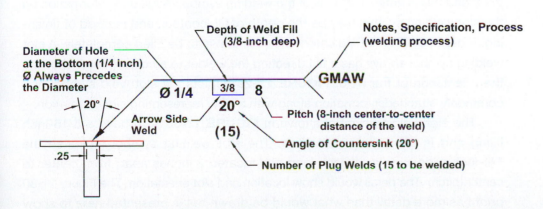

FIGURE 11-26 The plug welding symbol

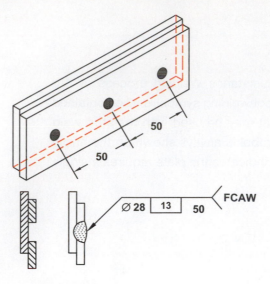

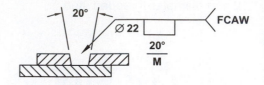

FIGURE 11-27 Plug weld with size, depth of fill, and pitch.

FIGURE 11-28 Plug welds may be countersunk to aid the welder in obtaining root penetration.

shown in **FIGURE 11-27** indicates that a plug weld to be made on the arrow side has a 28 mm diameter hole and a pitch of 50 mm. The 13 indicates the hole is to be filled 13 mm deep. If no depth of fill is given, the hole is filled completely. The welding process to be used is FCAW. The plug weld size is always preceded by the diameter symbol (Ø).

The plug weld shown in **FIGURE 11-28** is countersunk to a 20° angle as opposed to having straight walls. The 22 mm diameter is the diameter at the *bottom* of the plate. If a depth of fill is not indicated, as in Figure 11-28, then the hole is to be filled all the way. The symbol also indicates the weld is to be machined flush after welding with the FCAW process.

SLOT WELD SYMBOL

The slot weld symbol uses the same symbol as the plug weld. However, the diameter symbol (Ø) does *not* precede the size and is one distinction when telling them apart. Orientation of the slot weld is illustrated in a detailed drawing on the print, and this is noted in the tail of the welding symbol. Additional information on the slot welding symbol may be the depth of fill, contour, and method of finishing. If depth of fill is not indicated, then the slot is to be filled completely. A slot welding symbol *always* has a tail directing the welder to a note or detail showing the orientation of the weld. **FIGURE 11-29** depicts the slot weld symbol and commonly attached information along with a view representing slot orientation.

The slot welding symbol shown in **FIGURE 11-30** is to have a 5/16-inch filling and is then hammered flush. The slot welding symbol indicates the 1/4-inch-wide slots are 1 inch long and located 2 inches apart from center to center (pitch). The detail would show location and slot orientation. The Figure 11-30 print has more detail than what would be drawn but is presented here to show clarity to the reader.

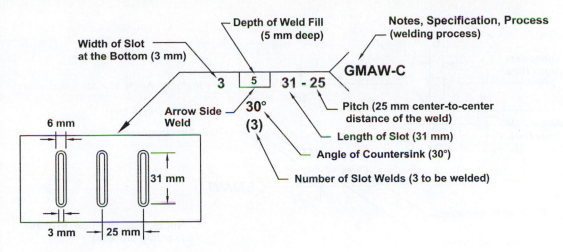

FIGURE 11-29 The slot welding symbol

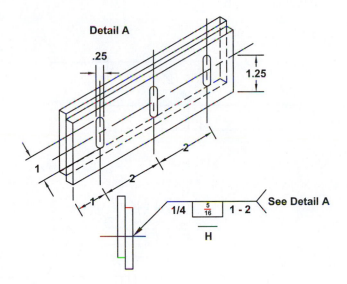

FIGURE 11-30 Because a slot weld may run vertically or horizontally, the orientation of the weld must be indicated on the print.

GROOVE WELD SYMBOLS

Groove welds are welds suitable for use on butt joints, where base metals to be joined butt up against one another on the same plane. They are also used on T-joints and corner joints when complete joint penetration is required. The common grove weld symbols are provided in **FIGURE 11-31**.

- **Square Groove Weld Symbol:** All groove welds follow the same format. The basic difference is the design of the joint. The square groove welding symbol shown in **FIGURE 11-32** has a 3 mm root opening on the arrow side. On the other side, a melt-through symbol is shown. The melt-through symbol is filled black. This means the welder should have the parameters set to melt through to the other side of the joint.
- **Single V-Groove and Double V-Groove Weld Symbols**: The V-groove welding symbol shown in **FIGURE 11-33** on the left is beveled at a 45° angle with a flush contour and has a 1/4-inch root opening. On the arrow side, a

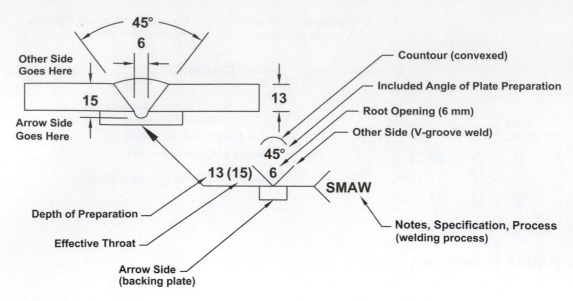

FIGURE 11-31 The groove welding symbol

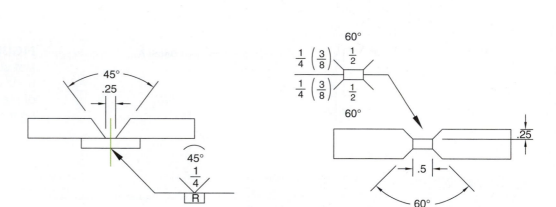

FIGURE 11-32 A square groove weld with root opening and weld penetration indicated.

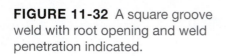

FIGURE 11-33 A backing may be on the opposite side of a weld or between groove welds in the form of a spacer if beveled on both sides.

backing plate symbol is shown; the R indicates a removal of the backing plate after welding.

The double V-groove welding symbol shown in Figure 11-33 on the right is beveled at a 60° angle and has a 1/2-inch root opening on both sides. A **spacer** is shown in the middle of the joint; the spacer acts as a backing plate for both sides and is made of the same material as the base metal. The depth of preparation is 1/4 inches for both sides, and the **effective throat** (depth of penetration) is 3/8 inches. This means there is complete penetration of the weld into the spacer and base metal.

■ **Single Bevel-Groove and Double Bevel-Groove Weld Symbols**: The single bevel-groove welding symbol shown in **FIGURE 11-34** on the left is

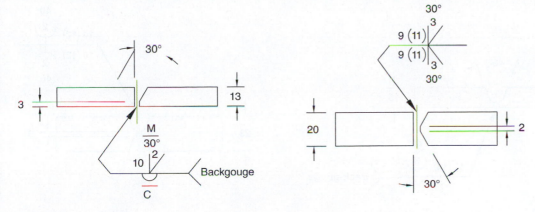

FIGURE 11-34 Bevel-groove welds have a broken arrow to indicate which side of the plate receives the bevel, as shown on these bevel groove welds.

beveled on the other side to a depth of 10 mm at a 30° angle; it has a 2 mm root opening, and the weld is machined flush. Note the broken arrow to indicate which side is to be beveled. It is then back-gouged and welded on the arrow side by any suitable means (i.e., oxy-fuel torch, plasma arc, etc.). The other side weld symbol is known as a **back weld** because it is a weld put on the root side of the joint *after* the groove weld is made. The back weld is then chipped flush.

The double beveled-groove welding symbol shown in Figure 11-34 on the right is beveled to a depth of 9 mm at a 30° angle on both sides; it has a 3 mm root opening. Again, the broken arrow indicates the side to be beveled. It is then welded with an effective throat (depth of penetration) to 11 mm on both sides to ensure 100% penetration.

- **Combination Weld Symbols**: Bevel-groove welds lend themselves to combination welds. The single bevel-groove welding symbol shown in **FIGURE 11-35** on the left is first beveled on the arrow side to a depth of 1/2 inches at a 30° angle. It is then followed up with a 3/8-inch fillet weld. The welding symbol is always read from the reference line outward. The order is important when producing the weld. When no effective throat (depth of penetration) is given, the joint is to be welded all the way (complete joint penetration).

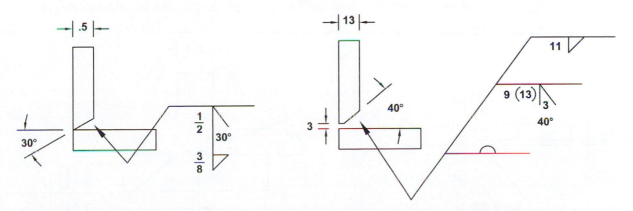

FIGURE 11-35 Combination welds may be shown on a single reference line or stacked on multiple reference lines. When a multiple reference line is used, the symbol closest to the arrow is the first thing to be done, the second closest is the second thing to be done, and so on.

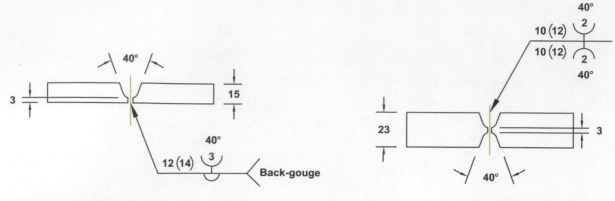

FIGURE 11-36 U-groove welding symbols

The welding symbol shown in Figure 11-35 on the right is first welded with a **backing weld** on the other side. It is called a backing weld because it is welded on the root side of the joint prior to welding the groove weld. It is then followed by a single bevel-groove weld on the arrow side prepared to a depth of 9 mm at a 40° angle and welded with a 3 mm root opening and a 13 mm effective throat (depth of penetration) to ensure complete penetration. Finally, an 11 mm fillet weld is placed on the arrow side. Again, the welding symbol is read from the reference line outward. In this case, there are three separate reference lines. To read multiple reference lines, start at the arrow and then read the first reference line encountered, continuing on to the next.

- **Single U-Groove and Double U-Groove Weld Symbols**: The single U-groove welding symbol shown in **FIGURE 11-36** on the left is beveled on the other side to a depth of 12 mm at a 40° angle; it has a 3 mm root opening. The arrow side weld is welded to an effective throat (depth of penetration) of 14 mm. It is then back-gouged and welded on the other side with a back weld.

  The double U-groove welding symbol shown in Figure 11-36 on the right has a 10 mm depth of preparation, with a 12 mm effective throat (depth of penetration) on both the arrow side and the other side. The root opening is 2 mm, and both U-grooves are beveled to a 40° angle.

- **Single J-Groove and Double J-Groove Weld Symbols**: The single J-groove welding symbol shown in **FIGURE 11-37** on the left is beveled on the other side to a depth of 9 mm at a 25° angle, with an effective throat

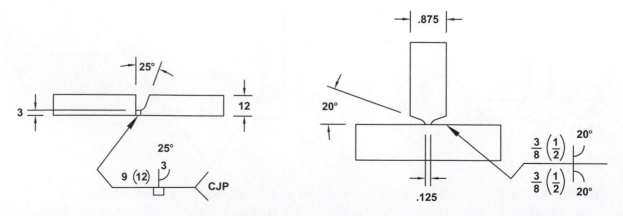

FIGURE 11-37 J-groove welding symbols.

(depth of penetration) of 12 mm. The little square on the arrow side is not a backing plate but rather a **consumable insert**. The name is what is implied: The insert is consumed, becoming part of the weld metal during welding. The root opening of 3 mm is also the size of the consumable insert. CJP stands for Complete Joint Penetration, and the name is self-explanatory. A class of consumable insert may be placed in the tail; for example, Class 3.

The double J-groove welding symbol shown in Figure 11-37 on the right is beveled to a depth of 3/8 inches at a 20° angle on both sides. It is then welded with an effective throat (depth of penetration) to 1/2 inches on both sides to ensure 100% penetration.

- **Single-Flare V-Groove and Double-Flare V-Groove Weld Symbols**: The arrow side single-flare V-groove welding symbol shown in **FIGURE 11-38** on the left has a size of 18 mm. The size on flare-groove joints is defined as the radius from the point of tangency to the outer edge of the member. The effective throat (depth of penetration) of the weld is 16 mm. In flare V- and flare bevel-groove joints, it is not always necessary to achieve complete joint penetration, and the groove weld size (effective throat) is often less than the preparation size (distance from the point of tangency to the outer member).

  The double-flare V-groove welding symbol shown in Figure 11-38 on the right has a size of 3/4 inches from the center of the shaft to the tangent to the outside edge. The effective throat (depth of penetration) is 1/2 inches. This is indicated for both sides.

- **Single-Flare Bevel-Groove and Double-Flare Groove Weld Symbols**: The arrow side single-flare bevel-groove welding symbol shown in **FIGURE 11-39** on the left has a size of 1/2-inch radius to the outside edge. The effective throat (depth of penetration) is 3/8 inches and is welded with the GMAW process. The weld has a convex contour by hammering.

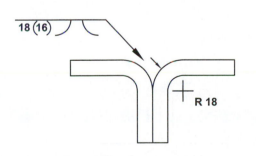

FIGURE 11-38 Flare V-groove welding symbols

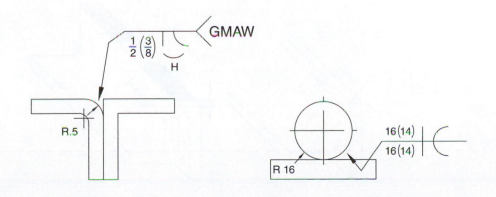

FIGURE 11-39 Flare bevel-groove welding symbols have a broken arrow to indicate which side of the part is flared.

The double-flare bevel-groove welding symbol shown in Figure 11-39 on the right has a size of 16 mm from the center of the shaft to tangent to the outside edge. The effective throat (depth of penetration) is 14 mm. This is indicated for both sides.

EDGE WELD SYMBOL

The edge welding symbol has replaced the (no longer used) flange weld symbol. A compilation of information that can be applied to the edge weld symbol is shown in **FIGURE 11-40**. Size is the height or buildup of the weld. The edge welding symbol shown in **FIGURE 11-41** has a size (height) of 5 mm.

The welding symbol shown in **FIGURE 11-42** has a 1/4-inch edge weld and is 2 inches long, with a 3-inch pitch (center to center). The size of the weld is always designated on the left-hand side of the symbol, and the length and pitch are on the right-hand side.

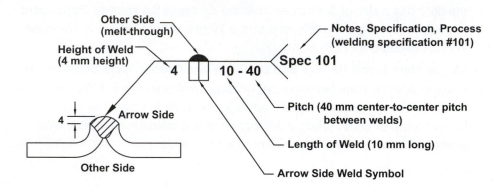

FIGURE 11-40 Edge welding symbol

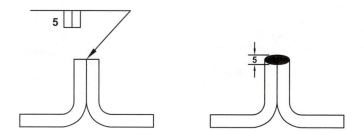

FIGURE 11-41 The size of an edge weld is the height

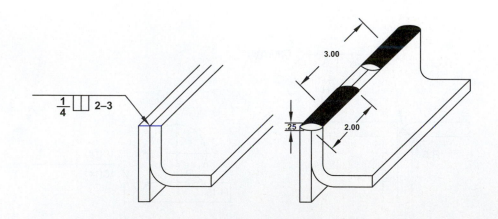

FIGURE 11-42 Edge welding symbols have the same characteristics as fillet welding symbols.

SURFACE WELD SYMBOL

The surface welding symbol is used to show a buildup, or cladding, on the surface of a part. It is often used with hard surfacing rods to both build up the thickness and to increase the hardness of the surface, such as on heavy road construction equipment.

The surface weld symbol will usually have the symbol only or may have the height (size) of the weld, as the 4 mm is shown in **FIGURE 11-43**. The length, or area, to be covered is shown in a detail on the print.

NONDESTRUCTIVE WELD SYMBOLS

Nondestructive testing of welds enables testing without affecting the service of welds on a product. The following abbreviations may be found on the reference line of a welding symbol:

VT = Visual Testing

RT = Radiographic Testing

NRT = Neutron Radiographic Testing

MT = Magnetic Particle Testing

AET = Acoustic Emission Testing

UT = Ultrasonic Testing

PT = Penetrant Testing

ET = Eddy Current Testing

PRT = Proof Testing

LT = Leak Testing

Examples are shown in **FIGURE 11-44**. The * is a symbol for the radioactive material, and its purpose is to show the angle, or direction, of radiation. Nondestructive testing symbols may be arrow side, other side, or no side significant.

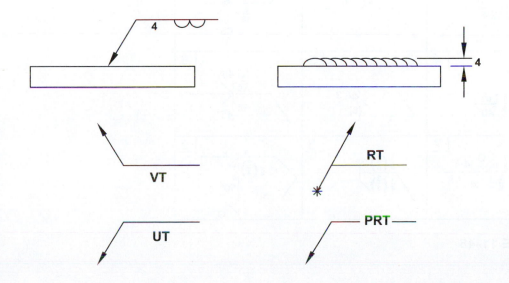

FIGURE 11-43 The surface welding symbol

FIGURE 11-44 Various nondestructive weld symbols

MULTIPLE CHOICE

For questions 1–9 refer to the symbols in the corresponding boxes 1–9 in **FIGURE 11-45**.

1. Refer to the welding symbol in the box numbered 1.
   a. What type of weld is to be made?
   b. What is the size of the weld on the other side?
   c. What is the weld pitch on the other side?
   d. What is the length of the weld on the arrow side?

2. Refer to the welding symbol in the box numbered 2.
   a. What type of weld is to be made?
   b. What is the size of the weld?
   c. How many welds are to be made?

3. Refer to the welding symbol in the box numbered 3.
   a. What type of weld is to be made on the arrow side?
   b. What is the weld symbol on the other side?
   c. What does the flag signify?
   d. What is the root opening?
   e. How is the weld to be finished on the other side?

4. Refer to the welding symbol in the box numbered 4.
   a. What type of weld is to be made?
   b. What is the diameter of the hole?
   c. What is the depth of fill?

5. Refer to the welding symbol in the box numbered 5.
   a. What type of weld is to be made?
   b. What is the pitch?
   c. What is the size?
   d. How many of these types of welds are to be made?

6. Refer to the welding symbol in the box numbered 6.
   a. What type of weld is to be made?
   b. What is the depth of preparation?
   c. What is the effective throat (depth of penetration)?
   d. What is the angle of preparation?

7. Refer to the welding symbol in the box numbered 7.
   a. What type of weld is to be made?
   b. What is the size of the weld on the arrow side?
   c. What is the length of the weld on the other side?
   d. How is the weld to be finished on the arrow side?
   e. How is the weld to be finished on the other side?

8. Refer to the welding symbol in the box numbered 8.
   a. What type of weld is to be made?
   b. What is the significance of the square on the reference line?
   c. What is the depth of preparation on the arrow side?
   d. What is the effective throat of the weld on the other side?

9. Refer to the welding symbol in the box numbered 9.
   a. What type of weld is to be made on the arrow side first?
   b. What type of weld is to be made on the arrow side second?
   c. What type of weld is to be made on the other side?
   d. What is the depth of preparation on the arrow side?
   e. How is the weld to be finished on the other side?

10. Pick one welding symbol out of the nine given in Figure 11-45 and then correctly perform a weld on metal with any welding process.

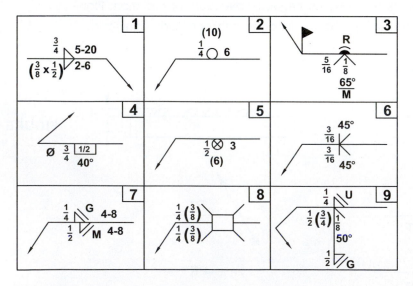

FIGURE 11-45

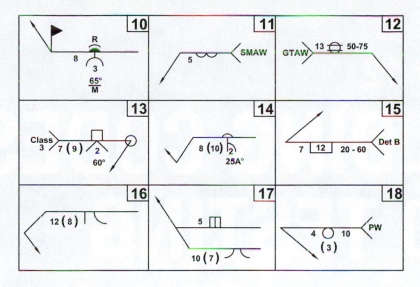

FIGURE 11-46

For questions 11–20 refer to the symbols in the corresponding boxes 10–18 in **FIGURE 11-46**.

11. Refer to the welding symbol in the box numbered 10.
    a. What type of weld is to be made on the arrow side?
    b. What is the size of the effective throat size?
    c. What is the root opening?
    d. What is the angle?
    e. What is weld symbol on the other side?
    f. What does the symbol directly below the R represent?
    g. What does the R mean?

12. Refer to the welding symbol in the box numbered 11.
    a. What type of weld is to be made on the arrow side?
    b. What does the number 5 mean?
    c. What welding process is used?

13. Refer to the welding symbol in the box numbered 12.
    a. What type of weld is to be made on the other side?
    b. What is the length?
    c. What is the width?
    d. What is the pitch?
    e. What is the welding process?

14. Refer to the welding symbol in the box numbered 13.
    a. What type of weld is applied to the arrow side?
    b. What symbol is on the other side of the welding symbol?
    c. What is the effective throat size?
    d. What is the depth of preparation?
    e. What is the groove angle?
    f. What is the root opening?
    g. What is in the tail of the welding symbol?
    h. What is meant by the circle at the junction of the arrow and reference line?

15. Refer to the welding symbol in the box numbered 14.
    a. What type of weld is to be made on the arrow side?
    b. What is the size of the effective throat size?
    c. What is the depth of preparation?
    d. What is the root opening?
    e. What is the angle?
    f. What is weld symbol on the other side?

16. Refer to the welding symbol in the box numbered 15.
    a. What type of weld is to be made?
    b. What is the width of the weld?
    c. What is the length of the weld?
    d. What is the pitch of the welds?
    e. What is the depth of fill?
    f. What does the "Det B" mean?

17. Refer to the welding symbol in the box numbered 16.
    a. What type of weld is to be made?
    b. What is the size of the effective throat size?
    c. What is the depth of preparation?

18. Refer to the welding symbol in the box numbered 17.
    a. What type of weld is applied to the first?
    b. Is the first weld to be applied arrow side or other side?
    c. What type of weld is applied to the arrow side second?
    d. What is the 5 represent?
    e. What is the 10 represent?
    f. What is the 7 represent?

19. Refer to the welding symbol in the box numbered 18.
    a. What type of weld is to be applied?
    b. What is the 4 represent?
    c. What is the 10 represent?
    d. What is the 3 represent?

20. Pick one welding symbol out of the nine given in *Figure 11-46* and correctly perform a weld on metal with any welding process.

Chapter 12

WELDING CODES AND TESTING

LEARNING OBJECTIVES

In this chapter, the reader will learn how to:

1. Report reasons for the formation of welding societies.
2. List the requirements to become a certified welding inspector.
3. Differentiate between a code, a standard, and a specification.
4. Write a welding procedure in a accordance with a code.
5. Determine if a weld coupon meets visual code requirements.
6. Extract specimens from a weld coupon according to code requirements.
7. Destructively test a weld coupon.
8. Evaluate weld specimens based on code requirements.
9. Examine a weld sample by using a nondestructive test method.

KEY TERMS

Coupons 429

Weldment 429

Welding Procedure Specifications (WPS) 433

Procedure Qualification Record (PQR) 433

Welder Qualification Test Record (WQTR) 434

Standards 434

Codes 434

Specifications 434

Essential Variables 439

Nonessential Variable 439

Discontinuity 447

Radiography 464

Ultrasonic Testing 466

Magnetic Particle Testing 469

Dye Penetrant Testing 473

INTRODUCTION

WHY TEST WELD COUPONS?   How does an individual know that the weld he or she made is free of defects? By definition, weld defects are severe enough to cause failure. There are welds that from outward appearances are picture-book perfect but fail a code-required test. There are weld tests, although acceptable to code criteria, that are not aesthetically pleasing yet pass qualification requirements. Situations exist where the failure of a weld may cost someone his or her life. It is for this reason that weld tests and the criteria for passing or failing were developed.

A welder can improve welding techniques and dispel previous beliefs by testing weld samples and examining the results. Testing could also be used to confirm or correct work angles and parameter settings. Weld tests are as simple as breaking a T-joint on a steel table with a hammer or having it thoroughly examined by radiographic testing. Between these two examples exist many different types of testing procedures.

Weld testing can be divided into two categories: destructive testing and nondestructive testing. Destructive testing is quite common because it is generally the least expensive means to evaluate weld quality. It is also a means to evaluate mechanical properties determining if filler metal is compatible with base metal. Although only a few samples, called **coupons**, are extracted from a weld test assembly, the samples taken are a fair representation of the entire weldment. A **weldment** is the finished product of two or more materials joined by welding. Destructive testing extracts coupons from a weld test assembly, while the nondestructive method allows the part to remain in service if applicable criteria are met.

Standards for industry predate welding. Prior to the turn of the 20th century, in the late 1800s, U.S. industries relied on steam boilers to harness the power to propel the Industrial Revolution. Unfortunately, there were no standards for boilers. As a result, boilers were blowing up on a daily basis. Some 10,000 boilers exploded between 1880 and 1900. By 1910, almost 1,400 boilers exploded annually. Naturally, there was a public outcry for someone to do something about this very dangerous workhorse of industry. The American Society for Mechanical Engineers (ASME) was created to establish standards for boilers.

Around 1919, when welding became an acceptable method to join metal and later deemed superior as compared to riveting, ASME created a section to put forth standards on this new technology. At the same time, both the American Welding Society (AWS) and the American Petroleum Institute (API) were organized to create welding standards in their respective fields. Many other organizations have sprung up over the years; however, ASME, AWS, and API are the standards much of the industry follows. It is so much so that if a person wishes to be a welding inspector, the ultimate credential that the inspector can carry is as an AWS Certified Welding Inspector (CWI).

WHAT DOES A WELDING INSPECTOR DO?   A welding inspector's job duty is one of quality control and inspection for a manufacturer, an insurance company, or a government agency. A welding inspector is responsible for judging the acceptability of a product according to a written welding specification. A welding inspector must understand the specification both as to its limitations and intent. However, it is not the duty of a welding inspector to analyze or interpret design of a product or weldment; that responsibility lies with the engineer or designer. The goal is to strive for the required quality without delaying completion and delivery of the product unless there is justifiable reason. A welding inspector's job is a serious occupation and must be conducted with integrity, ethics, and fairness.

Welding inspectors find themselves working in many different industries. Each position within a particular company may have a slightly different job responsibility. Some industries that employ welding inspectors are:

- Nuclear Plants
- Chemical Processing
- Petroleum Product Refining
- Transportation
- Building Construction
- Bridge Construction

WHAT ARE THE QUALIFICATIONS TO BE A WELDING INSPECTOR?   To become a welding inspector, an individual must meet certain requirements. The published standard for welding inspector requirements is AWS QC1. The only physical requirement list in AWS QC1 is eyesight. However, the *Certification Manual for Welding Inspectors*, published by the Educational Department of AWS, lists the necessity of physical conditioning. The physical requirements listed in the *Certification Manual for Welding Inspectors* are there because an inspector may be required to crawl around all sorts of structures while inspecting welds. This includes places with extreme heights. AWS QC1 outlines visual acuity and knowledge as necessary in the inspection of welds. Although a person needs experience in the welding field to become a certified welding inspector, the ability to weld is not a criterion. It is the knowledge of welding itself that is required.

Regardless of which welding code the inspector follows, there is a three-part 6-hour exam administered through the American Welding Society (AWS) to become a certified welding inspector (CWI). If the individual fails to meet the minimum criteria for any one of the three exams, the entire three-part exam must be retaken. The three parts consists of 1) General Knowledge, 2) Inspection, and 3) Welding Code Interpretation. In the Welding Code Interpretation section, the most common codes used during the exam are AWS D1.1 and API 1104, which are shown in **FIGURE 12-1**. Another common and useful code, ASME, may also

FIGURE 12-1 AWS D1.1 and API 1104 welding codes

be used. The available code clinics preparing for the exam generally cover only AWS D1.1 and API 1104. However, when applying for the Welding Inspector Exam Application, a person checks which code he or she wishes to use during his or her exam. Each code deals with the qualification of welding personnel and welding procedures. The choices are:

- AWS D1.1 – Structural Steel
- API – 1104 Pipeline
- AWS D15.1 – Railroad
- AWS D1.5 – Bridges
- ASME Section IX

The Welding Code Interpretation exam is an open book test, and the intent is to determine if an individual can interpret the intended meaning of the code as opposed to the memorization or general knowledge of a particular code. To become a Certified Welding Inspector, integrity is one of the most important qualifications. Any person who wishes to do the inspection job conscientiously and professionally must take these qualifications seriously. An inspector's license can be revoked if an issue regarding integrity arises.

Visual Acuity: Good vision is vital. The ability to examine weld surface conditions and judge their acceptability according to a written requirement is the primary duty of a welding inspector. An AWS Certified Welding Inspector (CWI) is required to have 20–40 vision, as determined by corrective eye charts and the Jaeger J-1 near vision acuity test, with or without corrective lenses. The required eye examination also includes a color perception test for red/green and blue/yellow differentiation. Although color perception for most visual welding inspection jobs is not a requirement for obtaining inspector certification, it is part of the renewal examination. It is up to the employer to determine and enforce any color perception requirement.

Knowledge: The improper use of welding terminology by a welding inspector could create more than just an embarrassment; it could lead to miscommunications that are detrimental to the outcome of the product. Consequently, the welding inspector must know and communicate correctly the language of welding,

FIGURE 12-2 AWS, *QC1 Standard for AWS Certification of Welding Inspectors*

and this is reflected in the testing to becoming a CWI. The inspector must communicate effectively with welders performing the work, individuals who make repairs, and supervisors and engineers who planned the work and accept the final structure. The vocabulary used—both spoken and written—must be in terms understandable to everyone involved.

ARE THERE DIFFERENT LEVELS OF CERTIFICATION?

LEVELS OF CERTIFICATION There are three levels of AWS certification of welding inspectors: Senior Certified Welding Inspector (SCWI), Certified Welding Inspector (CWI), and Certified Associate Welding Inspector (CAWI). The AWS specification QC-1, shown in **FIGURE 12-2**, outlines the areas of knowledge the applicant must be familiar with to successfully pass the examination. The SCWI must have a more extensive knowledge of nondestructive weld examination. The CAWI must be familiar with and understand fundamentals of the same joining and cutting processes as the CWI. **TABLE 12-1** summarizes the qualifications for each level.

CODES, STANDARDS, AND SPECIFICATIONS

WHY HAVE CODES, STANDARDS, OR SPECIFICATIONS? As noted earlier in this chapter, at one time, there were a great many hazards when working around boilers or other such equipment, with grave consequences to the individual but little or no liability for the company using or producing the hazardous product. Today, there is a much greater emphasis on protecting the public and employees, with monetary consequences to the company if it fails to do so. However, there are still areas of interpretation in how to best protect the individual. Where general safety or situations that could threaten a life are involved, the amount and precision of welding inspection usually must conform to a code, standard, or specification. These codes, standards, or specifications may come from a government agency or a private organization, such as AWS or ASME. The choice

	Requirements to Become a Certified Welding Inspector		
	SCWI	**CWI**	**CAWI**
Title	**Senior Certified Welding Inspector**	**Certified Welding Inspector**	**Certified Associate Welding Inspector**
Work Experience	No fewer than 15 years related to welding	No fewer than 5 years related to welding *or* 10 years if no high school diploma *or* 15 years if less than 8th grade education	No fewer than 2 years related to welding *or* 4 years if no high school diploma *or* 6 years if less than 8th grade education
Work Experience Substitution	*Substitute* 2 years with associate degree in engineering technology or physical science; *Substitute* 1 year with trade/vocational diploma; *Substitute* 3 years with welding teaching experience	*Substitute* 2 years with associate degree in engineering technology or physical science; *Substitute* 1 year with trade/vocational diploma; *Substitute* 3 years with welding teaching experience	8th grade *plus* 1 year of vocational schooling *plus* 3 years of occupation experience: Associate degree in engineering technology or physical science *plus* 6 months of occupation experience
Type of Jobs/Duties	Design, production, construction, repair, weld examination, quality control, quality assurance	Design, production, construction, repair, weld examination	
Education	High school graduate *or* approved high school equivalency diploma	High school graduate *or* approved high school equivalency diploma	*(See Work Experience Substitution)*
Visual Acuity	20–40 vision and Jaeger J-1 near vision acuity, with or without corrective lenses	20–40 vision and Jaeger J-1 near vision acuity, with or without corrective lenses	20–40 vision and Jaeger J-1 near vision acuity, with or without corrective lenses
Examination Results	Must have minimum 75% correct*	Must have minimum 70% correct**	Must have minimum 50% correct**
Other	Must be a CWI for at least 6 years	—	—

TABLE 12-1

*The examination will include questions relating to inspection, quality assurance, nondestructive examination, and welding processes.
**The examination will include questions relating to the understanding of codes, welding processes, and destructive and some nondestructive examination.

of which of these codes, standards, or specifications must apply is made by the design engineer of record and is interpreted by the insurance and customer inspectors on the job.

WHAT ARE THE DIFFERENCES AMONG CODES, STANDARDS, AND SPECIFICATIONS?

Codes and specifications are similar types of standards. It might be easier to comprehend the difference by understanding that a specification describes a specific part or material that may become an integral part of a product fabricated in accordance with a specific code. Almost all codes and standards require written **Welding Procedure Specifications (WPS)** that outline such requirements as preheat and postheat, amperage, voltage, travel speed, and electrode requirements. These specifications provide the welder and inspector with details about the welding processes to be applied. When a welding inspector is on the job, the inspector refers to the codes, standards, specifications, or a combination of the three to learn the requirements for the fabrication or construction. A number of documents exist that relate to weld quality. Depending on the type of welding used during construction, different requirements are enforced.

A WPS establishes the procedure the welder must use during welding. It is also used to qualify or test the welder. A **Procedure Qualification Record (PQR)** is a documentation of the results of the tests performed to make sure the WPS

Standards, Codes, and Specifications			
	Standards	**Codes**	**Specifications**
Definition	Any document or object used as a basis of comparison Once a standard has been defined, an inspector must then judge the adequacy of a product based on how it compares with that established standard.	Codes are mandatory when specified by the government or when a contract is made that requires them. Codes are generally applicable to a process. A code is a body of rules for the construction or fabrication of a product and describes a much larger scope of construction or fabrication.	Specifications are generally associated with a product. They describes the requirements for a particular object, material, service, etc.
What They Include	These documents include *Codes*, *Specifications*, *Procedures*, *Recommended Practices*, *Guides* and *Groups of Graphic Symbols*.	Codes use the words *shall* and *will*; this is the part that makes them mandatory.	Specifications use the words *shall* and *will*; this is the part that makes them mandatory.

TABLE 12-2

stands up to destructive and non-destructive tests required by the applicable code or standard. A **Welder Qualification Test Record (WQTR)** is documentation of the test results the welder obtained after following the WPS. **TABLE 12-2** summarizes **standards**, **codes**, and **specifications**.

As noted in Table 12-2, *Recommended Practices* and *Guides* are a type of *Standard*. Both recommended practices and guides are nonmandatory and are standards that are offered primarily as aids. They use words such as *may* or *should*. They could become mandatory if parties involved agree in a contract that they are to be included. An example would be a method of cutting plates. A code book might state that "plates may be sawed or flame cut." If by contractual agreement no cutting method producing heat is allowed, then it is mandatory the plates be sawed or cut in some fashion not producing heat.

Classifications and *Methods* include a list of established *Recommended Practices* of categories for the processing of products. Certain electrodes are classified in a group, and a method for testing a weld for qualification testing is something that would be established as a recommend practice.

HOW ARE THESE CODES USED IN INDUSTRY?   Many codes exist—each with a specific yet broad-ranging purpose. A code may be written to spell out rules governing how to build structures, pressure vessels, gas lines, rotating or earthmoving equipment, and the like. Other codes focus on specific materials, such as stainless steels or aluminum.

Anyone who uses a code should become familiar with its intended use. The intended use is usually written within the Scope or Introduction section of the book. For example, the first sentence in Section 1 of AWS D1.1 Structural Welding Code-Steel states, "This Code contains welding requirements for fabricating and erecting welded structures." It goes on to explain that the code is not intended for higher strength steels with a tensile strength greater than 100 KSI (690 MPa), steels less than 1/8 inches (3.2 mm), pressure vessels or pressure piping, or steels other than carbon or low alloy.

	Welding Codes
AWS	AWS D1.1–Structural Welding Code-Steel
	AWS D1.2–Structural Welding Code-Aluminum
	AWS D1.3–Structural Welding Code-Sheet Steel
	AWS D1.4–Structural Welding Code—Reinforcing Steel
	AWS D1.5–Bridge Welding Code
	AWS D1.6–Structural Welding Code-Stainless Steel
	AWS D15.1–Railroad Welding Specification-Cars and Locomotives
ASME	Section IX– Qualification Standard for Welding and Brazing Procedures, Welders, Brazers, and Welding and Brazing Operators
API	1104–Welding of Pipelines and Related Facilities
	620–Design and Construction of Large, Welded, Low-Pressure Storage Tanks
	650–Welded Tanks for Oil Storage

TABLE 12-3

If the desired base metal to be qualified is less than 1/8 inches (3.2 mm), a different code must be used, such as the AWS D1.3 Structural Weld Code–Sheet Steel. The AWS D1.3 code states, "This specification is applicable to the arc welding of sheet steels or strip steels, or both, including cold-formed members, 0.18 inch (4.6 mm) or less in thickness. Such welding involves sheet steel, strip steel, or both, being welded to supporting structural members."

For another example, if a fabricator just wants to set a standard for his or her workers to follow and is not welding anything that requires a structural welding code, AWS has Sheet Metal Welding Code, D9.1, for that purpose. This code states, "This Code provides qualification, workmanship, and inspection requirements for both arc welding (Part A) and braze welding (Part B) as they apply to the fabrication, manufacture, and erection of nonstructural sheet metal components and systems. It covers the thickness range for sheet metal up to as thick as 3 gauge, or 0.239 in. (6.07 mm). Where vacuum or pressure exceeds 120 inches of water (30 kPa) or where structural requirements are concerned, other standards shall be used."

It is also helpful to read the Foreword to determine the committee's intent of the code. In any event, the first part of the code book determines if the welders are qualifying and testing to the correct code. The next question that comes to mind is, "Who is responsible for tests required by a code?" The short answer is the manufacturer or contractor. Each code book spells out precisely who is responsible, which may or may not include the engineer. **TABLE 12-3** lists some of the welding codes available and their intended purposes.

QUALIFICATIONS

HOW DO I BECOME A CERTIFIED WELDER? The phrase "certified welder" or wanting to obtain a "weld certification" is inaccurate. A welder cannot get certified or obtain his or her certification in welding per se. There are many examples in the welding industry where terms or phrases are misused. A common example is saying "Heli-arc" instead of Gas Tungsten Arc Welding.

A person can become a Certified Welding Inspector (CWI) or obtain an AWS Certification of Automated Process Operators and Technician (Robotic Welding). As for "certified welder," it is more correct to say "qualified welder." The weld test conductor certifies that the welder has correctly followed a written welding procedure and, after the testing of the weld test assembly, meets the acceptance criteria specified in the code. The "certified welder" phrase may have been derived from wording requiring welders to be "certified" to a state test or other such requirements.

To become a qualified welder, a welder must:

1. Obtain a written welding procedure.
2. Have someone qualified to certify that the procedure is followed and witness the welding test.
3. Have the weld test assembly tested and evaluated. Depending on code requirements, the test assembly may be destructively tested, nondestructively tested, or both.
4. Have the results of the test written in a weld test report.

IS THERE ONE TEST THAT I CAN TAKE TO COVER EVERYTHING FOR WELDING?

The short answer to this question is no; many welder qualification tests exist. The first step is to determine what welding code is to be followed. The ASME Section IX Qualification Standard for Welding and Brazing Procedures, Welders, Brazers, and Welding and Brazing Operators is one of the most useful codes in existence. Another code is the AWS D1.1 Structural Welding Code-Steel. A reason the ASME code is popular is because the code encompasses so many base materials. Even if a base material is not covered in the ASME code, there is a provision, QW403, to use any base material provided that the chemical analysis and mechanical properties are specified. AWS and ASME are similar in their specifications, testing, and evaluation of the weld test assemblies. However, there are enough differences that one cannot assume that what one code allows, the other also allows; they are two different codes.

The second step is to determine what the limitations are in joint design and position. For example, qualifying for fillet welds does not qualify the welder to weld on groove welds or pipe. Qualifying on groove welds does not qualify the welder to weld on pipe. However, qualifying on pipe qualifies the welder to weld on pipe and plate both grooves and fillet welds provided they are within the limits of the type of process, metal thickness, and other essential variables. For example, welding a fillet weld in the horizontal position only qualifies the welder to weld in the flat, horizontal position on fillets. However, qualifying on a pipe in the 6G position (45-degree fixed position) qualifies the welder to weld in all positions and all types of joints.

The third step is to determine the minimum and maximum thickness of material to be qualified to weld. For example, AWS D1.1 covers the qualification on steels 1/8 inches (3.2 mm) and greater in thickness. The maximum thickness, depending on the particular test, can be unlimited. A thickness of less than 1/8 inches (3.2 mm) requires qualification from a different code. The ASME Section IX code qualifies base metals as thin as 1/16 inches or 3/16 inches (1.6 mm or 4.8 mm). For performance qualification, the maximum thickness qualified for ASME is usually twice the weld test coupon thickness (2t) unless the test thickness is over 3/4 inches (19 mm); then, the maximum thickness qualified is unlimited thickness.

The last step is to determine the welding process. Each process is qualified separately. Just because a welder is extremely competent at welding with the SMAW process does not automatically make that welder competent at welding with the GMAW or GTAW welding processes. Other factors to be determined are the type and diameter of the electrode, shielding gas composition and flow rate, voltage and amperage ranges, gas cup or nozzle diameter, and so on. With all these variables, it becomes quite clear why a written welding procedure is required.

WHAT ARE WELDER PERFORMANCE LIMITATIONS?   **TABLE 12-4** through **TABLE 12-7** compare performance qualification limitations for two commonly used codes. Table 12-4 through Table 12-7 are not intended to supplement the tables already in AWS and ASME but are able to provide a quick comparison. They show plate and pipe weld type, metal thickness, pipe and tubing diameter, and weld positions qualified. The API 1104 code is not included in the tables

Welder Qualification with ASME Section IX and AWS D1.1		
	ASME	**AWS**
Welder Qualifies on Fillet Weld	Qualified on fillet welds	Qualified on fillet welds
Welder Qualifies on Plate Groove	Qualified on fillet welds	Qualified on fillet welds
	Treats plate thickness and pipe diameter as separate issues	Qualified on groove plates[1]
Welder Qualifies on Pipe Groove	Qualified on fillet welds	Qualified on fillet welds
	Treats plate thickness and pipe diameter as separate issues	Qualified on groove plates
		Qualified on groove pipe

TABLE 12-4

[1]Qualifies on pipe over 24 inches in diameter with backing, back-gouging, or both.

Welder Plate Qualification with ASME Section IX and AWS D1.1						
	ASME			**AWS**		
Plate Thickness Qualification[1]	**Qualified**	**Min**	**Max**	**Qualified**	**Min**	**Max**
	1/16 to 3/8 inches (1.6 to 9.5 mm)	1/16 (1.6 mm)	2T[2]	$1/8 \leq T \leq 3/8$ $(3.2 \leq T \leq 9.5)$	1/8 (3.2 mm)	2T[2]
	$+ 3/8 < 3/4$ inches $(+ 9.5 < 19$ mm$)$	3/16 (4.8 mm)	2T[2]	$3/8 \leq T \leq 1$ $(9.5 \leq T \leq 25)$	1/8 (3.2 mm)	2T[2]
	Over 3/4 inches (Over 19 mm)	3/16 (4.8 mm)	Any	1 inch and over (25 mm & over)	1/8 (3.2 mm)	Unlimited

TABLE 12-5

[1]ASME treats grooves and fillets the same for qualification. AWS has two fillet weld options: Option 1 is a 1/2-inch (13 mm) T-joint. Option 2 is a double lap joint on a 3/8-inch (9.5 mm) plate. Both qualify from 1/8 inches (3.2 mm) to unlimited thickness.
[2]2T = Twice the metal thickness

Welder Pipe Qualification with ASME Section IX and AWS D1.1

Pipe Diameters (Ø)	ASME		AWS		
	Outside Ø Tested	**Min Qualified**	**Inside Ø Tested**	**Min**	**Max**
	< 1 inch (< 25 mm)	Size welded	2-inch (51 mm) Schedule 80 *or* 3-inch (76 mm) Schedule 40	3/4 inches to 4 inches Ø (19 mm to 102 mm Ø)	
	1 inch to < 2⅞ (25 to < 73 mm)	1 inch (OD) (25 mm)		1/8[1] (3.2 mm)	3/4[1] (19 mm)
	2⅞ inches and over (73 mm and over)	2⅞ inches (OD) (73 mm)	6-inch (152 mm) Schedule 120 *or* 8-inch (203 mm) Schedule 80	4 inches Ø and over (102 mm Ø and over)	
				3/16[1] (4.8 mm)	Unlimited

TABLE 12-6

[1]Pipe wall thickness

Welder Position Qualification with ASME Section IX and AWS D1.1

Position Qualified	ASME				AWS			
	Plate		**Pipe**		**Plate**		**Pipe**	
	Groove	**Fillet**	**Groove**	**Fillet**	**Groove**	**Fillet**	**Groove[2]**	**Fillet**
Plate[1] 1G	1G	1F	1G	1F	1G	1F, 2F	1G	1F, 2F
2G	1G, 2G	1F, 2F	1G, 2G	1F, 2F	1G, 2G	1F, 2F	1G, 2G	1F, 2F
3G	1G, 3G	1, 2, 3F	1G	1F, 2F	1, 2, 3G	1, 2, 3F	1, 2, 3G	1, 2, 3F
4G	1G, 4G	1, 2, 4F	1G	1, 2, 4F	1G, 4G	1, 2, 4F	1G, 4G	1, 2, 4F
3G and 4G	1, 3, 4G	All	1G	All	All	All	All	All
Pipe[1] 1G	1G	1F	1G	1F	1G	1F, 2F	1G	1F, 2F
2G	1G, 2G	1F, 2F	1G, 2G	1F, 2F	1G, 2G	1F, 2F	1G, 2G	1F, 2F
5G	1, 3, 4G	1 to 4F	1G, 5G	All	1, 3, 4G	1, 2, 34	1G, 5G	1F, 5F
6G	All	All	All	All	All	All	All	All
2G and 5G	All	All	All	All	All	All	All	All

TABLE 12-7

[1]Qualifying on fillet welds does not qualify to weld grooves. For qualifying position, see the ***Fillet*** columns.
[2]Qualifies on pipe over 24 inches in diameter with back-backing, back-gouging, or both.

Note: Positions = 1-Flat, 2-Horizontal, 3-Vertical, 4-Overhead, 5-Horizontal Fixed Pipe, and 6-45° Fixed

because the criteria are different from AWS and ASME. Each welding code must be purchased and studied separately in order to write a weld procedure and perform the appropriate welder performance test.

WELDING PROCEDURES

WHAT IS THE WRITTEN WELDING PROCEDURE THAT HAS BEEN MENTIONED? A written welding procedure is a document outlining all the variables that are used to produce a sound weld. The Procedure Qualification tests whether the filler material and other variables used on a particular base metal are compatible and can withstand the rigors of a destructive test. For example,

Company X uses a flux-cored electrode to weld ASTM A36 (mild steel) plates and structural shapes. The manufacturer's literature of the electrode shows all mechanical properties and welding parameters in its brochure with a blend of 75% argon/25% CO_2 shielding gas composition. But Company X also welds with the GMAW process with steel electrodes by using a 90% Argon/10% CO_2 shielding gas and wants to stock only one type of shielding gas. The manufacturer's brochure states that a maximum of 80% argon is recommended for flux-cored electrodes. If the 90/10 shielding gas is used, will the extra 10% argon cause overalloying on the flux-cored welds and possibly fracture when the product is sold and in service? The way to find out is to destructively test the weld, and this compatibility testing is an example of a purpose of procedure qualification.

The first step is to write a Welding Procedure Specification (WPS). The WPS states all the variables necessary to produce a sound weld. Welding codes list **essential variables** that, if changed, require the company to retest and document a new procedure. There are other variables that, if changed, do not affect the mechanical properties of the weld and need only be documented and dated. These variables are called **nonessential variables**. Many of the variables in the WPS have a range; for example, amperage, voltage, contact-to-tip work distance, and shielding gas flow rate. These ranges should be such that if the welder stays within the written WPS, the mechanical properties are unaffected and/or discontinuities do not occur.

Once the WPS is written, a procedure qualification test is performed by using suitable weld test coupons. The exact parameters while testing are written down and recorded on a Procedure Qualification Record (PQR). The weld test assembly is cut into smaller samples according to the location and size specified in the welding code. After testing, the samples are evaluated and the results are recorded on the PQR. Once it is determined that the WPS is written correctly by testing through the PQR, the WPS is qualified. When the welder qualifies, that person follows the variables and parameters on the WPS. The test results of the welder are recorded on a document called a Welder Qualification Test Record (WQTR). The testing of the weld coupon for a WQTR, with most codes, may be destructively tested or nondestructively tested with radiography (x-ray). Radiographic testing for welder qualification can be used because the test determines the ability of the welder to produce a sound weld and not essential variable compatibility. The essential variable compatibility was proven through the PQR.

WHAT DO THESE PROCEDURES LOOK LIKE?

Each welding code contains a sample form that identifies all the essential variables necessary to comply with that specific welding code. In addition, there are many fine software programs designed to produce welding procedures. Some of these software packages are written specifically for a particular welding code, while others are more generic. Even a software program written for a specific code can be adapted to be written for another welding code provided that all the essential variables are addressed. **FIGURE 12-3**, **FIGURE 12-4**, and **FIGURE 12-5** show examples of a WPS, PQR, and WQTR.

WELDING METALS, INC.
WELDING PROCEDURE SPECIFICATIONS

WPS No. _WPS-1_   Revision _____   Date _____   By _Dave Hoffman_

Authorized By _Kevin Dahle_____   Date _08/29/08_   Prequalified ☒

Welding Process(es) _GMAW_   --_____   Type: Manual ☐   Machine☐   Semi-Auto☒   Auto ☐

Supporting PQR's _PQR-1_____   _____   _B-U2a-GF_

JOINT

Type _Butt_____

Backing   Yes ☒   No ☐   Single Weld ☒   Double Weld ☐

Backing Material   _A36_____

Root Opening _1/4 in_   Root Face Dimension   _None_

Groove Angle _45 Deg_   Radius (J-U) _N/A_

Back Gouge   Yes ☐   No ☒

Method _None_____

BASE METALS

Material Spec. _ASTM A36_____   To   _ASTM A36_

Type or Grade _N/A_____   To   _N/A_

Thickness: Groove (in)   _3/8_   --   _3/8_

Fillet ( )   _N/A_   --   _____

Diameter (pipe,)   _N/A_   --   _____

POSITION

Position of Groove   _1G_____   Fillet _--_

Vertical Progression: Up ☐   Down ☐

FILLER METALS

AWS Specification   _A5-18_   --

AWS Classification   _ER70S-X_   --

_____   _____

ELECTRICAL CHARACTERISTICS

Transfer Mode (GMAW):

Short-Circuiting ☐   Globular☐   Spray ☒

Current: AC ☐   DCEP ☒   DCEN ☐   Pulsed ☐

Other _--_

Tungsten Electrode (GTAW): _ASTM A36_

Size _N/A_____   Type _N/A_

SHIELDING

Flux   Gas _Argon/CO2_

_N/A_   Composition _90% Argon/10% CO2_

Electrode-Flux (Class)   Flow Rate _35-45 CFH_

_N/A_   Gas Cup Size _5/8 or 3/4 in_

TECHNIQUE

Stringer or Weave Bead _Both_

Multi-Pass or Single Pass (Per Side) _Multiple_

Number of Electrodes _____1_____

Electrode Spacing: Longitudinal _N/A_

Lateral _N/A_

Angle _N/A_

Contact Tube to Work Distance _1/2 - 1 in_

Peening _None_____

Interpass Cleaning _Brush_____

PREHEAT

Preheat Temp., Min. _70 Deg F._

Thickness Up to 3/4 in   Temperature _70 Deg F._

Over 3/4 to 1 1/2 in   _150 Deg F._

Over 1 1/2 to 2 1/2 in   _225 Deg F._

Over 2 1/2 in   _300 Deg F._

Interpass Temp., Min. _70 Deg F._   Max. _500 Deg F._

POSTWELD HEAT TREATMENT PWHT Required ☐

Temp. _None_   Time _N/A_

WELDING PROCEDURES

Layer/Pass	Process	Filler Metal Class	Diameter	Cur. Type	Amps or WFS	Volts	Travel Speed	Other Notes
All	GMAW	ER70S-X	0.045	DCEP	300-350 A	26-30 V	12-16 IPM	400-500 WFS

FIGURE 12-3 An example of a completed Welding Procedure Specification

WELDING METALS, INC.
PROCEDURE QUALIFICATION RECORD

PQR-1

WPS No. __PQR-1__   Revision _____   Date _____   By __Dave Hoffman__

Authorized By __Kevin Dahle__   Date __08/29/08__   Type: Manual ☐   Machine ☐

Welding Process(es) __GMAW__ -- _____   Reference WPS No. __WPS-A1__   Semi-Auto ☒   Auto ☐

JOINT

Type __Butt__

Backing   Yes ☒   No ☐   Single Weld ☒   Double Weld ☐

Backing Material __A36__

Root Opening __1/4 in__   Root Face Dimension __None__

Groove Angle __45 Deg__   Radius (J-U) __N/A__

Back Gouge   Yes ☐   No ☒

Method __None__

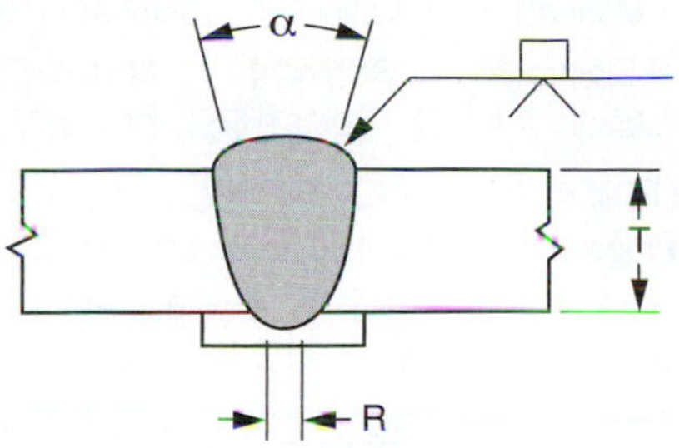

BASE METALS

Material Spec. __ASTM A36__   To __ASTM A36__

Type or Grade __N/A__   To __N/A__

Thickness: Groove (in) __3/8__   Fillet (in) __3/8__

Diameter (pipe,) __N/A__

POSITION

Position of Groove __1G__   Fillet __--__

Vertical Progression: Up ☐   Down ☐

FILLER METALS

AWS Specification __A5.18__   --

AWS Classification __ER70S-6__   --

ELECTRICAL CHARACTERISTICS

Transfer Mode (GMAW):

Short-Circuiting ☐   Globular ☐   Spray ☒

Current: AC ☐   DCEP ☒   DCEN ☐   Pulsed ☐

Other --

Tungsten Electrode (GTAW): _____

Size __N/A__   Type __N/A__

SHIELDING

Flux   Gas __Argon/CO2__

__N/A__   Composition __90% Argon/10% CO2__

Electrode-Flux (Class)   Flow Rate __40 CFH__

__N/A__   Gas Cup Size __3/4 in__

TECHNIQUE

Stringer or Weave Bead __Stringer__

Multi-Pass or Single Pass (Per Side) __Multiple__

Number of Electrodes __1__

Electrode Spacing: Longitudinal __N/A__

Lateral __N/A__

Angle __N/A__

Contact Tube to Work Distance __3/4 in__

Peening __None__

Interpass Cleaning __Brush__

PREHEAT

Preheat Temp., Min. __70 Deg F.__

Interpass Temp., Min. __70 Deg F.__   Max. __500 Deg F.__

POSTWELD HEAT TREATMENT   Required ☐

Temp. __None__

Time __N/A__

WELDING PROCEDURES

Layer/Pass	Process	Filler Metal Class	Diameter	Cur. Type	Amps or WFS	Volts	Travel Speed	Other Notes
1-6	GMAW	ER70S-6	0.045	DCEP	325 A	28 V	14 IPM	450 WFS

FIGURE 12-4 An example of a completed Procedure Qualification Record

WELDING METALS, INC.
WELDER QUALIFICATION TEST RECORD

WQTR-1

WQTR No. _WQTP-1_   Welder Name _Joe Welder, Jr._   Welder Id. _JWJ_

WPS No. _WPS-1_   Revision No. _____   Date _09/28/08_

VARIABLES: Record actual values used in qualification | **QUALIFICATION RANGE**

Process (Table 4.10, Item (1)) _GMAW_ | _GMAW_

Transfer Mode (GMAW): Short-Cir ☐   Globular ☐   Spray ☒   Short-Circuiting ☐   Globular ☐   Spray ☒

Type   Manual ☐   Machine ☐   Semi-Auto ☒   Auto ☐   Manual ☐   Machine ☐   Semi-Auto ☒   Auto ☐

Number of Electrodes   Single ☒   Multiple ☐   Single ☒   Multiple ☐

Current/Polarity   AC ☐   DCEP ☒   DCEN ☐   Pulsed ☐   AC ☐   DCEP ☒   DCEN ☐   Pulsed ☐

Position (Table 4.10, Item (4)) _GMAW_ | _1G, 1F, 2F_

    Weld Progression (Table 4.10, Item (6))   Up ☐   Down ☐   Up ☐   Down ☐

Backing (Table 4.10, Item (7))   Use Backing ☒   With Backing ☒   Without Backing ☐

Consumable Insert (GTAW)   Use Insert ☐   With Insert ☐   Without Insert ☐

Material/Spec. _ASTM A36_   to _ASTM A36_   _Group 1_

Thickness (Plate): Groove (in) _3/8_	_1/8_ -- _3/4_ in
Fillet ( ) _N/A_	_Any_ -- _Unlimited_ in
Thickness (Pipe/Tube): Groove ( ) _N/A_	_1/8_ -- _3/4_ in
Fillet ( ) _N/A_	_Any_ -- _Unlimited_ in
Diameter (Pipe): Groove ( ) _N/A_	_24_ -- _Unlimited_ in
Fillet ( ) _N/A_	_Any_ -- _Unlimited_ in

Notes _____

Filler Metal (Table 10, Item (2))

     Spec. _Spec. A5.18_   _A5.18_

     Class. _Class ER70S-6_   _ER70S-X_

     F-No. _F-No. F6_   _F6_

Gas Flux Type (Table 4.10, Item (3)) _Argon/CO2_   _Argon/CO2_

Other _90/10 Composition_   _90/10 Composition_

VISUAL INSPECTION (4.8.1)   Acceptable

GUIDED BEND TEST RESULTS (4.30.5)

Type	Result	Type	Result
Face Bend	Pass		
Root Bend	Pass		

FILLET TEST RESULTS (4.30.2.3 AND 4.30.4.1)

Appearance _N/A_   Fillet Size _N/A_   Macroetch _N/A_

Fracture Test Root Penetration _N/A_   Description _____

Inspected By _Dave Fisher_   Test No. _15_   Organization _DF Testing Agency_   Date _09/28/08_

RADIOGRAPHIC TEST RESULTS (4.30.3.1)

Film Identification No.	Result	Remark	
N/A	N/A	N/A	Interpreted By --
			Organization --
			Test No. --
			Date --

We, the undersigned, certify that the statements in this record are correct and that the test welds were prepared, welded, and tested in accordance with the requirements of section 4 of ANSI/AWS D1.1 (2006) Structural Welding Code-Steel.

Manufacturer _Welding Metals, Inc._   Authorized By _____   Date _09/28/08_

FIGURE 12-5 An example of a completed Welder Qualification Test Record

ESSENTIAL AND NONESSENTIAL VARIABLES

WHAT ARE ESSENTIAL AND NONESSENTIAL VARIABLES? Essential variables require requalification if a change is made. Requalification means that a new weld test assembly is completed and tested according to the code followed. Nonessential variables require only a written change be made on the WPS. In situations that include notch toughness testing, supplementary essential variables list what changes require requalification. **TABLE 12-8** lists many of the essential variables for AWS, AMSE, and API. Not all the essential variables that are applicable for each code are listed. The variables included in Table 12-8 are selected for the sake of comparison only. The applicable code needs to be studied to provide more detail on specifics.

Most codes require procedure qualification on any joint, base metal, and process regardless of how many other companies may have qualified similar procedures. Some of the AWS codes—AWS D1.1 in particular—contain prequalified weld joints. A process to note with AWS D1.1 is GMAW-S. Short-circuit transfer requires procedure qualification with AWS D1.1 (4.15.1). ASME Section IX allows GMAW short circuiting, but the welder is only qualified to 1.1 times the thickness tested (QW-403).

Looking over Table 12-8, it may seem odd that weld position, for example, is an essential variable for AWS but is a nonessential variable for ASME. The variables in Table 12-8 are for Procedure Qualification. The ASME code is concerned that filler metal, base metal, and all the other variables are compatible and pass a destructive test. The individual welder's ability to produce that particular weld out of position is addressed in the Welder Performance variables. **TABLE 12-9** compares and contrasts welder qualification variables for the AWS D1.1 and ASME Section IX codes.

WHAT ARE OTHER GROUPS OR CATEGORIES?

GROUPING Qualification testing can become quite expensive. To alleviate costs, societies that develop codes group certain aspects of the code to cut down on the number of qualifications required. An example is grouping similar composition base metals. AWS D1.1 has three prequalified base metal groups. If a welder passes a qualification test on a Group I base metal, the welder is qualified to weld on all other base metals in Group I. API 1104 has a category called API Specification 5L. API also allows welding any applicable ASTM (American Society of Testing Materials) specification. ASME Section IX uses P-Numbers and Group Numbers. There are over 100 ASTM metals assigned to P-1. A qualification from one P-1 base metal to another P-1 base metal will qualify the welder to weld any other base metal listed as a P-1 base metal.

Another grouping is electrodes and welding rods. A specific electrode is assigned a classification number. Electrodes with different classification numbers but with similar characteristics are grouped under the same specification number. If electrodes with different specification numbers share similar characteristics,

PQR Essential Variables Changes–AWS and ASME

	SMAW	GMAW	FCAW	GTAW	SAW	OFW
Filler Metal						
Classification Strength	AWS	AWS	AWS			
± Filler				AWS/ASME		
Chemical Composition	ASME	ASME	ASME	ASME	AWS/ASME	ASME
F-Number or Groups	ASME	ASME	ASME	ASME	ASME	ASME
Specification						ASME
± Insert				ASME	AWS	
± Granular Flux					AWS/ASME	
Granular Flux Composition					AWS/ASME	
Solid to Tubular		ASME				
Weld Joint						
Groove Design	AWS	AWS	AWS	AWS	AWS	AWS
Omission of Backing	AWS	AWS	AWS	AWS	AWS	
Base Metal						
P-Number or Group Number	AWS/ASME	AWS/ASME	AWS/ASME	AWS/ASME	AWS/ASME	AWS/ASME
Thickness Not Qualified	AWS/ASME	AWS/ASME	AWS/ASME	AWS/ASME	AWS/ASME	AWS/ASME
Diameter Not Qualified	AWS/ASME	AWS/ASME	AWS/ASME	AWS/ASME	AWS/ASME	ASME
Electrical Characteristics						
Amperage	AWS	AWS	AWS	AWS	AWS	
Voltage	AWS	AWS	AWS	AWS	AWS	
Type of Current (AC or DC)		AWS	AWS		AWS	
Polarity		AWS	AWS		AWS	
Mode of Transfer		AWS/ASME				
Gas						
Shielding Composition		AWS/ASME	AWS/ASME	AWS/ASME		
Shielding Flow Rate		AWS/ASME	AWS/ASME	AWS/ASME		
Shielding or Trailing		ASME	ASME	ASME		
Type of Fuel						ASME
Weld Position						
Weld Position Not Qualified	AWS	AWS	AWS	AWS	AWS	AWS
Vertical Up or Down	AWS	AWS	AWS	AWS		
Heat Treatment						
± Preheat Temperature	AWS/ASME	AWS/ASME	AWS/ASME	AWS/ASME	AWS/ASME	
Add or Omit Postheat	AWS/ASME	AWS/ASME	AWS/ASME	AWS/ASME	AWS/ASME	ASME
Inter-Pass Temperature	AWS	AWS	AWS	AWS	AWS	
General						
Number of Electrodes		AWS	AWS		AWS/ASME	
Number of Passes (welds)	AWS	AWS	AWS	AWS	AWS	
Tungsten Electrode				AWS		
Travel Speed		AWS	AWS	AWS	AWS	
Method of Cleaning	ASME					

TABLE 12-8

Welder Performance Essential Variable Changes Requiring Requalification Essential Variables Changes–AWS and ASME						
	SMAW	**GMAW**	**FCAW**	**GTAW**	**SAW**	**OFW**
Base Metal						
Thickness Not Qualified	AWS/ASME	AWS/ASME	AWS/ASME	AWS/ASME	AWS/ASME	ASME
Diameter Not Qualified	AWS/ASME	AWS/ASME	AWS/ASME	AWS/ASME	AWS/ASME	
Position Not Qualified	AWS/ASME	AWS/ASME	AWS/ASME	AWS/ASME	AWS/ASME	ASME
Vertical Up or Down	AWS/ASME	AWS/ASME	AWS/ASME	AWS/ASME	AWS	
Omission of Backing	AWS/ASME	AWS/ASME	AWS/ASME	AWS/ASME	AWS/ASME	
Joint Design						
Addition of Backing						ASME
Filler Metal						
F-Number Limits	ASME	ASME	ASME	ASME	ASME	ASME
± of Consumable Inserts				ASME		
To Multiple Electrodes		AWS	AWS	AWS	AWS	
Classification Within Group						
Unapproved Filler and Shielding Gas Combination		AWS	AWS	AWS		
Gas						
Type of Fuel Gas						ASME
Change in Shielding Gas		AWS	AWS	AWS		
Omission of Backing Gas		ASME	ASME	ASME		
Electrical Characteristics						
Mode of Transfer		AWS/ASME				
Current or Polarity				ASME		
Other						
Welding Process not Qualified	AWS/ASME	AWS/ASME	AWS/ASME	AWS/ASME	AWS/ASME	ASME

TABLE 12-9

they are grouped with the same F-Number. **TABLE 12-10** illustrates an F-Number grouping to specification to individual electrode classification. Note specifications for ASME begin with SFA and specifications for AWS begin with A (i.e., SFA 5.18, A 5.18). **TABLE 12-11** expands on Table 12-10.

If a filler metal is proprietary, it more than likely has not been assigned an AWS classification, specification, or a F-number. ASME has made a provision for filler metals not included in one of the F-Number groups by substituting with an A-Number. A designated A-Number is a classification of ferrous weld metal analysis or, more simply, its chemical composition. A table is provided in the ASME code that provides a percentage range of alloying elements for electrode composition. The A-Number groups range from 1 to 12 beginning with mild steel and include alloys, such as chromium-nickel and nickel-chromium-molybdenum.

Although there are a large number of base metals grouped or categorized in welding codes, there are many that are not. ASME has made provisions for creating procedures to accommodate base metals not categorized (QW 403). An example is welding the various chrome-moly (chromium-molybdenum) base metals or 400 series

Example of Grouping F-6 Electrodes	
F-6	
A 5.18	A 5.20
ER80S-2	E70T-1
ER70S-3	E71T-1
ER70S-6	E71T-5
Etc.	Etc.

TABLE 12-10

F-NUMBERS Grouping of Electrode & Welding Rods

F-Number	Specification	Classification	Category
1	A–5.1 and 5.5	EXX20, EXX22, EXX24, EXX27, EXX28	SMAW electrodes
1	A–5.4	EXX25, EXX26	
2	A–5.1 and 5.5	EXX12, EXX13, EXX14, EXX19	
3	A–5.1 and 5.5	EXX10, EXX11	
4	A–5.1 and 5.5	EXX15, EXX16, EXX18, EXX48	
5	A–5.4	EXX15, EXX16, EXX17	SMAW stainless electrodes
6	A–5.2	RX	Oxy-fuel solid steel rods
6	A–5.17	FXX-EXX, FXXECX	SAW – steel
6	A–5.9	ERXXX, ECXXX, EQXX	Stainless steel solid/composite electrode
6	A–5.18	ERXXS-X, EXXC-X, ECXXC-XX	Steel solid and composite electrode
6	A–5.20	EXXT-X	Flux-cored electrodes
6	A–5.22	EXXXT-X	Stainless steel flux-cored electrodes
6	A–5.23	FXX-EXXX-X, FXX-ECXXX-X, FXX-EXXX-XN, FXX-ECXXX-XN	SAW – solid and metal cored flux and electrode
6	A–5.28	ER-XXS-X, EXXC-X	Low-alloy rods – solid and metal cored
6	A–5.29	EXXT1-X	Low-alloy steel flux-cored electrodes
6	A–5.30	INXXXX	Consumable inserts
21	A–5.3	E1100, E3003	Aluminum alloy SMAW electrodes
21	A–5.10	ER1100, ER1188	Aluminum alloy electrodes
22		ER5554, ER5356, ER5556, ER5183, ER5654	
23		ER4009, ER4010, ER4043, ER4047,	
25		ER4643	
		ER2319	
31	A–5.6 and 5.7	RCu, ECu	Copper alloy solid electrodes
32		RCuSi-A, ECuSi	
33		RCuSn-A, ECuSn-A, ECuSnC	
34		RCu, ECuNi	
35		RBCuZn-A, RCuZn-C	
36		ER-CuAl-Al, ERCuAl-A2, ECuAl-A2, ER-Cu-Al-A3, E-CuAl-B	
37		E-CuNiAl, ECuMnNiAl, ERCuNiAl, ERCuMnNiAl	
41	A–5.11	ENi-1	Nickel and nickel-based alloy electrodes
	A–5.14	ERNi-1	
42	A–5.11	ENiCu-7	Nickel and nickel-based alloy electrodes
	A–5.14	ERNiCu-7	
43	A–5.11	ENiCrFe-1, ENiCrFe-2, ENiCrFe-3, ENiCrFe-4, ENiCrMo-2, ENiCrMo-3	Nickel and nickel-based alloy electrodes
43	A–5.14	ERNiCr-3, ERNiCrFe-1, ERNiCrFe-5, ERNiCrFe-6, ERNiCrMo-2, ERNiCrMo-3	Nickel and nickel-based alloy electrodes
44	A–5.11	ENiMo-1, ENiCrMo-4, ENiCrMo-5	Nickel and nickel-based alloy electrodes
44	A–5.14	ERNiMo-1, ERNiMo-2, ERNiMo-7, ERNiCrMo-4, ERNiCrMo-5, ERNiCrMo-7	Nickel and nickel-based alloy electrodes
45	A–5.11	ENiCrMo-1	Nickel and nickel-based alloy electrodes
	A–5.14	ERNiCrMo-1, ERNiFeCr-1	
51	A–5.16	ERTi-1, ERTi-2, ERTi-3, ERTi-4	Titanium electrodes
52	A–5.16		
53	A–5.16		
54	A–5.16		
61	A–5.24	ERZr1, ERZr2, ERZr3	Zirconium and zirconium alloy electrodes
71	A–5.13	RXXX-X, EXXX-X	Hard-facing weld metal overlay
72	A–5.21	RXXX-X	

TABLE 12-11

stainless steels. One need only provide the chemical and mechanical properties of each alloying element that make up the base metal. The Procedure Qualification Record (PQR) lists the *exact* percentage of alloying elements and the *exact* mechanical properties of the base metal. The Welding Procedure Specification (WPS) contains a minimum and maximum *range* of chemical and mechanical properties.

WELD TESTING

The Welding Procedure Specification (WPS) is a recipe in that it lists the range of variables to be used in welding. Unless AWS D1.1 Structural Steel Welding Code prequalified procedures are used, the WPS must first be proven by performing a procedure qualification test and documenting the test results on a Procedure Qualification Record (PQR). A weld test assembly with a minimum size specified in the code is welded and tested.

After the PQR has successfully passed, the individual welder need only follow the WPS, staying within the range of variables listed on the WPS. For welder performance qualification, a weld test assembly is welded and tested. Where under ASME Section IX the PQR is required to be destructively tested, the welder performance qualification may be either destructively tested or nondestructively tested. AWS D1.1 requires that the PQR documents passing both destructive and nondestructive test methods. Because the cost of nondestructive equipment can be quite high and the testing centers have already invested in destructive testing equipment, most testing centers use the same equipment to evaluate both procedure and welder qualification weld test assemblies.

WHAT IS THE FIRST STEP AFTER THE COUPON SAMPLE HAS BEEN WELDED?   Prior to destructive or nondestructive testing, the weld coupon must first be visually inspected and meet minimum or maximum criteria of the code. Some of the criteria include the amount of melt-through, buildup on the cap pass, and root concavity. A certain amount of irregularities are permissible and referred to as discontinuities. If a **discontinuity** exceeds the code criteria, it becomes a defect and the weld test assembly fails prior to further testing.

WHAT TYPES OF VISUAL DISCONTINUITIES ARE EXAMINED PRIOR TO DESTRUCTIVE TESTING?   While there are a number of discontinuities that are examined prior to destructive testing, the three codes covered in this chapter all list penetration and fusion between base metal and filler metal as a criteria to successfully passing a weld test. Some root concavity is acceptable; incomplete root penetration is not (see **FIGURE 12-6** and **FIGURE 12-7**).

Weld reinforcement (convexity)—or the amount of buildup on the cap pass—must be a minimum of flush with the base metal regardless of the code. The amount of buildup varies from code to code and is often a suggestion as opposed to mandatory (see **FIGURE 12-8**). Undercut must also be kept to a minimum (see **FIGURE 12-9**). Undercut could create stress risers that may lead to cracking. Most codes permit up to 1/32 inches undercut. As a rule of thumb, if you can feel the undercut with your fingernail, the undercut exceeds minimum standards.

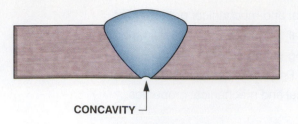

FIGURE 12-6 A groove weld joint with concavity

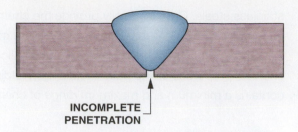

FIGURE 12-7 A groove weld joint with incomplete joint penetration

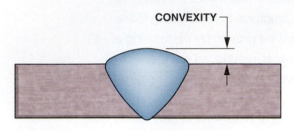

FIGURE 12-8 A groove weld joint showing convexity

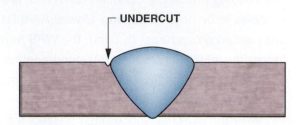

FIGURE 12-9 A groove weld joint with undercut

While the inspector's responsibility is not to judge aesthetics of a weld, the weld profile must meet minimum criteria. The cross-sectional area of a fillet weld must fit within the profile of the weld or it is undersized. In **FIGURE 12-10**, the bead profile on the right—while not the most aesthetically pleasing—does fit within the triangle of the weld cross-sectional area. The fillet weld on the left side of Figure 12-10 does not meet minimum criteria.

TABLE 12-12 lists acceptance criteria for discontinuities that may exist and fail a weld sample prior to destructive examination.

A number of external discontinuities may be visible that result in weld test failure prior to destructive testing. In addition, there are a number of internal discontinuities that would fail the coupon during or after destructive testing. **FIGURE 12-11** identifies a number of discontinuities that may be found in a weld sample.

Appendix A provides a longer list of discontinuities and photographs of defects that help identify them.

HOW ARE THESE WELD COUPONS DESTRUCTIVELY TESTED?

EXTRACTION OF TEST SAMPLES   While the location of specimens extracted from weld coupons appear to be arbitrary, they are representative of what to expect regardless

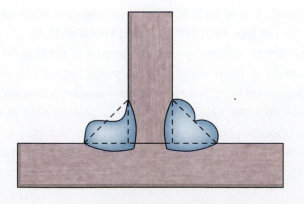

FIGURE 12-10 The weld on the left is undersized; the weld on the right is not undersized.

Visual Examination Criteria Before Destructive Testing[1]	
AWS (4.8)	
General	**For Fillets**
1. Free of cracks 2. Craters filled to full cross-sectional area 3. Weld face – minimum of flush 4. Undercut – not to exceed 1/32 inches 5. No evidence in root of incomplete fusion, cracks, or inadequate penetration 6. Maximum root concavity – 1/32 inches 7. Maximum melt-through – 1/8 inches	1. Complete fusion 2. Fillet leg size is specified minimum 3. No cracks 4. Undercut – not to exceed 1/32 inches
ASME (QW-194 and QW-184)	
General	**For Fillets**
1. Complete joint penetration 2. Complete fusion of weld and base metal	1. No more than 1/8-inch leg size variance for procedure qualification 2. No more than 1/16-inch convexity for performance qualification
API (3.4)	
1. No cracks 2. Inadequate penetration – shall not exceed 1 inch in 12 inches of weld or 8% if welds are less than 12 inches long 3. Incomplete fusion – shall not exceed 1 inch in 12 inches of weld or 8% if welds are less than 12 inches long 4. Burn-through – 1/4 inch maximum or the sum of separate burn-throughs is less than 1/2 inches in 12 inches of weld 5. Undercut – no more than 2 inches in a 12-inch weld, or the aggregate length exceeds 1/6 of the weld length	

TABLE 12-12

[1]Criteria is minimum acceptable by code; more stringent criteria may always be set by an inspector.

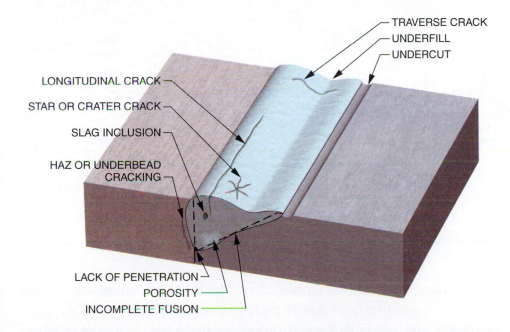

FIGURE 12-11 Undesirable weld discontinuities

of where the samples are taken from. AWS and ASME requirements are similar on weld coupon size and extraction location of weld specimens, while API is slightly different. The ASME Procedure Qualification fillet weld test for a T-joint requires 5 etched specimens, whereas AWS requires only 2 etched specimens. **FIGURE 12-12** shows where specimens are extracted for an ASME fillet plate and pipe test. **FIGURE 12-13** illustrates the similarities that AWS has with ASME on a T-joint performance fillet weld test. AWS also has an Option 2 fillet performance weld test that entails two lap joints with 15/16-inch spacing. The test requires 2 root bend tests.

V I D E O

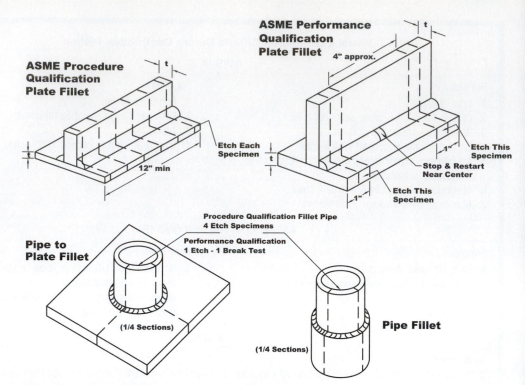

FIGURE 12-12 Locations of ASME procedure and performance weld test samples

Source: Reprinted from ASME 2001 BPVC Edition, Section IX, by permission of The American Society of Mechanical Engineers. All rights reserved.

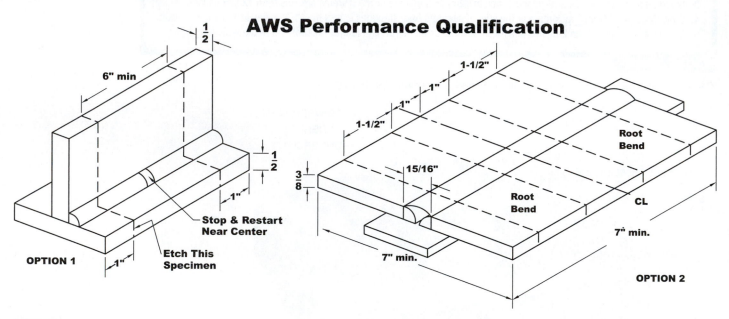

AWS Performance Qualification

FIGURE 12-13 Examples of AWS performance qualification weld test samples

Source: AWS D1.1/D1.1M:2008. Figure 4.9, reproduced with permission of the American Welding Society (AWS), Miami, Florida.

A typical AWS and ASME plate coupon for PQR testing requires 2 tensile tests and 4 bend tests. A welder performance qualification for both codes requires 2 bend tests or radiographic inspection of the test plate, pipe, or tubing. The bend tests are either side bends or alternatively face and root bends depending on the thickness of the welded plate. **FIGURE 12-14** illustrates where specimens are extracted from the plate for Procedure Qualification.

FIGURE 12-15 illustrates where specimens are extracted from the plate for Performance Qualification. For ASME, depending on procedure requirements, the backing may or may not be required. Groove angle and root opening are specified in the PQR and WPS.

Procedure Qualification - Plate

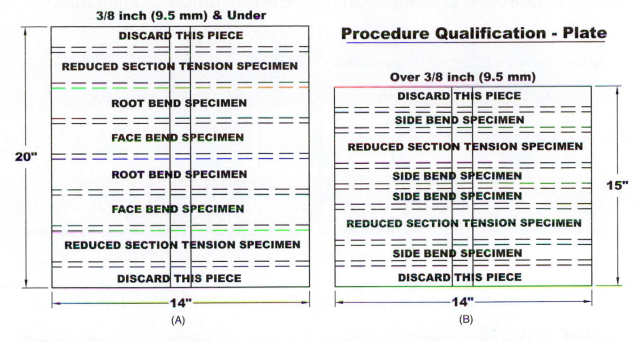

3/8 inch (9.5 mm) & Under

| DISCARD THIS PIECE |
| REDUCED SECTION TENSION SPECIMEN |
| ROOT BEND SPECIMEN |
| FACE BEND SPECIMEN |
| ROOT BEND SPECIMEN |
| FACE BEND SPECIMEN |
| REDUCED SECTION TENSION SPECIMEN |
| DISCARD THIS PIECE |

20"

14"

(A)

Over 3/8 inch (9.5 mm)

| DISCARD THIS PIECE |
| SIDE BEND SPECIMEN |
| REDUCED SECTION TENSION SPECIMEN |
| SIDE BEND SPECIMEN |
| SIDE BEND SPECIMEN |
| REDUCED SECTION TENSION SPECIMEN |
| SIDE BEND SPECIMEN |
| DISCARD THIS PIECE |

15"

14"

(B)

FIGURE 12-14 Location of weld test samples for procedure qualification on plate

Performance Qualification - Plate

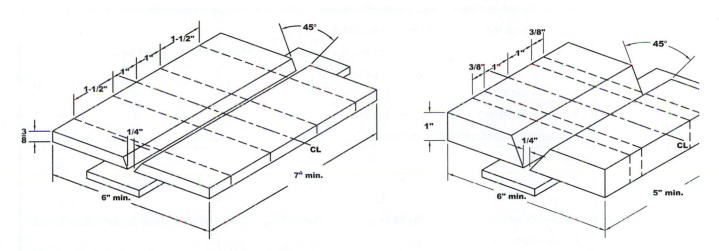

FIGURE 12-15 Location of weld test samples for performance qualification on plate

Specimens extracted from the pipe for ASME and AWS are the same. Side bend tests or face and root bends depend on the base metal pipe wall thickness. **FIGURE 12-16** illustrates where the specimens are extracted for Procedure Qualification and Performance Qualification.

API tests differ in where the specimens are extracted as well as the type of tests required. The type and quantity of specimens extracted depend on the diameter and pipe wall thickness. An API code book is required to determine how many and where to take the samples. API also requires nick break tests in addition to bend and tensile tests.

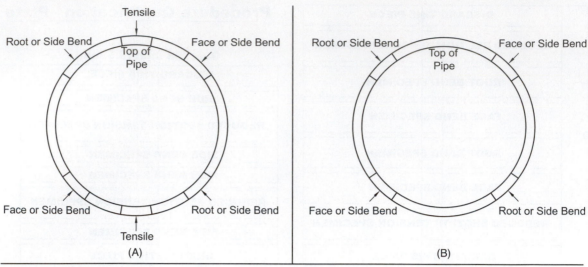

Procedure Qualification

Tensile

Root or Side Bend — Top of Pipe — Face or Side Bend

Face or Side Bend

Root or Side Bend

Tensile

(A)

Performance Qualification

Root or Side Bend — Top of Pipe — Face or Side Bend

Face or Side Bend

Root or Side Bend

(B)

Root/Face or Side Bend Depends on Metal Thickness

FIGURE 12-16 Location of weld test samples for procedure and performance qualification on pipe
Source: (A) AWS D1.1: 2008 p. 184, Figure 4.34, reproduced with permission of the American Welding Society (AWS), Miami, Florida. (B) Reprinted from ASME 2001 BPVC Edition, Section IX, by permission of The American Society of Mechanical Engineers. All rights reserved.

TABLE 12-13 and **TABLE 12-14** summarize the tests required for AWS and ASME by listing the number and types of test specimens. Lengths of pipe

Weld Testing for Plate and Pipe – AWS D1.1

Fillet	Thickness	Break Test	Macro-Etch	Root Bend	–
Procedure Qualification[1]	Varies (See Code)	–	3	–	–
Welder Qualification	Option 1 (Tee)	1	1	–	–
Welder Qualification	Option 2 (Lap)	–	–	2	–
Procedure Qualification[1]	**Thickness**	**Tensile**	**Root Bend**	**Face Bend**	**Side Bend**
Groove – Plate	$1/8 \leq T \leq 3/8$	2	2	2	–
Groove – Plate	$3/8 < T < 1$	2	–	–	4
Groove – Plate	1 inch and over	2	–	–	4
Welder Qualification	**Thickness**	**Tensile**	**Root Bend**	**Face Bend**	**Side Bend**
Groove – Plate	3/8	–	1	1	–
Groove – Plate	$3/8 < T < 1$	–	–	–	2
Groove – Plate	1 inch and over	–	–	–	2
Pipe	**Diameter**	**Thickness**	**Tensile**	**Root Bend**	**Face Bend**	**Side Bend**
Procedure Qualification[1]	2 or 3 inches	Sch 80 or 40	2	2	2	–
Procedure Qualification[1]	6 or 8 inches	Sch 120 or 80	2	–	–	4
Welder Qualification	≤ 4 inches	Any	–	$5G/6G^2 - 2$	$5G/6G^2 - 2$	–
Welder Qualification	> 4 inches	< 3/8 inches	–	$5G/6G^2 - 2$	$5G/6G^2 - 2$	–
Welder Qualification	> 4 inches	= 3/8 inches	–	–	–	$5G/6G^1 - 4$

TABLE 12-13

[1]Before preparing mechanical test specimens, weld test assembly shall be nondestructively tested by either Radiographic Testing or Ultrasonic Testing.
[2]Note: For 1G and 2G positions, the number of tests, is half that of the 5G and 6G positions.

Weld Testing for Plate and Pipe – ASME Section IX

Fillet Weld Test - Plate	Thickness	Break Test	Macro-Etch	–	–
Procedure Qualification	Varies (See Code)	–	5	–	–
Performance Qualification	3/16 to 3/8 inches	1	2	–	–
Fillet Weld Test - Pipe	**Thickness**	**Break Test**	**Macro-Etch**	**–**	**–**
Procedure Qualification	Wall Thickness		4[1]	–	–
Performance Qualification	Wall Thickness	1[1]	1[1]	–	–
Groove Weld - Plate	**Thickness**	**Tensile**	**Root Bend**	**Face Bend**	**Side Bend**
Procedure Qualification	< 1/16 to < 3/4	2	2[2]	2[2]	–
Procedure Qualification	3/4 & Over	2	–	–	4
Performance Qualification	< 1/16 to < 3/4	–	1[2]	1[2]	–
Performance Qualification	3/4 & Over	–	–	–	2

Pipe	Diameter	Thickness	Tensile	Root Bend	Face Bend	Side Bend
Procedure Qualification	Any	< 1/16 to < 3/4	2	2[2]	2[2]	–
Procedure Qualification	Any	3/4 & Over	2	–	–	4
Performance Qualification	Any	< 1/16 to < 3/4	–	2[2a]	2[2a]	–
Performance Qualification	Any	3/4 & Over	–	–	–	4
Procedure Qualification	Small Diameter[3]		Full Specimen	–	–	–
Performance Qualification	Small Diameter[3]		Full Specimen	–	–	–

TABLE 12-14

[1]Pipe to be cut into 1/4 sections.
[2]For thickness over 3/8 inches, 4 side bend tests may be substituted for the 2 root bend and 2 face bend tests.
[3]Pipe 3 inches OD and less may be tested by tensile-pulling the full specimen or by tensile, face bend, or root bend methods.

specimens depend on equipment used to perform the tests. Six inches is a minimum for specimen length; however, 8- to 9-inch lengths are more typical.

HOW ARE WELD COUPONS PREPARED, TESTED, AND EVALUATED?

There are a great deal of similarities between the codes. ASME and AWS have the greatest similarities, with only a slight variation in the preparation of the weld test samples. Fillet weld tests are measured, etched, and broken to determine uniformity and penetration. Bend specimens are tested by placing them in fixtures intended to exert a enough pressure on the face of the specimen to stretch and expose the surface and subsurface discontinuities for examination. Testing tensile specimens are pulled in a special machine that records the amount of stress exerted to fracture the specimen. API nick-break specimens may be pulled or fractured by repeated blows from a hammer but are examined and evaluated differently.

After testing, the specimens are evaluated. Evaluating test specimens can be difficult but not subjective. The specimen is either within specification or it is not. The codes specify minimum requirements. Any company may, at its discretion, impose more stringent requirements. The minimum standards set by the code enable the welded product to meet minimum engineering standards while

VIDEO

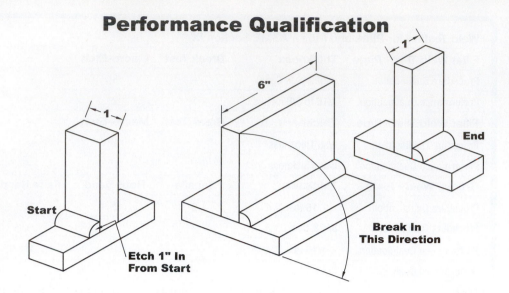

FIGURE 12-17 Locations and preparation of weld test samples on fillet welds.

Source: AWS D1.1: 2008 p 186, Figure 4.37, reproduced with permission of the American Welding Society (AWS), Miami, Florida.

in service. Establishing a more stringent requirement to pass a weld test does not jeopardize the set engineering standards.

FILLET WELD TESTS Fillet weld test assemblies consist of three parts: measuring the leg size, cut and etch, and break test. **FIGURE 12-17** illustrates how both AWS and ASME specimens are prepared. While only the start of the AWS specimen is acid-etched, the ASME code requires both the start and end specimens to be etched.

First, both legs of the fillet weld are measured to determine if the leg sizes conforms to the size requirements and are of (relative) equal size (see **FIGURE 12-18**). Next, the 1-inch (25 mm) end specimen(s) are etched to determine if proper fusion and penetration have been achieved. The etched specimen in **FIGURE 12-19** has adequate depth of fusion on the upper leg but lacks sufficient fusion on the lower leg to pass the test. Finally, the 6-inch (154 mm) midsection of the test assembly

FIGURE 12-18 Measuring fillet weld leg size

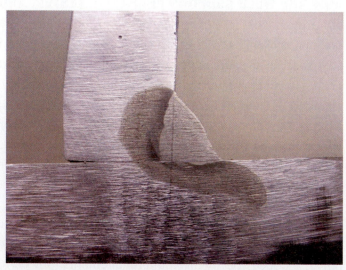

FIGURE 12-19 Etched sample showing depth of weld penetration

FIGURE 12-20 Placement of the middle section on a press for breaking

FIGURE 12-21 Fillet break showing incomplete fusion

is broken with the weld face up (**FIGURE 12-20**) to reveal the root and examined for complete fusion. The specimen in **FIGURE 12-21** has a section that reveals inadequate fusion.

TABLE 12-15 shows the allowable discontinuities for visual inspection after destructive testing. If any part of the three-part fillet weld test fails, the entire test fails. Any discontinuity greater than those listed is defined as a defect; the specimen and weld test fail. If a welder fails a test, most codes require further practice or training prior to retesting. If a code allows immediate retesting—AWS, for example—two weld test assemblies must be made and passed.

BEND TESTS   Bend tests are extracted from weld test assembles, as shown in Figure 12-13 and Figure 12-14. Extraction of the specimens is usually sawed or flame cut. The width of a side bend test is typically 3/8 inches (9.5 mm)—see **FIGURE 12-22**—depending on code requirements. The welds and backing bars,

AWS & ASME Bend Specimens

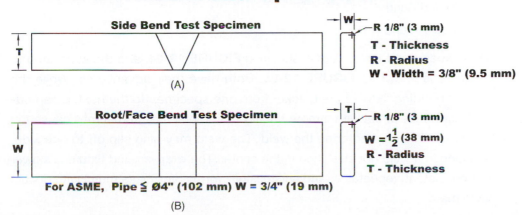

FIGURE 12-22 Bend test specimen dimensions

Source: (A) AWS D1.1: 2008 p.165–166—Figure 4.12, reproduced with permission of the American Welding Society (AWS), Miami, Florida. (B) Reprinted from ASME 2001 BPVC Edition, Section IX, by permission of The American Society of Mechanical Engineers. All rights reserved.

Grind Marks Must Run With The Specimen

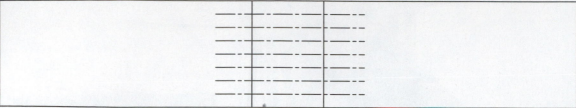

Grind Marks Running Across The Specimen Might Tear While Bend Testing

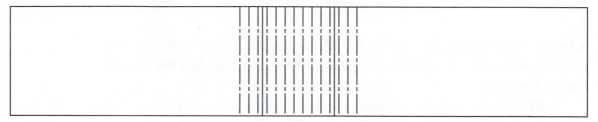

FIGURE 12-23 Orientation of grinding on bend test specimens

if used, are removed. Bend test specimens must have the grinding marks running lengthwise with the direction of the specimen (**FIGURE 12-23**). If grind marks are cross-wise, a tear may result along these marks, and it would be difficult to ascertain if a tear resulted from a weld discontinuity or from the grind marks.

After polishing the welds flush with the base metal, the specimen is formed around a mandrel to expose the face, root, or side of the weld. Bend test machines are generally in two forms:

- Guided Bend
- Wrap-Around

The guided bend test fixture is shown in **FIGURE 12-24**, and the wrap-around machine is shown in **FIGURE 12-25**. Both have their advantages. While the guided bending fixtures are quicker from one specimen to the next, a particularly hard base metal with a more ductile weld metal may sometimes cause the specimen to kink, fracturing the weld. The weld may also slip off to one side, resulting in a weld not stretched in the center. The wrap-around fixture is slower but consistent, repeatable, and successfully forms hard base metal with ductile weld metal.

WHAT ARE THE ACCEPTANCE CRITERIA FOR BEND WELD TEST SPECIMENS?

After bending, the outer, convex surface is examined for discontinuities. Weld bend test specimens are evaluated by measuring discontinuities in their greatest

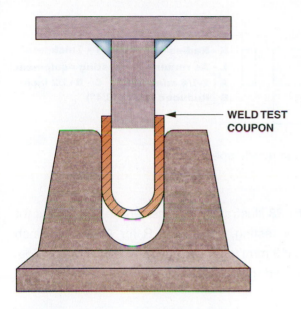

WELD TEST
COUPON

FIGURE 12-24 Fixture type guided bend test machine

FIGURE 12-25 Wrap-around guided bend test machine

dimension (**FIGURE 12-26**). Measurable acceptable discontinuities for AWS start at 1/32 inches (0.8 mm), with a maximum of 1/8 inches (3 mm), and a total of all discontinuities cannot exceed 3/8 inches (9.5 mm) per specimen. ASME states only that a discontinuity shall not exceed 1/8 inches (3 mm) per specimen. It is up to an inspector to determine the total amount of discontinuities allowed. It is often helpful to quote other codes to reinforce minimum requirements. Again, a company may designate stricter requirements than what a code states.

Cracks at the corner are not applicable to the 1/8-inch (3 mm) maximum requirement unless the crack occurred from an obvious discontinuity; if so, then the 1/8-inch (3 mm) rule applies (**FIGURE 12-27**). For AWS and ASME, Table 12-15 describes the allowable discontinuities for visual inspection after destructive testing.

FIGURE 12-26 Measuring the greatest dimension of a discontinuity

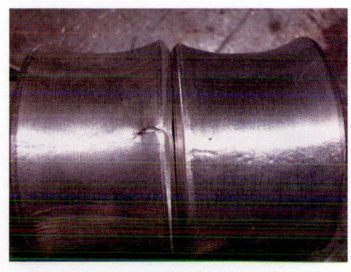

FIGURE 12-27 Example of a corner crack

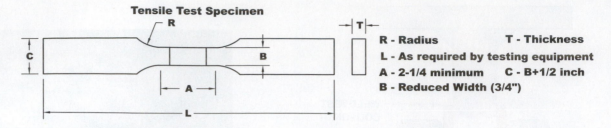

Tensile Test Specimen

R - Radius   T - Thickness
L - As required by testing equipment
A - 2-1/4 minimum   C - B+1/2 inch
B - Reduced Width (3/4")

FIGURE 12-28   Dimensions of a reduced-section tensile specimen

TENSILE SPECIMENS   **FIGURE 12-28** illustrates how the tensile specimens for AWS and ASME are prepared for testing. Dimension R for ASME is 1 inch (25 mm); for AWS, it is 1/2 inches (13 mm) for plate and 1 inch (25 mm) for pipe. API differs somewhat in that the weld reinforcement is not removed for tensile testing.

Prior to tensile testing, the specimen's width and thickness are measured to determine the cross-sectional area (CSA). For example, if the sample in Figure 12-28 has a "T" value of 0.438 inches (11 mm) and a "B" value of 0.750 inches (19 mm), the resulting cross-sectional area is 0.3285 inch² (209 mm²).

$$0.438 \times 0.750 = 0.3285 \text{ inch}^2$$
$$(11 \times 19 = 209 \text{ mm}^2)$$

The tensile specimen is placed in a tensile testing machine and subjected to a load. The load continues to increase until the specimen reaches the maximum load and then fractures. The machine measures the applied load, and that value is recorded. These steps are depicted in **FIGURE 12-29**, **FIGURE 12-30**, and **FIGURE 12-31**.

The ultimate tensile strength (UTS) is determined by dividing the cross-sectional area into the maximum load recorded on the tensile machine. Using the previous CSA example, if the recorded maximum load is 24,640 pounds

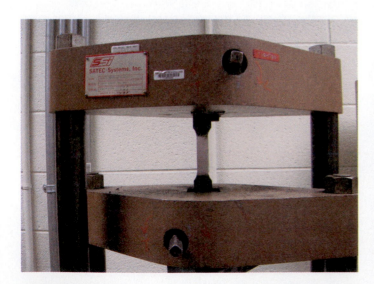

FIGURE 12-29 Tensile specimen setup before the load is applied

FIGURE 12-30 Tensile specimen after completed test

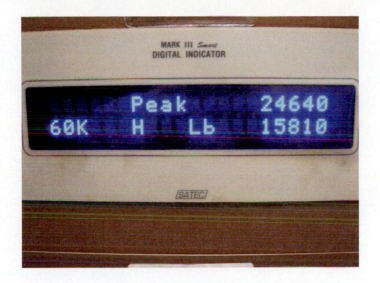

FIGURE 12-31 Tensile machine readout showing maximum load (24640) and breaking load (15810)

(109,535 newtons) and the cross-sectional area of the tensile specimen is 0.3285 inch^2 (209 mm^2), the UTS is 75,008 psi (524 MPa):

$$24,640 \text{ lbs} \div 0.3285 \text{ in}^2 = 75,008 \text{ psi}$$

$$109,535 \text{ N} \div 209 \text{ mm}^2 = 524 \text{ MPa}$$

One difference that ASME offers is an alternative tensile specification for a small diameter pipe (i.e., 3-inch (76 mm) diameter and under). With the ASME small diameter pipe, tensile and/or bend specimens are extracted or the entire weld test assembly is pulled. The fractured small diameter specimen is shown in **FIGURE 12-32**. To obtain the UTS of this specimen, the mean diameter would first be determined, and that value is then multiplied by the pipe wall thickness.

For example, a round tube measures 1 inch (25 mm) outside diameter (O.D.) with a 0.125 inch (3 mm) wall thickness. The mean diameter is 0.875 inches (22 mm). If the load fractured the specimen at 26,820 pounds (119,217 N), what would be the UTS?

$$1.000 \text{ in} - 0.125 \text{ in} = 0.875 \text{ in (O.D.} - \text{wall thickness} = \text{mean diameter)}$$
$$(25 \text{ mm} - 3 \text{ mm} = 22 \text{ mm})$$

$$3.1416 \times 0.875 \text{ in} = 2.7489 \text{ in } (\pi \times \text{mean diameter} = \text{circumference})$$
$$(3.1416 \times 22 \text{ mm} = 69 \text{ mm})$$

$$2.7489 \text{ in} \times 0.125 \text{ in} = 0.3436 \text{ in}^2 \text{ (circumference} \times \text{thickness} = \text{area)}$$
$$(69 \text{ mm} \times 3 \text{ mm} = 207 \text{ mm}^2)$$

$$26,820 \text{ lbs} \div 0.3436 \text{ in}^2 = 78,056 \text{ psi (load} \times \text{area} = \text{UTS)}$$
$$(119,217 \text{ N} \div 207 \text{ mm}^2 = 576 \text{ MPa})$$

To evaluate tensile test coupons, the minimum ultimate tensile strength of that particular base metal must be known. ASTM A36 (low carbon) steel, for example, has a minimum ultimate tensile strength of 58,000 psi (400 MPa). If the specimen's UTS is greater than 58,000 psi (400 MPa), the specimen passes. Table 12-15 summarizes the evaluation criteria for AWS, ASME, and API.

FIGURE 12-32 Fractured small diameter pipe tensile specimen

NICK-BREAK TESTS Nick-break tests are treated similarly to tensile pulls in that they are pulled in the same machine to break the specimen. Another method to test nick-break specimens is to secure the specimen in a vise and strike it with a hammer repeatedly until the specimen breaks. The 1/16-inch (1.6 mm) depth cut along the face of the weld for nick-break tests is intended as an option for automatic and semiautomatic welding processes. While a properly welded API tensile test should not break in the weld zone, the nick-break specimen's purpose is to examine the interior of the weld and is meant to break where sawed. The notch in the face enables the specimen to break more cleanly. **FIGURE 12-33** shows a nick-break specimen extracted from a weld pipe test assembly with a large discontinuity. **TABLE 12-15** outlines the acceptable criteria for successfully passing tensile, bend, fillet, and nick-break specimens.

A final weld test examining only the weld metal is called an All Weld Metal Test. It is used to identify the metallurgical mechanical properties of a filler

FIGURE 12-33 Fractured nick-break test specimen

Examination Criteria After Destructive Testing[1]

AWS

Tensile Test	Bend Test	Fillet Test
1. The tensile strength of the reduced specimen shall not be less than the minimum tensile strength range of the specified base metal.	No defect shall exceed: 1. 1/8 inches measured in any direction. 2. 3/8-inch sum of all defects exceeding 1/32 inches. 3. 1/4-inch corner crack unless crack is from obvious defect (then, 1/8-inch rule applies).	1. Macro-etch specimens shall show fusion to the root of the joint; full fusion between layers of weld metal and between base metal and weld metal; acceptable profiles and leg size; undercut not to exceed 1/32 inches; no cracks. 2. Break test to show full penetration into the root; no inclusions or porosity larger than 3/32 inches; the sum of any defect not to exceed 3/8 inches.

ASME

Tensile Test	Bend Test	Fillet Test
1. The tensile strength of the reduced specimen shall not be less than the minimum tensile strength of the base metal or the minimum tensile strength or the weaker of the two if different base metals are welded together.	1. The face of the bend-guided test coupon shall have no open defects greater than 1/8 inches. 2. Open defects on the corners are allowed provided they are not caused by a definite defect, such as slag inclusions, lack of fusion, or another internal defect.	1. Break tests shall show no evidence of cracks or incomplete root fusion; the sum of any defect shall not exceed 3/8 inches or 10% of a quarter section. 2. Macro-etch specimens shall show complete fusion and be free from cracks; Performance Qualification allows linear root indications up to 1/32 inches.

API

Tensile Test	Nick-Break Test	Bend Test
1. The tensile strength of the specimen shall be greater than or equal to the minimum specified tensile strength of the base metal. 2. If the specimen breaks in the weld metal, meets the minimum strength criteria, and meets the nick-break test criteria, the specimen passes.	1. Exposed surface shall show complete penetration and fusion. 2. No gas pocket to exceed 1/16 inches. 3. Sum of all gas pockets shall not exceed 2% of the exposed surface. 4. Slag inclusion not to exceed 1/32 inches in depth and 1/8 inches in length or one-half the nominal wall thickness (whichever is smaller). 5. There shall be at least 1/2 inches of sound metal between slag inclusions.	1. No cracks or other defects exceeding 1/8 inches or one-half the nominal wall thickness (whichever is smaller). 2. Cracks at the corner that are less than 1/4 inches are acceptable unless caused by an obvious defect.

TABLE 12-15

[1]Criteria is minimum acceptable by code; more stringent criteria may always be set by an inspector.

metal used in production. This test is often done when a desired filler metal does not meet any AWS classification or specification.

A groove joint is prepared and filled in with the filler metal to be tested. The base metal is cut away, and the remaining weld metal is turned in a lathe to produce a round test specimen (see **FIGURE 12-34** and **FIGURE 12-35**). The test sample is pulled in a tensile testing machine, and the mechanical properties are determined, as would a weld test sample previously described. Size of the specimen may vary, but the reduced section is of sufficient length to include the two marks for calculating the percentage of elongation.

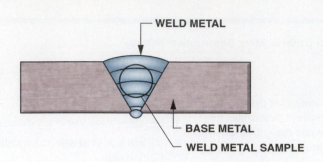

FIGURE 12-34 End-view of an All Weld Metal test showing from where the test sample is taken.

WELD METAL

BASE METAL

WELD METAL SAMPLE

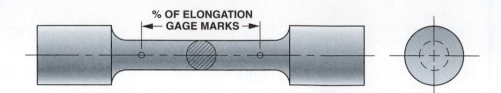

FIGURE 12-35 Example of general layout for an All Weld Metal Test specimen

% OF ELONGATION
GAGE MARKS

NONDESTRUCTIVE TESTING

WHAT NONDESTRUCTIVE TESTING METHODS ARE USED ON WELDED PARTS? The vast majority of inspection is visual, and there are many gauges and tools available to aid in the process. **FIGURE 12-36** displays an inspection tool kit with many measuring instruments and gauges.

Fillet welds are common in welding, and many different gauges are used to measure leg size, convexity, and concavity. **FIGURE 12-37** shows two different gauges used to measure the leg size of a fillet weld. **FIGURE 12-38** shows a gauge used to measure the convexity or concavity of a fillet weld.

Some gauges have multiple functions. The same combination gauge used to measure the leg size of a fillet weld may also be used to measure convexity.

FIGURE 12-36 A complete inspection tool kit

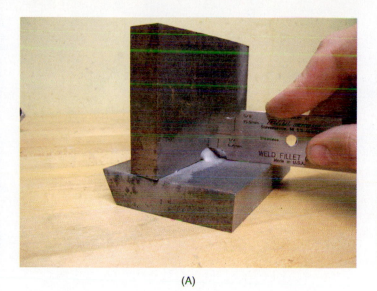

(A)

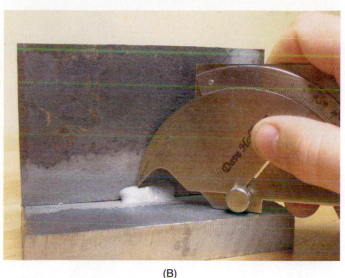

(B)

FIGURE 12-37A Measuring fillet weld leg size with a standard fillet weld gauge

FIGURE 12-37B Measuring fillet weld leg size with a Bridge Cam Gauge

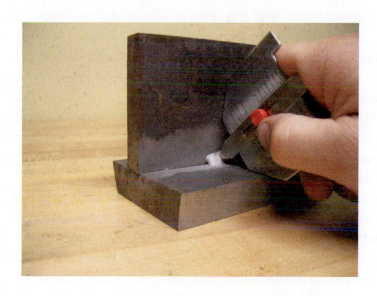

FIGURE 12-38 One type of AWS gauge measuring convexity of a fillet weld

And the same gauge may also be used to measure undercut or check the angle of beveled plate prior to welding (**FIGURE 12-39**).

While visual examination is utilized the majority of the time in weld inspection, it cannot diagnose what lies beneath the surface. There are four nondestructive methods commonly used to examine and evaluate welds other than visual inspection:

- Radiographic (x-ray)
- Ultrasonic
- Magnetic Particle
- Liquid Penetrant

Radiographic inspection is most often specified as an alternative to destructive testing. The ASME code requires destructive testing for procedure qualification but allows either destructive testing or radiographic testing for welder qualification. There is a certain amount of logic to this. In Procedure Qualification, the concern is

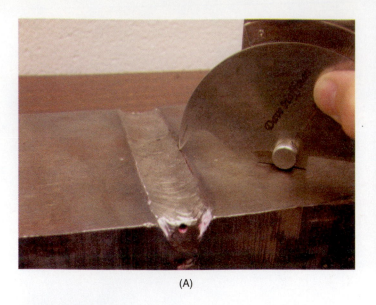

(A)

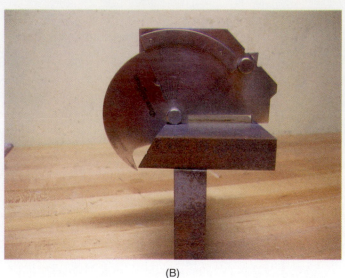

(B)

FIGURE 12-39A Using a Bridge Cam Gauge to measure undercut

FIGURE 12-39B Using a Bridge Cam Gauge to measure bevel angle

that the base metal, filler metal, and all the other variables that make up the welding procedure are compatible. If an inappropriate base metal/filler metal combination is used or there is no preheat or postheat treatment where there should have been, there is a good chance that even though the x-ray shows it is free of discontinuities, the weld may crack or even completely fracture during a simple bend test.

Likewise, API requires destructive testing for Procedure Qualification but will permit radiographic testing for welder qualification for butt welds in lieu of destructive testing. API refers to radiographic testing throughout the code and has a section devoted to those requirements.

The D1.1 Structural Steel Welding code is the most often used code when referring to AWS. For all welding processes this code specifies, and for the majority of the weld joints used to fabricate a product, AWS has Prequalified Procedures. In other words, a company does not have to write and test a welding procedure in order to qualify for this code. If an unconventional procedure is desired—for example, welding with short-circuit transfer on a plate steel—then a procedure qualification is necessary. However, AWS requires not only destructive testing for procedure qualification but also nondestructive testing. The code devotes several pages to radiographic requirements. However, the most extensive coverage is devoted to ultrasonic testing. AWS D1.1 also makes some reference to and requirements for magnetic particle and liquid penetrant (dye penetrant) testing.

RADIOGRAPHIC TESTING

WHAT IS RADIOGRAPHIC TESTING? **Radiography** (radiographic testing/ x-ray) is commonly used to inspect welds that are fabricated as specified by a welding code. Radiography is reliable but has a few major limitations. All work in the inspection area has to be stopped and the workers vacated to avoid exposure. This can impair work progress and delay the job. Also, the exposed film is only two-dimensional. Depth of discontinuities cannot be determined unless a second view can be exposed. Because of the configuration of the part being tested, this may not be possible. The development of real-time imaging computer processing techniques for radiographic testing helps overcome the time-consuming process that was involved with film recording.

Although there are many types of radiography, such as CT scan, reverse geometry x-ray, and micro-focus x-ray microscopy, the two types generally used for testing of weldments are x-ray and gamma ray radiography. In the x-ray or gamma ray method, a film is placed behind the part. A radioactive source is then exposed to the part. The radioactive particles bombard the part to reveal materials of various densities, which show up as lighter and darker areas on the film (**FIGURE 12-40**).

HOW ARE SIZES OF A DISCONTINUITY DETERMINED FROM AN X-RAY?

A standard film identification marker is usually included in every radiograph for exposure verification and film identification. The film identification marker is called an Image Quality Indicator (IQI), also referred to as a Penetrameter. IQI is made of the same material or a similar material as the specimen being radiographed and is of a simple geometric form. It contains some small structures (holes, wires, etc.), the dimensions of which bear some numerical relation to the thickness of the part being tested. The image quality of the IQI on the radiograph is permanent evidence that assures proper radiographic procedures have been followed.

Codes or agreements between customer and vendor may specify the type of IQI, its dimensions, and how it is to be employed. Even if Image Quality Indicators are not specified, their use is advisable because they provide an effective check of the overall quality of the radiographic inspection. The common IQI consists of either a simple geometric shape with specific size holes or a series of

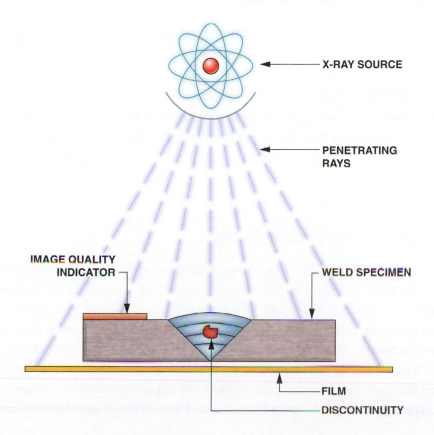

FIGURE **12-40** An illustration of radiographic testing

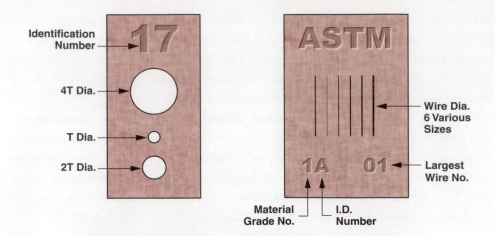

Identification
Number

4T Dia.

T Dia.

2T Dia.

Wire Dia.
6 Various
Sizes

Largest
Wire No.

Material
Grade No.

I.D.
Number

FIGURE 12-41 Examples of Image Quality Indicators

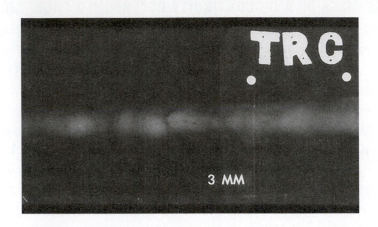

FIGURE 12-42 A radiographic image of a groove weld

wires of different diameters encased in clear plastic. **FIGURE 12-41** shows the two types of Image Quality Indicators.

WHAT DOES A RADIOGRAPHIC PICTURE LOOK LIKE?
The X-ray or gamma-ray beam penetrates through a test specimen striking a film placed on the side of the test specimen opposite from the source. The resulting film is a negative with heavy dense materials, such as tungsten, developing as bright white areas and porosity or slag inclusions developing as dark areas. **FIGURE 12-42** is a radiographic picture of a weld. Notice the dark line in the center of the weld indicating a less dense structure than the weld, such as a crack, slag inclusion, or linear porosity.

ULTRASONIC TESTING

WHAT IS ULTRASONICS?
Ultrasonic testing induces a high-frequency sound into a material by using a piezoelectric transducer searching for acoustical impedance mismatch. The sound frequency is much higher than human hearing. Humans can generally hear up to 20,000 Hz (hertz). Transducers, which are used in ultrasonic testing, commonly range from 1.0 MHz to 10.0 MHz (megahertz). On straight beam testing and calibration, when a sound travels through a material that is free from defects, it will ping as it strikes the opposing side and bounce back. The ultrasonic testing equipment is calibrated to read this thickness on a screen.

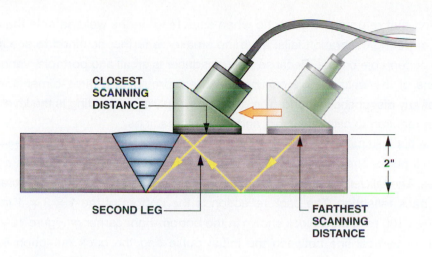

CLOSEST
SCANNING
DISTANCE

SECOND LEG

FARTHEST
SCANNING
DISTANCE

2"

FIGURE 12-43 An illustration of angle beam ultrasonic testing

When the sound is interrupted by something like slag inclusions (acoustical impedance mismatch), the ping is bounced back and read on the screen at a specific depth between the initial pulse (top of the plate) and the back reflection (bottom of the plate). The back reflection may or may not be the bottom of the plate depending on part geometry and requirements. It is usually calibrated as the greatest thickness scanned.

AWS specifies transducer size and frequency for longitudinal (straight-beam) and angle-beam search units. Angle-beam angles are to be 45°, 60°, or 70°. Straight-beam testing is easier to calibrate and for interpreting results but requires a smooth surface. Angle-beam testing requires a different procedure for calibration by using a special angle block with specific size holes. Angle-beam testing does not require the weld reinforcement to be removed when searching for discontinuities (**FIGURE 12-43**).

HOW DO YOU CALIBRATE AND READ AN ULTRASONIC UNIT? Prior to searching for discontinuities with ultrasonics, the unit must be calibrated on metal that has the same sound velocity as the weld sample (**FIGURE 12-44**). Calibration predetermines the size of discontinuities found while searching.

FIGURE 12-44 An ultrasonic readout showing initial pulse, discontinuity, and back reflection

Anything that returns a larger echo when searching on the weld sample than a level set during calibration fails; anything smaller is further scanned to see if it meets acceptable criteria. Because the transducer is small and portable, various sections of the weld sample can be searched, providing a three-dimensional scan of any discontinuity found. The downside of ultrasonic testing is the level of training required to calibrate, read, and interpret readings.

The first vertical line to the far left shown on the LCD screen in Figure 12-44 is the initial pulse. The initial pulse is the sound that the transducer produces as it vibrates. The tall vertical line that is approximately in the center of the LCD screen is the back reflection. The back reflection is the bottom of the $1 \times 3 \times 4$ inch ($25 \times 75 \times 100$ mm) steel block shown in the bottom-right corner of Figure 12-44. The short vertical line between the initial pulse and the back reflection is a 3/32-inch (2.4 mm) hole in the steel block directly below the transducer approximately 2 inches (50 mm) from the top surface. The hole in the steel block is approximately two-thirds from the top surface of the plate, and the corresponding short vertical reflection on the LCD screen is approximately two-thirds from the initial pulse to the back reflection. Thus, the depth of the discontinuity would be known by where a vertical line is displayed on the ultrasonic LCD screen. Any discontinuity in a steel sample to be examined larger than the calibrated 3/32-inch (2.4 mm) hole would show up as a taller vertical line than the short line in Figure 12-44.

Angle-beam search units pick up discontinuities on the second leg of the reflected sound wave, as shown in Figure 12-43. While the straight beam uses a longitudinal wave, the angle beam uses a shear wave. The shear wave initial pulse reflection on the ultrasonic LCD screen shows up as noise and is ignored (**FIGURE 12-45**).

AWS specifies a rectangular 2.5 MHz transducer and 45° angle lucite wedge when weld reinforcement removal is not desired, as shown in **FIGURE 12-46**.

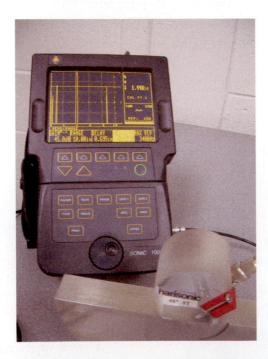

FIGURE 12-45 An ultrasonic readout showing shear wave initial pulse noise

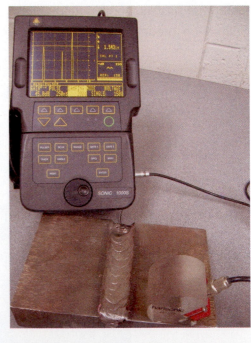

FIGURE 12-46 A 2.5 MHz transducer with 45° angle lucite wedge

FIGURE 12-47 An International Institute of Welding (IIW) ultrasonic reference block

AWS also specifies an International Institute of Welding (IIW) ultrasonic reference block for angle-beam search unit calibration (**FIGURE 12-47**).

MAGNETIC PARTICLE TESTING

WHAT IS MAGNETIC PARTICLE TESTING?   **Magnetic particle testing** is the process of inducing a ferromagnetic material (a material that will attract a magnet) with a current that produces lines of flux similar to what is produced in a typical magnet. Metal filings applied to the magnetized test specimen gather along discontinuities. The advantages of magnetic particle testing are:

- It is fast and simple.
- It is a reliable method for surface cracks. Near subsurface discontinuities can also be detected.
- There are few limitations on size or shape of a specimen.

The limitations of magnetic particle testing are:

- It can only be used on ferromagnetic materials.
- It utilizes large amounts of current.
- The orientation of magnetic fields must be correct to detect discontinuities.

The process can be applied wet or dry. The wet method is superior in detecting surface flaws, fine cracks, and for production applications. The powders used are less dense, smaller in size, and have a more uniform shape. It is also used where the test specimens have lower permeability and higher retentivity. When flourescent powders and a black light are used, the greatest visibility of discontinuities is obtainable. The dry method is more flexible and is more sensitive to subsurface discontinuities when used with direct current. Because the dry method is used with portable and mobile equipment, size is not a factor. For soft iron or low-carbon steel, the dry method is best. With either the wet or dry method, the test specimen must be ferromagnetic.

Ferromagnetic materials include iron, steel, cobalt, and their alloys. Common examples of nonferromagnetic materials include aluminum, magnesium, copper, brass, bronze, and titanium.

MAGNETIC FIELDS   There are two types of induced magnetic fields: circular and longitudinal. Regardless of the how the magnetic field is induced, the lines of force must cut across the discontinuity to create a leakage field. If the lines of force do not cut across the discontinuity, the leakage field will not be created and the discontinuity will not be revealed.

WHAT TYPE OF EQUIPMENT IS USED IN MAGNETIC PARTICLE TESTING?

MAGNETIC PARTICLE EQUIPMENT   The equipment may be AC, DC, or half-wave DC. AC is best for detecting surface discontinuities, and DC is good for detecting surface and subsurface discontinuities. Subsurface discontinuities must be fairly close to the surface. Categories of equipment are:

- Small portables capable of 700 amperes and some up to 2,000 amperes
- Large portables or mobiles capable of up to 3,000 amperes
- Stationary such as direct contact, central conductor, or magnetizing coil
- Special purpose for unusual applications like on disk-like parts with an I.D.

The magnetic yoke is often used because of its portability. Portable equipment usually utilizes the dry technique in which dry powder metal filings are sprayed on the test specimen while current is applied. If the test specimen has low permeability and high retentivity, the test specimen is first magnetized by inducing the current. Then, powder metal filings are applied after the current is turned off, while the specimen retains its magnetism. Dry metal filing powders (**FIGURE 12-48**) may be gray, red, yellow, or black.

SOME WORDS LIKE PERMEABILITY AND RETENTIVITY WERE USED. WHAT DO THEY MEAN?

PROPERTIES OF MAGNETIC PARTICLES   When magnetizing a material, certain properties must be considered:

- **Permeability:** The ease with which a magnetic flux can be established in a given material
- **Reluctance:** The opposition of a magnetic material to the establishment of a magnetic flux
- **Flux Density:** The number of lines of force per unit area

FIGURE 12-48 Dry metal powder filling in red, black, and gray

- **Residual Magnetism:** The amount of magnetism a magnetic material retains after the magnetizing force is removed
- **Retentivity:** The ability of a material to retain magnetism after the magnetizing force has been removed
- **Coercive Force:** The reverse magnetizing force necessary to reduce residual magnetism to an acceptable level. This force is opposite in direction and lower in value than the original force that produced the magnetism
- **Demagnetization:** The process by which residual magnetism is reduced to an acceptable level in a test specimen

WHAT ARE THE STEPS IN MAGNETIC PARTICLE TESTING?

- **Step 1: Check the retentivity before magnetizing.** Before applying a magnetic current to a specimen, it is beneficial to determine if any magnetism resides in the test specimen. Later, after testing, if demagnetization is needed or desired, it could be compared as a before-and-after magnetization scenario.
- **Step 2: Magnetize the test specimen.** Flip the switch on yoke to DC. DC is used to magnetize the specimen. The knob above the switch, if so equipped, increases or decreases the amount of current induced into the specimen. Press the *on* button on the top of the magnetic yoke to induce magnetism; see the small portable magnetic yoke in **FIGURE 12-49**.
- **Step 3: Check for retentivity.** After removing the magnetic yoke, it is necessary to see if there is an adequate amount of magnetism in the test specimen in order to conduct a test for discontinuities (**FIGURE 12-50**).
- **Step 4: Apply the iron powder.** The powder is applied after the part has been magnetized. The permeability and the retentivity of the material will determine if remagnetizing the specimen is needed (**FIGURE 12-51**).

FIGURE 12-49 A portable magnetic yoke magnetic particle tester

FIGURE 12-50 An indicator measuring residual magnetism

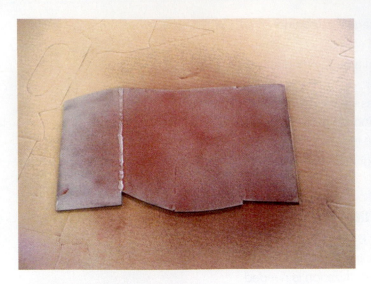

FIGURE 12-51 A magnetized part with powder applied

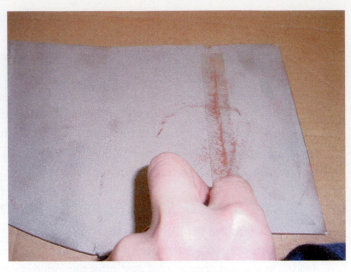

FIGURE 12-52 Clear tape used to record the magnetic leakage field for reporting

- **Step 5: Remove excess powder.** Removing the excess powder by gently blowing it away with a squeeze ball will reveal if a discontinuity exists and where it is located in the test specimen. Low retentivity will result in the iron filings moving quite easily and in the removal of evidence of a discontinuity.

- **Step 6: Record discontinuities.** Discontinuities are easily seen and are often included in an inspection report by using clear Scotch tape to collect the iron particles across the leakage field and taping them on the report (**FIGURE 12-52**).

- **Step 7: Demagnetize the test specimen.** Flip the switch on the magnetic yoke to AC. Hold the yoke just above the test specimen. Depress the "*on*" button on the top of the magnetic yoke. Slowly pull away from the test specimen while still depressing the on button. See the AC/DC switch in Figure 12-49.

- **Step 8: Check for retentivity after demagnetizing.** After demagnetizing the test specimen, check to see if it is necessary to repeat step 7 (**FIGURE 12-53**).

FIGURE 12-53 Measuring residual magnetism after a test

DYE PENETRANT TESTING

WHAT IS DYE PENETRANT TESTING? **Dye penetrant testing** is a method in which a penetrant—a high color contrast visible dye or a fluorescent spray—is applied to a specimen in conjunction with a developer to bring out the dye, there by testing for such things as cracks and porosity. A discontinuity *must* be open to the surface for penetrant testing to be successful. The advantages of dye penetrant testing are simplicity and economy, versatility, direct indication, and high sensitivity.

The limitations of dye penetrant testing are:

- The discontinuity must be open to the surface. Subsurface discontinuities cannot be detected.
- It is difficult to test porous specimens or those with rough surfaces from welding or grinding. These conditions may provide false discontinuities by holding the penetrant.
- Some penetrant materials can harm the test specimen. Sulfur and chlorine affect certain metals such as the nickel in stainless steel, and some plastics and rubber may be affected. Oil- and/or water-based penetrants can be difficult to remove from some areas and can cause defects if subsequent welds are to be made.

WHY USE DYE PENETRANT TESTING? When deciding to use the dye penetrant process, the type of penetrant and the process used depend on the discontinuities being sought and the material being tested. Some factors to consider include:

- Sensitivity
- Specimen size and shape
- The number of specimens being tested
- The availability of equipment and facilities

Sensitivity refers to the ease and certainty of seeing small amounts of penetrant and being able to locate discontinuities of a particular type in a particular specimen. Examples of these are surface defects in a rough casting or fine cracks in an exotic material.

There are basically two classifications of penetrant: visible dye (also known as color contrast) and fluorescent. Visible dye and fluorescent penetrant have about the same penetration ability. Visible dyes come in different colors, and enough contrast is needed to see a defect. Portability and simplicity are advantages of visible dye. Fluorescent penetrants seen under black light are brighter and are easier to see than visible dye, but a dark room and electricity are needed to perform the test. For spot-checking in the field, the visible dye is the preferred method.

WHAT ARE THE DIFFERENT TYPES OF PENETRANT PROCESSES? The three basic visible dye penetrant processes are:

- **Water Washable:** This is used for high-production testing of small parts or rough surfaces, such as threads or keyways. The downside of this process is that it is easily washed away on shallow discontinuities.

- **Postemulsified:** There are two types: lipophilic (oil-based) and hydrophilic (water-based). This is used in the high production of large parts. There is less danger of washing the penetrant from cracks than with the water washable type.
- **Solvent:** Solvents are used for spot checks or testing in the field. This is one of the more common processes due to its ease of use and portability. Solvents can be naphtha, mineral spirits, or a special formula.

WHAT ARE THE DIFFERENT TYPES OF DEVELOPERS?

The use of developers depends on required sensitivity, available equipment, and specimen surface condition. There are four basic types of developers:

- **Water Suspended:** Both water suspended and water soluble are most effective when applied to smooth surfaces.
- **Water Soluble:** This has the greatest sensitivity when applied as a spray.
- **Dry Powder:** This method is better adaptable for rough surfaces and automatic processing. It is the easiest to remove.
- **Nonaqueous:** This is most widely used for in-field testing and spot checks. It is also used when the greatest sensitivity is required.

WHAT MAKES UP THE MOST POPULAR COMPONENTS OF DYE PENETRANT TESTING?

The three main components used in dye penetrant testing are:

- **Penetrant:** A visible dye (**FIGURE 12-54**) that seeps into the cracks and voids of the specimen to be tested. The penetrant is a color easily visible. Red is a common color for penetrant.
- **Developer:** Used to draw out the penetrant from the voids and cracks. The developer (**FIGURE 12-55**) is a liquid that turns bright white after it dries. The penetrant bleeds through the developer, displaying a vivid contrast.
- **Cleaner:** Used to clean off the area to be inspected prior to penetrant application and to remove the penetrant and developer from the surface after testing (**FIGURE 12-56**).

FIGURE 12-54 A container of visible dye penetrant

FIGURE 12-55 A container of developer

FIGURE 12-56 A container of penetrant and developer cleaner/remover

WHAT ARE THE STEPS OF DYE PENETRANT TESTING? All Nondestructive Testing methods must have qualified procedures, just as welding methods do. There are some general guidelines, however, that apply to each method. Dye penetrant testing involves six basic steps:

1. **Surface Preparation (Precleaning):** The single most common cause of failure to test correctly is poor surface preparation. Contaminants can seal off discontinuities, cause nonrelevant indications, and obscure true discontinuities. Paint, scale, loose dirt, slag, glass deposits, oil, grease, spatter, and rust can all cause nonrelevant indications. Chemical cleaning is preferred because shot blasting, sand blasting, or wire brushing can peen, fill, or smear discontinuities. Solvents or water-based cleaners must be thoroughly removed and allowed to evaporate for 10–20 minutes or more depending on the temperature to ensure they are not retained in discontinuities during penetrant application. Slag, glass deposits from wire feed welding, and spatter will have to be removed mechanically. Whatever method is used, the surface must be clean and dry prior to penetrant application.

2. **Penetrant Application:** One of the most common methods of penetrant application, especially for remote locations, is spraying. Temperatures between 60°F and 120°F lead to the best results. The distance to the specimen should be 10 to 12 inches (25 to 30 cm). The time allotted to soak in penetrant should be 5 to 30 minutes. Although the penetrant needs to be applied thoroughly, excessive application is undesirable due to the extended time it takes for penetrant removal (**FIGURE 12-57**).

   The penetrant is usually applied to the exterior and cleaned off. The penetrant will bleed through the developer. However, the penetrant may also be applied to the interior and drawn through areas when a product is required to be airtight or watertight. The question as to whether the penetrant is applied internally or externally often depends on if it is undesirable to have the penetrant remain on the part or not. In the case of porosity or cracks, the discontinuity may only be open to the outside surface, and the penetrant must be applied to the external surface.

FIGURE 12-57 Application of visible penetrant

3. **Excessive Penetrant Removal:** The removal of excess penetrant depends on if the penetrant is water-soluble or emulsified. Dipping is best for postemulsified processes from 1 to 3 minutes; brushing or wiping is not recommended. If the penetrant is water-soluble, large water droplets are best applied at temperatures between 60°F and 110°F, at 30 to 40 psi at a 45° angle. These methods are often employed in a lab as opposed to a remote location. In a remote location, such as in the case of weldments or when searching for cracks on cast surfaces, the solvent-based penetrant is removed by spraying the chemical cleaner on a clean shop rag and wiping the surface clean. This step must be repeated until the surface is clean of all penetrant to prevent false indications. The chemical cleaner must never be sprayed on the surface, as it may remove penetrant that has seeped into cracks or porosity. Spraying the chemical cleaner directly on the surface could remove too much penetrant and therefore prevent finding an indication (**FIGURE 12-58** and **FIGURE 12-59**).

Correct Penetrant Removal:
Spray cleaner on a shop rag first.

FIGURE 12-58 Correct penetration removal involves spraying remover on a towel.

Incorrect Penetrant Removal:
Do not spray directly on the part.

FIGURE 12-59 Incorrect penetration involves spraying remover directly on the part.

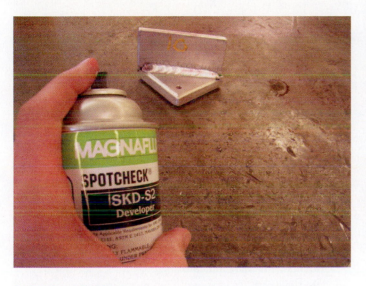

FIGURE 12-60 Application of developer

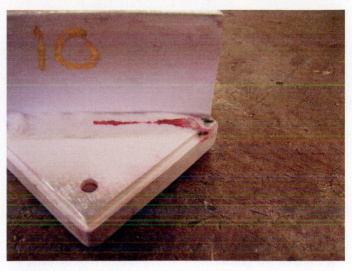

FIGURE 12-61 Developer dries and indicates defects by leakage of colored penetrant.

4. **Developer Application:** Water-based developers are applied immediately following the rinse. Temperatures of 150°F to 225°F are recommended, and the length of drying time should be just long enough for a part to dry. In the case of chemical cleaners, the surface should be completely dry prior to the application of the developer. Spray a thin, even coating. Overspraying may wash out the penetrant or obscure indications. Dry and nonaqueous developing time should be from 7 to 30 minutes (**FIGURE 12-60**).

5. **Inspection:** The inspection process is usually for cracks, porosity, and small inclusions. False indications result from the poor and incomplete washing of penetrant from the test surface. Also, rough surfaces, sharp fillets, blind holes, lint, dirt, threads, keyways, press fit or joints, and mating parts can retain penetrant and give false indications. **FIGURE 12-61** shows a part with indications caused from cracking. **FIGURE 12-62** shows an image with indications using fluorescent dye penetrant.

6. **Postcleaning:** Usually, postcleaning is necessary to rid penetrant and developer from the part. If the part is to be painted, plated, or finished in any way, postcleaning is necessary. Penetrant and developer residue can attract moisture and may cause corrosion. The best method to remove penetrant and developer is the same as precleaning.

FIGURE 12-62 Fluorescent penetrant testing

MULTIPLE CHOICE

1. What welding societies were formed around 1919?
   a. CWI, CAWI, and SCWI b. QC1, ASME, and AWS
   c. CWI, QC1, and AWS d. ASME, AWS, and API

2. How much work experience does a person need to become a Certified Welding Inspector?
   a. 15 years b. 5 years
   c. 2 years d. 1 year

3. What is the publication that outlines the requirements for becoming a Certified Welding Inspector?
   a. AWS QC1 b. ASME QC1
   c. API QC1 d. CWI QC1

4. What codes allowed for testing on an AWS CWI exam?
   a. Only AWS b. AWS or API
   c. AWS, ASME, and API d. AWS or AMSE

5. There are three parts of the written exam for certified welding inspectors: General Welding Knowledge, Welding Code Interpretation, and _____.
   a. Boiler and Pressure Vessel b. Welding Inspection
   c. Structural Steel d. Nondestructive Testing

6. The three levels of weld inspector qualification are CWI, SCWI, and _____.
   a. AWS b. ACWI
   c. CWI-A d. CAWI

7. What is a code?
   a. Any document or object used as a basis of comparison
   b. Mandatory when specified by the government or when a contract is made that requires them
   c. Generally associated with a product. It describes the requirements for a particular object, material, service, etc.

8. Can a welder take just one weld test to become qualified (certified) for everything?
   a. No b. Yes

9. The four most common nondestructive testing methods are radiographic, magnetic particle, ultrasonic, and _____.
   a. Eddy current b. Tensile testing
   c. Dye penetrant d. Bend testing

10. Another name for a penetrameter is
    a. Image Quality Indicator
    b. Radiography
    c. X-ray
    d. Film

11. The four categories of magnetic particle testing equipment are small portable, stationary, special purpose, and _____.
    a. Fixed b. Portable
    c. Medium portable d. Large portable

12. The metal filing colors used in magnetic particle testing are gray, black, yellow, and _____.
    a. Orange b. Green
    c. Red d. Blue

13. The two types of dyes used in dye penetrant testing are visible dye and _____.
    a. Invisible dye b. Fluorescent dye
    c. Transparent dye d. Opaque dye

MATCHING

14. Match the correct acronym with the following definitions.
    a. Is a recipe specifying all the variables to successfully complete a weldment
    b. Tests the mechanical properties of the welding procedure
    c. Is a record of the welder's successfully completed weld
    _____ WQTR
    _____ WPS
    _____ PQR

15. Match the testing methods used in ultrasonic with another name for them.
    a. Longitudinal _____ Angle Beam
    b. Shear _____ Straight Beam

16. Match the correct term with the definition for magnetic particle testing.
    a. If a crack is discovered, the name given where the metal filings gather around
    b. The ease with which a magnetic flux can be established in a given material
    c. The ability of a material to retain magnetism after the magnetizing force has been removed
    d. The amount of magnetism a magnetic material retains after the magnetizing force is removed
    _____ Retentivity
    _____ Leakage field
    _____ Residual Magnetism
    _____ Permeability

17. Identify the following as a standard, code, or specification:
    - SFA 5.1 _____
    - Weld Symbol _____
    - QC1 _____
    - API 1104 _____
    - A 36 _____
    - Procedures _____

SHORT ANSWER

18. List one advantage and one disadvantage of each nondestructive testing method.

19. Why must an ultrasonic testing unit be calibrated?

20. How is a record kept of the discontinuity for Magnetic Particle testing?

21. List the three basic visible dye penetrant processes.

22. List the four different types of developers for dye penetrant testing.

23. Why should cleaner never be sprayed directly on the specimen to be tested in dye penetrant testing?

24. How long should the soak time be for dye penetrant testing?

25. What are some reasons to remove the developer after testing is complete?

26. Calculate the cross-sectional area and the ultimate tensile strength from the following data:
    - Plate tensile specimen measures 0.752 inches × 0.370 inches (19.1 mm × 9.5 mm)
    - Hypothetical load is 23,950 pounds (106,467 N)
    - 1.315 inches (33.4 mm) outside diameter pipe with a 0.133-inch (3.4 mm) wall thickness
    - Hypothetical load is 35,320 pounds (157,006 N)

27. Write a welding procedure (WPS) with the following requirements:
    - Code Requirement: AWS
    - Process: FCAW
    - Material: mild steel
    - Weld Type: plate
    - Joint: groove, 45° included angle
    - Backing/Root Opening: yes - 1/4 inches (6.4 mm)
    - Qualification Range: 1/8 inches to 0.75 inches (3 mm to 19 mm)
    - Weld Position: all positions
    - Electrode Type: E71T-1M
    - Electrode Diameter: 0.045 inches (1.1 mm)
    - Shielding Gas: argon/CO_2 – 75/25

28. Write a welding procedure (WPS) with the following requirements:
    - Code Requirement: AWS
    - Process: SMAW
    - Material: mild steel
    - Weld Type: plate
    - Joint: fillet (use Option 1)
    - Qualification Range: 1/8 inches to unlimited (3 mm to unlimited)
    - Weld Position: flat/horizontal
    - Electrode Type: E7018
    - Electrode Diameter: 1/8 inches (3 mm)

29. Using the procedure created in problem #27 and/or problem #28, weld, examine, and destructively test to create a PQR. *Note:* Software for the writing procedure can be obtained from the Prentice Hall website.

30. Write a Welder Qualification Test Record (WQTR) for problem #37 and/or problem #28.

Chapter 13
WELDING DESIGN

LEARNING OBJECTIVES

In this chapter, the reader will learn how to:

1. Distinguish the difference between thermal and residual stress.
2. Identify how stresses are introduced into metal.
3. Discuss the consequences of localized heat in metal.
4. Calculate longitudinal and transverse shrinkage.
5. Calculate groove depth to avoid angular shrinkage.
6. List generic weld design considerations to minimize heat input.
7. Recommend fillet weld joint design.
8. Recommend groove weld joint design.
9. Assign groove joint accessibility.
10. List changes to weld joint design when welding aluminum.
11. Determine minimum fillet and groove weld size based on material thickness.
12. Calculate the strength of welds.

KEY TERMS

Static 482	Restrained 483
Dynamic 482	Shear Strength 490
Thermal Stress 483	Peening 497
Residual Stress 483	Doubler 498

INTRODUCTION

Weld joint design is critical to minimize heat input and distortion, and increase productivity. Weld joint design is also critical to ensure the welder has proper access to the joint to produce a weld free of defects. This chapter deals with considerations when developing weld joints, what to look for or avoid, and some basic calculations to aid in the understanding of weld size, strength, and distortion.

As an example, a typical included angle for V-grooves is 45° with a 1/4-inch (13 mm) root opening and a backing plate or a 60° included angle with a 0-inch to 1/8-inch (0.32 cm) open root without a backing plate. Some companies opt for a 70° to 75° included angle on groove joints. This is a fairly large angle requiring more weld metal to completely fill the joint. However, this large included angle also allows greater access to the joint for the welder. The companies using these larger included angles are more concerned with the welders creating a defect-free weld than the extra weld metal and time it takes to complete the weld. Other companies are more concerned with how much heat input and possible distortion a weld would create on a product. Joints, such as J-grooves and U-grooves, are utilized for their economics and to minimize heat input.

There are many factors to consider when selecting joints for welding. Selection may be based on reducing the amount of weld metal in a joint, resulting in a decrease of internal stress as well as reducing costs. Strength and joint accessibility may be other considerations. The information in this chapter is *not* a substitute for good engineering practices or calculations but rather provides a practical guideline when welding.

WELDING PROCEDURE CONSIDERATIONS

WHAT DO WELDING PROCEDURES HAVE TO DO WITH WELD JOINT DESIGN?

Weld joint design, costs, metallurgy, and welding procedures are intertwined. Decreasing a V-groove included angle, for example, decreases the amount of weld metal required to complete a joint. A decreased included angle also produces less distortion because the heat input is less. The cost is decreased because it takes less time, filler metal, and shielding gas to complete the joint. However, a decreased included angle is more difficult to weld and is prone to discontinuities. The following are some guidelines to consider when designing weld joints and developing welding procedures:[1]

- The use of backing strips or backing bars increases the speed of welding on the first pass in groove welds because it takes time to manipulate the electrode to produce a sound weld without a backing or the time it would take to back-gouge and reweld the other side. However, there is also an increase in the overall weld metal required to complete the joint.

---

[1] Cynthia L. Jenney and Annette O'Brien, editors, 2001, *AWS Welding Handbook, Volume 1,* 9th edition, American Welding Society, Chapter 5, "Design for Welding."

- To eliminate or reduce the need for preheating, consider the use of low-hydrogen electrodes on plain carbon steels. Material thickness also plays a roll in preheating. For example, A36 plates less than 3/4 inches (19 mm) require no preheat.

- If plates are 1 inch (25 mm) in thickness or less, a groove joint may be welded from only one side. This would eliminate welding in the overhead position, as would be the case in a double V-groove joint. However, angular distortion (warpage) must be carefully monitored when welding on only one side on thick plates. Distortion can be minimized with the use of U-grooves, low heat input processes, stringer beads instead of weave beads, and other techniques.

- Many times, weld reinforcement—weld filled above the base metal—is unnecessary to obtain a full-strength joint. Excessive weld reinforcement increases residual stress and may lead to distortion and/or cracking.

- Be conscious of grain direction when welding thick sections and the subsequent cooling effects on the base metal. If grain direction of the base metal is not considered, cracking may occur next to the weld during or after cooling.

These tips may help reduce costly repairs. However, there are other factors when determining how to weld a particular product. Questions to ask: "What will this part have to hold?" and "Where will this part go when or if a load is applied?" Fabricated parts are often overwelded. Excessive weld size creates large heat-affected zones (HAZ), often resulting in a failure in the base metal next to the weld. Excessive weld size increases residual stress, possibly causing delamination of the metal or lamellar tearing below the surface near the root of the weld.

WHAT ARE THERMAL AND RESIDUAL STRESS?
Different types of failures can occur in metal after a load of some kind is applied. In the case of compression members, as the compression load increases to the point where the load is excessive, the affected area is no longer able to resist the tendency to buckle, and failure results. **Static** members (**FIGURE 13-1**), such as those which are not meant to move—for example, trusses—can experience compression loads. In the case of members subjected to repeated loading and unloading—cyclic loads or **dynamic** loads—stresses have an effect on fatigue life. In this case, the repeated loading and unloading may trigger fatigue cracks and ultimately lead to a failure.

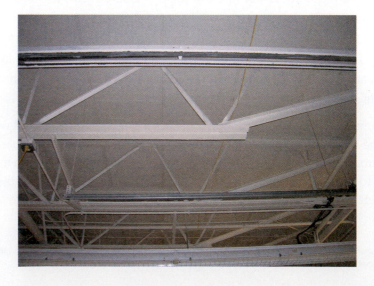

FIGURE 13-1 A truss is an example of a static load

Thermal stress is a result from the nonuniform heating and subsequent cooling of a welded joint. **Residual stress** develops after a joint is welded and then allowed to cool. While thermal stress is induced by heat, residual stresses develop from many manufacturing processes, including rolling, forming, casting, and welding. When metals are heated, they expand, and when they cool, they contract. Restrained welded parts are not allowed to freely expand and contract while cooling. This restraint imposes stress on the metal. In severe cases, the metal cracks. Unrestrained members shrink and warp as cooling imposes varying degrees of stress. These stresses cannot be prevented, but they can be reduced or minimized by proper techniques and procedures.

Welding creates significant and localized heat. If the base metal is ductile and the weld is of correct size, the strain of cooling is minimized and the resulting stress is also minimized. However, if the base metal is not very ductile, preheating and subsequent postweld heating are required to lessen the amount of residual stress. Still, even if the entire structure is preheated, the amount of heat in the weld area is considerably greater than the surrounding area, and some residual stress is present. If a preheat and postweld heat treatment are not practiced in less ductile metals, the residual stresses from welding may cause cracking, or lamellar tearing, or severely limit fatigue life. It is unlikely that any load applied to a member is responsible for initiating a lamellar tear, but, rather, the weld shrinkage stresses are responsible[2] (**FIGURE 13-2**).

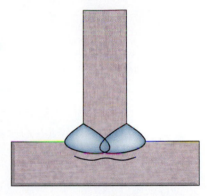

FIGURE 13-2 Shrinkage from welds can cause cracks.

The following precautions should help minimize the problems of lamellar tearing in welds having restricted connections while fabricating. The assumption is that a low-hydrogen electrode is being used.[3]

- The groove weld size should be as small as possible depending on the design; overwelding should be avoided.
- The welding sequence should be such as to minimize overall shrinkage, especially in areas where the members are highly **restrained**. Unrestrained welded members are allowed to expand and contract freely, so weld cracking occurs infrequently. Restrained members are not allowed to contract freely, and the possibility of cracking occurs more frequently.
- Selecting an electrode with the lowest tensile strength and the highest ductility is used to allow the weld metal rather than that in the base metal to strain. This may contradict common welding practice since it is often a requirement that welds be as strong as or stronger than the base metal. However, a filler metal with low ductility and high tensile strength does not have the ductility to stretch significantly when cooling. This results in base metal cracks adjacent to the weld metal if the base metal also lacks ductility.
- The specification of material with improved ductility should be considered for critical connections. In this situation, the base material as opposed to the filler material is considered to handle the strain while the weld metal cools.

Lamellar tearing is only one consideration in product failure. Welds fail for various reasons. It must be determined to what extent weld metal imperfections

[2] Cynthia L. Jenney and Annette O'Brien, editors, 2001, *AWS Welding Handbook, Volume 1,* 9th edition, American Welding Society, Chapter 7, "Residual Stress and Distortion."
[3] Cynthia L. Jenney and Annette O'Brien, editors, 2001, *AWS Welding Handbook, Volume 1,* 9th Edition, American Welding Society, Chapter 5, "Design for Welding."

can be tolerated without jeopardizing structural integrity if the blemish should be repaired. Repairs made by air arc gouging and subsequent welding will increase heat input and shrinkage, leading to further residual stresses, and could result in cracking and lamellar tearing. Some welding code books provide a guideline to follow when determining weld metal quality, but in many cases, the inspector is still the judge. Many factors must be weighed in cost savings, liability, and customer satisfaction.

TRANSVERSE AND LONGITUDINAL SHRINKAGE

ARE THERE CALCULATIONS FOR SHRINKAGE AND WARPAGE? Calculations for shrinkage help to understand how weld size and joint design affect metal distortion. When any part is welded, the base metal and weld metal will first expand and then shrink as they solidify. But which way will they shrink and how much? In unrestrained members, the entire part is allowed to move, and warpage is visible. In restrained members, because the welded members are not allowed to move, warpage is not visible, and the metal strains in the heated area. Something has to give. The ductility of the base metal and weld metal will determine if cracking occurs. There are different ways that welded parts shrink. They can be identified as transverse shrinkage, longitudinal shrinkage, and angular shrinkage. These changes take place whether the joint is a groove or a fillet. **FIGURE 13-3** illustrates these changes in a groove joint.

TRANSVERSE SHRINKAGE IN GROOVE JOINTS[4] Transverse shrinkage is measured transverse to the weld axis and across the weld members being joined (**FIGURES 13-4** and **13-5**).

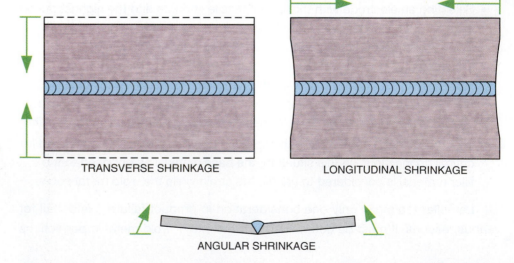

FIGURE 13-3 Shrinkage can occur transversely, longitudinally, and angularly.

Source: Redrawn from the American Welding Society (AWS) Handbook Committee, 2001, *AWS Welding Handbook*, 9th ed., vol. 1, Welding Science and Technology, Miami: American Welding Society.

TRANSVERSE SHRINKAGE

LONGITUDINAL SHRINKAGE

ANGULAR SHRINKAGE

---

[4] Cynthia L. Jenney and Annette O'Brien, editors, 2001, *AWS Welding Handbook, Volume 1,* 9th edition, American Welding Society, Chapter 7, "Residual Stress and Distortion."

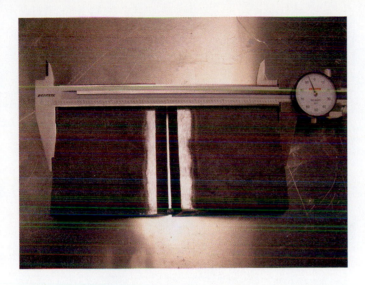 

FIGURE 13-4 Measuring for transverse shrinkage before welding

FIGURE 13-5 Measuring for transverse shrinkage after welding

Shrinkage depends on metal thickness and the root opening because both affect the cross-sectional area. As the cross-sectional area of the weld increases, shrinkage also increases. However, as the plate thickness increases, the percentage of shrinkage decreases. The formula for calculating transverse shrinkage in groove weld joints is:

$$0.2 \frac{A_w}{t} + 0.05d = S$$

Where: A_w = Cross-sectional area of weld, in inches^2 (cm^2)

t = Thickness of plates, in inches (cm)

d = Root opening, in inches (cm)

S = Transverse shrinkage, in inches (cm)

In this example, the plate thickness is 1/2 inches (13 mm). A single bevel groove is made with a 60° included angle and a 1/8-inch (3 mm) root opening and root face. The calculated cross-sectional area of the weld is 0.144 in^2 (93 mm^2).

U.S. Standard	*SI Standard*
$0.2 \dfrac{0.144}{0.5} + 0.05\,(0.125) = 0.064 \text{ in}$	$0.2 \dfrac{93}{13} + 0.05\,(3) = 1.6 \text{ mm}$

The groove will shrink in the transverse direction approximately 1/16 inches (1.6 mm). This is a fairly significant amount. The shrinkage does not occur 100% in the weld metal. Some shrinkage also occurs in the weld/base metal interface and in the HAZ. If there are multiple groove welds parallel to each other in a production part, the cumulative effect may pose difficulties in part tolerance and warpage. If the part is restrained, cracking may occur. If the grain direction is parallel to the groove joint, whether restrained or not, cracking may also occur.

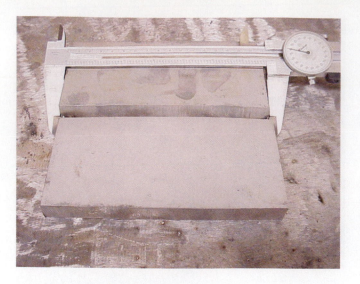

FIGURE 13-6 Measuring for longitudinal shrinkage before welding

FIGURE 13-7 Measuring for longitudinal shrinkage after welding

LONGITUDINAL SHRINKAGE OF GROOVE JOINTS[5] Longitudinal shrinkage is measured parallel to the weld axis and down the length of the weld members being joined (**FIGURES 13-6** and **13-7**).

$$\frac{C_3 I L}{t} 10^{-7} = \Delta L$$

Where: C_3 = 12 (305 if calculating in metric, mm)
  I = Welding current
  L = Length of weld, in inches (mm)
  t = Plate thickness, in inches (mm)
  ΔL = Longitudinal shrinkage

For convenience, the same groove joint in the transverse groove weld shrinkage example is used here. An additional parameter—amperage—is required. In this example, an E7018 electrode is welded at 125 amps. The length of the weld is 10 inches (250 mm).

U.S. Standard

$$\frac{(12)(125)(10)}{0.5} 10^{-7} = 0.003 \text{ in}$$

SI Standard

$$\frac{(305)(125)(250)}{13} 10^{-7} = 0.08 \text{ mm}$$

The groove weld shrinks 0.003 inches (0.08 mm). This appears minor in comparison to transverse groove weld shrinkage. However, longitudinal weld metal cracks are the result of transverse shrinkage in base metals. Longitudinal weld metal cracks occur in base metals that are not ductile if proper preheat and postweld heat procedures are not followed. On thicker restrained sections, transverse cracks may even occur in ductile base metals if the proper heat treatment procedures are not followed.

---

[5] Cynthia L. Jenney and Annette O'Brien, editors, 2001, *AWS Welding Handbook, Volume 1,* 9th edition, American Welding Society, Chapter 7, "Residual Stress and Distortion."

ANGULAR SHRINKAGE[6]   When discussing angular shrinkage, the discussion of how to prevent warpage would be more productive than in calculating or measuring angular distortion. In order to prevent angular shrinkage in single-groove welds, the member must be restrained. However, the weld area will experience residual stress that may or may not lead to cracking depending on base metal composition, base metal grain direction, weld metal composition, metal thickness, and preheat and/or postweld heat treatment.

Another method to combat angular shrinkage in groove welds is the use of the double V-groove joint. Experiments on steel conclude that a double V-groove approximately 60% larger on the first side than the second side eliminates shrinkage. The included angle of the V-groove is a typical 60° angle (**FIGURE 13-8**). **FIGURE 13-9** shows a simple method of restraining weld members using a series of vise-gripped C-clamps.

$$(T - \tfrac{1}{2}RF) \times 0.6 = FD$$

Where: T   = Thickness of plate

RF = Root face

0.6 = Constant (60%)

FD = First pass side and depth of preparation

Root opening and root face are proportional. As the root face (and root opening) decreases, angular distortion also decreases. The following example,

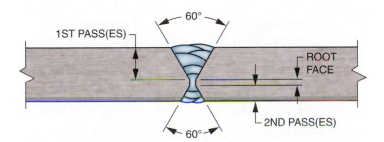

FIGURE 13-8 Weld sequence to avoid angular shrinkage

Source: Redrawn from the American Welding Society (AWS) Handbook Committee, 2001, *AWS Welding Handbook*, 9th ed., vol. 1, Welding Science and Technology, Miami: American Welding Society.

FIGURE 13-9 Aluminum frame for a fire truck

Source: Photo courtesy of E.J. Metals, Inc.

[6] Cynthia L. Jenney and Annette O'Brien, editors, 2001, *AWS Welding Handbook, Volume 1,* 9th edition, American Welding Society, Chapter 7, "Residual Stress and Distortion."

using a 1/8-inch (3.2 mm) root face on a 1-inch (25 mm) thick plate, minimizes distortion if the depth of the first groove is 9/16 inches (14 mm) and the depth of the second groove is 3/8 inches (9.5 mm).

U.S. Standard	SI Standard
$(1 \text{ in} - (\frac{1}{2} \times 0.125 \text{ in})) \times 0.6 = 0.5625 \text{ in}$	$(25 \text{ mm} - (\frac{1}{2} \times 3.2 \text{ mm})) \times 0.6 = 14 \text{ mm}$

When the root face and root opening are at or near zero, the depth of the first groove may be increased to as much as 70% of the base metal thickness. The formula is simplified without the root face to $T \times 0.7 = FD$.

TRANSVERSE SHRINKAGE IN FILLET WELDS   Transverse shrinkage in fillet welds is similar to groove welds, but the formula for calculating shrinkage differs.

CONTINUOUS FILLET WELDS[7]

$$C_1\left(\frac{D_f}{t_b}\right) = S$$

Where: D_f = Fillet leg length, in inches (mm)
t_b = Thickness of bottom plate, in inches (mm)
C_1 = 0.04 (1.02 if calculating in metric)
S = Shrinkage, in inches (mm)

In this example, the plate-to-plate thickness is 1/2 inches (13 mm). The fillet weld leg size is 3/8 inches (9.5 mm).

U.S. Standard	SI Standard
$0.04\left(\frac{0.375}{0.5}\right) = 0.03 \text{ in}$	$1.02\left(\frac{9.5}{13}\right) = 0.76 \text{ cm}$

The transverse shrinkage on a single fillet weld, as in the last example, is roughly half that of the groove weld on the same plate thickness. This is due to the cross-sectional area of the weld being much smaller on fillet welds than on groove welds.

FIGURE 13-10 illustrates the effects of an unrestrained and a restrained member and the resulting shrinkage and distortion from fillet welds. As the size of the welds increases, the associated shrinkage issues also increase.

Longitudinal shrinkage for fillet welds is proportional based on cross-sectional size of the weld deposited. Weld size plays a pivotal role in controlling distortion, warpage, and residual stresses in welded parts. To combat transverse and longitudinal shrinkage in fillet welds, intermittent welds are made as opposed to continuous welds. Intermittent welding helps alleviate problems by putting less weld metal in any given welded length. Intermittent fillet welds usually surpass the tensile and shear strength minimum requirements and decrease the residual stresses resulting from shrinkage. Labor and welding costs are also decreased with intermittent fillet welds.

Size is always a consideration when designing weld joints. Residual stress, tensile strength, brittleness, shrinkage, and distortion increase as the size of the weld increases and ductility decreases. Welds will often meet design

---

[7] Cynthia L. Jenney and Annette O'Brien, editors, 2001, *AWS Welding Handbook, Volume 1,* 9th edition, American Welding Society, Chapter 7, "Residual Stress and Distortion."

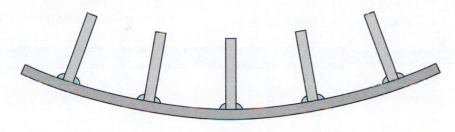

UNRESTRAINED MEMBER

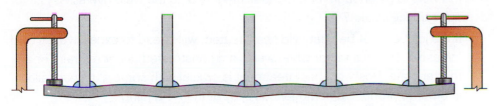

RESTRAINED MEMBER

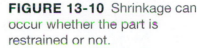

FIGURE 13-10 Shrinkage can occur whether the part is restrained or not.

Source: Redrawn from the American Welding Society (AWS) Handbook Committee, 2001, *AWS Welding Handbook*, 9th ed., vol. 1, Welding Science and Technology, Miami: American Welding Society.

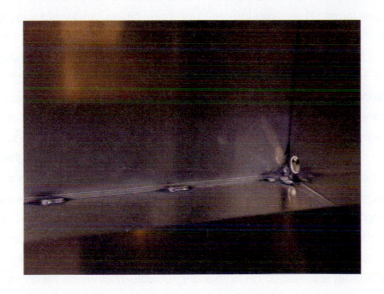

FIGURE 13-11 Welding on a fire truck

Source: Photo courtesy of E.J. Metals, Inc.

specifications even though they appear too small for the sake of appearance (**FIGURE 13-11**). In the final analysis, overwelding creates many problems not easily solved. Careful consideration in joint design and weld size will circumvent many problems.

WELD JOINT DESIGN

WHAT ARE SOME SPECIFIC DESIGN CONSIDERATIONS IN WELD JOINTS?

Some basic considerations must be observed:

- Is the joint accessible to welders and inspectors? What can be drawn on a print and what can be performed on the shop floor can be dramatically different. Consider the drawing in **FIGURE 13-12**. If the rod is 1/2 inches (13 mm), how is this accomplished? A simple change from welding the end

FIGURE 13-12 Example of an impossible-to-reach weld joint design

of the rod to the left end of the pipe makes the weld accessible to both the welder and the inspector.

- Does the design consider the economic factors of welding? Is the size of the weld large enough for structural integrity or is the weld oversized just for appearance's sake?
- If the joint cannot be postweld heat-treated, will it lead to excessive residual stresses? This is a factor when welds are constructed under restraint or on thicker plate. Postweld heat treatment is crucial with some aluminum alloys in order to produce a homogenous alloy at the weld joint.

FILLET WELDS   Fillet welds are much more common and more economical than groove joints. They are simpler to prepare, fit-up, and weld than groove joints. When designing fillet welds, how big is big enough? A good rule of thumb on steels using filler metals of matching strength is to design fillets weld leg sizes to be 75% to 80% of the base material thickness. This is large enough to sustain any reasonable load as far as cost is concerned and does not appear too small for anyone to question the weld strength. Doubling the size of a fillet weld doubles the strength but requires four times the weld metal[8] **(FIGURE 13-13)**.

The size of the fillet weld is always designed on the basis of shear strength on the throat regardless of the direction of the load or force. **Shear strength** is the amount of force required to fracture a weld. In the case of steel, the allowable shear strength is limited to 30% of the classified tensile strength of the filler metal.[9] T-joints and lap joints welded only on one side cannot stand opposing loads and will fracture easily when a load is applied. To ensure full strength, both lap joints and T-joints are welded on both sides. Double-fillet welds having a leg size equal to 75% of the base plate being welded would be sufficient for full strength.

Depending on the size of the fillet weld, it may be more economical to produce a single bevel-groove or a double bevel-groove weld to ensure adequate effective throat. When a fillet weld 5/8 inches (16 mm) or larger is specified, a bevel-groove joint is recommended in order to obtain adequate strength and an adequate effective throat.[10] This recommendation is for the size of the fillet weld

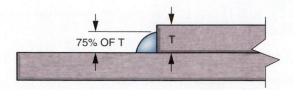

FIGURE 13-13 Welding slower produces larger welds but often lacks proper root penetration.

---

[8] Howard B. Cary, 1998, *Modern Welding Technology*, 4th edition, Prentice Hall, Chapter 19, "Design for Welding."
[9] American Welding Society, *Structural Welding Code—Steel, AWS D1.1:2006*, Tables 2.3 and 2.5.
[10] Cynthia L. Jenney and Annette O'Brien, editors, 2001, *AWS Welding Handbook, Volume 1,* 9th edition, American Welding Society, Chapter 5, "Design for Welding."

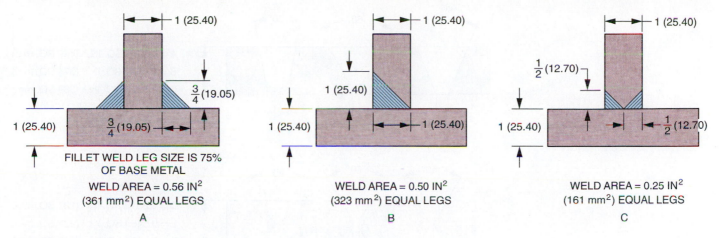

FIGURE 13-14 Various designs for thicker T-joints on plate members

Source: Redrawn from the American Welding Society (AWS) Handbook Committee, 2001, *AWS Welding Handbook*, 9th ed., vol. 1, Welding Science and Technology, Miami: American Welding Society.

and not the plate size. For example, for a 3/4-inch (19 mm) plate, a 75% fillet weld leg size equals 9/16 inches (14 mm).

The T-joints in **FIGURE 13-14** illustrate various methods to create a fillet weld on a 1-inch (25 mm) plate with adequate strength. Note the corresponding weld area for each of the three T-joints. In terms of weld metal volume alone, the benefits of beveling are revealed. In addition, the smaller single bevel and fillet weld combination reduces residual stresses. However, joints B and C require additional labor to create the bevel as well as possible use of a smaller diameter electrode to access the joint.

When a partial joint penetration bevel-groove weld is reinforced with a fillet weld, as would be the case in joints B and C in **FIGURE 13-15**, the effective throat is the combination of both the bevel-groove and fillet weld and not the sum of each individual weld made.

When the depth of preparation exceeds 3/4 inches (19 mm) on double bevel-groove welds, the double J-groove joint may be more economical because of reduced groove area. Also, J-grooves are used to minimize heat input.

V-GROOVE WELDS Single/double V-groove welds are more common than U-groove joints. However, U-grooves may be used to minimize heat input because there is generally less weld metal in a U-groove than a comparable V-groove joint. U-groove joints (and J-groove joints) require more skill to weld because of the straighter side walls. More skill is also required on V-groove

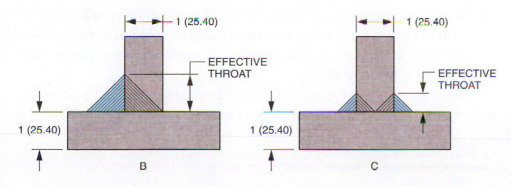

FIGURE 13-15 Joint C has the same effective throat as B but will have less adverse effects.

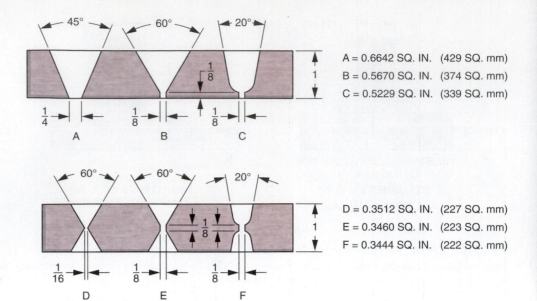

A = 0.6642 SQ. IN.   (429 SQ. mm)
B = 0.5670 SQ. IN.   (374 SQ. mm)
C = 0.5229 SQ. IN.   (339 SQ. mm)

D = 0.3512 SQ. IN.   (227 SQ. mm)
E = 0.3460 SQ. IN.   (223 SQ. mm)
F = 0.3444 SQ. IN.   (222 SQ. mm)

FIGURE 13-16 Various groove weld joint designs

welds when the included angle is decreased. For example, an AWS D1.1 prequalified procedure for FCAW (B-U2a-GF) allows the use of a 45° or 30° included angle V-groove. The 30° is more economical to produce but requires more skill to weld.

Double V-groove joints are economical when the individual groove depth does not exceed 3/4 inches (19 mm) and overhead welding does not present a problem. When plates exceed 11/2 inches (38 mm) or when the depth of preparation exceeds 3/4 inches (19 mm), the double U-groove joints may be more economical.[11] The groove joints in **FIGURE 13-16** reflect typical designs. Note the cross-sectional area for each groove joint.

The groove joints A, B, and C in Figure 13-16 are common joint designs. They are compared to the weld volume of D, E, and F if both thicknesses are 1 inch (25 mm). The economics of a double V-groove from a single V-groove is apparent as the base metal increases in thickness. It is quite substantial in terms of costs as well as the reduction of heat input and resulting residual stresses. The total reduction of weld area in a single or double U-groove depends on the radius at the bottom of the groove joint.

The groove joints in **FIGURE 13-17** illustrate changes in groove angle, root opening, and the root face of the groove joint. Each change in design also changes the time it takes to complete the weld and the resulting residual stresses. Another consideration is how easy or, conversely, how difficult it is for the welder to complete the joint. It does no good to design a joint with a narrow groove angle to minimize welding time and residual stresses if the welder cannot obtain proper fusion along the groove face. The groove angles in Figure 13-17 are common and can be found in the *AWS D1.1 Structural Welding Code – Steel* book as prequalified weld joints.

The groove joints in Figure 13-17, according to the AWS D1.1 prequalified section, require a backing plate or backing weld/back weld. Open-root welding

[11] Cynthia L. Jenney and Annette O'Brien, editors, 2001, *AWS Welding Handbook, Volume 1,* 9th edition, American Welding Society, Chapter 5, "Design for Welding."

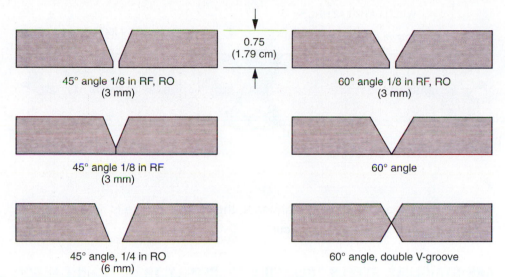

0.75
(1.79 cm)

45° angle 1/8 in RF, RO
(3 mm)

60° angle 1/8 in RF, RO
(3 mm)

45° angle 1/8 in RF
(3 mm)

60° angle

45° angle, 1/4 in RO
(6 mm)

60° angle, double V-groove

FIGURE 13-17 Warpage and weld joint access play an important role in joint design

increases the possibility of weld defects and discontinuities. The specific cross-sectional area for the joints in Figure 13-17 are:

- The 45° groove with the 1/8-inch (3 mm) root face and 1/8-inch (3 mm) root opening contains a total weld area of 0.256 in² (165 mm²).
- The 45° groove with the 1/8-inch (3 mm) root face and no root opening contains a total weld area of 0.162 in² (105 mm²).
- The 45° groove with a 1/4-inch (6 mm) root opening contains a total weld area of 0.421 in² (271 mm²). This is the most common joint when welding with a backing plate.
- The 60° groove with 1/8-inch (3 mm) root face and 1/8-inch (3 mm) root opening contains a total weld area of 0.319 in² (206 mm²). This is a common joint when welding with no backing plate but takes more skill than with a backing plate.
- The 60° groove with the no root face and no root opening contains a total weld area of 0.325 in² (210 mm²), but most likely will require back-gouging and back welding.
- The 60° double V-groove with no root face and no root opening contains a total weld area of 0.162 in² (105 mm²). It is the most efficient joint of the six joints shown. However, this joint requires more preparation time and may require back-gouging before welding the second side, thus adding to the total filler metal used and time consumed. As the thickness of the base plate increases, the decrease in weld time increases cost efficiency.

Proper joint design minimizes overwelding and reduces the ramifications of residual stresses. Overwelding on thicker plate steel or on hardenable steels not preheated and postheated may draw carbon out of the base metal and into the admixture of the weld. Steels not preheated and postheat treated will go through a thermal cycle in the HAZ, possibly creating bainite or martensite formation. The thermal cycle occurs because the unheated base material actually quenches the metal next to the weld. **FIGURE 13-18** illustrates the effects of hardenability from welding. The material in Figure 13-18 is AISI 403 stainless steel preheated and postweld heat-treated prior to welding with an ER410 NiMo filler metal.

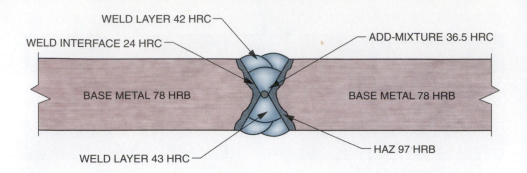

FIGURE 13-18 The effects of hardness due to welding on 400 grade stainless steel

WELD LAYER 42 HRC
WELD INTERFACE 24 HRC
ADD-MIXTURE 36.5 HRC
BASE METAL 78 HRB
BASE METAL 78 HRB
WELD LAYER 43 HRC
HAZ 97 HRB

Without preheat and postheat treatments, the weld interface is harder than the weld metal, and cracking could occur.

ARE STAINLESS STEELS TREATED DIFFERENTLY THAN PLAIN CARBON STEELS?

Joint design on stainless steel does not differ from that of plain carbon steel. However, weld size is a consideration. The most weldable type of stainless steel is the austenitic, chrome-nickel 300 series stainless (**FIGURES 13-19a** and **13-19b**). The chromium combines with iron (ferrite) and protects it from corrosion, while the nickel keeps it from being too hard. The chromium-nickel stainless steels contain small amounts of carbon. Carbon is undesirable, particularly in the .8% to 18% nickel group. Carbon will combine with chromium to form chromium carbides. The carbon pulls the chromium away from the ferrite, leaving it unprotected so it may be subject to corrosion. Chromium carbides are formed when the steel is held in the temperature range of 800°F (425°C) to 1600°F (870°C) for prolonged periods. The formation of chromium carbides is called carbide precipitation. Carbide precipitation occurs more readily when large welds are produced followed by slow cooling. Carbide precipitation is minimized or eliminated when small-size welds are made and cooling is more rapid.

(A)

(B)

FIGURE 13-19 Header for a racing boat
Source: Photo courtesy of Custom Marine, Inc.

IS ALUMINUM TREATED DIFFERENTLY THAN STEELS IN A WELD JOINT DESIGN? The welded joints used in aluminum fabricated parts—butt, lap, and T—for structural applications are acceptable; however, the edge and corner should be avoided if possible. The edge and corner are weaker joints, and aluminum is more prone to fatigue and failure than steel. If a corner joint is unavoidable and GTAW is the welding process, the use of a filler metal as opposed to welding the joint without filler metal, autogenously, increases strength.

LAP JOINTS Corner joints can be quite common in the fabrication of parts, but the lap joint (forming a lip to create a lap joint as opposed to a corner) has greater strength than a corner joint (**FIGURE 13-20**). The increase in strength can range from 70% to 100%, depending on base metal composition, temper, and postweld heat treatment. For aluminum, AWS recommends a minimum overlap of three times (3×) the thickness of the thinnest member. Fillet welds on thicknesses up to and including 1/2 inches (13 mm) provide adequate strength. It may be more economical with greater strength obtained if a bevel groove is used on thicknesses greater than 1/2 inches (13 mm).

FIGURE 13-21 shows additional methods for achieving a corner on aluminum fabricated parts. The choice of which style to use is made based on economics, strength, or appeal. Note that corrosion-resistant aluminum alloys in the 5XXX series may be more subject to corrosion if hot-forming is used.

T-JOINTS When metal thickness exceeds 3/4 inches (18 mm), it is beneficial to incorporate the bevel-groove joint for economics and strength. Welding from both sides is recommended when welding on aluminum. A joint welded from one side on aluminum is relatively weak. Also, small continuous welds are recommended rather than larger intermittent fillet welds. With steels, intermittent fillet welds are often preferred to decrease residual stress and costs. In addition, continuous fillet welds on aluminum are recommended over intermittent welds because of better fatigue life, reduced crevice corrosion, and decreased the possibility of crater cracks.

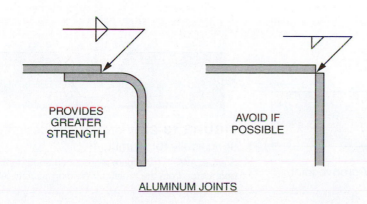

PROVIDES
GREATER
STRENGTH

AVOID IF
POSSIBLE

ALUMINUM JOINTS

FIGURE 13-20 Joint design for aluminum is more important than with other base metals.

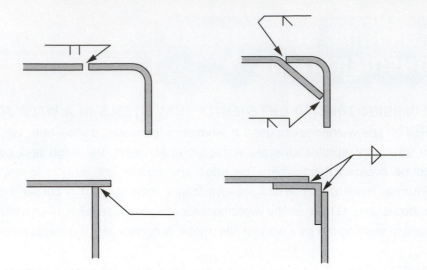

FIGURE 13-21 Various fillet and groove weld designs for aluminum

Aluminum is more susceptible to crater cracks than steels. A crater crack initiated on aluminum can quickly continue through the rest of the weld and the base metal. It is imperative that craters be filled or back-welded by reversing direction at the end of the joint 1/2 to 1 inch (13 to 25 mm). If only one side of a joint is accessible, then a large continuous fillet weld is recommended. The strength of a fillet weld is directly proportional to its size, but the cost—based on the amount of metal deposited—is in proportion to the square of its size.

GROOVE JOINTS While thinner plates and sheet metal can be welded with a square-groove joint, thicker plates are best served with V-grooves. Standard V-grooves (**FIGURE 13-22**) should use a backing plate or should be back-gouged and welded on both sides.

When pipe welding or any time the weld is accessible from only one side, the extended-land bevel joint is recommend (**FIGURE 13-23**). This joint design is supported by *AWS D1.2 Structural Welding Code – Aluminum*. The effectiveness of this joint design depends on surface tension supporting the molten metal in the weld pool. It is more difficult to weld an open root with aluminum than with steels. The extended-land bevel joint allows the base metal to melt without melting through. Preventing melt-through can be accomplished with a backing plate and an adequate root opening. However, a backing plate inside a pipe or tubing might pose an obstruction from whatever may be flowing through. It is recommended that the extended-land material thickness be 1/16 inches to 3/32 inches (2 mm

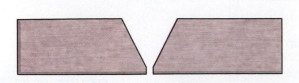

FIGURE 13-22 A typical V-groove joint design

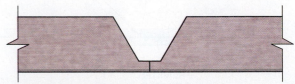

FIGURE 13-23 V-groove joint design specifically for aluminum.
Source: AWS D1.2. Figure reproduced and adapted with permission from the American Welding Society (AWS), Miami, Florida.

to 2.5 mm). The bottom of the groove should be wide enough to allow the welder to place a full root bead.

The advantages of the extended-land design are:

- Smooth, complete penetration control
- No "suck-back"
- No backing required
- Good for all fixed pipe positions
- Preheating not required

MAKING ALUMINUM WELDS STRONGER When welding aluminum, there are always trade-offs. For example, when welding heat-treatable aluminum alloys, nearly full strength can be obtained by solution heat-treating. Postweld heat treatment creates a more homogenous alloy between the weld metal and base metal. The problem with solution heat treating is that it is often not economical or practical.

In non-heat treatable aluminum alloys, the welded joint and the surrounding HAZ are softer than the base metal. A method to increase strength in non-heat treatable aluminum is to cold-work the parts after welding, and this is often not practical or economical. Without cold-working the weld zone, the weldment is the weakest part in an assembly.

One other method to develop as much strength in an aluminum weld as possible is to use the highest strength filler metal possible. The highest strength filler metal is ER5556. The use of higher strength filler metal permits smaller weldments, leading to smaller HAZ and further enhancing weld strength. A comparison of popular filler metals—ER5356 and ER4043—reveals that if a 1/4-inch (6.5 mm) fillet weld was made with ER5356, then a 3/8-inch (9.5 mm) fillet weld must be made with the ER4043 to achieve the same weld strength. On the other hand, when welding higher strength alloys such as 6061 T6, especially where postweld heat treatment will not be used, welds made with ER4043 filler metal would be less susceptible to cracking. Increased ductility of welds made with 5356 make it a popular choice for welding large aluminum fabrications, such as boats.

In aluminum groove welds, the fatigue strength can be increased by **peening** the weldment. Peening is accomplished by repeated blows to the weld metal. Also, it is a good procedure to ensure that the weld reinforcement blends smoothly onto the base metal to avoid abrupt changes in thickness. Weld reinforcement with a smooth transition decreases the risk of stress risers developing, which could lead to cracking. Welding procedures should be written in a way where little or no spatter develops. Spatter marks also create stress risers in the base metal adjacent to the weldment. While residual stresses do not affect the static strength of aluminum, they can be detrimental to fatigue strength.

Shot peening or other forms of peening are beneficial to reduce residual stress. Heat treatment also relieves residual stresses by increasing fatigue resistance and dimensional stability. For aluminum alloys that are not strengthened by heat treatment, such as the 5000 series, postweld heat treatment relieves the majority of the residual stresses, but may also reduce strength beyond the HAZ, since these alloys are strengthened by strain hardening when they are formed. Although heat-treatable alloys such as 6061 can be stress-relieved by thermal treatment, strength is also lost. This is because when temperatures are high enough to reduce residual stresses, they are also high enough to substantially diminish strength properties. However, if a loss of strength can be tolerated, a reduction of residual stresses can be diminished by approximately one-half, with only a slight loss in strength.[12]

The 5000 series aluminum can tolerate thermal treatment to reduce residual stresses, but they are not considered suitable for service in areas of high temperatures above 300°F (149°C).

On the other hand, most aluminum alloys perform extremely well down to the extreme cold ranges. It has been found that the strength in aluminum alloys will increase as temperatures decrease. Ultimate tensile strength, compressive strength, and shear strength increase as the temperature decreases. At temperatures of −300°F (−185°C) and colder, ultimate strengths are 35%–50% greater than at room temperature and yield strengths are 15%–25% higher. At −459°F (−273°C), strengths are even higher. (Note: Absolute 0 is −459°F/273°C).[13]

OTHER DESIGN CONSIDERATIONS Aluminum has four to six times the thermal conductivity of steel, depending on the carbon content of the steel. Aluminum dissipates heat readily, and cold starts are common. One method to combat cold starts is to taper the ends of angles and channels. Examples are shown in **FIGURE 13-24**.

A **doubler** is common for strengthening steel plates or beams. This is a simple method of welding a square or rectangular plate over the welded plate or beam. One method to combat cold starts on aluminum and to reduce stress is to use a triangular plate. A three-to-one ratio (3:1) or greater length-width should accomplish the desired results (**FIGURE 13-25**).

FIGURE 13-24 Tapering structural shapes helps alleviate stress on aluminum.

Source: Redrawn from the American Welding Society (AWS) Handbook Committee, 2001, *AWS Welding Handbook*, 9th ed., vol. 1, Welding Science and Technology, Miami: American Welding Society

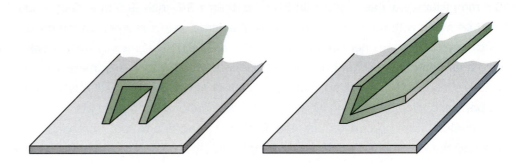

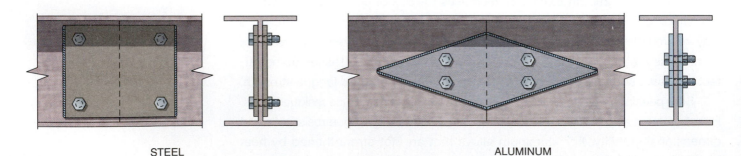

STEEL

ALUMINUM

FIGURE 13-25 Doublers are designed differently for steel than aluminum.

---

[12] Cynthia L. Jenney and Annette O'Brien, editors, 2001, *AWS Welding Handbook, Volume 1,* 9th edition, American Welding Society, Chapter 5, "Design for Welding."
[13] http://www.secat.net/answers_view_article.php?article=Alloys_for_Cryogenic_Applications.html

STRENGTH OF WELDS

HOW IS THE STRENGTH OF WELDS DETERMINED? With carbon steels in mind complete joint penetration groove welds are generally considered full strength welds because they are capable of withstanding the full stress of the welded member. The allowable stress is the same as the base metal. For partial penetration groove joints and fillet welds, there are other methods to determine how much stress a weld can endure. The strength of a weld can be mathematically calculated. Shear strength is useful in determining fillet weld size and the filler metal used to make the weld.

One method for calculating shear stress is the Allowable Stress Design (ASD) procedure. The allowable shear stress in steel weld metal is 30% of the nominal tensile strength of the weld metal. The ASD method is useful for determining fillet welds and partial penetration groove welds. The formula is: (Allowable Shear Stress) $\times$ (Shear Plane in Weld) = Allowable Unit Load.) The allowable shear stress is 30% of Electrode Tensile Strength. For fillet welds, the shear plane in weld is: throat dimension $\times$ (leg) length. This is for equal leg sizes where the effective throat is 70.7% of the weld size. In the example that follows, the length is 1 unit. As an example, if a 3/8-inch (9.5 mm) fillet weld is produced with an E7018 electrode:

Allowable Shear Stress =
US 0.30(70,000 psi) = 21,000   **SI** 0.30(483 MPa) = 145

Shear Plane in Weld =
US 0.707(3/8 in) = 0.265   **SI** 0.707(9.5 mm) = 6.72

Allowable Unit Load =
US 21,000 $\times$ 0.265 = 5565 lb/in   **SI** 145 $\times$ 6.72 = 975 N/mm

Based on these calculations, the *AWS D1.1 2006 Structural Welding Code – Steel* has determined the minimum fillet weld leg size (*AWS Table 5.8*) and effective throat on partial penetration groove welds (*AWS Table 3.4*) on steel (**TABLES 13-1 and 13-2**). Arguably, any weld strong enough for a building structure is strong enough for any welded connection. The recommended welds may seem small, but these tables support the argument for the rule of thumb for fillet leg size welds to be 75% to 80% of the base material thickness.

Fillet Welds–Steel	
Base Metal Thickness	**Minimum Fillet Weld Leg Size**
Over 1/8 to 1/4 inches (Over 3 mm to 6 mm)	1/8 inches (3 mm)
Over 1/4 to 1/2 inches (Over 6 mm to 13 mm)	3/16 inches (5 mm)
Over 1/2 to 3/4 inches (Over 13 mm to 19 mm)	1/4 inches (6 mm)
Over 3/4 inches (Over 19 mm)	5/16 inches (8 mm)

TABLE 13-1

Source: AWS D1.1. Reproduced and adapted with permission from the American Welding Society (AWS), Miami, Florida.

Partial Penetration Groove Welds–Steel	
Base Metal Thickness	Minimum Effective Throat
Over 1/8 to 3/16 inches (Over 3 to 5 mm)	1/16 inches (2 mm)
Over 3/16 to 1/4 inches (Over 5 to 6 mm)	1/8 inches (3 mm)
Over 1/4 to 1/2 inches (Over 6 to 13 mm)	3/16 inches (5 mm)
Over 1/2 to 3/4 inches (Over 13 to 19 mm)	1/4 inches (6 mm)
Over 3/4 to 1 1/2 inches (Over 19 to 38 mm)	5/16 inches (8 mm)
Over 1 1/2 to 2 1/4 inches (Over 38 to 57 mm)	3/8 inches (10 mm)
Over 2 1/2 to 6 inches (Over 57 to 152 mm)	1/2 inches (13 mm)
Over 6 inches (Over 152 mm)	5/8 inches (16 mm)

TABLE 13-2

Source: AWS D1.1. Reproduced and adapted with permission from the American Welding Society (AWS), Miami, Florida.

The ASD calculations do not account for the depth of penetration of the actual throat but rather use a theoretical throat. Additionally, manufacturers of filler metals always exceed minimum tensile strength. What this means is the actual unit load on a weld will exceed the calculated Allowable Unit Load.

A shear strength test will provide data on how much stress it will take to fracture a fillet weld. The fillet weld shear test uses a tensile testing machine to determine the shear strength of the welds. The test is intended to represent actual production welds. Of the two types of tests shown in **FIGURE 13-26**, the transverse shear test is the more common. When performing this test, it is crucial to maintain the exact size and spacing from one specimen to the next to ensure accurate test results.

To aid in uniformity and eliminate cold starts and crater fill issues, a larger specimen is welded. The width of the entire specimen is a minimum of 6 inches (150 mm) (**FIGURE 13-27**).

If the dimensions in Figure 13-27 are followed and the filler metal is ductile, the weld will shear. To ensure the test will result in the fracture of the welds, the base metal is larger than the fillet weld leg size. The dimensions for the transverse shear

FIGURE 13-26 Shear strength test specimens

Source: Redrawn from the American Welding Society (AWS) Handbook Committee, 2001, *AWS Welding Handbook*, 9th ed., vol. 1, Welding Science and Technology, Miami: American Welding Society.

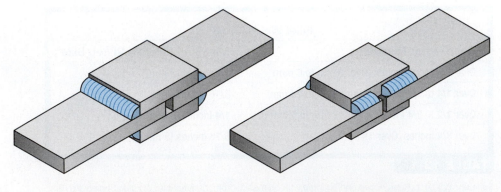

TRANSVERSE SHEAR TEST   LONGITUDINAL SHEAR TEST

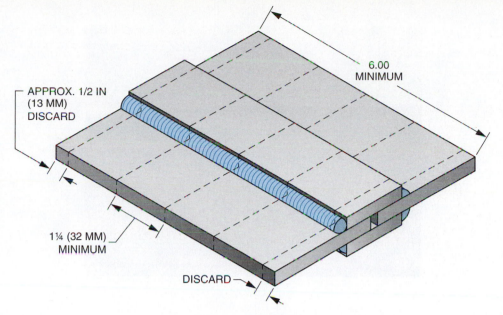

APPROX. 1/2 IN
(13 MM)
DISCARD

1¼ (32 MM)
MINIMUM

6.00
MINIMUM

DISCARD

FILLET SHEAR TEST

FIGURE 13-27 Size and location of shear strength test specimens

Source: AWS B2.1:2005, Fig. 2.6, is reproduced and adapted with permission from the American Welding Society (AWS), Miami, Florida.

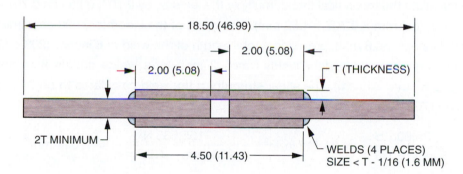

18.50 (46.99)

2.00 (5.08)

2.00 (5.08)

T (THICKNESS)

2T MINIMUM

4.50 (11.43)

WELDS (4 PLACES)
SIZE < T - 1/16 (1.6 MM)

FIGURE 13-28 Detail specifications of shear strength test specimen

Source: Cary, Howard, B., *Modern Welding Technology*, 4th edition, ©1998. Reproduced by permission of Pearson Education, Inc., Upper Saddle River, New Jersey.

test are illustrated[14] in **FIGURE 13-28**. **FIGURE 13-29** shows an actual transverse shear test weld test sample before pulling is complete. A direct reading from the tensile machine will provide a value. The reading on the tensile machine—after testing is complete—divided by the total length of the four welds will provide a tensile strength per unit of measurement comparable to the ASD calculations. The reading from the actual test should be greater than the calculated Allowable Unit Load because the ASD calculations use theoretical throat as opposed to actual throat.

A formula can be used to determine shear strength in lb/in^2 (N/cm^2). The results of the test are reported as the shear strength of the fillet welds. The formula to calculate the shear strength of a fillet weld uses the theoretical throat and the length of the weld because the actual throat dimension is impossible to measure throughout the length of the weldment. The formula is:

$$\frac{P}{L \times T} = S$$

P = Load, lb (N)

L = Total length of the fillet weld sheared, in (cm)

T = Theoretical throat dimension, in (cm)

S = Shear strength of the weld, lb/in^2 (N/cm^2)

---

[14] American Welding Society, *Specification for Welding Procedure and Performance Qualification*, AWS B2.1:2005, Section 2.

FIGURE 13-29 Shear strength test specimen

Using a laboratory example, the load applied to the transverse shear test was 35,000 pounds (155,585 N). The equal leg fillet weld size was 1/4 inches (6.4 mm). To calculate the theoretical throat, multiply the leg size by 0.707; 0.25 in × 0.707 = 0.177 in (6.4 mm × 0.707 = 4.52 mm). The width of the laboratory test specimen was 2 inches (50.8 mm), providing a total length of the weld of 8 inches (203 cm), welded on four sides; four welds times 2 inch long welds equals 8 inches (4 × 50.8 cm = 203 cm). The shear strength of the weld calculates to be 24,718 lb/in² (170 N/mm² or 16,956 N/cm²):

$$\frac{35,000 \text{ lb}}{8 \text{ in} \times 0.177 \text{ in}} = 24,718 \text{ lb/in}^2 \qquad \frac{155,585 \text{ N}}{203 \text{ mm} \times 4.52 \text{ mm}} = 170 \text{ N/mm}^2$$

Compared to the ASD calculations made earlier, this laboratory Unit Load is 4375 lb/in (766 N/mm). This value is achieved by 35,000 lbs ÷ 8 in = 4,375 lb/in (155,585 N ÷ 203 mm = 766 N/mm) or by 24,717 lb/in² × 0.177 in = 4,375 lb/in (170 N/mm² × 4.52 mm = 766 N/mm). The Unit Load value in this laboratory example is higher than the previous calculated ASD Allowable Unit Load because the machine load pulled a fillet weld with actual throat as opposed to a calculated theoretical throat.

CHAPTER QUESTIONS/ASSIGNMENTS

MULTIPLE CHOICE

1. The typical included angles for groove welds are 45° and _____.
   a. 22 1/2°
   b. 30°
   c. 60°
   d. 75°

2. The typical root openings for groove welds are 0 to 1/8 inches for 60° groove angles and _____ for 45° groove angles?
   a. 1/8 inches
   b. 1/4 inches
   c. 1/2 inches
   d. 3/4 inches

3. The thickness for steel where preheat prior to welding begins to be recommended is _____.
   a. 3/4 inches
   b. 1 inches
   c. 1½ inches
   d. 2 inches

4. What is the potential problem if grain direction is not considered?
   a. Porosity
   b. Distortion
   c. Warpage
   d. Base metal cracking

5. The types of stresses induced from welding are thermal and _____.
   a. Residual
   b. Primary
   c. Secondary
   d. Tertiary

SHORT ANSWER

6. Calculate groove depth to prevent warpage for the following plate thicknesses:
   a. 3/4-inch (19 mm) plate and 3/32-inch (2.5 mm) root face
   b. 1½-inch (38 mm) plate and 1/8-inch (3.2 mm) root face
   c. 2-inch (51 mm) plate and no root face

7. What is a good rule of thumb for fillet weld size?

8. What is preferred for aluminum; intermittent or continuous welds?

9. When decreasing the groove angle from larger to smaller, what are the positive and negative effects?

10. Calculate allowable unit load for a 1/2 inches (13 mm) fillet weld using 06011 and a 3/4-inch (19 mm) fillet weld using ER70S-6 electrode.

CLASS PROJECTS

11. Class Project: Calculate the shrinkage of an A36 plate. Weld a plate and then compare the calculations from actual part. Procedure:
   a. Obtain two pieces of A36 plate:
      i. Any thickness from 3/8 inches to 3/4 inches
      ii. 8 inches long by 4 inches wide
      iii. Cut a double V-groove.
      iv. Root face and root opening may be 0 inches to 1/8 inches.
   b. Tack the weld coupon together on the ends.
   c. Measure the coupon with calipers in both the transverse and longitudinal directs, *before* you start welding the plate.
   d. Obtain or calculate the cross-sectional area from **TABLE 14-5** in Chapter 14.
   e. Calculate the expected transverse and longitudinal shrinkage.
   f. Select any welding process.
   g. Alternate welding side 1 and side 2 or temporarily clamp the sample to a table. Erroneous readings will result if the plate warps. Allow cooling when unrestrained.
   h. Let it completely cool (allow 24 hours).
   i. Measure the cooled coupon in both transverse and longitudinal directions with calipers.

12. Class Project: Warpage
   a. Using either a structural or a formed angle, weld a T-joint and a channel joint (**FIGURE 13-30**).
   b. Size of angles may be 2 × 2 × 1/4 inches (50 × 50 × 6.5 mm) by 24 inches (610 mm) long or any suitable size and length.
   c. Weld all lengthwise joints continuously.
   d. Observe the results.

13. Class Project: Weld Size
   a. Choose a thickness of metal to weld a T-joint.
   b. Set appropriate welding parameters for a given welding process and electrode diameter.
   c. Select as many different welding processes as allowed or instructed.
   d. Make four T-joints for each welding process. Use the same weld technique for each joint to produce the fillet weld size by decreasing or increasing travel speed.
      i. Make the first weld larger than the base metal thickness.
      ii. Make the second weld as larger as the base metal thickness.
      iii. Make the third weld 75% of the base metal thickness.
      iv. Make the fourth weld 50% of the base metal thickness.
   e. Cut, polish, and etch each sample to observe root penetration. Cut each sample in the same location to ensure impartiality.

14. Class Project: Joint Accessibility
   a. Create three V-groove joints with a 0- to 1/8-inch (3.2 mm) root opening/root face.
      i. One with a 30° included angle
      ii. One with a 45° included angle
      iii. One with a 60° included angle
   b. Weld in the flat and vertical-up position.

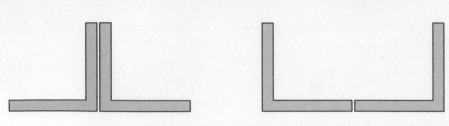

FIGURE 13-30

c. Cut the samples in thirds to examine for discontinuities.

d. Discuss the ease or difficulty in producing the welds.

15. Class Project: Shear Test

   a. Create a transverse shear strength test. Follow the dimensions of the example in Figures 13-15 and 13-16.

   b. Produce a 1/4-inch (6.5 mm) fillet weld.

   c. Use suitable metal thicknesses to develop the shear strength test.

   d. Using the ASD formula, calculate how much it will take to fracture the specimen. Note: The unit load must be multiplied to obtain the full length of the four welds.

   e. Weld the sample and then cut as illustrated in Figure 13-15.

   f. Pull the sample in the tensile testing machine. Record the reading upon fracture.

   g. Compare the actual results to the calculated results. Is there a difference? If so, why is there a difference?

   h. Calculate the shear strength of the weld in lb/in^2 (N/cm^2) by using the following formula:

$$\frac{P}{L \times T} = S$$

Chapter 14
WELDING COSTS

KEY TERMS

Yield 513

Deposition Efficiency 514

Deposition Rate (DR) 515

Operator Factor 519

KEY ACRONYMS

CSA (Cross-Sectional Area)

DDM (Density of Deposited Metal)

DR (Deposition Rate)

EC (Electrode Cost)

EP (Electrode Price)

FC (Flux Cost)

FDE (Filler Metal Deposition Efficiency)

FLR (Flux Ratio)

FP (Flux Price)

FR (Flow Rate)

GS (Gas Cost)

I (Amperage)

LC (Labor Cost)

LWW (Length of Wire Weight)

OC (Overhead Cost)

OF (Operator Factor)

OH (Overhead)

PC (Power Cost)

PE (Power Source Efficiency)

PG (Price of Gas)

TEL (Total Electrode Loss)

TS (Travel Speed)

V (Voltage)

W (Local Power Rates–Watts)

WFM (Weight of Filler Metal)

WFS (Wire Feed Speed)

WMD (Weight of Metal Deposited)

WPR (Welder Pay Rate)

WT (Weld Time)

Many things must be considered when determining what a product's final cost to the customer should be. A list consisting of material handling, fabrication, welding, painting, labor, overhead, etc., could be quite extensive to include all expenditures. It is no wonder a product can cost so much or a wonder when something apparently costs so little when every detail is included and a profit is still realized.

Three factors often determine welding practices: costs, quality, and operator appeal. Operator appeal is important because it increases the probability of better quality, which can ultimately lower costs and increase sales. When calculating costs between two processes, shielding gases, electrodes, etc., consider operator appeal if final costs are inconsequential.

Writing a good procedure will reduce costs while still maintaining quality. The welding procedure is critical not only to a specific product but also when establishing general shop practices, which ultimately establish welding costs. Changing the process, the shielding gas, filler material, etc., will affect the cost of welding. Sometimes, changing to a more expensive shielding gas, for example, leads not only to an increase in welding efficiency but also allows the use of a less expensive electrode. Both lead to a reduction in welding costs. Just because a shop practice has always worked before does not mean there is not a better way to decrease welding cost while still maintaining—or perhaps increasing—the quality of the welding process. For all these reasons, it is imperative that a welding procedure contain all the essential components making up a good weld. Voltage, amperage, gas flow rates, joint design, travel speed, etc., must be written down in a procedure and adhered to so quality and cost controls can be ensured.

There are many equations for calculating welding costs, and entire textbooks have been devoted to this subject. What this chapter presents is by no means all-inclusive. However, it should be sufficient to determine the outcome of the final costs when changing variables. Critical information is given in both U.S. Standard and SI Standard measures.

JOINT DESIGN

HOW DOES JOINT DESIGN AFFECT WELDING COST?   When examining the weld cost equations, keep in mind that the amount of metal deposited is determined by the joint design. When selecting the joint design, also consider the welding processes, the position of the workpiece, accessibility, and strength requirements. Joint design depends on the process used as well as on the ability to access the joint.

As an example for how cost is affected, consider an overwelded fillet joint. The fillet weld in **FIGURE 14-1** labeled (A) has a 1/4-inch (6.4 mm) leg size, yielding a 0.031 square inch (0.205 cm^2) weld area per inch (25.4 mm) of weld metal. The fillet weld labeled (B) has a 5/16-inch (8 mm) leg size, yielding a 0.049 square inch (0.312 cm^2) weld area per inch (25.4 mm) of weld metal. When the weld leg

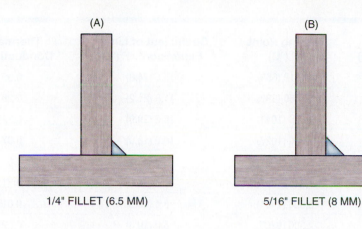

(A) 1/4" FILLET (6.5 MM)

(B) 5/16" FILLET (8 MM)

FIGURE 14-1 A larger weld costs more to produce than a smaller weld.

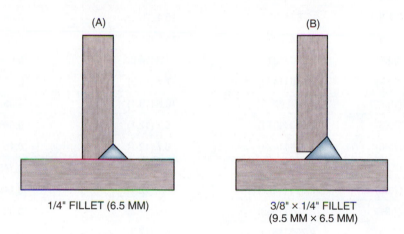

(A) 1/4" FILLET (6.5 MM)

(B) 3/8" × 1/4" FILLET (9.5 MM × 6.5 MM)

FIGURE 14-2 In order to obtain the same strength on a poor fit-up, a larger weld must be produced.

sizes are plugged into a weld cost equation, it can be seen how much more the larger weld cost.

Poor fit-up can also produce excessive total weld area. The single bevel/fillet weld labeled (A) in **FIGURE 14-2** has a 45° bevel angle with a 1/4-inch depth of preparation and a 1/4-inch (6.5 mm) leg size on the fillet weld. The total weld area per inch of weld metal would yield 0.062 inches squared (0.41 cm²). The poor fit-up on the joint labeled (B) has a 1/8-inch (3.2 mm) gap, and the weld area yields 0.014 square inches (0.90 cm²). Because of the gap, to obtain the same strength on joint (B) as joint (A) would require more than double the weld metal deposit.

FILLER METAL WEIGHTS AND PROCESS EFFICIENCIES

WHAT INFORMATION IS NEEDED TO CALCULATE WELDING COSTS? In order to perform cost calculations on weldments, some vital information is needed to insert into the equations. The needed information in **TABLE 14-1** is the density in units of lb/in³ highlighted in the third column. The additional information in Table 14-1 is provided for perspective on the differences between one metal and another.

Metal or Alloy	Specific Gravity	Density lb/in³ (gm/cm³)	Melting Point °F (°C)	Coefficient of Linear Expansion* °F (°C)	Thermal Conductivity
Aluminum	2.70	0.096 (2.65)	1,218 (659)	13.8 (24.8)	0.59
Brass	8.60	0.308 (8.52)	1,650 (899)	11.8 (21.2)	0.28
Aluminum bronze	7.69	0.278 (7.70)	1,905 (1041)	16.6 (29.9)	0.128
Silicon bronze	8.72	0.314 (8.70)	1,880 (1027)	10.0 (18.0)	0.07
Copper	8.89	0.322 (8.92)	1,981 (1083)	9.8 (17.6)	1.0
Gold	19.3	0.697 (9.30)	1,945 (1063)	7.8 (14.0)	0.71
Inconel	8.25	0.307 (8.50)	2,608 (1431)	6.4 (11.5)	0.015
Cast iron	7.50	0.260 (7.19)	2,300 (1260)	6.0 (10.8)	0.029
Wrought iron	7.80	0.281 (7.78)	2,750 (1510)	6.7 (12.1)	0.15
Lead	11.34	0.410 (11.35)	621 (327)	16.4 (29.5)	0.08
Magnesium	1.74	0.063 (1.75)	1,202 (650)	14.3 (25.7)	0.37
Monel	8.47	0.319 (8.83)	2,400 (1316)	7.8 (14.0)	0.036
Nickel	8.8	0.322 (8.92)	2,650 (1454)	7.4 (13.3)	0.23
Silver	10.45	0.380 (10.52)	1,764 (962)	10.6 (19.1)	1.06
Steel, high C	7.85	0.284 (7.86)	2,500 (1371)	6.7 (12.1)	0.095
Steel, med C	7.84	0.284 (7.86)	2,600 (1427)	6.7 (12.1)	0.15
Steel, low C	7.84	0.284 (7.86)	2,700 (1482)	6.7 (12.1)	0.145
Steel, low alloy	7.85	0.284 (7.86)	2,600 (1427)	6.7 (12.1)	0.145
SST, austenitic	7.9	0.286 (7.91)	2,550 (1399)	9.6 (17.3)	0.03
SST, ferritic	7.7	0.281 (7.78)	2,750 (1510)	9.5 (17.1)	0.03
SST, martensitic	7.7	0.281 (7.78)	2,600 (1427)	9.5 (17.1)	0.03
Tin	7.29	0.263 (7.28)	449 (232)	12.8 (23.0)	13.5
Titanium	4.5	0.163 (4.51)	3,031 (1666)	4.0 (7.2)	0.011
Tungsten	18.8	0.689 (19.07)	6,170 (3410)	2.5 (4.5)	0.31
Zinc	7.13	0.256 (7.08)	788 (420)	22.1 (39.8)	0.30

TABLE 14-1

*Coefficient of Linear Expansion is × 10⁻⁶ per

*Note: Electrical Conductivity is similar to Thermal Conductivity; however, the decimal must be moved over two places to the right. Example: Copper's Electrical Conductivity is **100***

Source: Cary, Howard: Modern Welding Technology, 4th edition, © 1998. Reproduced by permission of Pearson Education, Inc., Upper Saddle River, New Jersey.

Additional information needed for some equations, shown in **TABLE 14-2**, is how many inches of electrode wire is in a pound. It also provides some insight as to how much welding electrode is in a spool of wire. For example, a typical 0.035 inches (0.9MM) ER70S-6, 35-pound (15.9 kg) spool has 127,750 inches of electrode wire. The distance equals to 10,645.8 feet (3,245 m) or just over 2 miles (3.24 km).

Efficiency of welding depends on many variables. Arc length, type of shielding gas used, and proper welding parameter settings all affect how efficient the weld is performing. The data in **TABLES 14-3** and **14-4** are relative values. A decision must be made as to which figures are to be entered into an equation. In order for the results to be relevant, the numbers must stay consistent. If a choice is made to use the lowest efficiency of one welding process, then use the lowest efficiency

Electrode Diameter			Inches of Electrode/Pound (Centimeters/Kilogram)				
Decimal	Fraction	Metric	Aluminum	Magnesium	Flux-Cored	Mild Steel	Stainless
0.023	-	0.6 mm	25,072 (133,463)	38,204 (202,102)	-	8,475 (44,997)	8,416 (44,713)
0.030	-	0.8 mm	14,737 (75,073)	22,456 (113,682)	5,944 (30,189)	4,981 (25,311)	4,947 (25,151)
0.035	-	0.9 mm	10,827 (59,317)	16,498 (89,823)	4,367 (28,853)	3,660 (19,999)	3,634 (19,872)
0.040	-	1.0 mm	8,289 (48,047)	12,631 (72,757)	3,344 (19,321)	2,802 (16,199)	2,782 (16,097)
0.045	3/64	1.2 mm	6,550 (39,708)	9,980 (60,129)	2,642 (15,968)	2,214 (13,388)	2,198 (13,303)
0.052	-	1.4 mm	4,905 (28,430)	7,474 (43,051)	1,978 (11,432)	1,658 (9,585)	1,646 (9,525)
0.062	1/16	1.6 mm	3,450 (18,768)	5,258 (28,421)	1,392 (7,547)	1,166 (6,328)	1,158 (6,288)
0.078	5/64	2.0 mm	2,180 (12,012)	3,322 (18,189)	879 (4,830)	737 (4,050)	732 (4,024)
0.093	3/32	2.4 mm	1,533 (8,341)	2,337 (12,631)	619 (3,354)	518 (2,812)	515 (2,796)
0.125	1/8	3.2 mm	849 (4,692)	1,293 (7,105)	342 (1,887)	287 (1,582)	285 (1,572)
0.156	5/32	4 mm	545 (3,003)	830 (4,547)	220 (1,208)	184 (1,012)	183 (1,006)
0.187	3/16	4.8 mm	377 (2,085)	575 (3,158)	152 (839)	128 (703)	127 (699)

TABLE 14-2

Length vs. Weight

Source: Cary, Howard: *Modern Welding Technology*, 4th edition, © 1998. Reproduced by permission of Pearson Education, Inc., Upper Saddle River, New Jersey.

Process	Efficiency %
Covered Electrodes:	
SMAW 14-inch Manual	55%–65%
SMAW 18-inch Manual	60%–70%
SMAW 28-inch Automatic	65%–75%
Solid Bare Electrodes:	
Gas Metal Arc Welding	90%–95%
Submerged Arc Welding	95%–100%
Electrogas/Electroslag Welding	95%–100%
Tubular Cored Electrodes:	
FCAW–Gas Shielded	80%–90%
GMAW–Metal Cored	80%–90%
FCAW–Self Shielded	55%–65%

TABLE 14-3

Efficiency of Electrode Type and Process.

Method of Application	Efficiency % (Duty Cycle)
Manual	5%–30%
Semiautomatic	10%–60%
Mechanized	40%–90%
Automatic	50%–92%

TABLE 14-4

Welder Operator Efficiency.

on every other process being compared. The data may be skewed if there are inconsistencies in choosing high efficiency for one process and low efficiency for another—unless there is a very good reason for doing so.

TABLE 14-5 contains information on cross-sectional area (CSA) and weld metal deposited for steel. If the joint geometry needed for a weld cost equation differs from what is presented in the table, an equation is provided to calculate cross-sectional area for each joint design presented. All the data presented in Table 14-5 for steel are in inches.

TABLE 14-5 Size, CSA, and Steel Weld Deposit[1]

Weld Type	Joint Design	T (or L) Size in Inches (cm)	CSA in² (cm²) Theoretical	Weld Deposit lb/ft (kg/m)
Fillet Equal Legs		1/8 (0.32)	0.008 (0.051)	0.027 (0.040)
		3/16 (0.48)	0.061 (0.115)	0.020 (0.905)
		1/4 (0.64)	0.031 (0.205)	0.106 (0.161)
		5/16 (0.79)	0.049 (0.312)	0.167 (0.245)
		3/8 (0.95)	0.070 (0.451)	0.238 (0.355)
		1/2 (1.27)	0.125 (0.806)	0.425 (0.634)
			$CSA = \frac{1}{2}(L)^2$	
Fillet Two Size Legs		1/8 × 1/4 (0.32 × 0.64)	0.016 (0.102)	0.053 (0.080)
		1/4 × 3/8 (0.64 × 0.95)	0.047 (0.304)	0.160 (0.240)
		3/8 × 1/2 (0.95 × 1.27)	0.094 (0.603)	0.319 (0.474)
		1/2 × 5/8 (1.27 × 1.59)	0.156 (1.010)	0.530 (0.794)
		5/8 × 3/4 (1.59 × 1.90)	0.234 (1.511)	0.795 (1.187)
		3/4 × 1 (1.90 × 2.54)	0.375 (2.413)	1.274 (1.897)
			$CSA = \frac{1}{2}(L_1 \times L_2)$	
Butt RO = 1/8		1/8 (0.32)	0.016 (0.102)	0.054 (0.080)
		5/32 (0.40)	0.019 (0.128)	0.065 (0.101)
		3/16 (0.48)	0.023 (0.154)	0.078 (0.121)
		1/4 (0.64)	0.031 (0.205)	0.105 (0.161)
		5/16 (0.79)	0.039 (0.253)	0.132 (0.199)
		3/8 (0.95)	0.0467 (0.304)	0.160 (0.239)
			$CSA = RO \times T$	
Single V RO = 1/8 RF = 1/8 A = 60°		1/4 (0.64)	0.040 (0.264)	0.136 (0.207)
		3/8 (0.95)	0.083 (0.533)	0.283 (0.419)
		1/2 (1.27)	0.144 (0.927)	0.491 (0.729)
		5/8 (1.59)	0.223 (1.440)	0.760 (1.132)
		3/4 (1.90)	0.319 (2.049)	1.087 (1.611)
		1 (2.54)	0.567 (3.658)	1.932 (2.875)
			$CSA = (T - RF)^2 \times \tan(A/2) + (RO \times T)$	
Double V RO = 1/8 RF = 1/8 A = 60°		3/4 (1.90)	0.207 (1.329)	0.705 (1.044)
		1 (2.54)	0.346 (2.236)	1.179 (1.757)
		1 1/2 (3.18)	0.522 (3.379)	1.779 (2.656)
		1 1/2 (3.80)	0.733 (4.712)	2.498 (3.704)
		1 3/4 (4.45)	0.981 (6.348)	3.343 (4.989)
		2 (5.08)	1.265 (8.166)	4.311 (6.419)
			$CSA = \frac{1}{2}(T - RF)^2 \times \tan(A/2) + (RO \times T)$	
Single Bevel RO = 1/8 RF = 1/8 A = 45°		1/4 (0.64)	0.039 (0.256)	0.133 (0.201)
		3/8 (0.95)	0.078 (0.502)	0.266 (0.395)
		1/2 (1.27)	0.133 (0.858)	0.453 (0.674)
		5/8 (1.59)	0.203 (1.315)	0.692 (1.034)

3/4 (1.90)	0.289 (1.856)	0.985 (1.459)
1 (2.54)	0.508 (3.277)	1.731 (2.576)

$$CSA = \tfrac{1}{2}(T - RF)^2 \times \tan(A) + (RO \times T)$$

Double Bevel

$RO = 1/8$
$RF = 1/8$
$A = 20°$
$R = 1/2\,T$

3/4 (1.90)	0.191 (0.835)	0.651 (0.656)
1 (2.54)	0.316 (1.261)	1.077 (0.991)
$1\tfrac{1}{4}$ (3.18)	0.473 (1.762)	1.612 (1.385)
$1\tfrac{1}{2}$ (3.80)	0.660 (2.318)	2.249 (1.822)
$1\tfrac{3}{4}$ (4.45)	0.879 (2.976)	2.996 (2.339)
2 (5.08)	1.129 (3.687)	3.848 (2.898)

$$CSA = \tfrac{1}{4}(T - RF)^2 \times \tan(A) + (RO \times T)$$

Single U

$RO = 0$
$RF = 1/8$
$A = 20°$
$R = 1/2\,T$

1/2 (1.27)	0.163 (1.051)	0.557 (0.826)
3/4 (1.90)	0.419 (2.684)	1.429 (2.110)
1 (2.54)	0.792 (5.105)	2.701 (4.013)
$1\tfrac{1}{2}$ (3.80)	1.890 (12.115)	6.441 (9.522)
2 (5.08)	3.456 (22.2813)	11.777 (17.513)
$2\tfrac{1}{2}$ (6.35)	5.490 (35.401)	18.710 (27.825)

$$CSA = (T - R - RF)^2 \times \tan(A/2) + 2R(T - R - RF) + \tfrac{1}{2}\pi R^2 + (RO \times T)$$

Double U

$RO = 0$
$RF = 1/8$
$A = 20°$
$R = 1/2\,T$

1 (2.54)	0.148 (4.263)	0.505 (3.351)
$1\tfrac{1}{2}$ (3.80)	0.514 (10.134)	1.752 (7.965)
2 (5.08)	1.076 (18.652)	3.668 (14.660)
$2\tfrac{1}{2}$ (6.35)	1.835 (29.646)	6.253 (23.302)
3 (7.62)	2.790 (43.174)	9.508 (33.935)
$3\tfrac{1}{2}$ (8.89)	3.941 (59.236)	13.431 (46.559)

$$CSA = \tfrac{1}{2}(T - 2R - RF)^2 \times \tan(A/2) + 2R(T - 2R - RF) + (R^2 + (RO \times T))$$

Single J

$RO = 0$
$RF = 1/8$
$A = 20°$
$R = 1/2\,T$

1/2 (1.27)	0.114 (0.525)	0.390 (0.413)
3/4 (1.90)	0.309 (1.342)	1.052 (1.055)
1 (2.54)	0.597 (2.553)	2.031 (2.007)
1 1/2 (3.80)	1.449 (6.057)	4.935 (4.761)
2 (5.08)	2.670 (11.140)	9.101 (8.756)
2 1/2 (6.35)	4.262 (17.701)	14.528 (13.912)

$$CSA = \tfrac{1}{2}(T - R - RF)^2 \times \tan(A) + R(T - R - RF) + \tfrac{1}{4}\pi R^2 + (RO \times T)$$

Double J

$RO = 0$
$RF = 1/8$
$A = 20°$
$R = 1/2\,T$

1 (2.54)	0.270 (2.132)	0.922 (1.675)
$1\tfrac{1}{2}$ (3.80)	0.699 (5.067))	2.381 (3.983)
2 (5.08)	1.324 (9.326))	4.512 (7.330)
$2\tfrac{1}{2}$ (6.35)	2.145 (14.823)	7.309 (11.651)
3 (7.62)	3.162 (21.587)	10.776 (16.968)
$3\tfrac{1}{2}$ (8.89)	4.376 (29.618)	14.913 (23.280)

$$CSA = \tfrac{1}{4}(T - 2R - RF)^2 \times \tan(A) + R(T - 2R - RF) + \tfrac{1}{2}\pi R^2 + (RO \times T)$$

[1]Note: Use the equation if the root face and/or the root opening is of a different value.

Source: Cary, Howard; *Modern Welding Technology*, 4th edition, © 1998. Reproduced by permission of Pearson Education, Inc., Upper Saddle River, New Jersey.

WELDING COST CALCULATIONS

WHAT EQUATIONS AND CALCULATIONS ARE THERE FOR DETERMINING WELDING COST?

WEIGHT Densities of various metals are found in Table 14-1. Density of metal in pounds per cubic inch (lbs/in^3) or grams per cubic centimeter (gm/cm^3) are used to obtain the weight of the weld metal deposited (WMD). The weight of the metal deposited is used for many of the equations in this chapter and can be useful for determining the cost of a linear weld. (Intermittent welds **FIGURE 14-3** are less costly than continuous welds.) Steel is the most common metal used for welding (**FIGURE 14-4** is an example using stainless steel), and the weight of the weld metal deposited for various weld joints is provided in Table 14-5.

To acquire the weight of the weld metal deposited other than steel the density of the metal must first be obtained from Table 14-1. For example, the filler metal is aluminum and the weld deposited is a 1/4-inch (0.64 cm) equal leg fillet. From Table 14-1, the density of aluminum is 0.096 lbs/in^3 (2.65 gm/cm^3). From Table 14-5, the cross-sectional area of this fillet weld is 0.031 in^2 (0.205 cm^2). The equation to calculate the weight of the weld metal deposited is:

$$CSA \times L \times DDM = WMD$$
$$0.031 \times 12 \times 0.096 = 0.036 \text{ lb/ft} \quad (0.205 \times 100 \times 2.65)/1000 = 0.054 \text{ kg/m}$$

Where: CSA = Cross-sectional area, (in^2) or (cm^2)
          L = Length is based on 12 in or 100 cm (1 meter) of welded metal
          DDM = Density of the deposited metal, (lb/in.^{3}) or (gm/cm^3)
          WMD = Weight of the weld metal deposited, pounds per foot
          (Divide by 1,000 to convert grams to kilograms per meter.)

Calculating the weight of the weld metal deposited is useful in comparing how much weld metal is deposited in a proper weld size versus an oversized weld. For example:

Compare a 1/4-inch (0.64 cm) fillet weld to a 5/16-inch (0.79cm) fillet weld:

$$1/4 \text{ inch: } 0.031 \text{ in}^2 \times 12 \times 0.284 \text{ lb/in}^3 = 0.106 \text{ lb/ft}$$
$$5/16 \text{ inch: } 0.049 \text{ in}^2 \times 12 \times 0.284 \text{ lb/in}^3 = 0.167 \text{ lb/ft}$$

or

$$0.64 \text{ cm: } (0.205 \text{ cm}^2 \times 100 \times 7.86 \text{ gm/cm}^3)/1000 = 0.161 \text{ kg/m}$$
$$0.79 \text{ cm: } (0.312 \text{ cm}^2 \times 100 \times 7.86 \text{ gm/cm}^3)/1000 = 0.245 \text{ kg/m}$$

Again, the weight of metal deposited for steel could have been taken directly from Table 14-5. However, the weight of the weld metal deposited per foot can be obtained for any metal by using this equation and the densities provided in Table 14-1. If the cross-sectional area is not provided in Table 14-5 because of variances in joint geometry, use the equation listed within each cell to calculate the desired CSA.

The cost of the weld metal is not relevant at this point. However, deriving the weight of the metal deposited is a necessary first step because it will be used in many of the equations to determine weld costs.

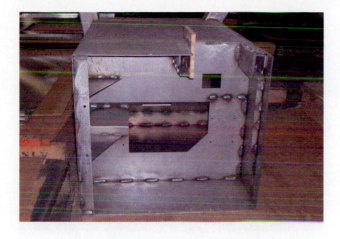

FIGURE 14-3 Intermittent welds are less costly to produce than continuous welds.
Source: Photo courtesy of Muza Metal Products, Inc.

EFFICIENCY Efficiency can be defined as the difference between the weight of the filler metal used and the weight of the actual weld metal deposited. When the GTAW process is performed properly, 100% of the filler metal goes into producing the weld. With SMAW, much of the electrode produces a slag that is removed as waste and is not part of the weld. This makes SMAW less efficient than other welding processes. Efficiency of an electrode depends on many factors. Incorrect voltage and amperage parameters produce excessive spatter, reducing the process efficiency. Modes of metal transfer and different blends of shielding gas can affect efficiency.

Efficiency may be difficult to determine, and obtaining the electrode manufacturers data is critical. Table 14-3 provides some idea of what to expect. The efficiency, or **yield**, can be determined by a simple experiment:

1. Obtain two pieces of metal of any substantial length to produce a lap joint.
2. Weigh the metal. Best results are obtained by using grams.
3. Clamp and weld the joint. Determine how much filler metal was used to produce the weld (i.e., calculate the weight of the filler metal used). With SMAW, it is simply a matter of weighing an unused electrode compared to the stub left after welding. With a process such as GMAW, it is necessary to know the WFS and the travel speed (or the time it took to weld the joint). Table 14-2 is helpful in calculating filler metal weight.
4. Remove anything that is not weld metal, such as spatter and slag.
5. Weigh the welded joint.
6. The difference between how much filler metal was used compared to the actual deposited weld metal is the process efficiency

The equation to define this efficiency (*Efficiency of Electrode*) may be expressed as follows:

$$\frac{Weight\ of\ Weldment}{Weight\ of\ Electrode} \times Efficiency$$

Where: Weight of Weldment is the weld metal deposited
Weight of Electrode is the electrode before welding
Efficiency is how the percentages (%) in Table 14-3 may be derived

Manufacturers have gone to great lengths to provide process and electrode efficiencies and efficiency data may be found in electrode manufacturer's literature. However, those wishing to determine the exact efficiency of their welding procedure may very well use the method describe.

FIGURE 14-4 Parts of a stainless steel racing boat header.
Source: Photo courtesy of Muza Metal Products, Inc.

Using data from Table 14-3 may be used to justify using one process over another, welding mode of transfer, or making a switch from one shielding gas mixture to another. The equation for determining how much electrode is actually required depending on the process efficiency is as follows:

$$\frac{WMD}{FDE} \times WFM$$

Where: WMD = Weight of the weld metal deposited, (lb/ft) or (kg/m)
   FDE = Filler metal deposition efficiency (%)
   WFM = Weight of filler metal required, (lb) or (kg)

Electrode manufacturers may aid in selecting filler metal deposition efficiency (FDE). **Deposition efficiency** is the actual amount of electrode going into creating only the weld metal. Brochures advertising electrodes from a specific manufacturer sometimes provide efficiency information. The results can be very skewed if the efficiency percentages are selected at the high end of one process and at the low end of another process—unless there is good justification.

Example: Compare 1/4-inch (0.64 cm) fillet welds by using 100% CO_2 with an efficiency of 80% versus a mixture of 90/10 (argon/CO_2) with an efficiency of 90%:

	U.S. Standard	*SI Standard*
80% efficient:	$\dfrac{0.106 \text{ lb/ft}}{0.80} = 0.133 \text{ lb/ft}$	$\dfrac{0.161 \text{ kg/m}}{0.80} = 0.201 \text{ kg/m}$
90% efficient:	$\dfrac{0.106 \text{ lb/ft}}{0.90} = 0.118 \text{ lb/ft}$	$\dfrac{0.161 \text{ kg/m}}{0.90} = 0.179 \text{ kg/m}$

This example calculates—using the same $0.75 per pound of electrode as in the previous examples—as $0.10/ft ($0.133 \times 0.75 = 0.10$) for CO_2 as opposed to $0.09/ft ($0.118 \times 0.75 = 0.09$) for the argon/$CO_2$ mixture. The cost per kilogram is approximately $1.65. The SI Standard efficiency calculation would cost $0.33/meter ($0.201 \times 1.65 = 0.33$) for CO_2 and around $0.30/meter ($0.179 \times 1.65 = 0.30$) for the argon/$CO_2$ mixture.

In these last two examples, there appears to be a significant change when creating a larger weld bead or using a less efficient shielding gas. The penny or two difference might seem insignificant until factoring in how much weld metal some companies produce on a weekly basis. If a company produces several thousand of feet or meters of weld per week, the cost is indeed significant.

Results can be altered depending on what efficiencies are used in the equation. Using the highest efficiency for the cheaper 100% CO_2 shielding gas and then using the lowest efficiency for the more expensive argon/CO_2 mixture would skew the results. The numbers in this case would favor the 100% CO_2 shielding gas, but the comparison is not realistic.

The efficiency equations presented so far may work well for custom fabrication, a short production run, or a small shop trying to get an idea on what to charge for a given product. The efficiency equations are a quick way to get a handle on welding costs. As more specific information is factored in—such as cost of electrodes, cost of shielding gas, electricity, and production method—more precise costs can be acquired.

PRICE VARIABLES   The next several equations deal with specific prices and costs. Costs fluctuate from year to year, so the values in the examples are hypothetical and should by no means be used for calculating final costs. Always obtain the latest cost of supplies before inserting them into a cost equation. The following equation is calculated by using the cost of the filler metal and comparing the efficiencies used in the previous example. The results show a cost savings if efficiency can be increased.

$$\frac{EP \times WMD}{FDE} = EC$$

Where: EP = Electrode price, ($/lb) or ($/kg)
WMD = Weight of the weld metal deposited, (lb/ft) or (kg/m)
FDE = Filler metal deposition efficiency (%)
EC = Electrode cost, ($/ft) or ($/m)

	U.S. Standard	*SI Standard*
80% efficient:	$\dfrac{\$1.36/\text{lb} \times 0.106\ \text{lb/ft}}{0.80} = \$0.18/\text{ft}$	$\dfrac{\$3.00/\text{kg} \times 0.161\ \text{kg/m}}{0.80} = \$0.60/\text{meter}$
90% efficient:	$\dfrac{\$1.36/\text{lb} \times 0.106\ \text{lb/ft}}{0.90} = \$0.16/\text{ft}$	$\dfrac{\$3.00/\text{kg} \times 0.161\ \text{kg/m}}{0.90} = \$0.54/\text{meter}$

The equations presented so far work well when all the welds for a product are measured. Sometimes this is easy—if the product is small or the welds are linear and are easily measured. However, on large scale production or when welds are not easily measured, another way to calculate the weight of the filler metal required for a job is based on the information written in a welding procedure; that is, the wire feed speed (or melt-off rate for other processes), type of wire, and travel speed.

To determine the type of wire needed, use the figures for the diameter and density of the electrode (e.g., steel, aluminum, etc.). The result is the length of the wire in a given weight (i.e., inches of a particular diameter electrode in a pound or centimeters in a kilogram). Table 14-2 supplies these figures for common electrode wire diameters and the more common filler metals used (e.g., steel, aluminum, stainless steel, and flux-cored electrode wires).

There are three equations to calculate the final costs, but the end result is the electrode cost per foot or meter of weld. These equations are for a continuous wire feed process and best for single-pass welding. The **deposition rate (DR)**, the

FIGURE 14-5 Sequencing of welds reduces distortion and costly rework.
Source: Photo courtesy of Muza Metal Products, Inc.

amount of weld metal deposited, in the following equation takes into account the filler metal efficiency. This equation alone is useful in determining the efficiency of various welding processes, which in itself determines welding costs. Sequencing of welds (shown in **FIGURE 14-5**) reduces rework and cost. The first equation is as follows:

$$\frac{WFS \times 60 \times FDE}{LWW} = DR$$

Where: WFS = Wire feed speed, in IPM or MPM or melt-off rate
60 = A constant, 60 minutes in an hour
FDE = Filler metal deposition efficiency (%)
LWW = Length of wire per weight, (in/lb) or (cm/kg)
DR = Deposition rate, (lb/hr) or (kg/hr)

In the next two examples, the comparison is between a 0.035 inches (0.9 mm) diameter mild steel electrode using short-circuit transfer and a 0.035 inches (0.9 mm) diameter metal core electrode. Assume the same efficiency for both welding electrodes and the same weight for solid steel wire in Table 14-2 as for the composite electrode. Wire feed speed on feeders set up for SI Standard may be in meters per minute (MPM) or millimeters per second (mm/sec.). For convenience in the formula, the wire feed speed is in centimeters per minute.

	U.S. Standard	*SI Standard*
Solid Steel Wire:	$\dfrac{250\ IPM \times 60 \times 0.92}{3660\ inches/lb} = 3.77\ lb/hr$	$\dfrac{635 \times 60 \times 0.92}{19{,}999\ cm/kg} = 1.75\ kg/hr$
Metal Cored:	$\dfrac{500\ IPM \times 60 \times 0.92}{3660\ inches/lb} = 7.54\ lb/hr$	$\dfrac{1270 \times 60 \times 0.92}{19{,}999\ cm/kg} = 3.51\ kg/hr$

The second equation in the three series set is used to calculate the welding travel speed and arrive at a rate in feet per hour or meters per hour. Typically, a U.S. Standard weld procedure is written in inches per minute (IPM) for travel speed, but for this equation, the travel speed in feet per hour is needed. Note that the travel speed for metal core is considerably faster than for short-circuit transfer. This is due to higher voltages and amperages used for the composite electrode. The equation is as follows:

$$\frac{TS_1 \times 60}{L} = TS_2$$

Where: TS_1 = Travel speed, (in/min) or (cm/min)

60 = A constant, 60 minutes in an hour

L = 12 inches in a foot or (100 cm in a meter)

TS_2 = Travel speed, (ft/hr) or (m/hour)

	U.S. Standard	*SI Standard*
Short Circuit:	$\dfrac{22\ inches/min \times 60}{12} = 110\ ft/hr$	$\dfrac{56\ cm/min \times 60}{100} = 33.6\ meters/hr$
Metal Cored:	$\dfrac{48\ inches/min \times 60}{12} = 240\ ft/hr$	$\dfrac{122\ cm/min \times 60}{100} = 73.2\ meters/hr$

With a simplified version of the equation the same results are obtained by using the following:

U.S. Standard	*SI Standard*
$TS_1 \times 5 = TS_2$	$TS_1 \times 0.6 = TS_2$
22 inches/min $\times$ 5 = 110 ft/hr	56 cm/min $\times$ 0.6 = 33.6 meters/hr
48 inches/min $\times$ 5 = 240 ft/hr	122 cm/min $\times$ 0.6 = 73.2 meters/hr

The third and final part of the three series equations is to determine the weight of filler metal required. The DR in this last equation comes from the first equation in the series. The TS_2 comes from the second equation in the series. The final equation follows:

$$\frac{DR}{TS_2} = WFM$$

Where: DR = Deposition rate, (lb/hr) or (kg/hr)

TS_2 = Travel speed, (ft/hr) or (m/hr)

WFM = Weight of filler metal required, (lb/hr) or (kg/hr)

	U.S. Standard	*SI Standard*
Short Circuit:	$\dfrac{3.77\ lb/hr}{110\ ft/hr} = 0.0343\ lb/ft$	$\dfrac{1.75\ kg/hr}{33.6\ m/hr} = 0.0521\ kg/m$
Metal Core:	$\dfrac{7.54\ lb/hr}{240\ ft/hr} = 0.0314\ lb/ft$	$\dfrac{3.51\ kg/hr}{73.2\ m/hr} = 0.0480\ kg/m$

All that remains is to obtain the electrode cost in $/ft ($/meter). This is done by multiplying the weight of filler metal required (WFM) by the electrode price ($/lb or $/kg). Note that the composite electrode is more expensive than a solid electrode wire.

	U.S. Standard	*SI Standard*
Short Circuit:	0.0343 lb/ft × $1.36/lb = $0.047/ft	0.0521 kg/m × $3.00/kg = $0.16/meter
Metal Cored:	0.0314 lb/ft × $1.79/lb = $0.056/ft	0.0480 kg/m × $3.95/kg = $0.19/meter

The results indicate that it is more expensive to purchase a composite electrode. However, with more than twice the travel speed, labor cost is dramatically reduced, and the overall cost is much lower when implementing the composite electrode. Pulse spray metal transfer is an example where deposition efficiency as well as travel speed can be increased while the cost of the electrode wire does not increase.

OTHER COST VARIABLES   The next several equations are useful if there is a need to calculate other variables, such as shielding gas or flux cost for SAW. Shielding gas costs may be calculated for a job in one of two ways. For typical welding operations, the equation is:

$$\frac{PG \times FR}{TS_2} = GS_1$$

Where: PG = Price of gas, ($/cubic foot or $/cubic liter)
FR = Flow rate, (cubic foot/hour or cubic liters/minute)
TS_2 = Travel speed, (ft/hr or m/hr)
GS_1 = Gas cost, ($/ft or $/m)

The following example compares a standard argon/CO_2 mixture and an argon/CO_2/O_2 blend.

U.S. Standard

75/25 Gas: $\dfrac{\$0.18 \times 30\ \text{CFH}}{110\ \text{ft/hr}} = \$0.05/\text{ft}$

Tri–Mix Gas: $\dfrac{\$0.31 \times 30\ \text{CFH}}{150\ \text{ft/hr}} = \$0.08\ \text{ft}$

SI Standard

$\dfrac{\$0.006 \times (14\ \text{LPM} \times 60)}{33.6\ \text{m/hr}} = \$0.16/\text{m}$

$\dfrac{\$0.011 \times (14\ \text{LPM} \times 60)}{45.7\ \text{m/hr}} = \$0.20/\text{m}$

There are times when the cost of spot welding with a wire feed welder is needed. For making a spot weld with a wire feed welder, the last equation is not practical because the weld time is so short. For spot welding, use this equation:

$$\frac{PG \times FR \times WT}{60} = GS_2$$

Where: PG = Price of gas, ($/cubic ft) or ($/cubic liter)
FR = Flow rate, (cubic ft/hour) or cubic liters/min
WT = Weld time, (min)
60 = A constant, 60 minutes in an hour
GS_2 = Gas cost, ($/weld)

U.S. Standard

$\dfrac{\$0.18 \times 30\ \text{CFH} \times 0.03\ \text{min}}{60} = \$0.003/\text{weld}$

SI Standard

$\dfrac{\$0.006 \times (14\ \text{LPM} \times 60) \times 0.03\ \text{min}}{60} = \$0.003/\text{weld}$

In a completely different scenario, if flux is being used, such as in the case of the ESW or SAW process, the flux cost must be accounted for. One factor in the equation is the flux ratio, which is the amount of flux used to the amount of steel wire used. In the case of the SAW process, this ratio is about 1 to 1. However, this ratio can go as high as 1.5 to 1. Tests should be run in order to determine the correct ratio.

The flux ratio for the ESW process may run from 5–10 pounds of flux for every 100 pounds of electrode. This gives a 0.05:1 to 0.1:1 ratio. Again, tests should be run to determine the correct ratio. The equation for determining flux costs is:

$$FP \times WMD \times FLR = FC$$

Where: FP = Flux price, ($/lb) or ($/kg)

WMD = Weight of weld metal deposited, (lb/ft) or (kg/m)

FLR = Flux ratio

FC = Flux cost, ($/ft) or ($/m)

U.S. Standard	SI Standard
$0.50/lb × 1.704 lb/ft × 1 = $0.85/ft	$1.10/kg × 2.535 kg/m × 1 = $2.79/m

This example uses a 1-inch (2.54 cm) equal leg fillet weld producing a WMD of 1.704 lb/ft (2.535 kg/m) and a SAW flux ratio of 1:1.

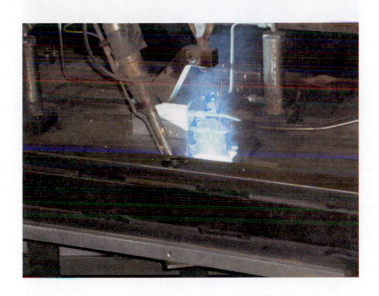

FIGURE 14-6 Robotic Arc Welding on a riding mower deck increases efficiency.
Source: Photo courtesy of the Ariens Company.

LABOR COSTS One way to determine labor costs is to calculate how much it costs for an individual to weld a 1-foot weld. This does not include filler metal. In the two following equations, the **operator factor** is the time spent actually welding. This operator factor varies depending on the process and is found in Table 14-4. The equation is:

$$\frac{WPR}{TS_2 \times OF} = LC$$

Where: WPR = Welder pay rate, ($/hr)

TS_2 = Travel speed, (ft/hr) or (m/hr)

OF = Operator factor, (%)

LC = Labor cost, ($/ft) or ($/m)

Using the metal-cored versus solid steel wire in a previous example, it was determined the composite electrode cost more per linear weld to produce. The faster travel speed of the composite filler metal and the labor—a part of this equation—reveal a considerable cost savings. Labor in this example includes wage, benefits, and other employer costs per employee.

	U.S. Standard	SI Standard
Short Circuit:	$\dfrac{\$50/hr}{(22\ IPM \times 5) \times 0.35} = \$1.30/ft$	$\dfrac{\$50/hr}{(56\ cm/min \times 0.6) \times 0.35} = \$4.25/m$
Metal Cored:	$\dfrac{\$50/hr}{(48\ IPM \times 5) \times 0.35} = \$0.60/ft$	$\dfrac{\$50/hr}{(122\ cm/min \times 0.6) \times 0.35} = \$1.95/m$

FIGURE 14-7 Robotic arc welding on a riding mower deck increases efficiency.

Source: Photo courtesy of the Ariens Company.

Operator factor is the arc-on time of the welder. When a welder moves from one weld to another, takes the time to clean or change a contact tip, chip slag, etc., this influences the percentage of time the welder is actually welding in a given period of time. Automated welding **(FIGURES 4-6 and 14-7)** improves the operator factor. Another useful way to calculate labor costs for procedures requiring multipass welding is:

$$\frac{WPR \times WMD}{DR \times OF} = LC$$

Where: WPR = Welder pay rate, ($/hr)
         WMD = Weight of weld metal deposited, (lb/ft) or (kg/m)
         DR = Deposition rate, (lb/hr) or (kg/hr)
         OF = Operator factor, (%)
         LC = Labor cost, ($/ft) or ($/m)

WMD, DR, and OF all depend on the process, size and type of joint, and the electrode used. Using information from the previous equations—i.e., $50/hr labor cost—welding on a 1-inch (2.54 cm) multipass fillet weld producing 1.704 lb/ft WMD (2.535 kg/m), a deposition rate of the metal-cored example (7.54 lb/hr or 3.51 kg/hr), and a 35% operator factor, an example is:

U.S. Standard

$$\frac{\$50/hr \times 1.704\ lb/ft}{7.54\ lb/hr \times 0.35} = \$32.28/ft$$

SI Standard

$$\frac{\$50/hr \times 2.535\ kg/m}{3.51\ kg/hr \times 0.35} = \$103.17/m$$

OVERHEAD COSTS To determine cost based on overhead, it is practical to substitute the welder's pay with an overhead rate instead. Overhead includes such things as the amount of electrical usage broken down into an hourly cost. The amount it costs to heat and cool the workplace (if used), ventilate fumes, uniforms provided, tools, welding gloves, welding helmet, etc., may all be included. In other words, all those costs incurred by the company need to be included in overhead. They are very similar to the last two equations, as follows:

$$\frac{OH}{TS_2 \times OF} = OC$$

Where:
OH = Overhead rate, ($/hr)
TS_2 = Travel speed, (ft/hr or m/hr)
OF = Operator factor, (%)
OC = Overhead cost, ($/ft or $/m)

For this example, using $100/hr for overhead rate, the travel speed for short-circuiting welding previously used (single-pass weld), and a 35% operator factor, it equates as follows:

U.S. Standard
$$\frac{\$100/hr}{(22\ IPM \times 5) \times 0.35} = \$2.60/ft$$

SI Standard
$$\frac{\$100/hr}{(56\ cm/min \times 0.6) \times 0.35} = \$8.50/m$$

It is sometimes difficult to calculate welding costs from some of the equations given if a large multipass weld is to be made. The last equation, using travel speed, is more beneficial for single-pass welding. For multipass welding, the weld metal deposited equation is more accurate. This equation would be:

$$\frac{OH \times WMD}{DR \times OF} = OC$$

Where:
OH = Overhead rate, ($/hr)
WMD = Weight of weld metal deposited, (lb/ft) or (kg/m)
DR = Deposition rate, (lb/hr) or (kg/m)
OF = Operator factor, (%)
OC = Overhead cost, ($/ft) or ($/m)

Using information from previous examples—i.e., $100/hr overhead, 1.704 lb/ft (2.535 kg/m) weld metal deposit, 7.54 lb/hr (3.51 kg/hr) deposition rate, and an operator factor of 35%—the example would be:

U.S. Standard
$$\frac{\$100/hr \times 1.704\ lb/ft}{7.54\ lb/hr \times 0.35} = \$64.57/ft$$

SI Standard
$$\frac{\$100/hr \times 2.535\ kg/m}{3.51\ kg/hr \times 0.35} = \$206.35/m$$

POWER COSTS   The last equation to consider includes the cost of power. This situation is used where the employees provide their own tools, gloves, welding helmet, etc. The company pays for electrical power and does not want to include any other overhead costs. This equation works very well for a company wishing to justify the purchase of newer, more efficient power sources to replace existing equipment. To find the final costs, calculate the power consumption for welding alone. The equation for power costs is:

$$\frac{W \times V \times I \times WMD}{1000 \times DR \times OF \times PE} = PC$$

Where:
W = Local power rates, ($/kWh)
V = Volts
I = Amps
WMD = Weld metal deposit, (lb/ft) or (kg/m)
1000 = A constant
DR = Deposition rate, (lb/hr) or (kg/hr)
OF = Operator factor, (%)
PE = Power source efficiency, (%)
PC = Power cost, ($/ft) or ($/m)

Hypothetically, this example uses the GMAW process with a 0.035 (0.9 mm) diameter electrode at 18.5 volts and 150 amps at 250 IPM (6.35 MPM). This produced a 1/4-inch (0.6 mm) fillet weld yielding a WMD of 0.106 lb/ft (0.161 kg/m) and a DR of 3.77 lb/hr (1.75 kg/hr). The on-peak electrical power rate is $0.20/kWh. The OF is 35%. Comparing an older transformer/rectifier power source of 60% efficiency to an inverter power source of 90% efficiency, the equations would be:

U.S. Standard

60% Efficient: $\dfrac{\$0.20/\text{kWh} \times 18.5 \times 150 \times 0.106\,\text{lb/ft}}{1000 \times 3.77\,\text{lb/hr} \times 0.35 \times 0.60} = \$0.07/\text{ft}$

90% Efficient: $\dfrac{\$0.20/\text{kWh} \times 18.5 \times 150 \times 0.106\,\text{kg/m}}{1000 \times 3.77\,\text{lb/hr} \times 0.35 \times 0.90} = \$0.05/\text{ft}$

SI Standard

60% Efficient: $\dfrac{\$0.20/\text{kWh} \times 18.5 \times 150 \times 0.161\,\text{kg/m}}{1000 \times 1.75\,\text{kg/hr} \times 0.35 \times 0.60} = \$0.24/\text{m}$

90% Efficient: $\dfrac{\$0.20/\text{kWh} \times 18.5 \times 150 \times 0.161\,\text{kg/m}}{1000 \times 1.75\,\text{kg/hr} \times 0.35 \times 0.90} = \$0.16/\text{m}$

And comparing the last example with off-peak electrical power rate at $0.05/kWh, the equations would be:

U.S. Standard

60% Efficient: $\dfrac{\$0.05/\text{kWh} \times 18.5 \times 150 \times 0.106\,\text{lb/ft}}{1000 \times 3.77\,\text{lb/hr} \times 0.35 \times 0.60} = \$0.02/\text{ft}$

90% Efficient: $\dfrac{\$0.05/\text{kWh} \times 18.5 \times 150 \times 0.106\,\text{lb/ft}}{1000 \times 3.77\,\text{lb/hr} \times 0.35 \times 0.90} = \$0.01/\text{ft}$

SI Standard

60% Efficient: $\dfrac{\$0.05/\text{kWh} \times 18.5 \times 150 \times 0.161\,\text{kg/m}}{1000 \times 1.75\,\text{kg/hr} \times 0.35 \times 0.60} = \$0.06/\text{m}$

90% Efficient: $\dfrac{\$0.05/\text{kWh} \times 18.5 \times 150 \times 0.161\,\text{kg/m}}{1000 \times 1.75\,\text{kg/hr} \times 0.35 \times 0.90} = \$0.04/\text{m}$

CHAPTER QUESTIONS/ASSIGNMENTS

TRUE OR FALSE

1. An increased travel speed often not only leads to increased root penetration but also reduces the final cost of a product.

2. Even though an increase in the percentage of argon in a shielding gas for welding on steels increases efficiency, the increased cost does not justify the final cost of welding.

3. Other than operator appeal, there is no advantage to using the pulse spray mode of transfer.

4. Poor fit-up is one of the largest culprits of increased costs in a welded part.

5. Because of so many variables, welder operator efficiency is one of the most subjective parameters in calculating welding costs.

6. Joint design of a weldment has very little impact on the final cost of a welded product.

SHORT ANSWER

7. Calculate the CSA of a single V-groove 3/8-inch plate with 45° angle, 1/4-inch root opening (having a backing plate), and no root face. Also, convert the result to SI Standard.

8. Calculate the WMD for steel, aluminum, and stainless steel with the CSA from problem #7 in both U.S. Standard and SI Standard.

9. Calculate the difference among the metals used in problem #8 by using a welding process with 75% efficiency to a welding process capable of 95% efficiency.

10. Calculate the cost of the aluminum filler metal in problem #8 by using a 90% process efficiency. Obtain the price per pound of each GMAW electrode wire from the Internet.

11. Compare the cost in problem #10 by using the three-step formula that uses wire feed speed, type of wire, and travel speed to arrive at price per foot and price per meter. Use a 3/64-inch (1.2 mm) ER4043 electrode at 325 IPM (825 cm/min) wire feed speed and 30 IPM (76 cm/min) travel speed. Use the same efficiency as in problem #10.

12. Calculate the difference it costs for welding by using 35 CFH (16.5 LPM) compared to using 45 CFH (21.2 LPM) of shielding gas for the ER4043 electrode used in previous problems. Use any information from previous problems to calculate costs per foot and costs per meter. Obtain the cost of argon shielding gas from the Internet.

13. Examine world economics. Using the aluminum filler metal in previous problems, calculate how much it cost per foot and per meter for a product welding in the United States to any foreign county. Use an operator factor (OF) of 30%. Use the Internet to determine foreign pay rate to compare to local pay rate. To make things more simplistic, just use an hourly wage.

14. Calculate the costs for welding a 3/16-inch (0.48 cm) square groove weld on steel with a 1/8-inch (0.32 cm) root opening by using an old electric motor generator power source at 20% efficiency to a more modern 85% efficiency inverter power source. Using an ER70-6, 0.035 diameter electrode, it is operating at 19 volts, 160 amps, 290 IPM (737 cm/min) wire feed speed, 90% deposition efficiency, and 35% operator factor. Obtain the $/kWh from your home bill. Calculate both U.S. Standard and SI Standard.

Chapter 15
POWER SOURCES

LEARNING OBJECTIVES

In this chapter, the reader will learn how to:

1. Understand the electron theory of current flow.
2. Understand how a welding power source functions.
3. Know the advantages and disadvantages of the different technology levels of welding power sources.
4. Determine when a constant current or a constant voltage power source is the correct choice.
5. Use the manufacturer's duty cycle specifications to determine the correct size power source for a specific welding process.

KEY TERMS

Valence Shell 526

Insulator 526

Conductor 526

Semiconductor 527

Voltage 528

Current 528

Resistance 529

Direct Current (DC) 530

Alternating Current (AC) 532

Frequency (Hz) 532

Rectification 535

Diode 535

SCR (silicon controlled rectifier) 538

IGBT (insolated gate bipolar transistor) 540

Wattage 549

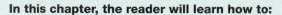

PEARSON
myweldinglab

INTRODUCTION

Arc welding has become one of the most important manufacturing processes in today's modern society. It is used to produce the appliances we use, build the buildings we work in, and manufacture the vehicles that take us there.

As the name implies, arc welding processes use an electrical arc. Because of modern electrical grid systems, electricity is available almost everywhere. When electrical current is not available, generators can be utilized to produce the electrical current needed.

Arc welding cannot be accomplished by just connecting two welding cables to the primary current supply. This causes a very large amount of uncontrolled alternating electrical current to be conducted. Attaching the same welding cables to the positive and negative leads of an automotive battery would also cause a very large amount of uncontrolled direct current to be conducted. The end results in both cases would be less than desirable and unsafe.

In order to produce the desired welds the arc—which from this point forward will be called the welding current—must be controllable. The purpose of the arc welding power source is to allow the welder to have control of the welding current.

A welder does not need to have a complete understanding of how each individual circuit of a power source operates. This type of information would be good to know but would not normally benefit the welder in his or her daily job. What does benefit the welder is an understanding of the differences among different types and technology levels of power sources. Knowing this information enables a welder to answer questions such as:

- What power source can be used with which arc welding process?
- Which power source will be best used to produce a specific product?
- What primary voltage can a power source operate on and how large should breaker and diameter primary conductors be?
- What duty cycle is required to meet production needs, and what power source meets this requirement?
- What size generator/welder will meet the requirements of a project?

ELECTRON THEORY

When two pieces of mild steel are welded together, the two individual pieces are permanently joined together at the smallest molecular level in a relatively small area. This joining is called coalescence and is done by using localized heating of the metal to the point where a portion of each individual piece melts and then joins together to become one. The source of heat can be from a chemical reaction (Oxy Fuel), friction (friction stir), or electrical current (Arc and Resistance Welding).

This chapter focuses on electrical arc welding and the power sources used to arc weld, so it is important to understand what an electrical arc is. The arc and its resulting noise and light are audible and visual signs of the welding current being conducted across the arc gap. At the same time current is being conducted

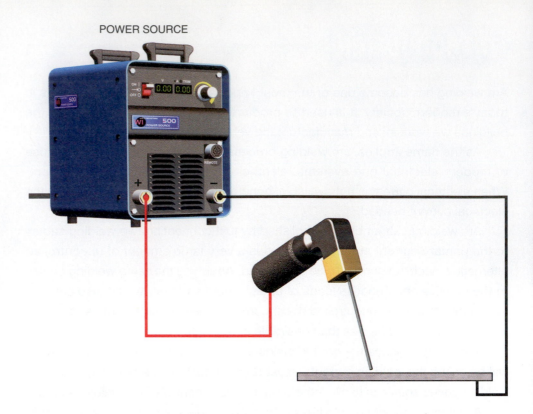

POWER SOURCE

FIGURE 15-1 A common SMAW welding circuit

across the gap, it is simultaneously flowing through the welding power source, output connections, electrode and work cable, work clamp and electrode holder, base metal, and consumable electrode. **FIGURE 15-1** shows a common SMAW welding circuit.

ELECTRICAL CURRENT

Electrical current is the movement (conduction or flow) of electrons through a conductive metal caused by the application of an electromotive force (voltage). To explain this definition, let us first look at the structure of an atom. The atom is the building block of all liquids, solids, and gases. All metals that are welded contain atoms. The majority of metals are alloys (mixes of several different pure elements). What all metals have in common is that they are all conductors and will allow the movement of electrons or current.

Every atom has a nucleus orbited by negatively charged electrons. The nucleus consists of positively charged protons and neutral (having no positive or negative charge) neutrons. The electron orbits are arranged in shells (rings). The negatively charged electrons remain in their orbits because they are attracted to the positively charged nucleus. The negatively charged electrons avoid each other's orbits because they each have the same electrical charge, thus causing them to repel each other. **FIGURE 15-2** shows an atom.

The outermost shell of an atom is called the **valence shell**. The number of electrons found in the valence shell determines if the atom is an **insulator** (a material that will not conduct electrical current) or a **conductor** (a material that will conduct electrical current).

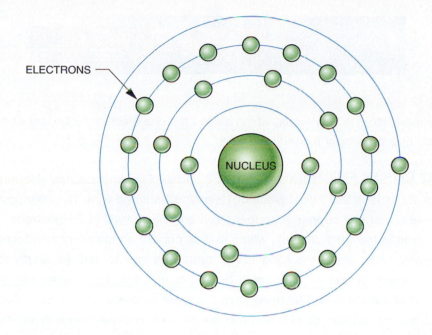

ELECTRONS

NUCLEUS

FIGURE 15-2 An atom is composed of electrons orbiting a nucleus.

A valence shell that is full will not accept electrons from or give up any electrons to other atoms. The material that is made of this specific atom would then be an insulator.

A valence shell that is not full may accept electrons from or give up any electrons to other atoms. The material that is made of this specific atom may be a conductor. How easily the atom gives or receives other electrons depends on the number of electrons in its valence shell. If an atom is just one electron short of full, it may not be a good conductor and may conduct electrical current only when very specific conditions are met. Materials made of these types of atoms are called **semiconductors**.

When an atom has one or very few electrons in its valence shell and the conditions are right, free atoms will begin to move from one atom to the next, producing electrical current. **FIGURE 15-3** shows a drawing of an oxygen atom and a copper atom. The oxygen atom, which is an insulator, has a valence shell with 6 electrons. With 6 electrons, the valence shell is almost full, making the bond of electrons too strong to allow electrons to be given up to other atoms. The copper atom has only 1 electron in its valance shell. The bond between electrons in

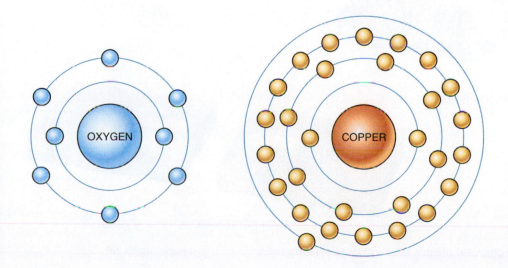

OXYGEN

COPPER

FIGURE 15-3 An oxygen atom (left) with many electrons in its outer shell compared to a copper atom (right) with only 1 electron in its outer shell

FIGURE 15-4 A conductor (the wire cable) with no movement or flow of electrons

the copper atom is very weak, which allows the one electron to be lost to other atoms, making copper a conductor.

WHAT CAUSES ELECTRONS TO MOVE?

To explain what causes electrons to move, let us look at a 2-foot piece of 12-gauge insulated wire. The conductor in the wire is made of copper. A copper atom only has a total of 29 electrons, with 1 electron in the valence shell, which makes copper an excellent conductor of electrons. With the wire lying on a table not connected to anything, a measurement of current flow could be taken, resulting in an extremely small amount of current flow. Electrons from a copper atom can move from atom to atom, but in this case, there is no pressure or force to make the electrons move. **FIGURE 15-4** shows a conductor with no movement or flow of electrons.

Now attach each end of the same insulated wire to the posts of a car battery. The battery has negative (–) and positive (+) posts, thus the wire now has positive and negative sides. Just like magnets, opposite electrical charges attract each other and like electrical charges repel each other. **FIGURE 15-5** shows how two magnets respond to like and to opposite charges.

The electrons of the atom are going to be naturally attracted toward the positive post of the battery. The car battery not only has terminals that establish the polarity, but it also provides the electromotive force (pressure) that causes electrons to move from one atom to the next. This electromotive force (pressure) is the **voltage** of the battery, which in this case is 12 volts. With an established polarity and voltage, the electrons in the atoms of the copper will begin to move from one atom to the next, producing electrical **current**. As long as the voltage is applied, the electrons will continue to move though the circuit (wire and battery).

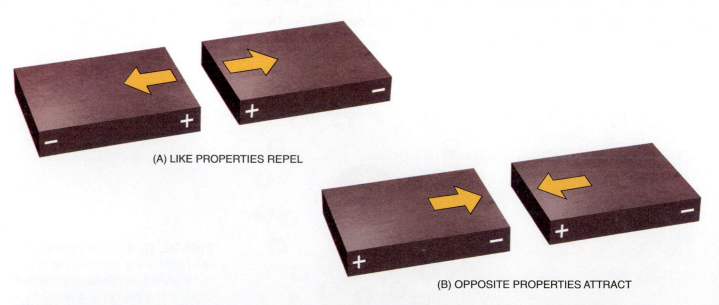

(A) LIKE PROPERTIES REPEL

(B) OPPOSITE PROPERTIES ATTRACT

FIGURE 15-5 Opposite electrical charges attract each other and like electrical charges repel each other.

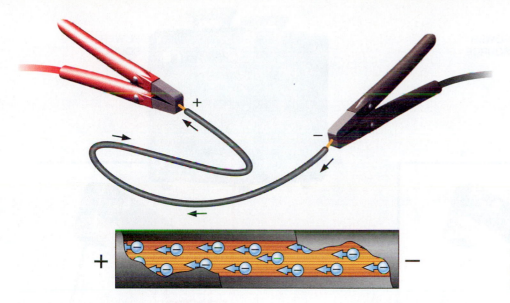

FIGURE 15-6 shows how electrons of a conductor respond when a polarity and voltage (pressure) are applied.

If either of the connections to the posts are disconnected, the current flow instantly stops. This is not unlike what happens when a welder is SMAW (stick welding) mild steel. The power source provides the voltage (electromotive force). But there is no welding arc or current flow until the electrode is touched to the mild steel base metal. When the electrode comes in contact with the mild steel, electrons begin to move from one atom to the next, being pushed by the voltage (pressure) supplied by the power source. If the welder wants to stop the arc, all that needs to be done is to increase the distance between the mild steel and the electrode, causing the conductive path to be lost and current flow to stop. **FIGURE 15-7** shows how current flow (welding output) can be stopped when the welding circuit is opened.

WHAT CAUSES THE MILD STEEL AND ELECTRODE TO MELT? Let us look

back at the 12-gauge wire connected between the posts of the car battery. As soon as the connection was closed (connected) between the posts, current began to flow through the copper wire. A car battery is designed to start a vehicle's engine by causing a large engine to begin to cycle through its movement and fire the spark plugs, causing combustion to start. In order to perform these tasks, a car's battery has to have a voltage potential and the ability to conduct a large amount of current. Twelve-gauge wire is not very large, with a diameter of 0.0808 in (2.053 mm). With the battery's ability to conduct a large amount of current, the small diameter wire would quickly begin to get hot. In a short amount of time, the wire would melt, and the connection between posts would be opened (disconnected), instantly ending current flow.

Copper is an excellent conductor of current, but in this case, the small diameter of the wire limits the number of electrons that can flow through it. This limitation or opposition to current flow is called **resistance**. As the current flows through the wire, the resistance to its flow produces heat. As the current continues to flow, the heat continues to build up, resulting in the wire melting.

If the 12-gauge wire was replaced with a 2-foot piece of 2/0 (pronounced 2 ought) welding cable, which has a diameter of 0.3648 inches (9.266 mm), the

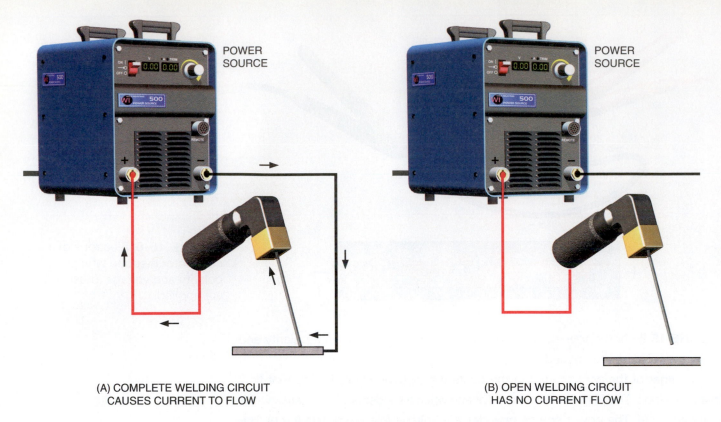

POWER
SOURCE

POWER
SOURCE

(A) COMPLETE WELDING CIRCUIT
CAUSES CURRENT TO FLOW

(B) OPEN WELDING CIRCUIT
HAS NO CURRENT FLOW

FIGURE 15-7 Welding output can be stopped when the welding circuit is opened.

result would be that the welding cable would become hot but would most likely not melt. The copper in the wire and the cable are the same type of copper but are different in size. The larger diameter cable is able to conduct more current because of its greater size, resulting in less resistance.

There is no perfect conductor of electrical current. Thus, every conductor has a certain resistance to current flow. One of the physical properties of all metals is its resistance to current flow.

In the case of mild steel being stick-welded, the steel and electrode resistances to current flow are what cause the base metal and electrode to melt. It is important in the case of arc welding that the base metal and electrode are the only conductors in the welding circuit to melt. Later in this chapter, we will discuss the importance of using the proper size welding cables, electrode holders, welding guns, and work clamps for the amount of welding current needed.

DIRECT CURRENT

Direct Current (DC) is the movement of electrons in one direction only. An everyday example of this is a two-cell flashlight. **FIGURE 15-8** shows a drawing of a common two-cell household flashlight.

Two batteries are arranged in order of positive to negative. In contact with the negative side is a wire that runs to the on/off switch. On the other side of the switch is a wire that runs to one contact point of the light bulb. In contact with the positive side is the other contact point of the light bulb. All this forms a very simple electrical circuit.

(A) SWITCH OFF

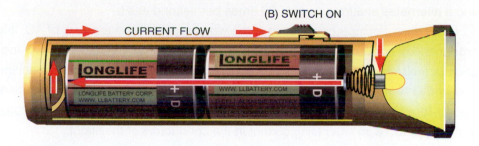

(B) SWITCH ON

CURRENT FLOW

FIGURE 15-8 A common two-cell household flashlight

When the on/off switch is open (off position), there is no electrical current flow and the light bulb will not light. When the switch is closed (on position), the voltage potential of the two batteries causes electrons in the entire circuit to move from negative to positive. This causes the light bulb to turn on.

The simple circuit used for the flashlight is very similar to a welding circuit. For example, let us again look at the circuit formed when stick welding mild steel. Instead of batteries, the power source provides the source of voltage potential. Connected to the negative terminal is the work lead, which is then clamped to the workpiece. The electrode lead is connected to the positive terminal. **FIGURE 15-9** shows the work lead connected to the negative terminal and the workpiece and the electrode lead to positive.

On the other end of the electrode lead is the electrode holder, which holds a stick electrode. When the stick electrode is not touching the workpiece, the welding circuit is open. As soon as the welder touches the electrode to the workpiece, current begins to flow throughout the welding circuit. The current again flows from negative to positive only.

FIGURE 15-9 DCEP current is applied when the electrode is positive and the work lead is negative.

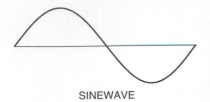

SINEWAVE

FIGURE 15-10 An AC sinewave form

Alternating Current (AC) is the movement of electrons in one direction, then changing direction and traveling in the opposite direction. The rate (times per second) in which AC electrical current changes direction is called the **frequency (Hz)**. **FIGURE 15-10** shows a common 230-volt AC waveform used in residential and industrial systems throughout North America. The waveform is called a sinewave, and it alternates at a frequency of 60 times per second (60Hz).

Almost all electrical power used in the world is supplied in the form of AC. It is produced by converting one form of mechanical energy (hydro, nuclear, wind, solar, or fuel) into electricity by using a generator. A generator is a rotating machine that converts mechanical energy into electrical energy. It uses the mechanical energy to rotate a magnetized rotor inside a barrel-shaped stator. As the rotor turns, it produces an alternating current in the conductors of the stator.

The voltage level of the electricity produced by the generator is generally around several thousand volts. Electrical generation plants are generally located away from the metropolitan areas that they serve. If the electrical power was transmitted at the lower thousand volt level, the current produced would be extremely high. High current flow requires large diameter conductors. Using large diameter conductors would be very expensive and inefficient.

In order to efficiently transmit electrical current from the plant to its customers, it is advantageous to increase the voltage to a very high voltage—around 110,000 volts—using a step-up transformer. As the voltage level goes up, the current level decreases, allowing smaller diameter conductors to be used. This makes it possible for the electrical generation plant to be located hundreds of miles from its furthest customer.

Because the use of high voltage is unsafe, the high voltage can then be reduced at its point of use. This is done with the use of a step-down transformer.

Alternating current is used for arc welding. The two most common examples are SMAW and GTAW. In **FIGURE 15-11**, the tungsten electrode is positive, the work is negative, and the current flow is from work to electrode. Then, the polarity switches, and current flow is from electrode to work.

FIGURE 15-11 When welding on AC, the polarity alternates between the electrode and the base metal.

As mentioned in the introduction, connecting welding leads directly to a DC battery or to the primary AC power would result in a very large amount of uncontrolled current being conducted. The resulting welds would not be the quality that is expected, and there would be a serious safety risk. In order to produce the welds that are expected by the industry today, the welder has to have a very controllable and consistent flow of welding current.

All welding power sources are designed to provide controlled and consistent current flow. If all power sources did this the same way, the choice of which one to purchase would be left to the flip of a coin or the welder's favorite color. Over the last 110 years, the manufacturers of welding power sources have used technical innovation to provide the welding industry with a variety of power sources that not only provide controlled and consistent current flow but also provide benefits such as increased portability, remote current control, increased power efficiency, and programmable memory. The purpose of this part of the chapter is to explain:

- How a welding power source functions
- The benefits and drawbacks of the different technology levels of welding power sources (tap select, magnetic amplifier, solid state, and inverter)

POWER SOURCE FUNCTIONS   How a welding power source functions can be broken down into sections. These sections are step-down, rectification, and output control. An easy way to visualize these sections is in a block form, as seen in **FIGURE 15-12**.

The left side of the diagram is where the primary input power is connected, and on the right side are the electrode and work leads. Everything between these two points is the welding power source.

Step-Down   The primary AC power that a power source is plugged into ranges in voltage level from 115–575 vac. Arc welding with voltage levels this high is uncontrollable and unsafe. The voltage has to be reduced to a lower level; this is done by using a step-down transformer. **FIGURE 15-13** shows the schematic symbol and a picture of a single-phase transformer.

A transformer is a very simple electrical device that consists of three parts: the primary coil, the core, and the secondary coil. Whenever electrical current flows through a conductor, a magnetic field (lines of flux) radiates out around the conductor. If another conductor crosses the lines of flux, the field will be imposed on the conductor.

FIGURE 15-12 A schematic of step-down, rectification, and output.

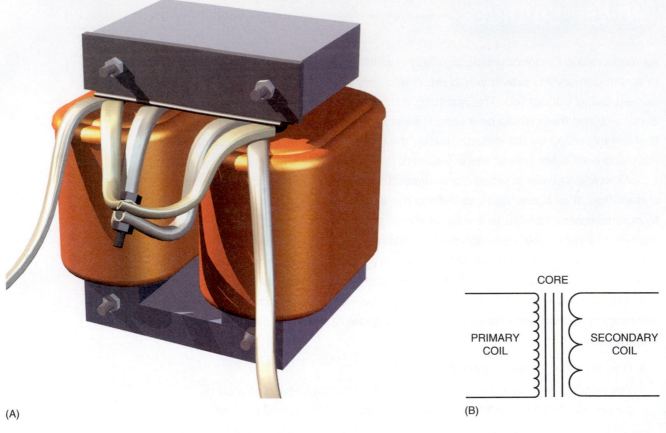

(A)

CORE

PRIMARY
COIL

SECONDARY
COIL

(B)

FIGURE 15-13 (A) Drawing of a single-phase welding transformer. (B) Schematic of a single-phase transformer.

When the welder turns on the power switch, AC current flows through the windings of the primary coil. As the current flows through the primary coil, lines of flux radiate out and cross the core. The core of a welding step-down transformer consists of laminated metal sheets pressed together and then welded along the sides. Because the core is conductive, current begins to flow though it when the lines of flux cross it. As current flows through the core, it also produces lines of flux that cross over the secondary coil, again imposing current flow through the coil. **FIGURE 15-14** shows the lines of flux.

Like the primary coil, the secondary coil is wrapped around the core. The difference between the two coils is in the number of times the secondary coil's conductor is wrapped around the core and in the diameter of the conductor.

The reason the transformer is being used is to change the high voltage/low amperage primary power into a usable welding power of low voltage/high

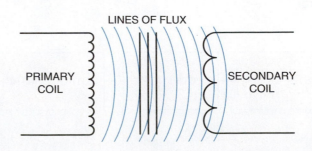

LINES OF FLUX

PRIMARY
COIL

SECONDARY
COIL

FIGURE 15-14 Lines of flux
between the primary and
secondary coils

amperage. In order to reduce the voltage, the secondary coil has less wraps (turns) of conductor around the coil.

Another difference is the diameter of the wire used in each winding. The primary is high voltage/low amperage, and the secondary is low voltage/high amperage. The increased current conducted in the secondary winding requires a larger diameter wire. The secondary of a welding transformer can always be identified by the larger diameter wire.

Rectification   The current that is stepped down by the step-down transformer is AC. The majority of arc welding processes today use DC; thus, the AC must be changed to DC. DC flows through a conductor in one direction only, not in both directions like AC.

The changing of AC to DC is called **rectification** and is done with a device called a rectifier. Rectifiers are very common device and are used in any electronic equipment that is plugged into AC power. Some power sources available on the market today are AC only; in this case, there is no rectification of the AC current needed. **FIGURE 15-15** shows the schematic symbol.

The rectifier above is a single-phase welding rectifier that consists of heat sinks and four diodes. A diode is a unique device in that it is not an insulator or a conductor; it is a semiconductor. A **diode** is a device that will conduct electrical current only when specific conditions are present and in only one direction. The main condition that must be met is that the diode has the correct polarity. **FIGURE 15-16** shows the schematic symbol and picture of a diode.

A diode has two sides: the anode and cathode. When the + is on the anode and − on the cathode (forward-biased), the diode will conduct current. When the polarity is reversed (reverse-biased), the diode will not conduct current. This ability allows the diode to act as a one direction gate only, allowing current to be conducted in only one direction. Its function is similar to a check valve used in a hydraulic system.

When AC is of the correct polarity, the diode conducts, and when the polarity is reversed, the diode does not conduct. When several diodes are

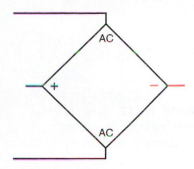

FIGURE 15-15 Changing alternating current to direct current is called rectification.

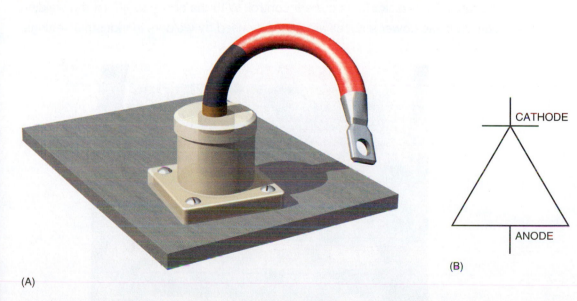

(A)

(B)

FIGURE 15-16 (A) Drawing of a high amperage diode. (B) Schematic symbol of a diode.

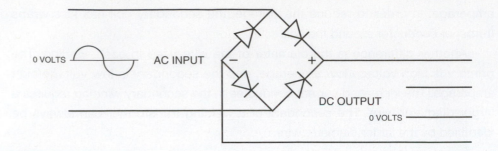

0 VOLTS ─ AC INPUT

DC OUTPUT

0 VOLTS

FIGURE 15-17 When several diodes are used in a rectifier, AC is turned into DC.

used in a rectifier, AC is turned into DC. **FIGURE 15-17** shows how a rectifier functions.

Output Control   The welding current is now stepped down and rectified, but there is no method of controlling the amount of current that is used for welding. There are four methods of controlling the amount of welding current: tap select, magnetic control, solid state, and inverter.

Tap Select   This method of control is inexpensive and easier for the welder to understand; thus, it is often used in power sources that are sold to the home hobbyist. The use of taps does not allow the welder to have fine adjustment of the welding output. There is no ability to remotely control the welding output.

When a tap select transformer is being manufactured, T connections or taps are installed at locations in the secondary winding. A switch is connected to the secondary of the transformer, which allows the welder to select different taps from which welding current will flow. The closer the selected tap is to the beginning of the secondary winding, the lower the welding voltage (less turns). The farther the tap is from the beginning, the higher the welding voltage (more turns). The control of the welding output is simply limited by the number of turns. This same method of adding taps can also be done with the primary coil of the transformer. **FIGURE 15-18** shows a tap select welding power source.

Magnetic Control   This type of control was first designed to compensate for the tap select's lack of fine current control. With the ability to adjust the welding output, these power sources are generally used by welders in industrial settings.

FIGURE 15-18 A tap select welding power source

Magnetic control power sources were the most advanced power source at one time and were extremely popular in the industry. Because of the way they were manufactured, the power sources were extremely robust, and it is not uncommon to find 30- to 40-year old power sources being used in shops today. This style of manufacturing is very labor-intensive; thus, in today's very competitive power source industry, the power source is generally not cost-competitive. There are still some magnetic control power sources sold today.

There are two types of magnetic control power sources: movable shunt and magnetic amplifier. Both types allow the welder to adjust the welding output throughout the entire range of the power source's welding range. The magnetic amplifier allows remote current control with the use of a foot control.

The movable shunt power source has a movable portion (shunt) of the step-down transformer. Moving the shunt manipulates the lines of flux that affect the secondary coil, changing the welding output. The shunt is a mechanical device that is adjusted by turning a handle to move the shunt. The welding output can be adjusted while welding on some power sources, while in others, the turning of the handle can cause extensive damage to the transformer (consult the owner's manual before attempting to adjust output while welding). These power sources do not have the ability to use a remote current control. **FIGURE 15-19** shows a movable shunt power source.

The magnetic amplifier is similar to the movable shunt in that it controls the welding output by manipulating a magnetic field. The difference is that the field that it manipulates does not affect the transformer but instead affects the reactor.

A reactor consists of a coil wrapped around an iron core. It is placed in line with one side of the secondary coil. As welding current flows through the coil, it builds a magnetic field, but this field does not positively affect the welding output. As the strength of the field is increased, the field reduces the flow of welding current.

The magnetic field is increased by the use of an extra circuit, which consists of extra winding around the reactor, DC power supply, and rheostat. The welder uses the adjustable rheostat to change the amount of DC current that flows

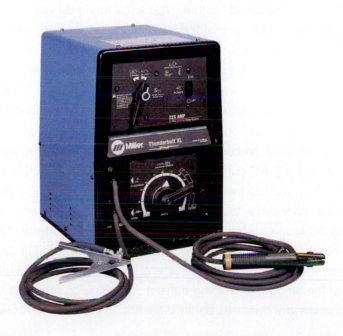

FIGURE 15-19 A movable shunt power source
Source: Photo courtesy of Miller Electric Manufacturing Co.

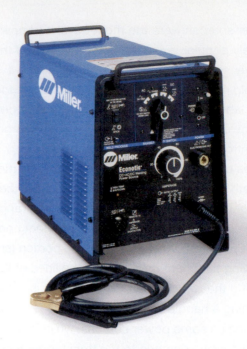

FIGURE 15-20 A magnetic amplifier power source

Source: Photo courtesy of Miller Electric Manufacturing Co.

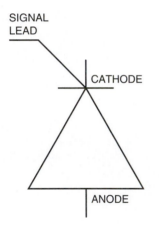

FIGURE 15-21 The schematic symbol of an SCR (silicon controlled rectifier)

through the extra winding. Welding output is directly affected by adjusting of the rheostat. Unlike the movable shunt, the welding output of a magnetic amplifier can be adjusted while welding and can be remotely controlled. **FIGURE 15-20** shows a magnetic amplifier power source.

Solid State   One of the drawbacks of the tap select and magnetic control power sources is that the switching on and off of the welding output (contactor control) is done by the opening and closing of the main power switch or a welding contactor. Both of these are mechanical devices that use contact points. With normal use, the points will become pitted and fail in time.

Instead of using a mechanical device, the solid-state power sources use electronic devices to turn the welding output on and off. One of the most common devices is the **SCR (silicon controlled rectifier)**. **FIGURE 15-21** shows the schematic symbol of an SCR.

An SCR is similar to a diode in that it has to be forward-biased in order to conduct current. Unlike a diode, even if the SCR is correctly biased, it will not conduct current until the SCR is turned on by a low-voltage signal. The low-voltage signal (gate signal) comes from a control board, which is able to receive commands from the welder. For example, the welder squeezes the trigger of the GMAW torch and the circuit board receives the command and then sends the signal to the SCR.

The SCR can be turned on, but even if the signal is removed, it will continue to conduct current until the current flow stops. Because of this, the SCR works best when conducting AC current. When the AC cycle forward-biases and the gate signal is sent, the SCR will conduct current. When the AC cycle begins to transition to the other cycle, the SCR is no longer forward-biased and turns off.

Because SCRs are controllable diodes, they replace diodes in the solid-state power sources bridge rectifier. Using SCRs allows the rectifier to be used to control the on and off of the welding current. The other benefit of using SCRs is the control of the amount of welding current. This type of control is called phase control. Phase control operates by timing when the SCRs are turned on during

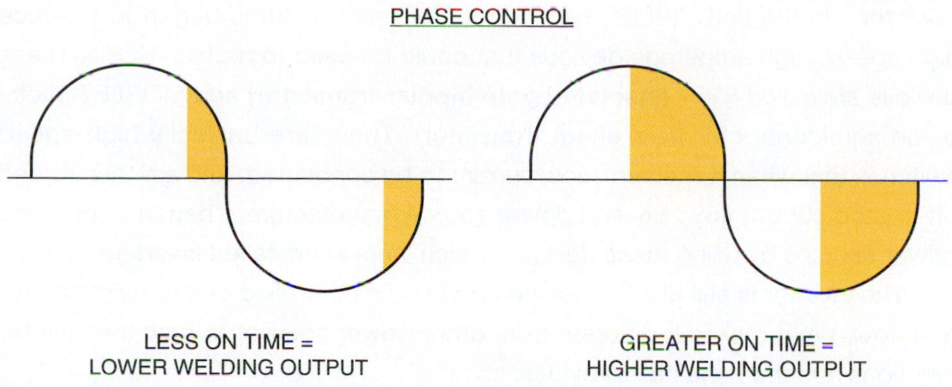

LESS ON TIME =
LOWER WELDING OUTPUT

GREATER ON TIME =
HIGHER WELDING OUTPUT

FIGURE 15-22 The sooner an SCR is turned on during a cycle, the greater the welding output.

each AC cycle. The sooner in an AC cycle the SCR is turned on, the higher the amount of welding current output. **FIGURE 15-22** represents how the sooner an SCR is turned on during a cycle, the greater the welding output.

The control board turns on the SCRs based on information from the power source's amperage or voltage control. An example of this would be a welder performing a FCAW vertical-up weld on a 3/8-inch groove weld. The welder finds that the welding voltage needs to be higher to fall within the range of the Welding Procedure Specification (WPS). To change the voltage, the welder increases the voltage setting on the power source. By adjusting the voltage control, the welder changes a signal that is sent to the circuit board. This signal tells the board how much welding voltage is needed. The board then turns the SCRs on at the correct time in the phase.

ARE THERE MORE BENEFITS WHEN USING A SOLID-STATE POWER SOURCE?

The benefits of using SCRs and a circuit board have made solid-state control one of the most popular means of controlling a power source's welding current. Not only does this style of control provide contactor and output level control, but it also can provide:

- Control contactor and welding current even in a power source that is used to perform an AC welding process, such as AC GTAW
- More accurate welding output with the use of feedback circuits that monitor and react to changes in the welding voltage or current
- Can react to changes every 120th of a second. An AC 60Hz single-phase cycle changes every 120th of a second. Changes can be even faster when the power source is operating on a three-phase primary.
- Welding current up-slope and down-slope control
- Limited welding current pulsing control
- Programmable welding parameters

ARE THERE ANY DISADVANTAGES WHEN USING A SOLID-STATE POWER SOURCE?

The technology that enables the production of solid-state power sources requires circuit boards and more components compared to the other types of power sources. With the addition of circuit boards and more components, there is a greater risk of power source malfunction or failure. Another disadvantage is the higher cost that comes with a solid-state power source.

Inverter   In the early 1980s, semiconductor manufactures began to produce high speed, high amperage devices that could be used to replace SCRs. These devices are called **IGBT (insolated gate bipolar transistor)** and MOSFET (metal oxide semiconductor field effect transistor). These are basically high-speed switches that when turned on allow current to be conducted, and are then turned off to stop current flow. Several power source manufacturers began to produce power sources by using these devices, which were soon called inverters.

The inverter is still just a machine providing a controlled and consistent current flow. What makes it different from other power sources is how the inverter functions and the benefits it provides.

As seen in FIGURE 15-12, a standard power source steps-down the primary voltage, rectifies the secondary current, controls the welding output, switches the welding current on and off, and conditions the welding current to be the most useful for the welding process in which it was designed. **FIGURE 15-23** shows the block diagram of a typical inverter.

Rectification   When the main power switch is turned on, the primary input current is rectified. This rectification produces DC voltage that can range from 500–800 Vdc. The rectifier used in this circuit is a three-phase rectifier but can also be used with a single-phase primary.

The ability to be used on a single- or three-phase primary circuit is one advantage of most inverter power sources. There are some inverters that require a three-phase primary only. This is because the three-phase primary is needed in order to meet the output range of the inverter.

Switching   The 500–800 Vdc current produced when the primary current is rectified is ready to be run through a set of IGBTs. The purpose of this circuit is to take the DC and change it back into AC. At first, this would seem to be a waste of time—turning AC into DC and then back into AC—but these steps produce several very important advantages.

The two conductors (buss) that carry the rectified current are connected to a set of IGBTs. Between the IGBTs is the primary coil of the step-down transformer. Contained in an IGBT package are two individual switches. When a signal is given by a circuit board, one switch from each package turns on, and current is conducted through one switch, across the primary coil, and then finally through the other switch. Next, the circuit board turns off the signal, stopping current flow across the primary coil. The circuit board then turns on the remaining switches,

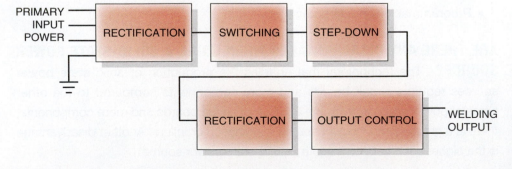

FIGURE 15-23 A block diagram of a typical inverter

repeating the process—except that current flow across the primary coil is in the opposite direction; thus, the circuit has just produced its own AC.

The advantage of making its own AC is that the step-down transformer of an inverter can be very small and lightweight. The step-down transformer of a standard power source accounts for a majority of the power source's size and weight. This makes the power source very heavy and not very portable. An inverter with the same welding range will be a fraction of the size and weight. This makes inverters very portable, and they can often be carried by one individual.

When an electrical engineer is designing a transformer, one of the variables that determines the size of the transformer is the frequency of the AC. The lower the AC transformer frequency, the larger the core needs to be. In North America, the primary current frequency is 60Hz. At this low frequency, the transformer must have a large core. Because the IGBTs used in the switch circuit are made to switch at high speed, the current across the transformer can be alternated at a very fast rate. Thus, the AC frequencies of the inverter's switching circuit can be in the millions (MHz) of hertz. This high frequency allows the transformer to be reduced in size and weight.

Step-Down   This part of an inverter is no different than a standard power source, other than that the frequency of the AC current is in the mega Hertz range (MHz) and the transformer is a fraction of the size and weight. The primary side of the transformer is still high voltage/low current and the secondary is low voltage/high current.

Rectification   This part of the inverter is in some cases very similar to standard power sources. The AC from the secondary of the transformer must now be turned into DC for use with the majority of the arc welding processes.

In the late 1990s, several power source manufacturers began producing inverter power sources capable of supplying both AC and DC welding current. Before AC/DC inverters were available, a welder could not use GTAW and SMAW on AC.

In the case of an AC/DC inverter, AC from the secondary is still turned into DC, but when an AC welding current is required, the DC must be turned into AC again. This process is again done with high-speed switches that are turned on and off. This time, AC current produced is alternated across the work and electrode.

The ability to produce AC welding current by using high-speed switches has several advantages. First, the AC welding current can be produced at frequencies lower and higher than the standard 60Hz. Second, the amount of time in which current is conducted in each direction can be different, changing the balance of electrode negative welding current versus electrode positive welding current. Both of these advantages give the welder more control over the welding arc and the resulting GTAW weld.

Output Control   Contactor and welding output are controlled by the main circuit board. This is done by controlling the on and off switching of the primary IGBTs. When the circuit board turns on a pair of IGBTs, it will allow the other pair to stay on until the required amount of current has been conducted. At this point, the circuit board turns off that pair and turns on the other pair. The longer each pair is turned on, the higher the welding output.

Because the inverter changes the primary AC curent to DC, changes to the welding output are not a limited by 120th of a second changes of a 60Hz primary. This allows an inverter to react very fast to changes.

WHAT ADVANTAGES DOES AN INVERTER POWER SOURCE HAVE OVER OTHER WELDING POWER SOURCES?

Because of the advanced technology and higher speed components, inverter power sources can have several advantages over other welding power sources. Not all inverters have all the following advantages:

- Decreased size and weight
- Increased primary voltage ranges
- Greater electrical efficiency
- Ability to be used as constant current and constant voltage processes
- Increased range of pulses per second when GMAW and GTAW pulsing. Some power sources produced today have pulsing ranges up to 5,000 pulses per second.
- Reacts quickly to changes in command from robotic and automated controllers
- Provides independent electrode negative and electrode positive current control for AC GTAW welding

DOES AN INVERTER POWER SOURCE HAVE ANY DISADVANTAGES?

Inverter power sources offer a wide range of advantages, but there are disadvantages. A welder needs to understand the advantages and disadvantages in order to make the best decision when looking to purchase a new power source. Not all inverters have both the following disadvantages:

- Increased cost compared to equal welding output noninverter power sources
- Greater number of electronic components; thus, a greater chance of failure

POWER SOURCE RATINGS

Manufacturers of welding power sources provide several ratings for their equipment. Each rating is designed to answer a question regarding the setup and operation of the welding power source. These ratings are very important in that they provide a guideline that a purchaser can use to help make the best decision when purchasing equipment.

It is important to understand what each of these ratings means. Just as important is for the welder to have a good understanding of what is required to successfully complete the welding for which the power source is being purchased. This means that even if the ratings are understood, not understanding the welding process and its required voltage, current, and time of welding can cause a power source to be purchased that does not have enough output. This can cause lost time by the power source overheating and shutting down. Not understanding the process can also cause the purchaser to buy too large of a power source. Because larger means higher cost, the welder may end up spending too much and never getting the full value of the purchase.

Several ratings pertain to the voltage and current required to operate the power source. It is never a good idea to assume that the voltage level and current capacity of your shop are adequate to operate a power source. If you are not absolutely sure of these values, it is recommended that an electrical contractor be contacted to provide this information.

HOW LOW AND HIGH ARE THE POWER SOURCE'S WELDING OUTPUTS?

WELDING OUTPUT RANGE   Ratings are listed in welding voltage, welding current, or both. AC/DC power sources may have separate ratings for AC and DC. **TABLE 15-1** shows an example of a power source's welding current and voltage range.

Welding Current Range	30–300A
Welding Voltage Range	12–42V

TABLE 15-1

Power source manufacturers are very precise in regards to the welding range. If the rating lists maximum amperage of 250A, the power source should not be expected to provide 275A. It is very important to purchase a power source that has a range that will meet welding requirements.

HOW LONG CAN THE POWER SOURCE WELD AT A CERTAIN WELDING OUTPUT CURRENT BEFORE IT OVERHEATS?

DUTY CYCLE   After the power source manufacturer has designed and built a power source, it must now test it. One of the tests determines how hot internally the power source gets while operating through its entire welding output range. The temperature information that is gathered is put into a graph that rates duty cycle. The duty cycle is a guideline providing recommended welding time durations for the entire welding output range of the power source. This guideline is to prevent overheating and possible damage to the power source. **FIGURE 15-24** represents a duty cycle chart seen in the industry.

The duty cycle graph shown in FIGURE 15-24 has two sets of values: welding current and duty cycle %. The welding current is the output range of the power source, and the duty cycle is a percentage of a 10-minute time period, with 100% representing 10 minutes and each 10% representing 1 minute. As the welding output current is increased from 0 to maximum output, the duty cycle percentage is decreased. The chart is used to determine the amount of time a

FIGURE 15-24 Duty cycle increases as welding output current decreases

Duty Cycle %	Welding Time (Minutes)	Rest Time (Minutes)
100	Continuous	0
90	9	1
80	8	2
70	7	3
60	6	4
50	5	5
40	4	6
30	3	7
20	2	8
10	1	9

TABLE 15-2

Rated Output Current/Voltage/Duty Cycle	Rated Output Current/Voltage/Duty Cycle
80A/20V/100%	300A at 36V, 20% Duty Cycle

TABLE 15-3

power source can be continuously used before welding needs to be stopped to allow the power source to cool off. See **TABLE 15-2** for conversion of % to welding time versus rest time.

WHAT IS THE WELDING OUTPUT VOLTAGE AND CURRENT AT 20, 40, 60, OR 100% DUTY CYCLE?

RATED OUTPUT   This rating is often used to compare competing power sources. For example, power source A has a rated output of 250A at 40% duty cycle, and power source B has a rated output of 225A at 40% duty cycle. This comparison can be used to help choose which power source to purchase. **TABLE 15-3** compares rated outputs for two power sources.

WHAT PRIMARY VOLTAGES WILL THE POWER SOURCE OPERATE AT?

PRIMARY VOLTAGE REQUIREMENTS   Power sources are designed to operate on specific primary voltages or within a voltage range. If the power source is designed to operate on a specific voltage, it is very important that the power source be operated on a voltage close to the listed operating voltage. Damage can be done to the power source if the voltage is not equal to the listed operating voltage.

Most industrial power sources can operate on several voltages. For example, 200/230/460 is a very common configuration for industrial power sources sold in the United States. For machines intended to be sold in Canada, the voltages may be 230/460/575.

Some power sources are now produced that operate within a voltage range. Specifications may range from 90–500 volts. This means that the power source will operate on any voltage from 90–500 volts.

WILL THE POWER SOURCE OPERATE ON SINGLE-PHASE POWER, THREE-PHASE POWER, OR BOTH?

SINGLE- OR THREE-PHASE POWER Some industrial power sources can only be operated on three-phase power. Trying to operate one of these power sources on single-phase power will generally result in a reduced welding range, failure to operate, or damage to the power source.

Some power sources can operate on single- and three-phase power. It is important to read the owner's manual for instructions regarding the correct wiring for single- or three-phase operation.

If a buyer were looking to purchase a power source to be used in his or her welding shop at home, it is very important that the power source be able to operate on single-phase power. Three-phase power in a residential area is very rare but may be found in rural areas, such as on a farm.

PRIMARY CIRCUIT REQUIREMENTS The next three ratings provide important information about the size of a primary circuit needed to operate the power source.

HOW MUCH PRIMARY CURRENT IS USED OR DRAWN AT THE POWER SOURCE'S RATED OUTPUT?

Primary Current Draw This is used to determine if the primary circuit that the power source is intended to be used on is has a high enough amperage capacity to safely operate the power source. If a welder does not know what his or her intended primary circuit's current capacity is, it is very important to find out before purchasing a power source.

WHAT SIZE BREAKER OR FUSE IS REQUIRED TO SAFELY OPERATE THE POWER SOURCE?

Recommended Primary Current Protection If the purchased power source has too high of a current draw, the welder will find that the overcurrent protection devices (breakers or fuses) will be tripped, causing the power source to shut down. Using overcurrent protection devices (breakers or fuses) that are too large may result in overheating of the primary conductors, which may result in a fire.

WHAT SIZE (DIAMETER) PRIMARY CONDUCTOR IS REQUIRED TO SAFELY OPERATE THE POWER SOURCE?

Recommended Primary Conductor Size The diameter of an electrical conductor determines the amount of current the conductor can carry before overheating. If the conductors are too small to safely carry the amount of current required, the conductors need to be changed in order to get the full welding range from the welding power source.

WHAT WELDING PROCESS CAN THE POWER SOURCE BE USED FOR WELDING?

CC, CV, or CC/CV If a power source is CC (consent current), it is designed to be used for welding GTAW and SMAW. If a power source is CV (constant voltage), it is designed to be used for welding GMAW and FCAW. A CC/CV power source has the ability to weld all the processes. **FIGURE 15-25** shows a CC/CV welding power source.

FIGURE 15-25 A CC/CV (constant current/constant voltage) welding power source

Source: Photo courtesy of Miller Electric Manufacturing Co.

Purchasing a CC/CV power source seems like the answer to every welding shop's dilemma of having to purchase one power source for CV and another for CC welding processes. Before purchasing a power source, it is important to examine how the shop functions and what the shop's needs are.

For example, suppose a welding shop has just signed a contract to manufacture a product by using GMAW. The shop also welds aluminum and stainless steel pipe by using GTAW. The owner decides that a CC/CV power source solves the problem at hand. What the owner did not consider is:

- GMAW is used for at least 30 hours each week. The shop also does sporadic GTAW, which at times can amount to 20 hours a week. During the overlap time, the power source is not able to be used for both GMAW and GTAW. It may have been more cost-effective to purchase one CV and one CC power source.

- In order to be used for GMAW and GTAW, the power source requires a wire feeder, GMAW torch, high-frequency control box, GTAW torch, GTAW remote control, and two different regulators/gas hoses. To switch between processes, the welder needs to change polarity, switch electrode cables, and change settings on the power source. If only one regulator is used, this also has to be switched to a different shielding gas tank. The system will then need to be purged to eliminate contamination of welds by the wrong shielding gas. In a busy shop, time wasted switching around a welding system results in a loss of money.

- Most CC/CV power sources are DC only; thus, the GTAW of aluminum is not available.

WHAT POLARITY CAN THE POWER SOURCE WELD IN?

AC, DC, or AC/DC   The welding process and metal being welded determine the polarity needed. The only time an AC or AC/DC power source is required is if the welder is going to GTAW aluminum and magnesium or SMAW with an AC welding rod.

Warranty   All the major manufacturers of power sources have good warranties on paper. Just as important as the warranty is how well the manufacturers and distributors back up the warranty.

There are several ways to find out how well the warranty is backed up. First, ask the repair technician at the welding distributor. Most will be happy to tell which manufacturers give the best response or what power sources require the most repairs. Local welding contractors and fabricators are also good sources of information.

The owner must make sure that the welding distributor is aware of any special circumstances in regards to your situation. For example, suppose your shop is 75 miles away from the repair shop, and if the power source needs to be repaired, who is responsible for bringing the power source in?

WELDING GENERATOR   In North America, access to electrical power is generally not a problem. When a welding shop buys a new power source, the power source simply needs to be connected to a circuit that has the correct size conductors and overcurrent protection. If the shop's electrical system is not large enough to meet the power source's requirements, the shop owner may need to have the power company increase the building's amperage capacity or hire an electrical contractor to install a larger circuit, but access to electrical power is generally available.

When that same welding shop was built, electrical power may not have been available. A generator may have been brought in to supply electricity for lights, tools, and the compressor. When iron workers arrived to build the frame, they may have brought a power source and plugged it into the generator. This may have worked, but most likely, the generator would not have been large enough to handle the power source and everything else. In this situation, the iron workers most likely brought their own welding generator.

Both the generator and the welding generator are electro-mechanical machines that are used to produce electrical power. This is the same technology used to produce electrical power at a large power plant, just on a much smaller scale. Both the generator and the welding generator use a gas-, diesel-, or propane-powered engine to rotate a magnetized rotor inside of a stator. As the rotor turns inside the stator, the passing magnetic fields produce current flow within the conductors of the stator. The current produced is AC, and its voltage level can depend on the number of windings and the rotors revolutions per minute (RPM). The difference between the two is that the welding generator has extra stator windings that produce current that is used for arc welding. **FIGURE 15-26** shows a welding generator.

Just like a welding power source, generators and welding generators are rated by their manufacturers. It is very important to understand the ratings and the operating limitations they represent. Generators are discussed in this chapter because of advances in generator and welding power source technology. It is not uncommon to see power sources being operated off of the generator's electrical power; thus, it is important to understand a generator's specifications.

Determining the size generator to purchase requires the buyer to not only look at the job requirements at hand but to also estimate what will be required in the future. The buyer must determine if the need is for a generator that can supply electrical power only or a welding generator that can supply electrical power and be used to arc weld. A quick decision may be to purchase the welding

FIGURE 15-26 A welding (motor) generator

Source: Photo courtesy of Miller Electric Manufacturing Co.

generator and have everything in one package. This may be adequate in some situations, but if today's need is for a continuous electrical demand of 25 kW at 230 and 115 Vac or 460 Vac three-phase power, the welding generator is not going to meet the requirements. On the other hand, if the need is for a system that could be mounted on the back of a ranch maintenance truck to be used to operate tools and pumps and to perform SMAW and FCAW, the welding generator is the best option.

In order to make the best decision, a welder or contractor looking to purchase a generator needs to understand how the equipment is rated and what the ratings mean. It is also very important to understand the individual voltage, amperage, and wattage requirements of each piece of equipment that will be operated off of the generator output. The welding portion of a welding generator is rated the same as any other power source. Review the power source rating section for details.

WHAT VOLTAGE OR VOLTAGES ARE AVAILABLE FROM THE ELECTRICAL OUTPUT OF THE GENERATOR?

GENERATOR AND WELDING GENERATOR RATINGS

Auxiliary Output Voltage Ranges   Most generators today have 115 and 230 Vac electrical output. In order to operate hand tools and lights, a minimum of 115 Vac is required. The higher voltage 230 Vac is used to operate larger equipment, such as compressors, pumps, and welding power sources. Some generators have a DC power output. Caution must be taken not to try to run AC equipment on this output.

WHAT IS THE SIZE OF CIRCUIT PROTECTION DEVICE (BREAKER) THAT EACH POWER OUTPUT CIRCUIT IS PROTECTED BY?

Output Circuit Breaker Size   The maximum amperage rating of each breaker is determined by the size of the conductors of that circuit. The breaker is there to protect the individual circuit from amperage being drawn at a value above what the conductors can handle and causing damage to the circuit.

It is important to know what equipment and amperage requirements will be run on each circuit. It is also recommended that amperage requirements be over-estimated in order to compensate for unforeseen requirements. For example, the generator is needed to run an 115 Vac water pump and flood lights. If a contractor only estimated in the amperage draw of the pump, the operator may be working in the dark.

WHAT IS THE MAXIMUM AMOUNT OF POWER THAT THE GENERATOR CAN SUPPLY FOR A SHORT PERIOD OF TIME?

Peak Wattage Output   This rating is in the form of wattage (wattage = volts × amps). **Wattage** is the measurement of power (energy) that is produced by the flow of electrical current. For example, a rating may be 10,500 watts or 10.5 kW. This value is important to know when running equipment such as pumps, compressors, and motors. Electro-mechanical equipment such as these will require more watts when first started. Once the mechanical portion (piston or rotor) of the equipment is moving, the wattage requirements are reduced. If the peak wattage demand is beyond the range of the generator, permanent damage in the form of overheating can occur.

Care must also be taken when trying to start high-draw equipment when other circuits of the power output are already being used. The total wattage is a sum of all circuits.

WHAT IS THE MAXIMUM AMOUNT OF POWER THAT THE GENERATOR CAN PUT OUT CONTINUOUSLY?

Continuous Wattage Output   This rating is given in watts (watts = volts × amps). For example, a rating may be 9,500 watts or 9.5 kW. This value is always lower than the peak wattage output. This is a very important rating when determining how large of a generator will meet the welder's or contractor's needs. The smaller the kW rating, the less equipment that can be run off of the generator, and the larger the kW rating, the larger in size and the more expensive a generator becomes. Before buying, a welder or contractor needs to evaluate what equipment will be run off of the generator. It is best to consider worst-case scenarios in which multiple circuits are being used.

It is not uncommon to have welding power sources being run off of a generator's output power. Even inverter power sources can draw larger amounts of current and require a good portion of the generator's power output. It is important to know the welding output of the power source needed and what that output requires from the primary power system.

HOW MANY WATTS OF POWER CAN BE PROVIDED WHILE THE WELDING GENERATOR IS BEING USED TO WELD? THIS RATING ONLY APPLIES TO A WELDING GENERATOR.

Continuous Wattage Output While Welding   A well-designed welding generator should be able to support welding at the same time power is being used from the generator portion. When purchasing a welding generator, it is important to understand how much current is required to successfully complete a welding project. Based on the current required, the welder should be able to determine (based on the manufacturer's information) the maximum amount of wattage that can be delivered while welding. If too small of a welding generator is purchased, the welder may find that a helper cannot grind welds while he or she is welding. This will negatively affect production. **TABLE 15-4** shows a manufacturer's rating of dual welding and primary power usage.

Welding Current	Total Available Primary Watts	120V Circuit Current Draw	240V Circuit Current Draw
350A	1,000W	12A	8A
250A	3,800W	38A	18A
150A	6,300W	60A	30A
50A	8,500W	75A	40A
0	11,000W	80A	48A

TABLE 15-4

CHAPTER QUESTIONS/ASSIGNMENTS

MULTIPLE CHOICE

1. What are the three different parts of an atom?
   a. Nucleus (neutrons and protons) and electrons
   b. Electrodes and nucleus (neutrons and protons)
   c. Neurons, ions, and electrons
   d. Protons, positive electrons, and negative electrons

2. What is a material called that allows electrical current to be conducted through it?
   a. Insulator
   b. Promoter
   c. Conductor
   d. Electron acceptor

3. DC moves in how many directions?
   a. Three
   b. Two
   c. Four
   d. One

4. What is voltage?
   a. An electromotive force or pressure that causes electrons to move between atoms
   b. A force that causes atoms to vibrate thus causing heat
   c. An electrical characteristic that is the opposite of current
   d. Something that is too hard to describe

5. What are all welding power sources designed to do?
   a. Control the heat produced by welding
   b. Be operated off of a building's primary power
   c. Provide a controlled and consistent current flow
   d. Allow a welder to melt metal

6. What is a duty cycle rating?
   a. A time-based welding guideline
   b. Something power source manufacturers provide but that does not need to followed
   c. Guideline that is provided to prevent overheating and possible damage to the power source
   d. A maximum welding current rating

7. Why does a welding generator typically have more than one auxiliary output voltage?
   a. To allow the generator to weld
   b. Having multiple voltages available makes the generator more versatile and able to provide power for more equipment.
   c. Less expensive than single-voltage generators
   d. Greater fuel use than single-voltage generators

8. All metals are composed of _____.

9. Electrical current is the movement of _____ from one atom to the next in a conductor.

10. When conducting welding current, a smaller diameter conductor will become hot or even melt because it has a greater _____ than a larger diameter conductor.

11. _____ is electrical current that travels in one direction and then changes direction at a specific frequency.

12. The block diagram of a solid state power source consists of step-down, _____, and output control.

13. The rating that shows how low and high a power source's welding outputs are called _____.

14. The recommended primary circuit _____ provides a guideline of what size primary protection is required to safely operate a welding power source.

15. What is electrical resistance?

16. Why is the voltage of electrical power being transmitted over long distance increased to a very high level?

17. What are the four types of welding power sources?

18. List two advantages of using SCRs in welding power sources.

19. List two advantages of inverter welding power sources.

20. Using the duty cycle chart below, what is the duty cycle at 200 amps of welding current?

21. When welding with a welding generator, when would knowing the generator's continuous wattage output while welding rating be useful?

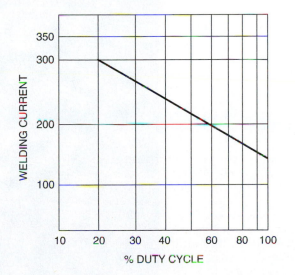

part 6

SUPPLEMENTARY INFORMATION

Appendix A Troubleshooting

WELD DISCONTINUITIES AND DEFECTS

AWS *A3.0 Standard Welding Terms and Definitions* defines a defect as "a discontinuity or discontinuities that by nature or accumulated effect (for example total crack length) render a part or product unable to meet minimum applicable acceptance standards or specifications. The term designates rejectability." And AWS defines a discontinuity as "an interruption of the typical structure of a material, such as a lack of homogeneity in its mechanical, metallurgical, or physical characteristics. A discontinuity is not necessarily a defect." There are many weld discontinuities and defects. Because some defects are closely related, some experts may argue that they should not be listed as a separate defect, while others may argue there is yet another closely related defect that should have been included. It is essential that terms and definitions have only one meaning. To ensure continuity, AWS *A3.0 Standard Welding Terms and Definitions* is used as a basis for authority in compiling a list and description of discontinuities.

This appendix attempts to aid the welder to not only identify problems encountered but also to help identify solutions to resolve these problems. To this end, there are other nonstandard AWS terms or situations outside the scope of AWS *A3.0 Standard Welding Terms and Definitions* included. Every possible problem that could exist would entail an exhausting list not every welder would encounter. Listing problems encountered in laser welding, ESW, EBW, etc., are specialized to a very smaller segment of the industry than, for example, SMAW or GMAW. This appendix attempts to deal with those problems encountered in the more common welding processes.

TERMS AND DEFINITIONS[1]   **FIGURES A-1a** and **A-1b** are pictorials identifying how these defects or discontinuities may appear in a weld and where they may appear.

1. Longitudinal Crack
2. Overlap
3. Inclusion (Slag)
4. Lack of Fusion
5. Undercut
6. Porosity (Piping)
7. Porosity (Worm Tracking)
8. Incomplete Joint Penetration

---

[1] Definitions reproduced from AWS *A3.0 2001 Standard Welding Terms and Definitions*, with permission from the American Welding Society (AWS), Miami, Florida.

9. HAZ or Underbead Cracking
10. Incomplete Fusion (Cold Lap)
11. Crater Crack (Star Crack)
12. Porosity (Linear)
13. Transverse Crack
14. Porosity (Cluster)
15. Underfill
16. Laminar Tear

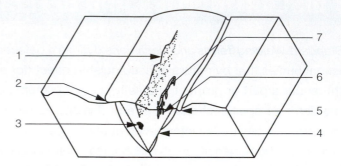

FIGURE A-1a Discontinuities 1-7

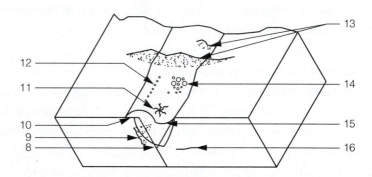

FIGURE A-1b Discontinuities 8-16

Arc strikes are a discontinuity resulting from an arc, consisting of any localized remelted metal, heat-affected metal, or change in the surface profile of any metal object. Arc strikes may develop localized stress risers, which could lead to cracking.

Groove Joint – SMAW Process

Possible Causes	Possible Solutions
■ Usually but not limited to the SMAW process when initiating an arc	■ Hold the electrode in the weld joint or over the weld metal prior to striking the arc. ■ Use a strike plate to strike the arc on, thus avoiding striking on the base metal. ■ Incorporate a self-darkening automatic welding helmet to see where the electrode is positioned prior to striking the arc.

A **concave root surface** results in an insufficient throat in a groove weld.

It is also known as suck-back (a nonstandard term).

Underside of Groove Joint – SMAW Process

Possible Causes	Possible Solutions
▪ The root face lacks sufficient size and/or the root opening is too wide.	▪ Increase the size of the root face and/or decrease the root opening.
▪ The puddle is too fluid due to excessive amperage or voltage depending on welding process.	▪ Decrease the amperage in welding processes such as SMAW. ▪ Decrease the arc length (voltage) in SMAW. ▪ Decrease the voltage in welding processes such as GMAW-S or overall welding parameters with GMAW-P.

▪ Incorrect electrode selection or process
  ▫ Welding the root pass can be difficult when using E7018 in an overhead position.
  ▫ Welding the root pass can be difficult to bridge the root gap when using FCAW.
	▪ Switch to a fast-freeze electrode.
	  ▫ E6010 or E6011 freezes quickly, providing a convex root surface.
	  ▫ GMAW-S provides a good root fusion with convex root surface.

Cracks are a fracture type discontinuity characterized by a sharp tip and high ratio of length and width to opening displacement. Cracks can be classified as resulting from being either cold or hot types.

The crack may be with the weld (longitudinal); across the weld from one toe to the other (transverse); multiple fractures producing a spider-like appearance (crater cracks); next to the weld in the base metal (HAZ, lamellar, or residual stress cracks); or beneath the surface (underbead cracking). Types of cracks include:

- Crater Crack
- Delayed Crack
- Face Crack
- Heat-Affected Zone Crack
- Hot Crack
- Longitudinal Crack

- Root Crack
- Root Surface Crack
- Stress-Relief Cracking
- Throat Crack
- Transverse Crack

- Stress-Corrosion Cracking
- Toe Crack
- Underbead Crack
- Weld Crack
- Weld Metal Crack

Longitudinal Crack – GMAW (Aluminum)

Crater Crack – FCAW Process

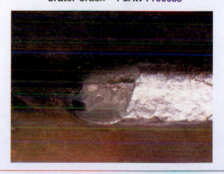

Possible Causes	Possible Solutions
▪ High restraint of the joint members	▪ On steels, use preheat to reduce the magnitude of residual stresses. ▪ Adjust the welding sequence to reduce restraint conditions.
▪ Poor joint fit-up or improper joint design	▪ Maintain proper groove dimensions to allow the deposition of adequate filler metal or weld cross-section to overcome restraining conditions.

continued

■ Too small a weld bead	■ Decrease the travel speed to increase the cross-section of the bead. ■ Increase the electrode size. ■ On steels, preheat the base material first or decrease the travel speed to increase the cross-section of the weld deposit.
■ Hot cracking ■ Heat input too high, causing excessive shrinkage and distortion (thin sections)	■ Reduce the heat input by reducing either the current or the voltage or both. ■ Minimize the joint restraints. ■ Increase the travel speed.
■ Hot-shortness	■ Increase the travel speed (aluminum). ■ Use an electrode with a higher manganese content. (Note: Use a shorter arc length to minimize the loss of manganese across the arc on steels.) ■ Adjust the pass sequence to reduce the restraint on the weld during cooling.
■ Cracking at the crater	■ Fill the crater before withdrawing the electrode. ■ Fill the crater or weld back into the weld approximately 3/4 inches to terminate the weld rather than at the end of the joint.
■ Hydrogen-induced cracking/Delayed cracking	■ Use a low-hydrogen welding electrode. ■ Use suitable preheat and postweld heat treatments. ■ Use a clean, dry electrode wire and a dry shielding gas. ■ Remove contaminants from the base metal. ■ Hold the weld at elevated temperatures for several hours before cooling to diffuse the hydrogen.
■ Presence of brittle phases in the microstructure of the base material (low ductility of the base material)	■ Use preheat. ■ Anneal the base metal. ■ Use ductile weld metal.
■ High residual stresses	■ Redesign the weld metal and reduce the restraints. ■ Change the welding sequence. ■ Use intermediate stress-relief heat treatment.
■ Hardening in the heat-affected zone	■ Preheat to the retard cooling rate.
■ Residual stresses too high	■ Use stress relief heat treatment (postheat).
■ Underbead cracking due to excessive heat input or brittle base metal	■ Weld a smaller bead. ■ Weld multiple stringer beads as opposed to one large weave bead. ■ Preheat the base metal prior to welding.

Drop-through is an undesirable sagging or surface irregularity, usually encountered when brazing or welding near the solidus of the base metal, caused by overheating with rapid diffusion or alloying between the filler metal and the base metal.

Mild Drop	Severe Drop

Possible Causes	Possible Solutions
■ Excessive amperage	■ Lower the amperage or use a smaller diameter electrode.
■ Improper root opening or insufficient root face on butt joints	■ Experiment with varying root openings and root faces with amperage settings. On thinner metal thicknesses, lower the amperage and increase the root opening rather than decrease the root opening. On thicker plates, decrease the root opening and increase the amperage.

Expulsion is the forceful ejection of molten metal from a resistance spot, seam, or projection weld usually at the faying surface. Spit is a nonstandard term when used for flash and expulsion.

Lap Joint – RSW Process

Possible Causes	Possible Solutions
■ Excessive amperage ■ Excessive tong pressure	■ Lower the amperage setting. ■ Decrease the tong pressure to slightly less than the thickness of the material being joined.

Inclusion is entrapped foreign sold material, such as slag, flux, tungsten, or oxide. It is a metallic or nonmetallic solid material entrapped in weld metal or between the weld metal and the base metal.

Lap Joint – SMAW Process

Possible Causes	Possible Solutions
■ Improper cleaning procedures	■ Clean work surfaces thoroughly by chipping, bushing, power wire "brushing" "grinding", and "chiselling" to ensure thorough removal of slag.
■ Current too low (SMAW/FCAW/SAW process)	■ Increase the current (amperage).
■ Narrow, inaccessible joints	■ Increase the groove angle.

■ Improper welding technique
  □ Long arc length
  □ Too high or too slow travel speed
  □ Slag flooding ahead of welding arc
	■ Improve the welding technique.
	■ Reposition work to prevent the loss of slag control wherever possible:
	  □ Decrease the travel speed in the flat/horizontal/ vertical-up position.
	  □ Increase the travel speed in the flat/horizontal/ vertical-down position.
	■ Restrict weaving to a minimum.

■ Too large of an electrode	■ Use a smaller electrode to permit a more controlled travel speed and access to the weld joint.
■ Tungsten spitting or tungsten sticking	■ Reduce the current or use a larger electrode to prevent tungsten spitting. ■ Adjust the tungsten to prevent the electrode from sticking in the weld puddle.

Incomplete fusion is a weld discontinuity in which fusion did not occur between the weld metal and the fusion faces or the adjoining weld beads. Nonstandard terms include lack of fusion, cold lap, over lap, and rollover.

Cold lap is a nonstandard term when used for incomplete fusion. It is the insufficient fusion of the weld reinforcement to the base metal. Cold lap is the total lack of fusion to the base piece. A cold lap bead does not have to be rounded; it could also be concave yet not be fused to the base metal.

T-Joint – GMAW-S Process

continued

Possible Causes	Possible Solutions
■ Improper work angle	■ Change to correct electrode angles and maintain them properly to the work surface to achieve maximum penetration.
■ Incorrect electrode extension or arc length causes poor joint fusion and penetration	■ Decrease the electrode extension for wire feed processes. ■ Decrease the arc length for SMAW and GTAW.
■ Contaminated work surface; weld zone surface is not free of film or excessive oxides	■ Clean the weld surfaces of mill scale impurities, paint, oils, galvanized coatings, etc., prior to welding.
■ Weld parameters set too low for the thickness of the metal welded prevents proper fusion ■ Weld parameters set too high	■ Increase the voltage for better fusion. Increase the amperage for greater penetration. ■ Increase the travel speed or lower the weld parameters.
■ Travel speed too slow for the parameters set, especially in the vertical-down position ■ Improper manipulation of arc ■ Too large a weld puddle	■ Keep the arc on the leading edge of the puddle. Increase the travel speeds to keep the electrode at the leading edge of the puddle. ■ Minimize excessive weaving to produce a more controllable weld puddle. ■ Increase the travel speed.

Overlap is the protrusion of weld metal beyond the toe, face, or root of the weld. It may also be known as roll over. Overlap is characterized by the weld bead rolling over the base metal and occurs at the toes of the weld. It differs from cold lap in that one toe of the weld (usually the bottom) is under the protruding weld face.

Lap Joint – GMAW-S (Stainless Steel)

Possible Causes	Possible Solutions
■ Improper work angle by concentrating the electrode to the top plate and/or too slow travel speed	■ Keep the electrode in the root of the joint and at the leading edge of the puddle. ■ Increase the travel speed.
■ Improper feed of filler metal by adding filler metal (GTAW or OFW) before a puddle is formed, resulting in the filler metal lying on top of the base metal and/or too large of an electrode	■ Always create a fluid puddle prior to dipping a filler metal onto the base metal by the arc or flame. ■ Decrease the electrode diameter.

Incomplete joint penetration is a joint root condition in a groove weld in which weld metal does not extend through the joint thickness, which is an insufficient weld metal depth into the base metal. Lack of penetration is a nonstandard term.

Underside of Groove Joint – GMAW-S Process

Possible Causes	Possible Solutions
■ Improper joint preparation ■ Excessively thick root face ■ Insufficient root opening ■ Bridging of root opening	■ Joint design must provide proper access to the bottom of the groove while maintaining the proper electrode extension. ■ Reduce the excessively large root face. ■ Increase the root gap in butt joints and increase the depth of the back gouge, if applicable.

■ Electrode diameter too large	■ Use a smaller electrode in the root.
	■ Increase the root opening.
■ Inadequate current	■ Follow the correct welding current and technique.
■ Improper weld technique	■ Change to correct electrode angles and maintain properly to the work surface to achieve maximum penetration.
	■ Keep the arc on the leading edge of the puddle. Travel speeds too fast or too slow may cause poor joint penetration.
	■ Keep the electrode extension to within an acceptable range (not too long).
■ Inadequate welding current	■ Increase the amperage.

Excessive melt-through is excessive joint penetration on a joint welded from one side on root joints (left photo) or the back side of fillet welds (right photo). Burn through is a nonstandard term when used for excessive melt-through or a hole through a root bead.

Underside of T-Joint – GTAW Process

Underside of Butt Joint – SMAW Process

Possible Causes	Possible Solutions
■ Excessive heat input	■ Reduce the wire feed speed and voltage.
	■ Increase the travel speed.
■ Poor fit-up	■ Decrease the root opening or fit-up.
	■ Increase the root face dimension.
■ Not enough electrode extension	■ Increase the stick-out.

Porosity describes cavity-type discontinuities formed by gas entrapment during solidification or in a thermal spray deposit. Blowhole or gas pocket is a nonstandard term when used for **Porosity**. Other nonstandard terms include:

- ■ Gas Mark
- ■ Hydrogen Porosity
- ■ Linear Porosity
- ■ Piping Porosity
- ■ Scattered Porosity

Gas mark—also known as worm tracking—is a condition that occurs in (gas shielded) FCAW. A bubble forms from the arc, which does not escape through the molten slag, becoming entrapped in the weld pool, forming a visible elongated defect on the surface of the weld bead.

T-Joint – FCAW Process

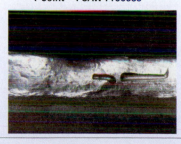

continued

Possible Causes	Possible Solutions
■ Surface base metal contamination ■ Rusty electrode wire or electrode has exceeded its shelf life ■ Insufficient shielding gas ■ Excessive amperage	■ Remove all grease, oil, moisture, rust, paint, and dirt from the work surface prior to welding. ■ Check the manufacturer's recommendation on electrode shelf life. ■ Increase the gas flow rate. ■ Decrease the wire feed speed.

Hydrogen porosity is a type of porosity occurring after a weld is completed, usually on the final pass of a multipass weld.

Possible Causes	Possible Solutions
■ Excessive heat input	■ Monitor the interpass weld temperatures on multipass welds. ■ Weld a stringer bead and avoid weave beads.

Linear porosity is a type of porosity characterized by holes running singularly with the weld bead.

Corner Joint – GMAW Process

Possible Causes	Possible Solutions
■ No/poor shielding gas	■ Turn the appropriate shielding gas on and adjust it to the proper flow rate. ■ Be conscious of the air flow from exhaust systems, fans, or wind near open doors.

Piping porosity is a type of porosity having a greater length running vertically into the weld metal from the root of the weld to the face.

Possible Causes	Possible Solutions
■ Typically caused from rapid solidification	■ Preheat the base metal to slow the solidification. ■ Run a weave or split-weave weld as opposed to stringer beads.

Scattered porosity is a type of porosity found interspersed throughout the weld.

Groove Joint – SMAW Process

Possible Causes	Possible Solutions
■ Inadequate shielding gas coverage	■ Increase the gas flow to displace all air from the weld zone. ■ Eliminate drafts, leaks in the gas line, or frozen regulators. ■ Reduce the travel speed or nozzle-to-work distance. ■ Welding gun nozzle with wire feed processes is plugged. Keep nozzle clear of excessive spatter. Check equipment for damage or fit.
■ Gas contamination or wrong shielding gas ■ Moisture in shielding	■ Use proper shielding gas or a better grade of shielding gas. ■ Replace the supply or cylinder. ■ Keep low-hydrogen electrodes dry.
■ Electrode contamination	■ Use only good, clean electrodes that are rust-free, lubricant-free, and have not exceeded shelf life. ■ Check for liner/conduit contamination.
■ Arc voltage too high ■ Excessive amperage	■ Reduce the voltage. ■ Reduce the amperage.

■ Workpiece contamination	■ Remove all grease, oil, moisture, rust, paint, and dirt from the work surface prior to welding.
	■ Use an electrode wire with higher percentages of deoxidizers.
■ Excessive shielding gas flow	■ Decrease the gas flow to avoid turbulence and the entrapment of air in the weld zone if the flow is excessive; reduce to the proper gas flow rate.
■ Travel speed too fast	■ Reduce the travel speed.
■ Improper gun angle	■ Use the correct gun angle.
■ High sulfur content	■ Use the basic coated base metal electrodes.
■ Excessive contact tube-to-work distance	■ Reduce the stick-out.
■ Arc length too long	■ Decrease the arc length.

Shrinkage void is a cavity-type discontinuity normally formed by shrinkage during solidification.

T-Joint – FCAW Process	T-Joint – FCAW Process

Possible Causes	**Possible Solutions**
■ Inadequate weld metal at the crater during solidification	■ Dwell at the end of the weld to fill the crater.

Spatter is the metal particles expelled during fusion welding that do not form a part of the weld.

T-Joint – GMAW Short Circuit

Possible Causes	**Possible Solutions**
■ Excessive voltage	■ Decrease the arc length for the SMAW process.
	■ Decrease the voltage for GMAW/FCAW processes.
■ Excessive amperage	■ Decrease the amperage.
■ Excessive electrode extension (stick-out)	■ Decrease the electrode wire extension from the contact tip to the base metal.
■ Wet, damaged, or rusty electrode	■ Ensure SMAW electrodes are dry and undamaged.
	■ Inspect GMAW and FCAW for oxidation (rust).

Undercut is a groove melted into the base metal adjacent to the weld toes or weld root and left unfilled by weld metal. Undercut can range from severe to mild.

Possible Causes	Possible Solutions
■ Undercut due to faulty electrode manipulation  Undercut-improper electrode manipulation	■ Pause (dwell) at the toes of the weld. ■ Decrease/eliminate weaving.
■ Undercut due to incorrect weld angle (too steep)  Undercut-improper work angle	■ Decrease the weld angle so electrode bisects weld the joint evenly. ■ Decrease the travel speed.
■ Undercut due to insufficient dwell time at weld toe  Undercut-insufficient dwell angle	■ Pause (dwell) at the toes of the weld when weaving in the vertical-up position.
■ Excessive welding current or voltage ■ Excessive travel speed	■ Decrease the welding current (usually SMAW or GTAW). ■ Decrease the voltage (arc length for SMAW or voltage control for GMAW/FCAW). ■ Decrease the travel speed.

Underfill is a condition in which the weld face or root surface extends below the adjacent surface of the base metal.

Lap Joint – GTAW (Stainless Steel)	T-Joint – GMAW Spray Transfer

Possible Causes	Possible Solutions
■ Inadequate filler metal	■ Add filler metal with a process such as GTAW.
■ Puddle is too fluid	■ Position the weld in the flat position as opposed to the horizontal position.
■ Weld is too large	■ Increase the travel speed to allow the puddle to freeze more quickly.

TROUBLESHOOTING PROCESSES

Weld defects and discontinuities occur during or after welding. There are a number of other problems the welder must troubleshoot and correct in order to continue to perform his or her primary job of welding. While some problems encountered are universal to every process—for example, no arc may mean no or poor work connections—others are specific to a particular welding process. The next section covers general and process-specific problems most encountered and offers why these problems occurred. Check the reasons these problems occur to correct the problem.

General	
Problems	**Reasons for Problems**

■ **No arc** means the arc cannot be established when the electrode contacts the base metal.	■ The work clamp is not connected to the workpiece.
	■ The work clamp is not making an electrical connection to the workpiece because:
	  ■ Painted surface prevents connections
	  ■ Excessive rust or oxidation prevents connection
	  ■ Work cable is not connected to the power source
	  ■ Work or power cable is severed

■ **Erratic arc** is an arc that is established but spits and sputters or performs poorly.	■ Poor work clamp connection due to paint, rust, or general oxidation ■ Poor electrical connection due to work clamp not connected to workpiece but lying on damp concrete flooring (electrical connection may still allow an arc to be established!)
■ **Insufficient amperage** means the power source is set correctly but the output amperage is too low.	■ Power cord is damaged ■ Power cord beyond its life cycle (after time and use, electrical cables cannot carry current as when they were new) ■ Incorrect input power (a three-phase 460V power source connected to a single-phase 230V outlet will not achieve the desired outcome) ■ Poor electrical connection due to work clamp not connected to workpiece but lying on damp concrete flooring (electrical connection may still allow an arc to be established!)

Oxy-Fuel

Problems	Possible causes/solutions
■ **Backfire** is the momentary recession of the flame into the welding tip, cutting tip, or flame spraying gun, followed by immediate reappearance or complete extinction of the flame, accompanied by a loud report.	■ Increase the fuel gas because it is starving the flame (the larger the welding or cutting orifice, the more pressure is required to maintain the flame). ■ Keep a proper distance from the workpiece and the end of the torch tip (back-fire often results when the tip of the welding or cutting tip is overheated).
■ **Flashback** is a recession of the flame into or in back of the mixing chamber of the oxy-fuel gas torch or flame spraying gun. This is a dangerous situation that could potentially melt the torch body if left unchecked. Without the installation of check valves, the flame could potentially work its way back to the cylinder, where spontaneous combustion would occur.	■ Keep a proper distance from the workpiece and the end of the torch tip (if the welding or cutting tip comes in contact with the base metal long enough to extinguish the flame, a flashback may result). ■ Clean the fuel gas orifice or orifices (dirty and/or plugged fuel gas orifices can result in a backfire occurring).
■ Black soot on base metal	■ Increase the oxygen to produce a neutral flame (black soot occurs with a carburizing flame—i.e., when too much fuel gas is in the mixture).
■ Puddle does not remain fluid when welding or cutting	■ Decrease the oxygen to produce a neutral flame (when too much oxygen is in the mixture—i.e., a oxidizing flame—the puddle appears to dry out while welding or cutting).

Shielded Metal Arc Welding

Problems	Possible causes/solutions
■ **Arc blow** is the deflection of an arc from its normal path because of magnetic forces. Arc blow occurs in DC arc welding but is especially problematic with Shielded Metal Arc Welding. ■ Backward blow occurs when welding toward a workpiece connection (ground clamp). Forward blow occurs when welding away from a workpiece connection. ■ Unbalanced magnetic lines of flux due to an electrical resistance change that occurs when welding toward the end of a plate, when using fixtures with low electrical resistance, or a variation in the conductive difference between hot and cold metal when multipass welding	
	■ Change from DC to AC if possible with the SMAW process. ■ Change the location of the work clamp. ■ Hold as short an arc as possible to help the arc force counteract the arc blow. ■ Reduce the welding current. ■ Use the back-step welding technique.
■ Electrode overheats	■ Current density has been exceeded. Electrodes of any given diameter can only carry so much current. When the amperage exceeds what the electrode can carry, the electrode overheats, turns red hot, and the coating breaks down.
■ Inadequate amperage or electrode continually sticks to the base metal when striking an arc; arc may also be erratic	■ Current is set too low ■ Poor work clamp connection ■ Long cable. For every 25 feet (7.6 meters), there is a loss of 5% of current (amperage). ■ Electrode cable needs to be replaced because it is worn out ■ Electrode is wet (coatings, especially low hydrogen, absorb moisture)

Gas Tungsten Arc Welding

Problems	Possible causes/solutions
■ No arc	■ Make sure the work clamp is attached. ■ Clean the tungsten. Tungsten is contaminated due to too low of a postflow shielding gas or base metal contamination. ■ Check the setup. Many GTAW power sources have the ability to set remote controls, such as foot pedals. ■ Turn the high frequency on if so equipped. ■ Check the shielding gas. No shielding gas, incorrect flow rates (too high or too low), or gas turbulence can adversely affect the starting of the arc.
■ Electrode over-heats and gets shorter while welding	■ Turn the amperage down; the current density has been exceeded. Electrodes of any given diameter can only carry so much current. When the amperage exceeds what the electrode can carry, the electrode overheats, turns red hot, and consumes. ■ Polarity is incorrect; set to DCEN when welding ferrous metals for all but special applications. Approximately 70% of heat is on the electrode when operating on DCEP. ■ Balance the control setting. A low balance setting (excessive DCEP) will cause the tungsten to overheat.
■ Electrode turns black after welding	■ Due to inadequate shielding gas. Set the shielding gas as recommended. ■ Due to too little or no postflow shielding gas. Set the postflow shielding time to 1 second for every 10 amps set on the power source. ■ Contaminated shielding gas. Loose connections or damage to the shielding gas system will cause atmosphere to be aspirated into the shielding gas.
■ Base metal turns black or dull (oxidized) after welding	■ Due to inadequate shielding gas. Set the shielding gas as recommended. ■ Due to too little or no postflow shielding gas. Set the postflow shielding time to 1 second for every 10 amps set on the power source. ■ A trailing gas is needed. Refractive and refractory metals require a blanket of an inert gas until they are completely solidified. ■ Contaminated shielding gas. Loose connections or damage to the shielding gas system will cause atmosphere to be aspirated into the shielding gas.
■ Porosity in the weld metal. GTAW is capable of producing X-ray quality welds completely void of any discontinuities.	■ Oils and other contaminats on base metal ■ A coolant leak on water-cooled torches ■ Excessive shielding gas flow. Set the shielding gas as recommended. ■ Contaminated shielding gas. Loose connections or damage to the shielding gas system will cause atmosphere to be aspirated into the shielding gas. ■ Moisture on base and/or filler metal. Base and filler metal should always be stored in a dry location. The temperature of the metal should always be warmer than the shop temperature to eliminate the dew point. Allow metal to dry after using water or liquid cleaners.

Wire-Feed Welding Processes (GMAW/FCAW)

Problem	Possible causes/solutions
■ No arc	■ Check the work clamp connection. ■ Check the power switch on the feeder unit. Separate wire feed units have an ON/OFF power switch. ■ Check the fuse on the wire feeder if so equipped. ■ Check to see if the power source is on.
■ Electrode wire does not feed but wire feeder and power source are turned on	■ Drive roll pressure is set too light ■ Electrode wire is stuck to contact tip ■ Electrode wire has bird's nest around drive rolls ■ Check the fuse on the wire feeder if so equipped.
■ Electrode overheats	■ Current density has been exceeded. Electrodes of any given diameter can only carry so much current. When the amperage exceeds what the electrode can carry, the electrode overheats, turns red hot, and the coating breaks down.

continued

Problem	Possible causes/solutions
■ Electrode wire bird's nest around drive rolls	■ Drive roll pressure is too tight; adjust the tension sufficiently to enough to feed the electrode but able to slip if electrode abruptly stops, such as when the electrode fuses to the contact tip. ■ Excessive gap between drive roll and wire guide to welding gun
■ Electrode balls up during or after welding	■ Wire feed speed is set too low or voltage is set too high for base metal welded
■ Electrode snaps back and fuses to contact tip	■ Contact tip is faulty or worn out needs to be changed ■ Drive roll pressure is insufficient; tighten drive roll pressure ■ Wrong size contact tip
■ Electrode stubs into base metal	■ Wire feed speed is set too high ■ Poor connection ■ Electrode cable needs to be replaced because it is worn out ■ Voltage is set too low ■ Inductance is set too high ■ Slope, if adjustable, is set to too many turns (too steep) ■ Poor work clamp connection
■ No shielding gas flow	■ Turn on the shielding gas. ■ Adjust the flow meter to the correct flow rate. ■ Gas shielding is not connected from flow meter to wire feeder ■ Nozzle plugged ■ Drive roll area where welding whip connects to wire feeder is loose or not sealed and tightened in correct location

Cast and Helix

When an electrode is produced, the manufacturer follows specifications on how much cast and helix is allowed. If a section of wire is cut from a spool of electrode wire and allowed to lay on the floor or tabletop, it will roll into a circle. The diameter of this circle is called the cast. While one end of the wire is laying flat, the other end is suspended in the air. The distance from the tabletop, for example, to the end of the suspended wire is the helix.

A piece of cut GMAW electrode showing cast and helix

The larger the cast, the more smoothly the wire will feed. A smaller cast creates more friction and can cause the tip of the electrode to wander. This is quite evident as the electrode wire nears the end of the spool with only a few rounds left. A larger helix can cause the tip of the electrode to spiral or change directions (left to right, up or down) as it exits the contact tip. Because they are not as stiff, smaller electrode diameters are more susceptible to cast and helix problems. This is especially true with small spools, such as 1-pound spools used for aluminum that wander and spiral more noticeably. Increasing the diameter of the electrode helps alleviate this problem. While manufacturing these electrodes, drawing compounds or lubricants may inadvertently work into the surface of the electrode. Manufacturers of electrodes are conscientious about removing these particulates prior to shipping; still, some residue may remain. These compounds or lubricants sometimes result in some otherwise unexplained weld discontinuities, such as porosity or cracking.

Decimal/Fractional Chart

Thread Size	Drill Size	Decimal	Thread Size	Drill Size	Decimal
	#80	0.0135	2–64	#50	0.0700
	#79	0.0145		#49	0.0730
	1/64	**0.015625**		#48	0.0760
	#78	0.0160	3–48	**5/64**	**0.078125**
	#77	0.0180		#47	0.0785
	#76	0.0200	3–56	#46	0.0810
	#75	0.0210		#45	0.0820
	#74	0.0225		#44	0.0860
	#73	0.0240	4–40	#43	0.0890
	#72	0.0250	4–48	#42	0.0935
	#71	0.0260		**3/32**	**0.09375**
	#70	0.0280		#41	0.0960
	#69	0.0292		#40	0.0980
	#68	0.0310	5–40	#39	0.0995
	1/32	**0.03125**		#38	0.1015
	#67	0.0320	5–44	#37	0.1040
	#66	0.0330	6–32	#36	0.1065
	#65	0.0350		**7/64**	**0.109375**
	#64	0.0360		#35	0.1100
	#63	0.0370		#34	0.1110
	#62	0.0380	6–40	#33	0.1130
	#61	0.0390		#32	0.1160
	#60	0.0400		#31	0.1200
	#59	0.0410		**1/8**	**0.1250**
	#58	0.0420		#30	0.1285
	#57	0.0430	8–32 and 8–36	#29	0.1360
	#56	0.0465		#28	0.1405
0–80	**3/64**	**0.046875**		**9/64**	**0.140625**
	#55	0.0520		#27	0.1440
	#54	0.0550		#26	0.1470
1–64 and 1–72	#53	0.0595	10–24	#25	0.1495
	1/16	**0.0625**		#24	0.1520
	#52	0.0635		#23	0.1540
2–56	#51	0.0670		**5/32**	**0.15625**

continued

Thread Size	Drill Size	Decimal	Thread Size	Drill Size	Decimal
	#22	0.1570		9/32	0.28125
10–32	#21	0.1590		L	0.2900
	#20	0.1610		M	0.2950
	#19	0.1660		19/64	0.296875
	#18	0.1695		N	0.3020
	11/64	0.171875	3/8–16	5/16	0.3125
12–24	#17	0.1730		O	0.3160
	#16	0.1770		P	0.3230
12–28	#15	0.1800		21/64	0.328125
	#14	0.1820	3/8–24	Q	0.3320
	#13	0.1850		R	0.3390
	3/16	0.1875		11/32	0.34375
	#12	0.1890		S	0.3480
	#11	0.1910		T	0.3580
	#10	0.1935		23/64	0.359375
	#9	0.1960	7/16–14	U	0.3680
1/4–20	#8	0.1990		3/8	0.375
	#7	0.2010		V	0.3770
	13/64	0.203125	7/16–20	W	0.3860
	#6	0.2040		25/64	0.390625
	#5	0.2055		X	0.3970
	#4	0.2090		Y	0.4040
1/4–28	#3	0.2130		13/32	0.40625
	7/32	0.21875		Z	0.4130
	#2	0.2210	1/2–12 and 1/2–13	27/64	0.421875
	#1	0.2280		7/16	0.4375
	A	0.2340	1/2–20	29/64	0.453125
	15/64	0.234375		15/32	0.46875
	B	0.2380	9/16–12	31/64	0.484375
	C	0.2420		1/2	0.5
	D	0.2460	9/16–18	33/64	0.515625
	1/4	0.25	5/8–11	17/32	0.53125
	E	0.2500		35/64	0.546875
5/16–18	F	0.2570		9/16	0.5625
	G	0.2610	5/8–18	37/64	0.578125
	17/64	0.265625		19/32	0.59375
	H	0.2660		39/64	0.609375
5/16–24	I	0.2720		5/8	0.625
	J	0.2770		41/64	0.640625
	K	0.2810	3/4–10	21/32	0.65625

continued

Thread Size	Drill Size	Decimal	Thread Size	Drill Size	Decimal
	43/64	0.671875		27/32	0.84375
3/4–16	**11/16**	**0.6875**		55/64	0.859375
	45/64	0.703125	1–8	**7/8**	**0.875**
	23/32	0.71875		57/64	0.890625
	47/64	0.734375		29/32	0.90625
	3/4	**0.75**	1–12	59/64	0.921875
7/8–9	49/64	0.765625	1–14	**15/16**	**0.9375**
	25/32	0.78125		61/64	0.953125
	51/64	0.796875		31/32	0.96875
7/8–14	**13/16**	**0.8125**		63/64	0.984375
	53/64	0.828125		**1**	**1.0**

Sheet Metal Gauge, Decimal Equivalent, and Weight Equivalent

Gauge	Decimal	Weight per Square Foot
26	0.01875 (0.022 galv)	0.75
25	0.021875 (0.025 galv)	0.875
24	0.025 (0.028 galv)	1
23	0.028125 (0.031 galv)	1.125
22	0.03125 (0.0336 galv)	1.25
21	0.034375	1.375
20	0.0375	1.5
19	0.04375	1.75
18	0.05	2
17	0.05625	2.25
16	0.0625	2.5
15	0.0703125	2.8125
14	0.078125	3.125
13	0.09375	3.75
12	0.109375	4.375
11	0.125	5
10	0.140625	5.625
9	0.15625	6.25
8	0.171875	6.875
7	0.1875	7.5
6	0.203125	8.125
5	0.21875	8.75
4	0.234375	9.375

continued

Plate (Fractional)	Decimal	Weight per Square Foot
1/4	0.25	10.20
5/16	0.3125	12.75
3/8	0.375	15.30
1/2	0.5	20.40
5/8	0.625	25.50
3/4	0.75	30.60
7/8	0.875	35.70
1	1.00	40.80

Millimeters to Decimal Conversion (1 inch = 25.4 mm and 1 mm = 0.0394 inches)			
Millimeters	Inches (Decimal)	Inches (Fraction)	Millimeters
1	0.039	1/64	0.40
2	0.078	1/32	0.79
3	0.117	3/64	1.19
4	0.156	1/16	1.59
5	0.195	5/64	1.98
6	0.234	3/32	2.38
7	0.273	7/64	2.78
8	0.312	1/8	3.18
9	0.351	5/32	3.97
10	0.390	3/16	4.76
11	0.429	7/32	5.56
12	0.468	1/4	6.35
13	0.507	9/32	7.14
14	0.546	5/16	7.94
15	0.585	11/32	8.73
16	0.624	3/8	9.53
17	0.663	7/16	11.11
18	0.702	1/2	12.7
19	0.741	9/16	14.29
20	0.780	5/8	15.88
21	0.819	11/16	17.46
22	0.858	3/4	19.05
23	0.897	13/16	20.64
24	0.936	7/8	22.23
25	0.975	15/16	23.81

Other Length Equivalents

Unit	mm	cm	in	ft	yd	m
1 mm =	1	0.1	0.0394	0.0033	0.0011	0.001
1 cm =	10	1	0.394	0.033	0.011	0.01
1 in =	25.4	2.54	1	0.0833	0.0278	0.0254
1 foot =	304.801	30.4801	12	1	0.3333	0.3048
1 yard =	914.402	91.4402	36	3	1	0.9144
1 meter =	1,000	100	39.37	3.2808	1.0936	1

Unit	Feet	Yards	Meters	Rods	Kilometers	Miles
1 rod =	16.5	5.5	5.02921	1	0.005029	0.003125
1 kilometer =	3,280.8	1,093.6	1,000	199	1	0.62137
1 mile =	5,280	1,760	1,609.34	320	1.60934	1

Area Equivalents

Unit	Square Inches	Square Feet	Square Yards	Square Meters
1 square foot =	144	1	0.1111	0.0929
1 square yard =	1,296	9	1	0.83613
1 square meter =	1,550	10.7639	1.19599	1
1 square rod =	39,204	272.25	30.25	25.293
1 acre =	6,272,640	43,560	4,840	4,046.86
1 square mile (640 acres) =	–	2,7878,400	3,097,600	2,589,999
1 square kilometer =	–	10,763,867	1,195,985	1,000,000

Weight Equivalents

Unit	Grains	Grams	Ounces	Pounds	Kilograms
1 grain =	1	0.064799	0.002286	0.000143	0.000065
1 gram =	15.4324	1	0.035274	0.002205	0.001
1 ounce =	437.5	28.3495	1	0.0625	0.028350
1 pound =	7,000	453.592	16	1	0.453592
1 kilogram =	15,432.4	1,000	35.274	2.20462	1

Unit	Kilograms	Pounds	Metric Tons	U.S. Tons
1 metric ton =	1,000	2,204.62	1	1.10231
1 U.S. ton =	907.185	2000	0.907185	1

Volume and Capacity Equivalents

Unit	Cubic Centimeters	Cubic Inches	Liters	Quarts	Gallons	Cubic Feet
1 cubic centimeter =	1	0.06102	0.001	0.00106	0.00026	0.00004
1 cubic inch =	16.387	1	0.01639	0.01732	0.00433	0.00058
1 pint =	473.18	28.875	0.47318	0.5	0.125	0.01671
1 liter =	1,000	61.023	1	1.0567	0.26417	0.03531
1 quart =	946.36	57.75	0.94636	1	0.25	0.03342
1 gallon =	3,785.4	231	3.7854	4	1	0.13368
1 cubic foot =	28,317	1,728	28.317	29.922	7.4805	1
1 bushel =	35,239.3	2,150.4	35.239	37.237	9.3092	1.2445
1 cubic yard =	764,559.4	46,656	764.56	807.9	201.97	27
1 cubic meter =	1,000,000	61,023.4	1,000	1,056.7	264.17	35.314

Fluid Capacity Equivalents

Unit	Teaspoon	Tablespoon	Fluid Ounce	Cup	Pint	Quart	Gallon	Milliliter	Liter
1 teaspoon =	1	0.333	0.167	0.021	0.010	0.005	0.001	4.929	0.005
1 teaspoon =	3	1	0.5	0.063	0.031	0.016	0.004	14.787	0.015
1 fluid ounce =	6	2	1	0.125	0.063	0.031	0.008	29.574	0.030
1 cup =	48	16	8	1	0.5	0.25	0.063	236.6	0.237
1 pint =	96	32	16	2	1	0.5	0.125	473.2	0.473
1 quart =	192	64	32	4	2	1	0.25	946	0.946
1 gallon =	768	256	128	16	8	4	1	3785	3.785
1 milliliter =	0.203	0.068	0.034	0.004	0.002	0.001	0.0003	1	0.001
1 liter =	202.8	67.628	33.8	4.227	2.113	1.057	0.264	1000	1

Welding-Related Conversion Formulas

To Convert From	To	Multiply by
inch (in)	meter (m)	0.0254
inch (in)	mm	25.4
inch squared (in²)	mm²	645.2
millimeter squared (mm²)	in²	0.00155
pounds (lb)	kg	0.454
kilograms (kg)	lb	2.2
ton	kg	907.2
kilogram	ton	0.0011
metric ton	kg	998.8
kilogram	metric ton	0.0010
pounds/hour (lbs/hr)	kg/hr	0.454
kilograms/hour (kg/hr)	lbs/hr	2.2

continued

To Convert From	To	Multiply by
liters/minute (LPM)	CFH	2.119
cubic feet/hour (CFH)	LPM	0.4719
pounds per square inch (psi)	kPa	6.895
kilopascal (kPa)	psi	0.145
megapascal (MPa)	psi	145
pounds per square inch (psi)	MPa	0.0069
inches per minute (ipm)	mm/sec	0.423
foot-pounds (ft-lbs)	J	1.356
Joule (J)	ft-lbs	0.737

To Get Electrode Length Per Unit of Weight (in/lb or cm/kg)

$(1 \div$ electrode density in lb/in$^3) \div 3.1416 \times \frac{1}{2}$ electrode diameter^2

$(1 \div$ electrode density in gm/cm$^3) \div (3.1416 \times \frac{1}{2}$ electrode diameter$^2) \times 1000$

Note: $3.1416 \times \frac{1}{2}$ electrode diameter^2 is πr^2.

To Convert From	To	Equation
°Fahrenheit	°Celsius	(°F − 32) ÷ 1.8
°Celsius	°Fahrenheit	(1.8 × °C) + 32

Common Formulas

Circumference of a Circle	$= \pi d$	(π is 3.1416, and d is diameter)
Area of a Circle	$= \pi r^2$	(π is 3.1416, and r is radius)
Area of a Triangle	$= (bh)/2$	(b is the base, and h is the height)
Area of a Square	$= a^2$	(a is one of the sides)
Area of a Rectangle	$= bh$	(b is the base, and h is the height)
Area of a Trapezoid	$= (h(a + b))/2$	(h is height, a is the longer parallel side, and b is the shorter)
Area of a Pentagon	$= 1.720a^2$	(a is one of the sides)
Area of a Hexagon	$= 2.598a^2$	(a is one of the sides)
Area of a Octagon	$= 4.828a^2$	(a is one of the sides)
Volume of a Cube	$= a^3$	(a is one of the edges)
Volume of a Prism	$= lwd$	(l is the length, w is the width, and d is the depth)
Volume of a Pyramid	$= (Ah)/3$	(A is the area of the base, and h is the height)
Volume of a Cylinder	$= \pi r^2 h$	(π is 3.1416, r is the radius of the base, and h is the height)
Volume of a Cone	$= (\pi r^2 h)/3$	(π is 3.1416, r is the radius of the base, and h is the height)
Volume of a Sphere	$= (4\pi r^3)/3$	(π is 3.1416, and r is the radius)

Kinetic Energy

U.S. Standard	Metric
KE = ½ × m × v²	1 Joule = 1 kg × (m² ÷ s²)

m = mass   v = velocity (speed)   kg = kilograms   m = meters   s = seconds

Ohm's Law

To Get Watts	To Get Volts	To Get Amps	To Get Resistance
$P = E^2 \div R$	$E = R \times I$	$I = \sqrt{P \div R}$	$R = P \div I^2$
$P = R \times I^2$	$E = P \div I$	$I = P \div E$	$R = E^2 \div P$
$P = E \times I$	$E = \sqrt{P \times R}$	$I = E \times R$	$R = E \div I$

P = Watts   E = Volts   I = Amps   R = Ohms (resistance)

AC/DC Formulas

To Find	Direct Current	AC/1-phase 115V or 120V	AC/1-phase 208V, 230V, or 240V	AC 3-phase All Voltages
amps when horsepower is known	$\dfrac{HP \times 746}{E \times Eff}$	$\dfrac{HP \times 746}{E \times Eff \times PF}$	$\dfrac{HP \times 746}{E \times Eff \times PF}$	$\dfrac{HP \times 746}{1.73 \times E \times Eff \times PF}$
amps when kilowatts is known	$\dfrac{kW \times 1000}{E}$	$\dfrac{kW \times 1000}{E \times PF}$	$\dfrac{kW \times 1000}{E \times PF}$	$\dfrac{kW \times 1000}{1.73 \times E \times PF}$
amps when kVA is known		$\dfrac{kVA \times 1000}{E}$	$\dfrac{kVA \times 1000}{E}$	$\dfrac{kVA \times 1000}{1.73 \times E}$
kilowatts	$\dfrac{I \times E}{1000}$	$\dfrac{I \times E \times PF}{1000}$	$\dfrac{I \times E \times PF}{1000}$	$\dfrac{I \times E \times 1.73\ PF}{1000}$
kilovolt-amps		$\dfrac{I \times E}{1000}$	$\dfrac{I \times E}{1000}$	$\dfrac{I \times E \times 1.73\ PF}{1000}$
horsepower (output)	$\dfrac{I \times E \times Eff}{746}$	$\dfrac{I \times E \times Eff \times PF}{746}$	$\dfrac{I \times E \times Eff \times PF}{746}$	$\dfrac{I \times E \times Eff \times 1.73 \times PF}{746}$

E = Voltage / I = Amps / W = Watts / PF = Power Factor / Eff = Efficiency / HP = Horsepower

Voltage Drop Formulas

Single Phase (2 or 3 wire)	$VD = \dfrac{2 \times K \times I \times L}{CM}$	K = ohms per mil foot (copper = 12.9 at 75°) (aluminium = 21.2 at 75°)
	$CM = \dfrac{2K \times L \times I}{VD}$	
Three Phase	$VD = \dfrac{.73 \times K \times I \times L}{CM}$	L = length of conductor in feet I = current in conductor (amperes)
	$CM = \dfrac{1.73 \times K \times L \times I}{VD}$	CM = circular mil area of conductor

		Average Properties of Standard Steels				
AISI or SAE No.	Condition of Steel	Tensile Strength psi	Yield Strength psi	% of Elongation in 2 Inches	Brinell Hardness	Rockwell Hardness
1010	Hot Rolled	50/60,000	30/40,000	30/40	115	64B
1212	Cold Rolled	75/90,000	60/70,000	10/15	150/210	85/95B
1018	Hot Rolled	55/70,000	35/50,000	30/40	120/140	67/80B
	Cold Rolled	70/85,000	60/75,000	18/25	150/180	80/90B
1020	Hot Rolled	55/70,000	35/50,000	30/40	120/140	67/80B
1117	Hot Rolled	65/75,000	40/50,000	25/35	135/155	74/82B
1040	Hot Rolled	85/95,000	50/60,000	20/30	175/200	88/93B
1045	Hot Rolled	90/105,000	55/65,000	15/25	190/220	90/98B
	Cold Rolled	90/110,000	75/90,000	12/20	195/230	90/99B
	Quench/Temp	120,000	92,000	20	255	25C
1137	Hot Rolled	90/105,000	55/70,000	15/25	180/220	89/98B
1144	Hot Rolled	95/110,000	55/70,000	15/25	190/220	90/99B
1050	Hot Rolled	95/110,000	55/70,000	15/25	200/235	93/99B
1095	Hot Rolled	130/150,000	75/95,000	7/17	260/300	26/32C
	Quench/Temp	170,000	110,000	14	290	36C
4140	Hot Rolled	95/150,000	60/95,000	17/25		
4150	Hot Rolled	90/110,000	65/75,000	20/30	185/215	90/96B
4340	Hot Rolled	100/120,000	65/85,000	20/30	210/250	16/25C
	Quench/Temp	170/180,000	140/155,000	12/20	350/375	36/39C
4620	Hot Rolled	80/95,000	60/70,000	20/30	165/200	85/93B
	Cold Rolled	95/105,000	80/90,000	15/25	200/220	93/97B
8620	Hot Rolled	80/95,000	60/70,000	18/25	165/200	85/93B
8642	Hot Rolled	100/130,000	65/85,000	15/25	210/275	17/29C
8742	Hot Rolled	100/130,000	65/85,000	15/25	210/275	17/29C
Type 302	Annealed Bar	80/90,000	30/40,000	55/65	150/180	80/90B
Type 304	Cold Rolled	80/90,000	30/40,000	55/65	150/180	80/90B
Type 308	Cold Rolled	80/90,000	30/40,000	45/55	130/150	75/85B
Type 309	Cold Rolled	90/100,000	35/45,000	40/50	160/180	80/90B
Type 310	Cold Rolled	90/100,000	35/45,000	40/50	160/180	80/90B
Type 316	Cold Rolled	80/90,000	25/40,000	55/65	150/180	80/90B
Type 321	Cold Rolled	80/100,000	30/40,000	50/60	140/165	75/85B
Type 410	Annealed Bar	70/100,000	35/45,000	30/40	150/200	80/90B
Type 416	Annealed Bar	80/100,000	55/65,000	15/25	190/220	90/95B
	Heat Treated	135/155,000	110/130,000	10/15	260/320	25/34C
Type 440F-Se	Annealed Bar	100/120,000	60/70,000	20/30	200/260	14/25C
PH 17-4	Preannealed	135/185,000	105/160,000	5/14	350/420	38/44C

Hardness Conversion Table for Carbon and Alloy Steels

Brinell Hardness No.	Rockwell Hardness Numbers			Tensile Strength	
	B Scale	A Scale	C Scale	ksi	kPa
–	–	84.5	66	–	–
722	–	83.4	64	–	–
688	–	82.3	62	–	–
654	–	81.2	60	–	–
615	–	80.1	58	–	–
577	–	79.0	56	313	2158
543	–	78.0	54	292	2013
512	–	76.8	52	273	1882
481	–	75.9	50	255	1758
455	–	74.7	48	237	1634
443	–	74.1	47	229	1579
432	–	73.6	46	222	1531
421	–	73.1	45	215	1482
409	–	72.5	44	208	1434
400	–	72.0	43	201	1386
390	–	71.5	42	194	1338
381	–	70.9	41	188	1296
371	–	70.4	40	181	1248
362	–	69.9	39	176	1214
353	–	69.4	38	171	1179
344	–	68.9	37	167	1151
336	–	68.4	36	162	1117
327	–	67.9	35	157	1083
319	–	67.4	34	153	1055
311	–	66.8	33	149	1027
301	–	66.3	32	145	1000
294	–	65.8	31	142	979
286	–	65.3	30	138	952
279	–	64.7	29	135	931
271	–	64.3	28	132	910
264	–	63.8	27	128	883
258	–	63.3	26	125	862
253	–	62.8	25	122	841
247	–	62.4	24	120	827
243	–	62.0	23	117	806
240	100	–	–	116	800
234	99	–	–	112	772
222	97	–	–	106	731

Hardness Conversion Table for Carbon and Alloy Steels

Brinell Hardness No.	Rockwell Hardness Numbers			Tensile Strength	
	B Scale	A Scale	C Scale	ksi	kPa
210	95	–	–	101	696
200	93	–	–	96	662
195	92	–	–	93	641
185	90	–	–	89	614
176	88	–	–	85	586
169	86	–	–	81	558
162	84	–	–	78	538
156	82	–	–	75	517
150	80	–	–	72	496
144	78	–	–		
139	76	–	–		
135	74	–	–		
130	72	–	–		
125	70	–	–		
121	68	–	–		
117	66	–	–		
114	64	–	–	55	379

SHEET METAL AND PLATE

In the production of sheet metal and plates, a slab is rolled until the desired thickness is achieved. The slab is at red heat, and the resulting metal is called hot-rolled steel (HRS). Hot-rolled metal has a telltale black mill scale covering its surface. If cold-rolled steel (CRS) is desired, the hot-rolled metal is pickled (soaked) in an acid bath to remove the mill scale and rolled again to the precise thickness. Because hot-rolled metal cools after it is rolled, the thickness may vary somewhat. Cold-rolled metal does not cool down and thus does not shrink, so an exact thickness can be achieved. Because the mill scale protects the surface of the metal to some extent, no special precaution against rusting is performed on hot-rolled metal. However, pickled cold-rolled steel is often covered with an oil to protect it. Hot-rolled metal that has been pickled is called pickle and oil and has a coating of thick sticky oil covering its surface; this sheet metal is known as HRPO (hot-rolled pickle and oil).

Sheet metal thickness measured in the U.S. Standard system (**FIGURE D-1**) is designated in gauge, whereas plate is designated in fractions of an inch. Sheet metal stops at 7 gauge (approximately 3/16 inch), and plate starts at 1/4 inch. Sizes vary depending on sheet thickness. Thin gauge metal, 28 gauge, 24 gauge, 22 gauge, etc., can be purchased in 3 feet by 8 feet sheets or in 4 feet by 10 feet sheets. As the thickness increases, 16 gauge and above, the sheet size also increases. Common sizes are 5 feet by 10 feet or 6 feet by 10 feet. Plate sizes, 1/4 inches, are also often purchased in 5 feet by 10 feet or 6 feet by 10 feet sheets. As the plate thickness increases, 1/2-inch plate, 1 inch plate, etc., the sheet size may start to decrease again to 4 feet by 10 feet or smaller sizes. Larger size sheets are possible with almost any thickness.

Sheet metal and plate are rolled into a coil when they are produced. A coil weighs 40,000 pounds. While it is cheaper to purchase steel in a coil, a company must have some means to straighten the metal. Larger companies have straighteners, but smaller companies usually have the distributor straighten and cut the metal to size. A sampling of how sheet metal and plate may appear in a steel book from a distributor is as shown in **TABLE D-1**.

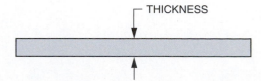

FIGURE D-1 Representation of sheet or plate thickness

Sheet Metal and Plate

Gauge or Thickness	Weight per Square Foot	Decimal	Standard Size
16 gauge	2.5 pounds	0.060 inch	72 inches × 120 inches
12 gauge	4.375 pounds	0.105 inch	72 inches × 120 inches
10 gauge	5.625 pounds	0.135 inch	72 inches × 120 inches
7 gauge	7.65 pounds	0.1875 inch	72 inches × 120 inches
1/4 inches	10.21 pounds	0.25 inch	72 inches × 120 inches
3/8 inches	15.32 pounds	0.375 inch	72 inches × 120 inches
1/2 inches	20.42 pounds	0.50 inch	48 inches × 120 inches
1 inches	40.84 pounds	1.00 inch	48 inches × 96 inches

TABLE D-1

BLACK PIPE

Pipes are usually produced by starting with a skelp (a flat bar), rolled to its final shape, and then seam-welded. Pipe (**FIGURE D-2**) is measured by the inside diameter (I.D.) for sizes up to and including 12 inches; greater than 12 inches, the pipe is measured by the outside diameter (O.D.). Wall thickness for most pipes is identified by schedules. The standard wall thickness is schedule 40; schedule 80 is a double wall thickness. There also exist schedules 10, 20, 120, and 160. Some manufacturers identify wall thickness as standard, double wall, triple wall, etc., or standard, XS, XXS, etc. Pipe is often sold in lengths of 21 feet; however, they are still sold by the pound per linear foot.

There are a few things of note. First, although pipe is measured I.D., the inside diameter of schedule 40 pipe does not measure precise; they actually measure larger, but 12-inch pipe measures exactly. Second, as the wall thickness increases, the inside diameter decreases and the outside diameter remains the same for a given diameter of pipe. The reason for this is that pipe is often threaded and coupled together. If the outside diameter would change when the wall thickness increased, there would have to be a set of threading dies made for each schedule of pipe for each diameter. Because pipe can have schedule 10, 20, 40, 80, 120, or 160 for pipe diameters ranging from 1/8 inch to 12 inches, there would be considerable more dies needed for threading. **TABLE D-2** is a sampling of how pipe may appear in a steel book from a distributor.

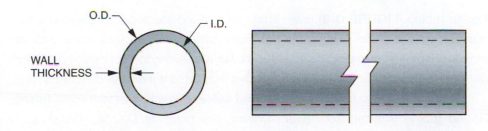

FIGURE D-2 Illustration of round pipe

Nominal Size (inch)	Schedule	Pipe Wall Thickness	Size I.D. (inch)	Size O.D. (inch)	Weight/Foot (pounds)	Length in Feet
1/2	40	0.109	0.622	0.840	0.85	21
1/2	80	0.147	0.546	0.840	1.09	21
3/4	40	0.113	0.824	1.050	1.13	21
3/4	80	0.154	0.742	1.050	1.47	21
1	40	0.133	1.049	1.315	1.68	21
1	80	0.179	0.957	1.315	2.17	21
1½	40	0.145	1.610	1.900	2.72	21
1½	80	0.200	1.500	1.900	3.63	21
2	40	0.154	2.067	2.375	3.65	21
2	80	0.218	1.939	2.375	5.02	21
2½	40	0.203	2.469	2.875	5.79	21
2½	80	0.276	2.323	2.875	7.66	21
3	40	0.216	3.068	3.500	7.58	21
3	80	0.300	2.900	2.375	10.25	21
4	40	0.237	4.026	4.500	10.79	21
4	80	0.337	3.826	4.500	14.98	21
5	40	0.258	5.057	5.563	14.62	16–24
5	80	0.375	4.813	5.563	20.78	16–24
6	40	0.280	6.065	6.625	18.97	16–24
6	80	0.432	5.761	6.625	28.57	16–24
8	40	0.322	7.981	8.625	28.55	16–24
8	80	0.500	7.625	8.625	43.39	16–24
10	40	0.365	10.020	10.750	40.48	16–24
10	80	–	–	10.750	–	16–24
12	40	0.375	12.000	12.750	49.56	16–24
12	80	0.500	11.750	12.750	65.42	16–24
14	40	0.375	13.250	14.000	54.57	16–24
14	80	0.500	13.000	14.000	72.09	16–24

TABLE D-2

ROUND TUBING

Round tubing (**FIGURE D-3**) often starts with round stock and is pierced with a mandrel. The mandrel pierces the hot round billet and the metal flows over the mandrel. Both the mandrel and the round billet are revolving in opposite directions during this piercing operation. **FIGURE D-4** illustrates a mandrel piercing a billet.

Measuring the outside diameter and the wall thickness identifies tubing. Round tubing is often C.D. (Cold Drawn) Seamless or D.O.M. (Drawn Over

FIGURE D-3 Round tubing

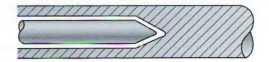

FIGURE D-4 Mandrel piercing billet

Mandrel). Because round tubing is cold drawn, the outside diameter and wall thickness are to exact tolerances. Round tubing is often sold in lengths of 20 feet. Square and rectangular tubing are formed by a series of rollers and then welded. Square and rectangular tubing are also measured by the outside diameter and wall thickness. Where round tubing wall thickness is usually identified by a decimal number, square and rectangular tubing are identified with the U.S. Standard system by gauge thickness for thinner walls and by fractional or decimal for heavier wall thickness. Square and rectangular tubing are often sold in lengths of 20 or 24 feet. Although round, square, and rectangular are 20 to 24 feet long, they are sold by the pound per linear foot. All metal is sold by weight. **TABLE D-3** shows a sampling of how tubing may appear in a steel book from a distributor.

	Round Tubing				
O.D. and Thickness	**Wall Diameter**	**I.D.**	**Weight Foot Per Pounds**	**C.D. Seamless**	**D.O.M. Welded**
3 inches × 18 gauge	.049	2.902	1.544		X
3 inches × 16 gauge	.065	2.870	2.037	X	
3 inches × 14 gauge	.079	2.834	2.586	X	
3 inches × 11 gauge	.120	2.760	3.691	X	X
3 inches × 1/8 inches	.125	2.875	3.838	X	X
3 inches × 3/16 inches	.187	2.625	5.646	X	X
3 inches × 7/32 inches	.219	2.687	6.505	X	X
3 inches × 1/4 inches	.250	2.500	7.343	X	X
3 inches × 5/16 inches	.313	2.375	8.982	X	X
3 inches × 3/8 inches	.375	2.250	10.51	X	

TABLE D-3

There are basically three different types of beams: Wide Flange Beams, H-Beams, and S-Beams. By far, the most common type of beam is the wide flange beam, also known as the W-beam (**FIGURE D-5**). The web the middle part is longer than the width of the top and bottom flanges. The difference between the W-beam and the H-beam is that the H-beam is more square; in other words, the web and flange are basically of equal size. The W-beam can also have equal flange and web sizes, and you would need a steel manufacturer booklet in order to tell the two apart.

The S-beam is what was commonly called the I-beam prior to the 1990s. The term I-beam is now used as a generic term for beams; however, when ordering beams, a person must distinguish which type of beam he or she wishes to purchase. On the S-beam, the flanges taper down from the web of the beam.

Beams used in the construction of houses and commercial building are W-beams. Beams used for log splitters may be an H-beam or a W-beam. Below is a sampling of how beams may appear in a steel book from a distributor.

Beams are specified on a print by the depth between flanges and the weight per linear foot. For example, the nominal size W10×8 listed in **TABLE D-4** would be specified on a print as W10×33 or W10@33. Note that the W6@20 (W6×6) and the H6@20 (H6×6) listed above have the same weight. The H-beam is slightly more symmetrical than the W-beam, but it would be difficult to tell the difference just by looking at them; it takes a steel booklet to distinguish the difference. Typical beam lengths are 40 feet, but they can also come in 60- and even 80-feet lengths.

FIGURE D-5 A wide-flange beam

Wide Flange and H-Beams					
Nominal Size	Weight/ Foot Pounds	Depth of Section	Flange Width	Flange Thickness	Web Thickness
W6 × 4	8.5	5.83	4.000	0.194	0.170
W6 × 6	15.5	6.00	5.995	0.269	0.235
W6 × 6	20.0	6.20	6.018	0.367	0.258
H6 × 6	20.0	6.00	5.938	0.379	0.250
W8 × 4	10.0	7.90	3.940	0.204	0.170
W8 × 4	15.0	8.12	4.015	0.314	0.245
W10 × 4	15.0	10.00	4.000	0.269	0.230
W10 × 8	33.0	9.75	7.964	0.433	0.292

TABLE D-4

CHANNEL IRON

Channel iron (**FIGURE D-6**) is very similar to beams in that it is specified on a print by the depth between the flanges and the weight per linear foot. For example, a channel would be specified on a print as C6×10.5 or C6@10.5. Channel iron is often sold in either 20- or 40-foot lengths. Longer lengths are also available. A sampling of how channel iron may appear in a steel book from a distributor is shown in **TABLE D-5**.

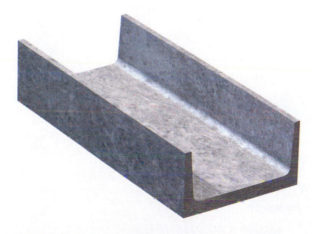

FIGURE D-6 Channel iron

Channel Iron			
Depth of Channel	Weight per Foot Pounds	Thickness of Web	Width of Flange
4	5.4	0.184	1.584
4	7.25	0.312	1.721
5	6.7	0.190	1.750
5	9.0	0.325	1.885
6	10.5	0.314	2.034
6	18.0	0.379	3.504
8	11.5	0.220	2.260
8	20.0	0.400	3.025

TABLE D-5

Fabricating such things as ATV ramps for trailers and other load-bearing projects is a great way to incorporate welding provided the welds are sound and liability is not an issue. The weight (or load) a fabricated project will endure without collapsing depends on a number of factors: the type of material, the structural shape of the material, and the length of the span.

The following five formulas are simply a starting point. These are standard engineering formulas that can be found by a search on the Internet, in the *Machinery's Handbook*, or other similar type books.

For calculating stress at the center:

$$\frac{Wl}{4Z} = S$$

Where: W = weight
l = length
Z = section modulus
S = allowable bending stress

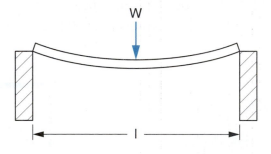

FIGURE E-1 Load at center of span

For calculating evenly distributed stress:

$$\frac{Wl}{8Z} = S$$

Where: W = weight
l = length
Z = section modulus
S = allowable bending stress

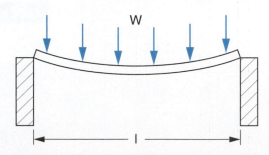

FIGURE E-2 Load distributed evenly along a span

For calculating uneven stress:

$$\frac{Wab}{Zl} = S$$

Where: W = weight

a = load point length (a)

b = load point length (b)

Z = section modulus

l = total length (a + b)

S = allowable bending stress

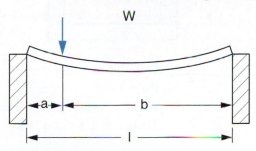

FIGURE E-3 Load distributed unevenly along a span

For calculating cantilever stress at end:

$$\frac{Wl}{Z} = S$$

Where: W = weight

l = length

Z = section modulus

S = allowable bending stress

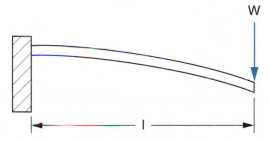

FIGURE E-4 Calculating Cantilever stress at the end

Calculating cantilever stress for load distributed evenly along a span:

$$\frac{Wl}{2Z} = S$$

Where: W = weight

l = length

Z = section modulus

S = allowable bending stress

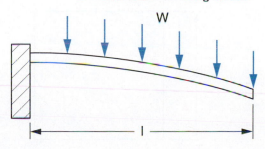

FIGURE E-5 Calculating cantilever stress for load distributed evenly along a span

The section modulus, Z, is calculated based on the structure of the shape. For example, to derive the section modulus for square tubing, the formula is:

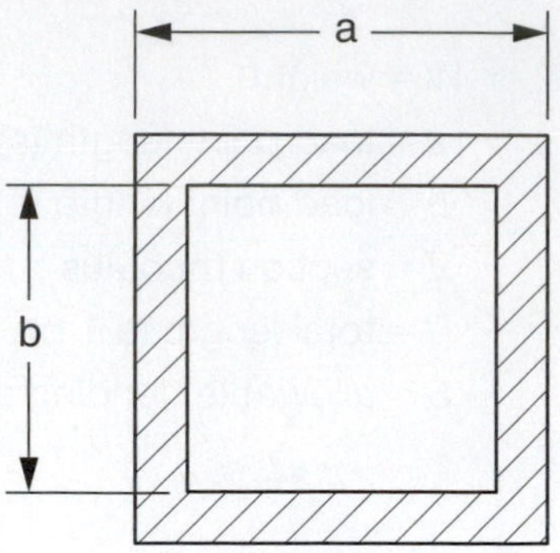

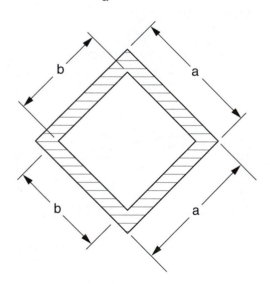

FIGURE E-6 Section moduls for square tubing

$$\frac{a^4 - b^4}{6a} = Z$$

Example:
If $a = 1$ inch
Wall thickness = 1/8 inches
Then, $b = 3/4$ inches

Some other common shapes are:

$$\frac{a^4 - b^4}{a} 0.118 = Z$$

FIGURE E-7 Section modulus for square tubing

$$\frac{a^3}{3} = Z$$

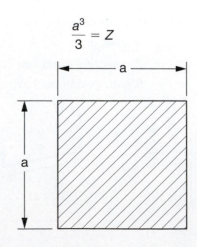

FIGURE E-8 Section moduls for a square

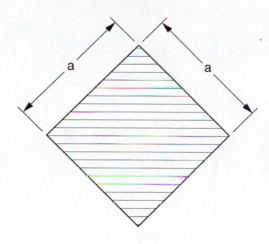

$$0.118a^3 = Z$$

FIGURE E-9 Section modulus for a square

$$\frac{bd^2}{3} = Z$$

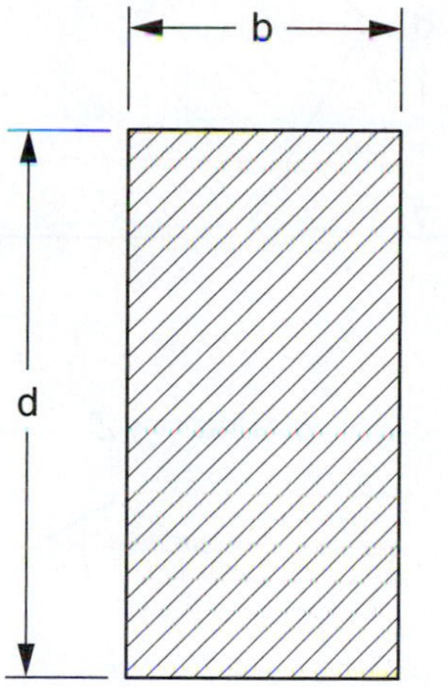

FIGURE E-10 Section modulus for a rectangle

$$\frac{bd^3 - hk^3}{6d} = Z$$

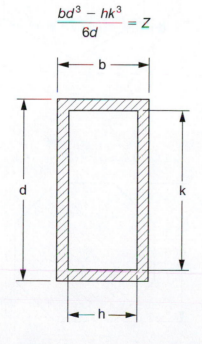

FIGURE E-11 Section modulus for rectangular tube

$$\frac{\pi(D^4 - d^4)}{32D} = Z$$

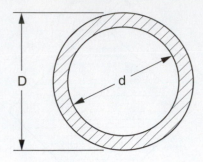

FIGURE E-12 Section modulus for a round tube or pipe

$$0.12d^3 = Z$$

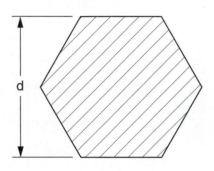

FIGURE E-13 Section modulus for a hexagon

$$0.104d^3 = Z$$

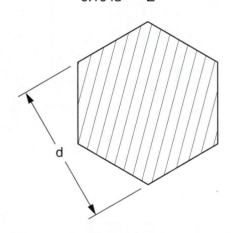

FIGURE E-14 Section modulus for a hexagon

$$0.109d^3 = Z$$

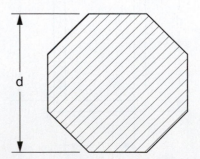

FIGURE E-15 Section modulus for an Octagon

$$\frac{\pi d^3}{32} = Z$$

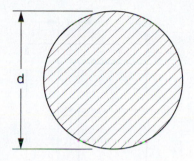

FIGURE E-16 Section modulus for a round

$$\frac{2sb^3 + ht^3}{6b} = Z$$

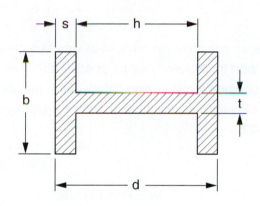

FIGURE E-17 Section modulus for a wide flange beam

$$\frac{bd^3 - h^3(b - t)}{6d} = Z$$

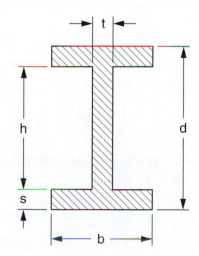

FIGURE E-18 Section modulus for a wide flange beam

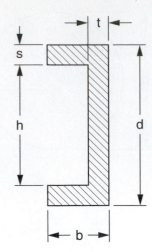

$$\frac{bd^3 - h^3(b - t)}{6d} = Z$$

FIGURE E-19 Section modulus for channel

The following pages calculate the amount of weight a given square or rectangular tube will support for a given length. For other structural shapes, consult the *Machinery's Handbook*. From a design safety point, the following examples are derived from calculating the stress at the center as opposed to a distributed load (see **FIGURE E-20).**

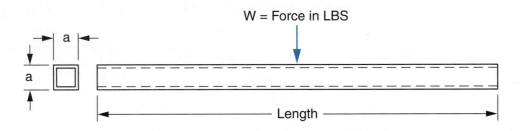

FIGURE E-20 Layout of square tubing for calculating for load bearing weight at center

The formula for calculating how much weight can be placed in the middle of a square tube before it buckles or fails is:

$$\frac{Wl}{4Z} = S$$

Where: W = force in pounds (lbs or kg)
l = length
Z = section modulus
4 = a constant
S = allowable bending stress

To get Z, the section modulus:

$$\frac{a^4 - b^4}{6a} = Z$$

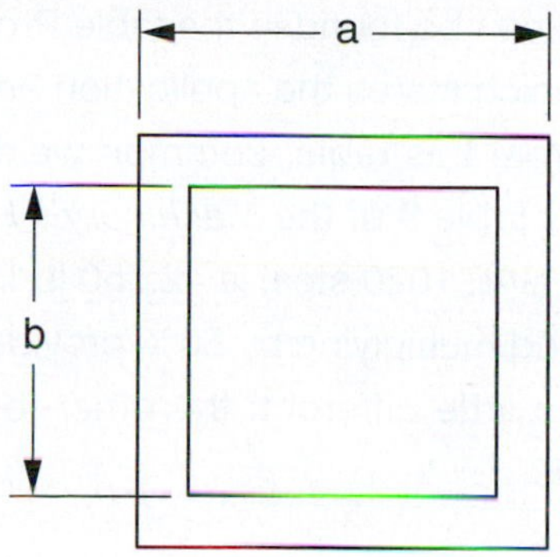

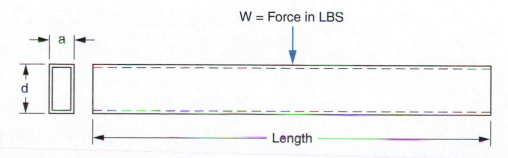

FIGURE E-21 Inside and outside dimensions of a square tube

Once the tubing material is identified and S is determined, the next step is to plug the known information into the formula. The desired information needed in the formula is W, the weight that can be loaded on the tubing in the middle. The following example uses a 1 inch × 1 inch × 14 gauge square steel tubing and is 30 inches long (14 gauge = 0.079 inches).

$$\frac{Wl}{4Z} = S \qquad \frac{W(30)}{4Z} = S$$

$$\frac{a^4 - b^4}{6a} = Z \qquad \frac{1^4 - 0.842^4}{6(1)} = Z \qquad \frac{1 - 0.5026}{6} = 0.0829$$

$$S = 24{,}000$$

$$\frac{Wl}{4Z} = S \qquad \frac{W(30)}{4(0.0829)} = 24{,}000 \qquad \frac{24{,}000 \times 0.3316}{30} = W = 265 \text{ lbs.}$$

A rectangular tube may be used to build a ramp for a trailer rather than using square tubing (see **FIGURE E-22**).

The S is the allowable bending stress. The value of S equals 60% of the elastic limit in tension but not more than 36% of the ultimate tensile strength.[1] If aluminum tubing is substituted, for example, the value of S will change. The value for the elastic limit in this example is the yield strength. "Yield Strength, S, is the

W = Force in LBS

Length

FIGURE E-22 Layout of rectangular tubing for calculating for load bearing weight at center

[1] *Machinery's Handbook*, 25th edition, Table 2, P. 280

maximum stress that can be applied without permanent deformation of the test specimen. This is the value of the stress at the elastic limit for materials for which there is an elastic limit."[2]

The yield strength changes for every grade of steel, aluminum, stainless, etc. The desired yield strength can be found in the table Properties of Materials in the *Machinery's Handbook*, which states the application and the corresponding SAE material specification.[3] From this table, common welded steel tubing specifies SAE 1020 as the material. Table 9 of the *Machinery's Handbook*[4] states that the annealed yield strength of SAE 1020 steel is 42,750 lb/in^2. Rounding this for engineering safety to 40,000 and multiplying by 60% provides a value of 24,000 for S.

The formula for Z is a little different; the other formulas would remain the same.[5] The formula for Z is:

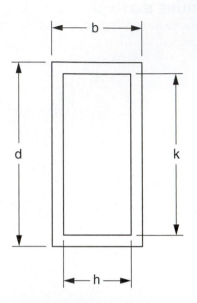

FIGURE E-23 Inside and outside dimensions of a rectangular tube

$$\frac{bd^3 - hk^3}{6d} = Z$$

Where (see **FIGURE E-4**):

b = outside width
d = outside length
h = inside width
k = inside length

Using the examples in **FIGURE E-2** and **FIGURE E-23** and a 2 inch × 1 inch × 14 gauge steel tubing with a 30-inch span (to compare a rectangle tube to a square tube), the formula is as follows:

$$\frac{Wl}{4Z} = S \qquad\qquad \frac{W(30)}{4Z} = S$$

$$\frac{bd^3 - hk^3}{6d} = Z \qquad \frac{1(2)^3 - 0.842(1.842)^3}{6(2)} = Z \qquad \frac{8 - 5.2625}{12} = 0.228$$

$$S = 24,000$$

$$\frac{Wl}{4Z} = S \qquad \frac{W(30)}{4(0.228)} = 24,000 \qquad \frac{24,000 \times 0.912}{30} = 730\ bs$$

Notice that the final weight for both examples of the square and rectangular tubing is rounded to the nearest whole number. Also, notice that placing the tubing as shown in **FIGURE E-23** more than doubles the strength of what the tubing can support compared to the square tube.

[2] *Machinery's Handbook*, 25th edition, Page 193
[3] *Machinery's Handbook*, 25th edition, Table 6, P. 422
[4] *Machinery's Handbook*, 25th edition, Table 9, P. 434
[5] *Machinery's Handbook*, 25th edition, Formulas, pp. 215–224

Periodic Table
of Elements

																		0
IA																		2 He
1 1 H	IIA											IIIA	IVA VA	VIA	VIIA			
2 3 Li	4 Be											5 B	6 C	7 N	8 O	9 F		10 Ne
3 11 Na	12 Mg	IIIB	IVB	VB	VIB	VIIB	—— VIII ——			IB	IIB	13 Al	14 Si	15 P	16 S	17 Cl		18 Ar
4 19 K	20 Ca	21 Sc	22 Ti	23 V	24 Cr	25 Mn	26 Fe	27 Co	28 Ni	29 Cu	30 Zn	31 Ga	32 Ge	33 As	34 Se	35 Br		36 Kr
5 37 Rb	38 Sr	39 Y	40 Zr	41 Nb	42 Mo	43 Tc	44 Ru	45 Rh	46 Pd	47 Ag	48 Cd	49 In	50 Sn	51 Sb	52 Te	53 I		54 Xe
6 55 Cs	56 Ba	57 *La	72 Hf	73 Ta	74 W	75 Re	76 Os	77 Ir	78 Pt	79 Au	80 Hg	81 Tl	82 Pb	83 Bi	84 Po	85 At		86 Rn
7 87 Fr	88 Ra	89 †Ac	104 Rf	105 Ha	106 106	107 107	108 108	109 109	110 110									

*Lanthanide Series	58 Ce	59 Pr	60 Nd	61 Pm	62 Sm	63 Eu	64 Gd	65 Tb	66 Dy	67 Ho	68 Er	69 Tm	70 Yb	71 Lu
†Actinide Series	90 Th	91 Pa	92 U	93 Np	94 Pu	95 Am	96 Cm	97 Bk	98 Cf	99 Es	100 Fm	101 Md	102 No	103 Lr

Legend - click to find out more...

H - gas

Li - solid

Br - liquid

Tc - synthetic

- Non-Metals
- Transition Metals
- Rare Earth Metals
- Halogens
- Alkali Metals
- Alkali Earth Metals
- Other Metals
- Inert Elements

Actual Throat The minimum distance from the weld root to the face of a fillet weld (**FIGURE 1**).

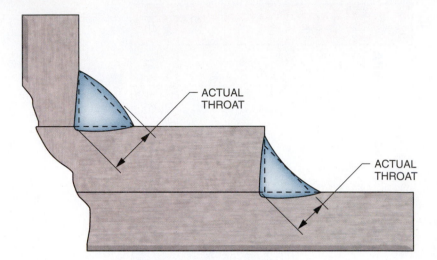

ACTUAL THROAT

ACTUAL THROAT

FIGURE 1

Adaptability Skills A desirable skill set for welding employees that includes demonstrating integrity, adapting to change, and demonstrating a positive attitude.

Air-Cooled Torch GTAW torch that uses the flow of shielding gas to help cool the torch.

Alloy A homogeneous mixture of two or more elements, of which one is a metal.

Alternating Current In welding, a current setting where the magnitude and direction of current flow switches at regularly recurring intervals. Electrons flow from negative to positive alternating from a state where the electrode is positive to one where the electrode is negative.

Amperage A measurement of the flow of electrical current. An electrical variable in welding that determines the melt-off rate of the electrode and strongly influences weld penetration.

Arc Blow The deflection of the welding arc from its intended path due to magnetic forces imposed.

Arc Burn A burn on the skin that results from exposure to intensive ultraviolet and infrared rays from the welding arc.

Arc Flash A burning and blistering of the sensitive membrane of the eye caused by exposure of the eyes to arc rays from welding.

Arc Length For arc welding processes, the distance from the end of the welding electrode to the weld pool (**FIGURE 2**).

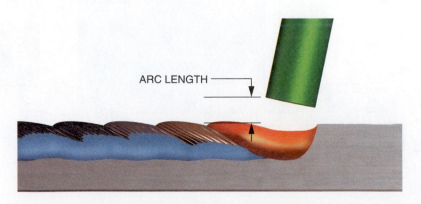

ARC LENGTH

FIGURE 2

594

Arc Strike Burn marks left outside the weld caused by the electrode in contact with the workpiece surface. These discontinuities can cause undesirable localized hardening and must be avoided.

Arc Voltage The voltage in the welding circuit as measured at the welding arc.

Arrow Side The bottom-side (below) information of the reference line on a welding symbol applied to the arrow side of the weld joint.

Asperites Surface imperfections found in Solid-State welding.

Asphyxiant A gas (argon or nitrogen) heavier than air that will cause oxygen to be pushed out of the lungs, causing a person to suffocate.

Austenite Also called gamma iron; is an interstitial solid solution of carbon in iron and occurs in steels above a critical temperature in the face-centered cubic form.

Axial Spray Transfer A mode of filler metal transfer deposit where small molten droplets are transferred across an open gap.

Back Cap Part of the GTAW equipment that pushes the collet into the collet body.

Back-Gouging Removal of metal from the back side of a weld joint in order to allow for complete joint penetration or complete fusion on a double groove weld.

Backhand (Drag) Technique Refers to a travel angle where the electrode is pointing back into the weld pool opposite the travel direction (**FIGURE 3**).

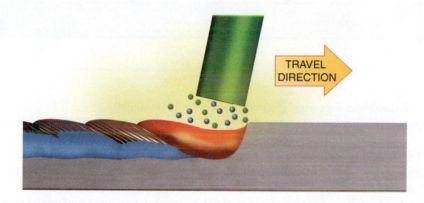

TRAVEL DIRECTION

FIGURE 3

Backing Plate A plate placed on the back side of a groove joint to support the weld metal and to prevent melt-through.

Backing Weld A weld on the opposite side from a groove weld before the groove weld is completed. It acts in the same manner as a backing plate.

Backstep Sequence A progression of welding where subsequent welds are made from one end of the joint to the other, such that each subsequent weld is added from the opposite direction as the travel direction. This is a method used to reduce weld distortion (**FIGURE 4**).

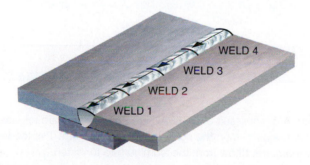

WELD 4
WELD 3
WELD 2
WELD 1

FIGURE 4

Back Weld A weld on the opposite side from a groove weld after the groove weld is completed.

Bainite A hard transformation structure in iron alloys—harder than pearlite but softer than martensite.

Base Metal (Base Material) The metal of the weld joint to be welded.

Bevel Angle The angle amount cut off from one base metal member to help form a groove joint (**FIGURE 5**).

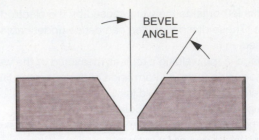

FIGURE 5

Bird's Nest Wrapping of the filler metal electrode in and around the drive roll assembly.

Block Sequence A welding sequence in a thick multipass groove joint where sections of the groove are filled with weld layers and other portions are left to be subsequently filled (**FIGURE 6**).

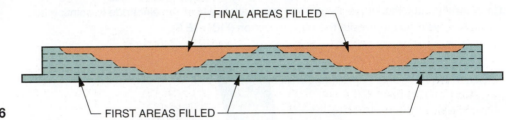

FIGURE 6

Boxing Extending a fillet weld around a corner of the weld joint.

Brittleness The property of materials that exhibit low fracture toughness and do not deform under load but shatter or break suddenly.

Burn Through A defect caused by excessive melting through a weld joint.

Butt Joint A weld joint between two base metal pieces aligned approximately in the same plane.

Carbide Precipitation An occurrence in stainless steels when weld and base metal temperatures are sustained during welding in a range of 800°F to 1,600°F (427°C to 871°C) for too long; chromium combines with carbon, forming chromium carbides.

Carbon Equivalence A measure of a steel alloy's chemical composition used to determine its weldability.

Carburrizing Flame A flame consisting of an unbalanced mix of oxygen and fuel gas, with an excess amount of fuel gas.

Cascade Sequence A groove weld made where progression is such that welds overlap longitudinally. The bottom weld layer being the longest and each subsequent layer being shorter (**FIGURE 7**).

FIGURE 7

Cathodic Etching A direct result of DCEP where positive ions are released from the electrode and strike the oxides on the surface of the aluminum. The loose oxides are then lifted away by the current flow from the aluminum to tungsten.

Cementite Also called iron-carbide; is the hardest structure represented on the iron-carbon diagram. It contains 6.67% carbon by weight and is found as small deposits in steels.

Chain Intermittent Fillet The placing of intermittent fillet welds directly opposite each other on a T-joint.

Code A body of mandated rules to be followed for the construction or fabrication of a product.

Cold Crack A crack that develops after the weld has cooled down.

Cold Lap Incomplete fusion or overlap of weld metal onto the base metal or another weld layer.

Collet Part of the GTAW torch to push against the inside wall of the collet body and squeeze around the tungsten, holding it in place.

Collet Body Part of the GTAW torch that along with the collet helps secure the tungsten in place and helps to diffuse shielding gas through the cup to the weld.

Complete Fusion Flowing together of filler metal and base metal between all weld passes into one basically homogeneous mass.

Complete Joint Penetration Extension of the weld bead fusion through the root (back) side of a groove weld (**FIGURE 8**).

COMPLETE JOINT PENETRATION

FIGURE 8

Compressive Strength The maximum stress that can be applied to a material in compression without permanent deformation or failure.

Concave Weld A weld with a concave weld face (**FIGURE 9**).

Concavity A measure from the center face of a concave weld to an imaginary line connecting the weld toes (**FIGURE 9**).

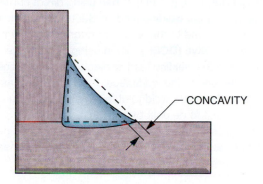

CONCAVITY

FIGURE 9

Conductor A material that will conduct electrical current.

Constant Current Power Source A power source commonly used for SMAW and GTAW with a relatively flat volt/amp curve. This results in small changes in welding current and an increasing amount of voltage when the electrode is moved away from the base metal.

Constant Voltage Power Source A power source commonly used for GMAW, FCAW, and SAW, with a relatively steep volt/amp curve. This results in a significant change in amperage for a small change in voltage when wire feed speed is increased.

Consumable Insert An insert set into the root of a weld joint that is consumed and becomes part of the weld metal during welding.

Contact Tip A copper alloy tube that guides a filler metal electrode wire from the welding gun to the workpiece.

Coupons Test specimens taken from a completed weld test assembly.

Convex Weld A weld with a convex weld face (**FIGURE 10**).

Convexity A measure from the center face of a convex fillet weld to an imaginary line connecting the fillet weld toes (**FIGURE 10**).

Crater A depression at the termination of a weld bead caused by solidification of the weld pool.

Creep Strength A measure of a material's tendency to relieve applied stresses by gradual deformation over time.

Cup Part of the GTAW torch that funnels shielding gas over the tungsten tip and weld puddle.

Current Movement of electrons from one atom to the next through a conductor.

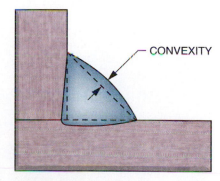

CONVEXITY

FIGURE 10

Current Density The measure of the working current used for a particular diameter electrode divided by its cross-sectional area.

Defect A flaw or combined total of flaws in a welded joint that cause it to be unfit for use by a code or standard, such as the various types of cracks. See Appendix A for more details.

Deoxidizers Alloying elements that help to remove oxides from the weld metal and weld pool.

Deposition Rate The weight amount of weld metal deposited in a time span.

Depth of Bevel (depth of preparation) The distance from the surface of a groove joint, perpendicular to the root of the weld or to the inside edge of the root face (**FIGURE 11**).

FIGURE 11

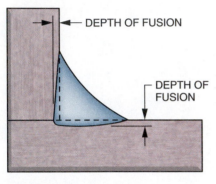

FIGURE 12

Depth of Fusion The distance that fusion extends, from the surface of the weld into the weld joint or previous weld deposit (**FIGURE 12**).

Diffusion The transfer of atoms between individual pieces of metals' outer surface layers.

Diode A device that conducts electrical current only when specific conditions are present and in only one direction.

Direct Current Electrode Negative (DCEN) When using direct current such that the negative terminal is attached to the welding lead or electrode holder. The positive terminal is attached to the work lead and to the workpiece; formerly known as straight polarity.

Direct Current Electrode Positive (DCEP) When using direct current such that the positive terminal is attached to the welding lead or electrode holder. The negative terminal is attached to the work lead and to the workpiece; formerly known as reverse polarity.

Discontinuity Undesirable flaws in a weld joint, such as porosity, undercut, and underfill, that may or may not cause it to be unfit by a code or standard.

Distortion Warping or displacement of base metal parts due to the nonuniform expansion and contraction forces caused by heat effects of welding and heating operations and the subsequent cooling cycles that follow (**FIGURE 13**).

FIGURE 13

Downhill Welding A welding progression where welding starts at the top of the joint and travels downward to the bottom.

Drag Lines For oxy-fuel or arc cutting processes, the ripples formed along the edge of the cut.

Ductility The property of a material to deform permanently or to exhibit plasticity without rupture while under tension.

Duty Cycle The amount of time (percent) a welding power source can be run at a rated output without overheating or being damaged. In the United States, it refers to a 10-minute time span. For example, a 40% duty cycle denotes usage for 4 minutes out of a 10-minute period.

Dye Penetrant Testing A method in which a penetrant—a high color contrast visible dye or a fluorescent spray—is applied to a specimen in conjunction with a developer—to bring out the dye—testing for such things as cracks and porosity.

Effective Throat The minimum distance from the weld root to the face of a weld, minus any convexity (**FIGURE 14**).

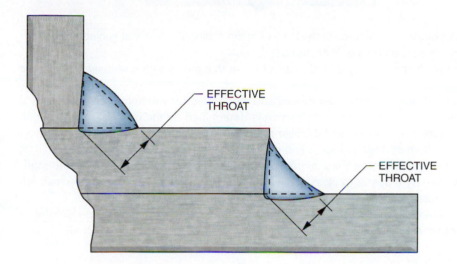

FIGURE 14

Elasticity A measure of metal's resilience or ability to resist deformation by stretching.

Electrode Extension (stick-out) With wirefeed processes (GMAW, FCAW, and SAW), the distance the electrode extends from the contact tip during welding (**FIGURE 15**).

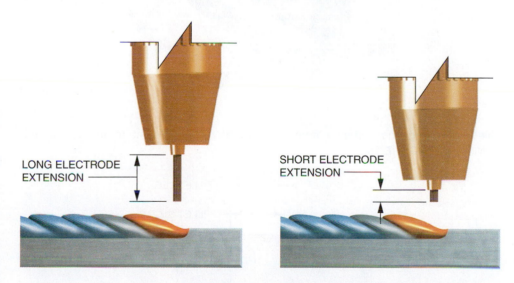

FIGURE 15

Electrode Holder A clamping device attached to the electrode lead used to hold the electrode in position for welding.

Electrode Lead The electrical cable used to attach the electrode holder to the power supply.

Elongation The increase in length that occurs before a material is fractured when subjected to tensile stress.

Essential Variables Welding variables that if changed require a retest and documentation of a new procedure.

Eutectoid Point In metals, a point where a single solid solution structure transforms directly to a mixture of more than one structure upon cooling.

Face Reinforcement Weld reinforcement on the side of the weld joint on which the weld is deposited (**FIGURE 16**).

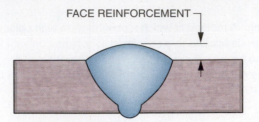

FIGURE 16

Fatigue Strength A measures of the load-carrying ability of a material subjected to loading, which is repeated a definite number of cycles.

Filler Metal The metal alloy added to a weld joint that mixes with and fuses to the base metals.

Filler Metal Classification Filler metals and electrodes for welding are identified by a classification number. The classification is determined by the chemical composition and mechanical properties of the filler metal. The classification number separates an electrode from others by process and usability.

Filler Metal Specification Filler metal specifications are composed of rules and standards used by industry to classify filler metals. They determine how filler metals are manufactured, packaged, and identified.

Fillet Weld A weld made between two base materials approximately perpendicular to each other. A fillet weld cross-section takes on a triangular shape (**FIGURE 17**).

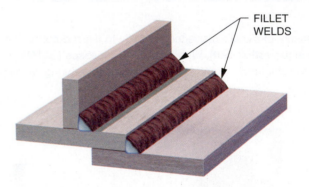

FIGURE 17

Fillet Weld Leg The distance from the root of the weld joint to the toe of the fillet weld (**FIGURE 18**).

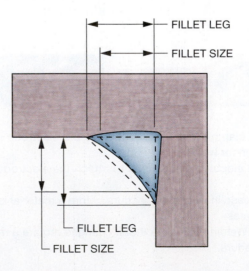

FIGURE 18

Fillet Weld Size The largest triangle that can be inscribed within a weld cross-section is used to determine the fillet weld size. For equal leg triangles, the leg length of an isosceles triangle's sides is the measure of the fillet weld size (**FIGURE 18**). For unequal leg fillet welds, the unequal leg triangle sides are used to determine the weld size.

Fire Watcher A person who observes the work area during and for half an hour after welding and cutting operations are completed to prevent occurrence of a fire.

Flashback A condition that can occur that will cause the oxy-fuel flame to move back into the tip, torch, hoses, or gas cylinders.

Forehand (Push) Technique Refers to a travel angle where the electrode is pointing ahead of the weld pool in the travel direction (**FIGURE 19**).

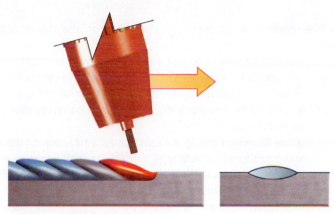

FOREHAND/PUSH TECHNIQUE   **FIGURE 19**

Frequency (Hz) The rate (times per second) in which AC electrical current changes direction.

Fusion The melting and mixing of two base metals together, with or without the melting and mixing with a filler metal.

Fusion Face A surface of base metal including previous weld deposits that will be melted into a weld.

Fusion Welding The melting together of base metals and filler metals to make a weld.

Gas Nozzle An apparatus used to direct gas flow to the workpiece from the welding torch or gun.

Gas Regulator An apparatus used to reduce cylinder gas pressures to a lower constant pressure suitable for welding.

Globular Transfer A mode of filler metal transfer deposit where large molten metal droplets transfer across an open arc.

Groove Angle (Included Angle) The total angle between two members to be joined (**FIGURE 20**).

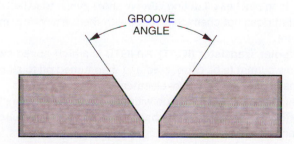

                                           FIGURE 20

Groove Face The surface of a butt weld forming the groove on the joint component (**FIGURE 21**).

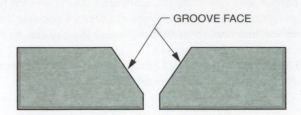

GROOVE FACE

FIGURE 21

GTAW Torch The conductor of welding current and the transmission line for shielding gas.

Hardenability The capacity of a metal to be hardened by heat treatment.

Hardness A significant property in metals that affects their ability to be formed. It is usually measured as resistance to penetration, resistance to scratching, or by the amount a steel ball will rebound off the metal's surface. Increased hardness strengthens metals and improves wear resistance.

Hazard Communication Standard (HCS) An OSHA standard requiring development of a hazard communication program whereby all employers with hazardous chemicals in their workplaces must have labels and Material Safety Data Sheets (MSDS) for their exposed workers and must train them to handle those materials.

Hazardous Materials Identification System (HMIS) A standard developed by the National Paint & Coatings Association (NPCA) to help employers comply with OSHA's Hazard Communication requirements. The system is composed of a color-code and numbering system, including labels or tags placed appropriately in the employee workspace to identify potential hazards and to inform workers of appropriate concerns and Personal Protective Equipment (PPE) required.

Heat-Affected Zone The area of base metal adjacent to the weld where grain structure and properties have been altered due to heat from welding.

High Frequency Often referred to as a high-frequency current; a high voltage (3,000 volts) oscillated at a high frequency (1MHz). The voltage level is high enough that it has the potential to conduct current across the gap from tungsten to base metal.

Hydrogen Cracking A form of cold cracking in steels caused by the absorption of hydrogen in the base metal.

Impact Strength An expression of resistance of a material to fracture resulting from impact loading.

Inclusion A discontinuity caused by the introduction of a foreign substance such as flux, oxides, rust, slag, or tungsten into the weld.

Incomplete Fusion A discontinuity caused by an absence or flowing together of filler metal and base metal and between all weld passes into one basically homogeneous mass.

Incomplete Joint Penetration A groove weld condition in which the weld metal does not flow completely through the joint root.

Inductance The rise in amperage (current) between the time the electrode contacts the base metal and then pinches off during GMAW short-circuit transfer.

Inert Gas A gas that does not chemically react with the welding arc, molten puddle, or finished weldment.

Insolated Gate Bipolar Transistor (IGBT) An IGTB is a high-speed switch that can be turned on to allow current to be conducted and then turned off to stop current flow.

Insulator A material that will not conduct electrical current.

Intermittent Welds A series of welds made with spaces between them.

Interpass Temperature A measure of the base metal temperature surrounding the weld during welding and between applications of subsequent weld deposits.

Ionization Potential The ionization potential or ionization energy of a gas atom is the energy required to strip it of an electron.

Iron-Carbon Phase Diagram A graph of phase changes in steels based on carbon percent by weight at temperatures from ambient to above the melting points of steels.

Joint The interface where edges of base metals to be joined by welding meet.

Joint Design The setup of the weld joint including shape, angles, and spacing (**FIGURE 22**).

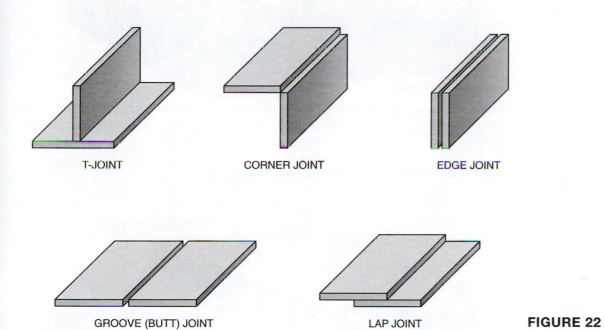

FIGURE 22

Joint Nomenclature Technical terms related to welds and weld joints. (**FIGURE 23–26**)

Joint Nomenclature–Fillet Welds

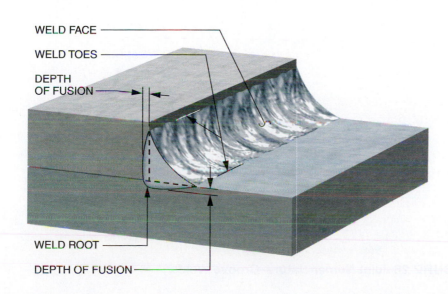

FIGURE 23

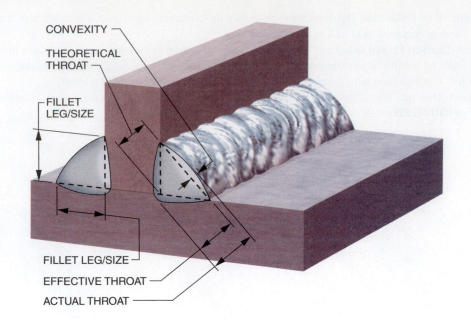

FIGURE 24 Joint Nomenclature–Convex Fillet Welds

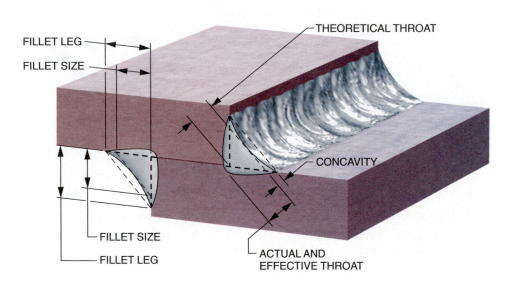

FIGURE 25 Joint Nomenclature–Convex Fillet Welds

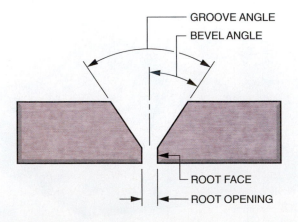

FIGURE 26 Joint Nomenclature–Groove Welds

Joint Penetration The maximum distance the weld deposit flows into the weld joint.

Joint Root The point in a weld joint in the middle or the back side where the two base metals come into closest proximity.

Kerf The width of the cut for any cutting process.

Kindling Temperature The temperature at which a material begins to burn or to oxidize when exposed to pure oxygen.

Knurled Drive Rolls Drive rolls that have serrations in the "v" portion, gripping the electrode for a more consistent feed.

Lap Joint A joint between two base metals lying in nearly the same plane that overlap.

Laser (Light Amplification by Stimulated Emission of Radiation) The laser's beam of light is produced by raising the energy level of specific matter to the point at which that matter releases photons of light (stimulated emission of radiation).

Lasing Material The matter in laser cutting that is used to produce high-energy photons.

Leakage Field In magnetic particle testing, the area where the magnetic particles concentrate, revealing the discontinuity.

Liner A conduit or tubular guide made of various materials, including steel, composite, and nylon, through which the filler metal electrode runs from the wire guides to the welding gun.

Macro-Etch A portion of a weld qualification test where a cross-section of a weld is cut out and polished, and then an acid mixture is applied to bring out and show the weld penetration under low magnification (**FIGURE 27**).

FIGURE 27

Magnetic Particle Testing The process of inducing a ferromagnetic material (a material which will attract a magnet) with a current that produce lines of flux similar to what is produced in a typical magnet.

Malleability The ability of a metal to deform permanently when loaded in compression.

Martensite A hard, brittle grain structure found in steels caused by cooling, usually rapidly, from austenite.

Material Safety Data Sheets (MSDS) A form containing data regarding the properties of a particular substance. An important component of workplace safety, it is intended to provide workers and emergency personnel with procedures for handling or working with that substance in a safe manner and includes information such as physical data (melting point, boiling point, flash point, etc.), toxicity, health effects, first aid, reactivity, storage, disposal, protective equipment, and spill-handling procedures.

Mechanical Properties Measures of the ease in which the metal can be shaped. Mechanical properties include strength, hardness, ductility, brittleness, elasticity, plasticity, malleability, and toughness.

Melt-Through The extension of weld metal through the joint root in a groove joint (**FIGURE 28**). Melt-through is a desired result on complete joint penetration welds as opposed to burn-through, which is excessive and unacceptable.

FIGURE 28

National Institute for Occupational Safety and Health (NIOSH) A research agency whose purpose is to determine the major types of hazards in the workplace and the ways of controlling them.

Neutral Flame A flame with an equal balance between the oxygen and fuel gas.

Nonessential Variables Welding variables that, if changed, do not require a retest and documentation of a new procedure.

Nozzle See **Gas Nozzle**.

Occupational Safety and Health Administration (OSHA) OSHA is a government regulatory agency whose mission is to prevent work-related injuries, illnesses, and deaths by issuing and enforcing rules (called standards) for workplace safety and health.

Open Circuit Voltage (OCV) The voltage measured across the welding terminals before the arc is started.

Other Side The information on the top-side (above) of the reference line on a welding symbol; it is applied to the opposite side of the joint from the arrow.

Overalloying The buildup of too high of an alloy content, such as a deoxidizing element, that can occur if using multipass welding with an electrode designed for single-pass welding only.

Overlap An incomplete fusion condition indicated by the flow of the weld metal at the weld toe or joint root onto the cold base metal (**FIGURE 29**).

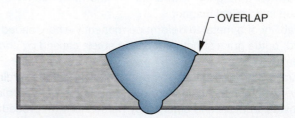

FIGURE 29

Oxidizing Flame A flame consisting of an unbalanced mix of oxygen and fuel gas, with an excess amount of oxygen.

Partial Joint Penetration Weld A groove weld design in which weld metal does not flow completely through the joint root (**FIGURE 30**).

FIGURE 30

Pearlite A lamellar grain structure in steels composed of alternating layers of ferrite and cementite.

Peening Mechanical working that plastically deforms metal with a hammer blow or similar means.

Personal Protective Equipment (PPE) Personal protective equipments are safety protections required under specific workplace conditions.

Physical Properties Characteristics of a metal that can be measured without changing the chemical nature, structure, behavior, or composition of the metal. These properties include magnetic properties, electrical properties, thermal properties, and the metal's atomic weight.

Plasma The state of matter that is produced when an electrical arc travels through the gas stream. The gas is heated to around 30,000°F (16,000°C) and is also ionized.

Plasticity The ability of a metal to sustain permanent deformation or to yield without rupture.

Plume A column of heat that rises from the welding arc containing particulate matter and vapors caused by the breakdown of filler metal, welding gas, and base metal constituents.

Polarity The selection of electrical lead settings for Direct Current welding of either Direct Current Electrode Positive (DCEP) or Direct Current Electrode Negative (DCEN).

Porosity Discontinuities caused by gases trapped during the solidification process in welds. Typically appears as round holes when visible from the weld face (**FIGURE 31**).

FIGURE 31

Position (Welding Position) The physical orientation of the workpiece, weld joint, and weld being completed. The American Welding Society has standardized welding positions for welder qualification. (**Welding Position/Welding Positions.**)

Postflow A timed shielding gas flow at the end of a weld.

Postheat Treating Heating a welded assembly after welding is completed; usually a stress relief heat treatment that reduces residual stresses in the weld assembly after welding is complete.

Power Source The electrical equipment used to provide welding current and voltage for welding.

Preflow A timed amount of shielding gas flow that begins before the start of welding.

Preheat The process of heating a base metal immediately before welding is initiated.

Problem-Solving Skills A desirable skill set for welding employees that includes the ability to think critically, apply problem-solving strategies, and apply mathematical reasoning.

Procedure Qualification Record (PQR) A documentation of the results of a weld test to make sure a Welding Procedure Specification (WPS) stands up to destructive testing and in-service use on a product; also called a Welding Procedure Qualification Record.

Productivity Skills A desirable skill set for welding employees that includes the ability to work productively, to follow directions, and to function in a manner that maintains a safe work environment.

Proportional Limit The greatest load a material can withstand and still return to its original length when the load is removed.

Pulsed Spray Transfer A mode of metal transfer that combines high-heat inputs of the spray transfer arc with slightly lower currents near the globular transfer range to provide a balanced average current low enough to allow for metal transfer in all positions.

Radiography Use of radiographic testing/x-ray to inspect welds that are fabricated as specified by a welding code and view internal components of the weld assembly that cannot be seen by visual examination.

Rectification The changing of alternating current (AC) to direct current (DC).

Reinforcement Weld metal beyond that required to fill a weld joint. There are two types or reinforcement. **Face reinforcement** is buildup on the face side from which welding is done. **Root reinforcement** is the amount of weld deposit or melt-through on the opposite side from which welding is done (**FIGURE 32**).

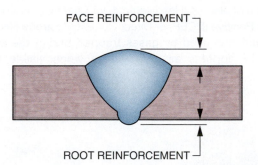

FIGURE 32

Residual Stress Stresses present in a welded joint caused by nonuniform heating and cooling.

Resistance Opposition to current flow.

Root Bead (Root Pass) A weld bead deposited in the joint root (**FIGURE 33**).

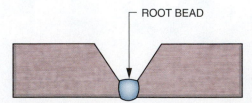

FIGURE 33

Root Edge The edge at the root of the joint.

Root Face The perpendicular part of the groove face within the joint root (**FIGURE 34**).

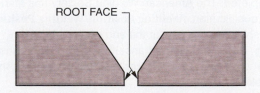

FIGURE 34

Root Opening The gap between the weld joint members at the joint root (**FIGURE 35**).

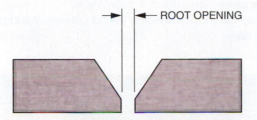

← ROOT OPENING →

FIGURE 35

Root Penetration The distance weld metal extends and fuses into the joint root.
Root Reinforcement The amount of weld deposit or melt-through on the opposite side from which welding is done (**FIGURE 36**).

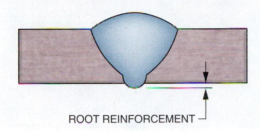

ROOT REINFORCEMENT

FIGURE 36

Root Surface The weld surface on the root side of the weld.
Seal Weld A weld made with the intent to prevent leakage.
Self-Shielded FCAW electrodes are those that use no external shielding gas because their inner core fluxes provide all the necessary shielding.
Semiconductor A conductor that conducts electrical current only when very specific conditions are met.
Shear Strength A measure exhibited by a load that tends to force materials apart by the application of forces pushing in opposite directions, pushing grains or crystals out of position by a slip-slide action.
Shielding Gas A gas used to prevent atmospheric contamination of the weld pool and filler metal electrode.
Short Circuit Transfer A method of filler metal transfer to the base metal where the filler metal is deposited from the electrode by short-circuiting to the workpiece surface.
Silicon Controlled Rectifier (SCR) A semiconductor similar to a diode that has to be forward-biased in order to conduct current.
Slag A nonmetallic residue left on the weld surface composed of burned fluxing agents and impurities from the weld (**FIGURE 37**).

FIGURE 37

Slope The proportion of the volt/amp curves in the welding output.

Spacer A spacer is centered in a groove joint and acts as a backing plate for both sides and is made of the same material as the base metal.

Spatter Metal particles from fusion-welding the fall outside of and do not become part of the weld (**FIGURE 38**).

FIGURE 38

Specifications A document that describes the manufacturing requirements for a particular object, material, service, and its use.

Spool Drag Excessive drag on the electrode wire being fed through the drive rolls caused by setting the spool tension too tight.

Spray Transfer A mode of metal transfer where small molten metal droplets deposited are transferred axially across an open arc.

Staggered Intermittent Fillet The placing of intermittent fillet welds such that the other-side welds are not directly across from the arrow-side welds but placed at a center distance from them and on the other side.

Standard Any document or object used as a basis of comparison. Once a standard has been defined, an inspector must then judge the adequacy of a product based on how it compares with that established standard.

Standoff The distance between a gas nozzle and the workpiece (**FIGURE 39**)

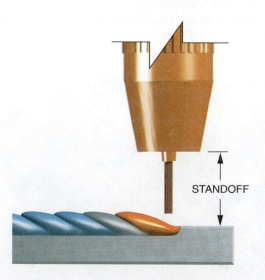

STANDOFF

FIGURE 39

Strength The measure of a material's resistance to external forces applied before it fails.

Stringer Bead A weld bead made with little or no weaving or oscillation.

Strongbacks A device attached to both joint members used to hold them in place and reduce distortion during welding (**FIGURE 40**).

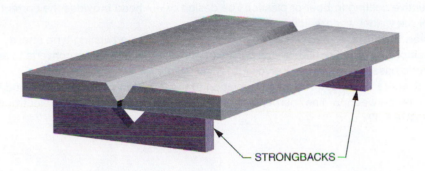

STRONGBACKS

FIGURE 40

Stub The unused portion of waste material left over from a welding electrode.

Surface Preparation Machining, brushing, or grinding used to clean and prepare the weld joint prior to welding.

Surface Tension The force or property that affects the liquid filler metal droplet attraction between the electrode and the workpiece surface. Surface tension can be recognized in everyday life in the household kitchen. When water is spilled on the kitchen countertop, we often see the water bead up. Surface tension holds the water from spreading out. Also, if we fill a glass of water to the top, we can often add just a bit more, and the water actually rises above the glass rim without spilling. Surface tension is what keeps the water from spilling. Wetting agents added to water break the surface tension. In welding, argon shielding gas works like a wetting agent in water to help wet-out the weld puddle.

Surfacing Applying a layer of metal deposit by welding, brazing, or soldering.

Tack Weld A fusion weld used to hold weld joints in place during welding.

Tack Welder A person who applies tack welds in order to hold components in place for welding.

Taper (bevel angle) For cutting processes, the angle of the cut edge when the kerf of the cut is wider at the top or bottom.

Team Skills A desirable skill set for welding employees that includes an ability to communicate clearly, to listen effectively, and to be capable of working cooperatively in teams.

Tempering A process of heating a metal to a precise temperature where a controlled cooling rate is that used to soften steels slightly, reducing the stressed condition in martensite.

Temporary Weld A nonpermanent weld that will be removed after welding is complete.

Tensile Strength The maximum load-carrying capability of a material based on a tensile test.

Theoretical Throat In a weld with no root opening, the distance from the joint root to the center of the hypotenuse that determines the fillet weld size (**FIGURE 41**).

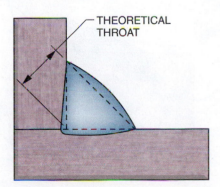

THEORETICAL THROAT

FIGURE 41

Thermal Conductivity The measure of the rate at which a quantity of heat travels through a metal per unit time, per unit cross-section, and per temperature gradient.

Thermal Expansion The tendency of a metal to grow in volume when heated.

T-Joint A weld joint where members to be joined are aligned approximately perpendicular to each other.

Torch Body Part of the GTAW torch that consists of a brass body covered by a nonconductive coating (rubber or plastic). The design of the head provides the correct angle for the welder when welding.

Torsion Strength A measure of a material's resistance to a twisting type stress.

Transition Current For GMAW, the point where one mode of transfer converts to another mode based on an amperage/voltage relationship.

Travel Angle The electrode angle in relation to the direction of travel determined by the angle between a line perpendicular to the weld axis and the electrode axis (**FIGURE 42**).

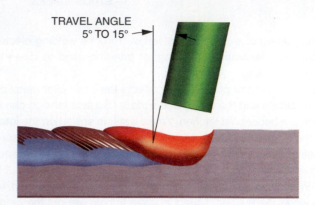

FIGURE 42

Travel Direction The path the electrode and weld pool progresses along the weld joint.

Travel Speed A measure of the distance the weld bead progresses along the weld joint in inches per minute or centimeters per minute.

Tungsten Electrode A nonconsumable electrode used as the conductor of welding current because it has the highest melting temperature of all pure metals at 6,192°F (3,422°C).

Ultrasonic Testing Use of a high-frequency sound projected into a material by using a piezoelectric transducer searching for acoustical impedance mismatch.

Undercut A depression at the edge of a weld toe or weld root caused by melting the base metal without adding sufficient fill (**FIGURE 43**).

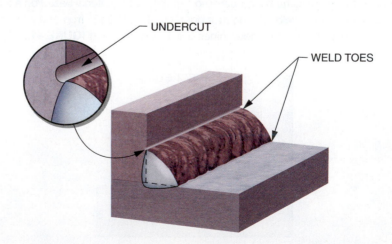

FIGURE 43

Underfill A depression in the weld face or weld root filled to a level below the adjacent base metal surfaces (**FIGURE 44**).

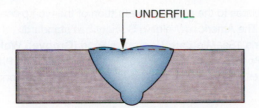

FIGURE 44

Uphill A welding progression where welding starts at the bottom of the joint and travels upward to the top.

Valence Shell The outermost most electron shell of an atom.

Velocity With ultrasonic testing, the rate at which sound travels through a material.

Voltage The electrical force that causes or pushes the flow of electrons through the welding leads.

Water-Cooled Torch GTAW torch that uses the flow of shielding gas and a continuous flow of liquid coolant to cool the torch.

Wattage The measurement of power (energy) that is produced by the flow of electrical current.

Weave Bead A weld bead made with a side-to-side motion or another other form of oscillation.

Weldability The capacity of a material to be welded under the imposed fabrication conditions into a specific, suitably designed structure and to perform satisfactorily in the intended service.

Weld Axis An imaginary line down the middle of the weld length.

Weld Bead A weld made from a single weld pass.

Welder A person who performs welding manually or with semiautomatic equipment.

Welder Qualification Test Record (WQTR) A documentation of the test results a welder obtains after following a Welding Procedure Specification (WPS).

Weld Face The surface of the weld on the side from which welding is done (**FIGURE 45**).

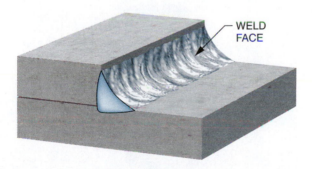

WELD FACE

FIGURE 45

Welding The joining of materials produced by heating them to the proper temperature with or without the application of pressure and with or without filler metal, or joining them together with the application of pressure alone, without the use of filler metal.

Welding Amperage The amount of current output to the electrode and therefore the welding arc.

Welding Arc The electrically charged gas plasma gap between a welding electrode and the base metal.

Welding Leads The electrical leads attached to the welding electrode and the workpiece.

Welding Operator A person who operates automatic welding equipment, ranging from simple mechanized track and rotational equipment to robotic welding equipment.

Welding Parameters Variables such as amperage, voltage, travel speed, shielding gas flow rates, and wire feed speeds used to control welding operations.

Weld Pitch Center-to-center distance from weld to weld.

Welding Position Relates to the physical orientation of the workpiece, and weld joint, weld being completed. The American Welding Society has standardized welding positions for welder qualifications. [See **Welding Positions Pipe (Groove), Welding Positions Pipe (Fillet), Welding Positions Plate (Groove), Welding Positions Plate (Fillet)**]. A combination number/letter designation is assigned to the welding positions:

1 = Flat
2 = Horizontal
3 = Vertical
4 = Overhead
F = Fillet
G = Groove

Welding Positions Pipe (Groove) (**FIGURES 46–50**)

FIGURE 46 Flat groove position 1G (rotated)

FIGURE 47 Horizontal groove position 2G

FIGURE 48 Vertical groove position 5G

FIGURE 49 All position groove 6G

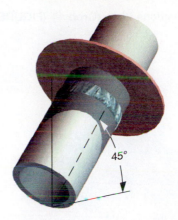

FIGURE 50 (All position groove 6GR (T, K, & Y Joints)

Welding Positions Pipe (Fillet) (FIGURES 51–56)

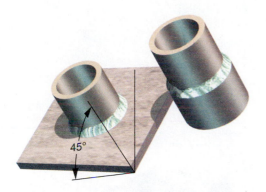

FIGURE 51 Flat fillet position 1F (rotated)

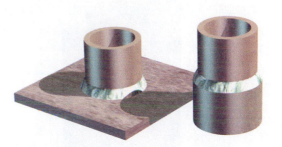

FIGURE 52 Horizontal fillet position 2F

FIGURE 53 Horizontal fillet position 2FR (rotated)

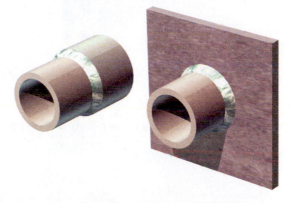

FIGURE 54 Vertical fillet position 5F (fixed position)

FIGURE 55 Overhead fillet position 4F

FIGURE 56 Fillet multiple position 6F

Welding Positions Plate (Groove) (FIGURES 57–60)

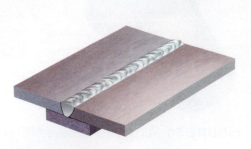

FIGURE 57 Flat groove position 1G

FIGURE 58 Horizontal groove position 2G

FIGURE 59 Vertical groove position 3G

FIGURE 60 Overhead groove position 4G

Welding Positions Plate (Fillet) (FIGURES 61–64)

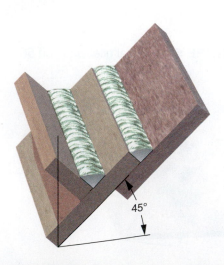

45°

FIGURE 61 Flat fillet position 1F

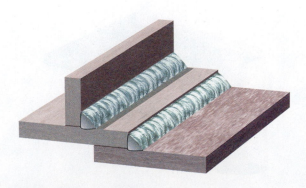

FIGURE 62 Horizontal fillet position 2F

FIGURE 63 Vertical fillet position 3F

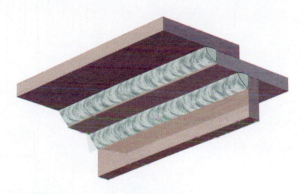

FIGURE 64 Overhead fillet position 4F

Welding Procedure The information detailing steps and welding parameters required for completion of a welding operation.

Welding Procedure Specification (WPS) A written welding procedure documenting specific welding variables for use by qualified welders and welding operators to ensure weld quality.

Welding Sequence The order that welds are applied to a welded assembly to control weld quality.

Welding Symbol An illustration used to communicate the specific type of weld to be used, including size, number, and extent of welding (**FIGURE 65**).

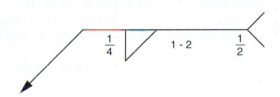

FIGURE 65

Welding Technique Methods a welder uses to produce a weld, such as arc length, electrode angle, and electrode manipulation.

Weldment The finish product of two or more materials joined by welding.

Weld Metal The melted portion of a fusion weld.

Weld Metal Crack Cracks within the weld or propagating from within the weld.

Weld Pass A single layer of weld deposited longitudinally within the weld joint.

Weld Pool The molten mass of metal in a fusion weld produced at the welding arc.

Weld Root The deepest point at the weld interface between two base metals where the weld extends through the weld joint (**FIGURE 66**).

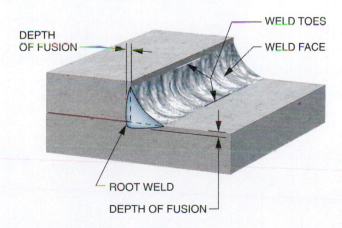

FIGURE 66

Weld Symbol An illustration that indicates the specific type of weld applied to a welding symbol. **FIGURE 67** shows 11 weld symbols.

FIGURE 67

Weld Tab A material added to the start or finish end of a weld assembly to facilitate weld starts and weld ends.

Weld Test Assembly Welded sample composed of individual coupons. Weld test specimens for bending, etching, and tensile pulls are extracted from a weld test assembly.

Weld Test Coupon Materials used for welder and procedure qualifications. Coupons are cut to dimensions specified in the welding code for use in weld test assemblies.

Whipping Technique The manipulation of a welding electrode in a back-and-forth motion in the travel direction.

Wire Feed Speed The rate that the filler metal electrode wire feeds out of the contact tip to the weld during welding in inches per minute or millimeters per second.

Work Angle The angle between the electrode axis and a line perpendicular to the workpiece surfaces.

Work Clamp A device attached to the work lead that makes the physical connection to the workpiece from the power source.

Work Lead The electrical cable connected to a clamp used to attach the workpiece to the power supply.

Workpiece The base metal parts to be welded together.

Yield Point The point during a tensile test at which a marked increase in deformation occurs without an increase in load stress.

Yield Strength The amount of stress in pounds per square inch (Mpa) required to cause a material to plastically deform.

Index

Please note: Page numbers in italics indicate figures, tables, and photographs.